# Topics in Fluorescence Spectroscopy

Volume 8
Radiative Decay Engineering

# Topics in Fluorescence Spectroscopy

Edited by JOSEPH R. LAKOWICZ and CHRIS D. GEDDES

# Topics in Fluorescence Spectroscopy

## Volume 8
## Radiative Decay Engineering

Edited by

CHRIS D. GEDDES

*The Institute of Fluorescence*
*Medical Biotechnology Center*
*University of Maryland Biotechnology Institute*
*Baltimore, Maryland*

and

JOSEPH R. LAKOWICZ

*Center for Fluorescence Spectroscopy and*
*Department of Biochemistry and Molecular Biology*
*University of Maryland School of Medicine*
*Baltimore, Maryland*

 Springer

The Library of Congress cataloged the first volume of this title as follows:

Topics in fluorescence spectroscopy/edited by Joseph R. Lakowicz.
   p. cm.
   Includes bibliographical references and index.
   Contents: v. 1. Techniques
   1. Fluorescence spectroscopy.   I. Lakowicz, Joseph R.

QD96.F56T66   1991                                            91-32671
543′.0858—dc20                                             CIP

Front cover—Surface Plasmon Coupled Emission (SPCE). See *Journal of Fluorescence* 14(1), 119–123, 2004, and chapter within

ISSN: 1574-1036
ISBN 0-387-22662-1 (HB)

Printed on acid-free paper



Printed in the United States of America

9  8  7  6  5  4  3  2  1        SPIN 11306818

springeronline.com

# Contributors

*Ricardo F. Aroca* • Materials & Surface Science Group, Department of Chemistry and Biochemistry, University of Windsor, N9B 3P4, Windsor, Ontario, Canada

*Kadir Aslan* • Institute of Fluorescence, University of Maryland Biotechnology Institute 725 West Lombard Street, Baltimore, Maryland 21201

*Donna Chen* • Department of Chemistry and Biochemistry, University of South Carolina, 631 Sumter Street, Columbia, South Carolina 29208

*Ashutosh Chilkoti* • Department of Biomedical Engineering, Duke University, Durham, North Carolina 27708

*Paula E. Colavita* • Department of Chemistry and Biochemistry, University of South Carolina, 631 Sumter Street, Columbia, South Carolina 29208

*Michael Doescher* • Department of Chemistry and Biochemistry, University of South Carolina, 631 Sumter Street, Columbia, South Carolina 29208

*Eric Dulkeith* • Photonics and Optoelectronics Group, Physics Department and Center for NanoScience, Ludwig-Maximilians-Universität München, Amalienstraße 54, 80799 München, Germany

*Robert C. Dunn* • Department of Chemistry, University of Kansas, Lawrence, Kansas, 66045

*Jochen Feldmann* • Photonics and Optoelectronics Group, Physics Department and Center for NanoScience, Ludwig-Maximilians-Universität München, Amalienstraße 54, 80799 München, Germany

*Chris D. Geddes* • Institute of Fluorescence, University of Maryland Biotechnology Institute, and Center for Fluorescence Spectroscopy, 725 West Lombard Street, Baltimore, Maryland 21201

*Joel I. Gersten* • Department of Physics, City College of the City University of New York, New York, New York 10031

*Paul J.G. Goulet* • Materials & Surface Science Group, Department of Chemistry and Biochemistry, University of Windsor, N9B 3P4, Windsor, Ontario, Canada

*Ignacy Gryczynski* • Center for Fluorescence Spectroscopy, University of Maryland, School of Medicine, 725 West Lombard Street, Baltimore, Maryland 21201

*Zygmunt Gryczynski* • Center for Fluorescence Spectroscopy, University of Maryland, School of Medicine, 725 West Lombard Street, Baltimore, Maryland 21201

*Amanda J. Haes* • Northwestern University, Department of Chemistry, 2145 Sheridan Road, Evanston, Illinois 60208-3113

*Christy L. Haynes* • Northwestern University, Department of Chemistry, 2145 Sheridan Road, Evanston, Illinois 60208-3113

*Arnim Henglein* • Hahn-Meitner Institut, 14109 Berlin, Germany

*Thomas A. Klar* • Photonics and Optoelectronics Group, Physics Department and Center for NanoScience, Ludwig-Maximilians-Universität München, Amalienstraße 54, 80799 München, Germany

*Wolfgang Knoll* • Max-Planck-Institute for Polymer Research, Ackermannweg 10, 55128 Mainz, Germany

*Joseph R. Lakowicz* • Center for Fluorescence Spectroscopy, University of Maryland, School of Medicine, 725 West Lombard Street, Baltimore, Maryland 21201

*Luis M. Liz-Marzán* • Departamento de Química Física, Universidade de Vigo, 36200, Vigo, Spain

*Joanna Malicka* • Center for Fluorescence Spectroscopy, University of Maryland, School of Medicine, 725 West Lombard Street, Baltimore, Maryland 21201

*C. Mayer* • Analytical Biotechnology, Technical University of Delft, Julianalaan 67, 2628 BC Delft, The Netherlands

*Adam D. McFarland* • Northwestern University, Department of Chemistry, 2145 Sheridan Road, Evanston, Illinois 60208-3113

*Paul Miney* • Department of Chemistry and Biochemistry, University of South Carolina, 631 Sumter Street, Columbia, South Carolina 29208

*Annabelle Molliet* • Department of Chemistry and Biochemistry, University of South Carolina, 631 Sumter Street, Columbia, South Carolina 29208

*David S. Moore-Nichols* • Department of Chemistry, University of Kansas, Lawrence, Kansas, 66045

*Michael L. Myrick* • Department of Chemistry and Biochemistry, University of South Carolina, 631 Sumter Street, Columbia, South Carolina 29208

*Nidhi Nath* • Department of Biomedical Engineering, Duke University, Durham, North Carolina 27708

*Thomas Neumann* • Max-Planck-Institute for Polymer Research, Ackermannweg 10, 55128 Mainz, Germany

*Lifang Niu* • Departments of Chemistry and of Materials Science, National University of Singapore, 10 Science Drive 4, Singapore 11754

*Steven J. Oldenburg* • Seashell Technology, La Jolla, California 92037

*Isabel Pastoriza-Santos* • Departamento de Química Física, Universidade de Vigo, 36200, Vigo, Spain

*Darren Pearson* • Department of Chemistry and Biochemistry, University of South Carolina, 631 Sumter Street, Columbia, South Carolina 29208

*Jorge Pérez-Juste* • Departamento de Química Física, Universidade de Vigo, 36200, Vigo, Spain

*John Reddic* • Department of Chemistry and Biochemistry, University of South Carolina, 631 Sumter Street, Columbia, South Carolina 29208

*Th. Schalkhammer* • Nanobioengineering, Vienna Biocenter, Universität Wien Dr. Bohrgasse 9, 1030 Wien, Austria. *Current address*: Schalkhammer KG, Klausenstrasse 129, 2534 Alland, Austria

*David A. Schultz* • David A. Schultz, University of California, San Diego, La Jolla, California 92093-0319

*Evelyne L. Schmid* • Departments of Chemistry and of Materials Science, National University of Singapore, 10 Science Drive 4, Singapore 11754

*Lindsay Taylor* • Department of Chemistry and Biochemistry, University of South Carolina, 631 Sumter Street, Columbia, South Carolina 29208

*Richard P. Van Duyne* • Northwestern University, Department of Chemistry, 2145 Sheridan Road, Evanston, Illinois 60208-3113

*Fang Yu* • Max-Planck-Institute for Polymer Research, Ackermannweg 10, 55128 Mainz, Germany

*Jing Zhou* • Department of Chemistry and Biochemistry, University of South Carolina, 631 Sumter Street, Columbia, South Carolina 29208

# Preface

Spatial control of photonic mode density is changing the practice of fluorescence spectroscopy. This laboratory has been active in fluorescence spectroscopy for nearly 30 years. During that time we have investigated many phenomena in fluorescence, including quenching, energy transfer and anisotropy, to name a few. Until recently we relied completely on the free-space emission properties of fluorophores observed in transparent media. The free-space quantities in fluorescence are determined by the values of the radiative and non-radiative properties of excited fluorophores. The observed changes in fluorescence intensities, lifetimes, etc. are due almost completely to changes in the non-radiative decay rates such as quenching. The rate of radiative decay is determined by the extinction coefficient or oscillator strength of the transition. This rate is essentially constant in most media.

In about 2000 we began to examine the effects of silver metallic particles on fluorescence. Examination of the literature revealed that proximity to silver particles could have dramatic effects on fluorescence quantum yields and lifetimes. Such changes are typically due to changes in the non-radiative decay rates. In contrast, the metal particles changed the radiative decay rate ($\Gamma$). These changes occur due to modifications of the photonic mode density (PMD) near the particle in $\Gamma$. This was the first time in 30 years that we saw an opportunity to modify this fundamental rate. Numerous opportunities became apparent as we considered the effects of PMD, including increased quantum yields, increased photostability and changes in resonance energy transfer. Additionally, we saw the opportunity to obtain directional rather than isotropic emission based on local changes in the PMD. We described these phenomena as radiative decay engineering (RDE) because we could engineer changes in the emission based on the fluorophore-metal particle geometries.

During these three years our enthusiasm for RDE has continually increased. Many of the early predictions have been confirmed experimentally. As one example we recently observed directional emission based on fluorophores located near a thin metal film, a phenomenon we call surface plasmon coupled emission (SPCE). We see numerous applications for RDE in biotechnology, clinical assays and analytical chemistry. The technology needed to implement RDE is straightforward and easily adapted by most laboratories. The procedures for making noble metal particles and surfaces are simple and inexpensive. The surface chemistry is well developed, and the noble metals are easily tolerated by biochemistry systems.

While implementation of RDE is relatively simple, understanding the principles of RDE is difficult. The concepts are widely distributed in the optics and chemical physics literature, often described in terms difficult to understand by biophysical scientists. In this volume we have presented chapters from the experts who have studied metal particle optics and fluorophore-metal interactions. We believe this collection describes the fundamental principles for the widespread use of radiative decay engineering in the biological sciences and nanotechnology.

Joseph R. Lakowicz and Chris D. Geddes
Center for Fluorescence Spectroscopy
Baltimore, Maryland
August 13, 2003

# Contents

**3.  Nanoparticles with Tunable Localized Surface Plasmon Resonances:
Topics in Fluorescence Spectroscopy**
Christy L. Haynes, Amanda J. Haes, Adam D. McFarland, and
Richard P. Van Duyne

## 4.  Colloid Surface Chemistry
Arnim Henglein

**5. Bioanalytical Sensing Using Noble Meal Colloids**
C. Mayer and Th. Schalkhammer

**6.   Theory of Fluorophore-Metallic Surface Interactions**
Joel I. Gersten



**7.   Surface-Enhancement of Fluorescence Near Noble Metal Nanostructures**
Paul J.G. Goulet and Ricardo F. Aroca



**8.   Time-Resolved Fluorescence Measurements of Fluorophores Close to
Metal Nanoparticles**
Thomas A. Klar, Eric Dulkeith, and Jochen Feldman

## 12. Noble Metal Nanoparticle Biosensors
Nidhi Nath and Ashutosh Chilkoti



## 13. Surface Plasmon-Coupled Emission: A New Method for Sensitive Fluorescence Detection
Ignacy Gryczynski, Joanna Malicka, Zygmunt Gryczynski
and Joseph R. Lalowicz

## 14. Radiative Decay Engineering (RDE)

Chris D. Geddes, Kadir Aslan, Ignacy Gryczynski, Joanna Malicka and
Joseph R. Lakowicz

# PREPARATION OF NOBLE METAL COLLOIDS AND SELECTED STRUCTURES

Isabel Pastoriza-Santos, Jorge Pérez-Juste, and
Luis M. Liz-Marzán[*]

## 1. INTRODUCTION

Metal colloids have been around for centuries because of their striking optical properties. A typical example of ancient colloidal metal nanoparticles is the famous Lycurgus cup, which dates back to the fourth century AD. The cup is still at the British Museum[1] and possesses the unique feature of changing color, since it is green when viewed in reflected light, but it appears red when a light is shone from inside and is transmitted through the glass. Analysis of the glass reveals that it contains a very small amount of tiny (ca. 70 nm) crystals of metal containing silver and gold in an approximate molar ratio or 14:1. It is the presence of these nanocrystals which imparts Lycurgus cup with its special color display. After such early origins, metal colloids, and particularly colloidal gold, were known and used in the Middle Ages by the alchemists for their health restorative properties, which are still popular nowadays.[2] Although the discovery that glass could be colored red by adding a small amount of gold powder is often credited to Johann Kunckel, a German glassmaker in the late seventeenth century,[1] it was not until 1857 that Michael Faraday performed a systematic study on the synthesis and color of colloidal gold.[3] Since that pioneering work, thousands of scientific papers have been published on the synthesis, modification, properties, and assembly of metal nanoparticles, using a large variety of solvents and other substrates. All this has led not only to reliable procedures for the preparation of metal nanoparticles of basically any size and shape, but also to a deep understanding of many of the physico-chemical features that determine the special behavior of these systems.

One of the very interesting aspects of metal colloids is that their optical properties strongly depend on particle size and shape. Bulk gold for instance has a high absorbance along the whole visible range, except for a slight dip around 400-500 nm, which makes Au thin films look blue in transmission. As particle size is reduced from say 500 nm down to ca. 3 nm, the color changes gradually to orange, through several tones of purple and red. These effects are due to changes in the so-called surface plasmon resonance,[4]

---

[*] Isabel Pastoriza-Santos, Jorge Pérez-Juste, and Luis M. Liz-Marzán, Departamento de Química Física, Universidade de Vigo, 36200, Vigo, Spain.

*Topics in Fluorescence Spectroscopy, Volume 8: Radiative Decay Engineering*
Edited by Geddes and Lakowicz, Springer Science+Business Media, Inc., New York, 2005

which is the frequency at which conduction electrons oscillate as a response to the alternating electric field of an incoming electromagnetic wave. Below 3 nm, the color changes to brownish and eventually colorless, which is mainly due to quantum effects.[5] Shape effects are even stronger, and slight deviations from the spherical geometry can lead to quite impressive color changes.

The plasmon resonance is one of the most characteristic properties of some noble metal colloids, and can be predicted using classical calculations of the scattering by small particles, which is usually known as Mie theory.[6] Although UV-visible spectra for all metals can be calculated from such theory,[7] only metals with free electrons (basically Au, Ag, Cu, and the alkali metals) possess plasmon resonances in the visible that provide them with intense colors. Elongated nanoparticles (ellipsoids and nanorods) display two distinct plasmon bands related to transversal and longitudinal electron oscillations, being the longitudinal one very sensitive to the aspect ratio of the particles.[8] Apart from single particle properties, also the environment in which the metal particles are dispersed is of relevance to the optical properties.[9] The refractive index of the dispersing medium[10] as well as the average distance between neighboring metal nanoparticles[11] have been shown to influence the spectral features, which has been exploited for applications such as DNA detection.[12]

Metal nanoparticles also have important applications in catalysis,[13,14] mainly because of their extremely large surface to volume ratio, which allows a more effective utilization of expensive transition metals, and additionally, since the metal nanoparticles can be either supported onto solid substrates or dispersed in a solvent, they can be used for both heterogeneous and homogeneous catalysis. We are not going to describe in detail the catalytic properties of metal colloids here.

In this chapter, we intend to introduce to the reader some of the most popular preparation methods of noble metal colloids, but do not intend by any means to thoroughly review the scientific literature on metal colloid synthesis. We have chosen examples of synthesis in aqueous and organic solvents, and differentiate between spherical and anisotropic nanoparticles (nanorods and nanoprisms). Additionally, in section 3 we shall describe one of the most popular recent procedures to assemble metal colloids into nanostructured materials (layer-by-layer assembly), as well as the properties of the resulting structures.

## 2. PREPARATION OF NOBLE METAL COLLOIDS

### 2.1. Spherical Nanoparticles in Water

#### 2.1.1. Citrate Reduction

Probably the most popular method to prepare Au nanospheres dispersed in water is the reduction with sodium citrate at high temperature. This method was first introduced by Turkevich and co-workers back in the fifties,[15] and further studied and modified in a number of later papers.[16-19] This preparation is basically performed by heating a dilute (typically $1\text{-}5 \times 10^{-4}$ M) aqueous solution of $HAuCl_4$ to its boiling point and then rapid addition of sodium citrate solution. The formation of the gold colloid is then rather slow, with a spectacular color change from light yellow (due to $HAuCl_4$) to colorless, then to gray, blue, purple, and finally (after some 10 minutes) deep wine red, revealing the formation of very uniform (standard deviation of ca. 10%) Au nanoparticles. Frens[17] demonstrated that the average particle diameter can be tuned over quite a wide range (ca. 10-100 nm) just by varying the concentration ratio between gold salt and sodium citrate. However, for particles larger than 30 nm, deviations from the spherical shape are observed, as well as a larger polydispersity. Lower temperatures can also be used, by which the reduction process is notably slower, and thus allowed to study the nucleation and growth mechanism.[19] Such studies revealed the presence of large and irregular aggregates of gold nuclei during the initial stages (these are responsible for the intermediate gray color) which eventually peptize to form the final monodisperse colloid. Examples of the initial aggregates and the final nanoparticles are shown in Figure 1.

The same procedure can be used to reduce a silver salt, but results in the formation of larger, non-uniform particles, with a rather high polydispersity. Using mixtures of Au and Ag salts, alloy nanoparticle colloids can be obtained, with surface plasmon bands centered at wavelengths intermediate between those for pure silver (400 nm) and pure gold (520 nm).[20,21] Citrate reduction has also been applied to the production of Pt colloids of much smaller particle sizes (2-4 nm), which could be further grown by hydrogen treatment.[22,23]

### 2.1.2. Borohydride Reduction

Sodium borohydride is quite a strong reducing agent and therefore it has been often used for the reduction of metal salts to produce small metal particles in solution. Klabunde and co-workers performed a large amount of work on the synthesis of magnetic particles based on iron and cobalt[24] using borohydride reduction. For these metals however, some complications can arise if the stoichiometry is not carefully adjusted, since otherwise one can easily end up with the corresponding metal borides.[25] Borohydride has also been applied for the reduction of noble metals, both in aqueous and in non-aqueous solution. We restrict ourselves in this section to reactions in aqueous media, while borohydride production of nonaqueous colloids is discussed in Section 2.2.1.

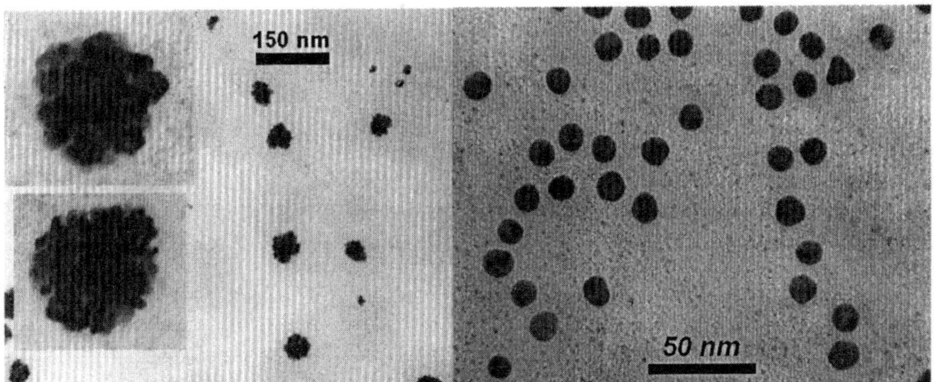


**Figure 1.** Typical TEM micrographs of aggregates form at the initial stages and the final Au nanoparticles prepared using the sodium citrate reduction method.

The formation of metallic platinum by borohydride was first reported by Brown and Brown[26] related to the study of hydrogenation reactions. These authors did not use any stabilizing agent, and therefore obtained what they termed "finely-divided black precipitates". A more detailed and systematic study using polyvinylpyrrolidone (PVP) as a protective agent (this was the traditional term for polymers, surfactants or other molecules used to stabilize colloids, and which has nowadays been replaced by other terms such as stabilizer or capping agent) and comparing borohydride with other reducing agents, was performed by Van Rheenen et al.[27]

Gold and silver, as well as bimetallic nanoparticles can also be easily formed using borohydride as a reductant, and suitable stabilizers. Liz-Marzán and Philipse[28] studied all possible combinations of Ag, Au and Pt, using clay fibers as supporting colloids. Optical analysis of the obtained colloids suggested that core-shell morphologies were obtained in some cases. Rather monodisperse silver nanoparticles can also be prepared by borohydride reduction in the presence of poly(acrylic acid),[29] and stabilization with citrate ions leads to small (ca. 3 nm) and rather uniform Au nanoparticles, which have been used as seeds for growth of larger spheres,[30] nanoshells,[31] and nanorods,[32] as detailed below.

### 2.1.3. γ-Radiolysis

When speaking about metal nanoparticles and their properties, Prof. Arnim Henglein's name should always be borne in mind. Henglein's pioneering contribution to understanding the special properties, not only of metal nanoparticles and clusters,[33] but also of semiconductor quantum dots[34] has inspired much of the work subsequently performed, as revealed in a recent festschrift issue in The Journal of Physical Chemistry.[35] One of these outstanding contributions has been the development of a very useful technique for the production of metal nanoparticles of controlled size, based on the formation of hydrated electrons and organic radicals in solution through γ-radiolysis, which then acted as reducing agents for metal ions, yielding metal atoms which subsequently coalesce to form larger particles.[33]

Henglein and co-workers were pioneers in the application of radiation chemistry to synthesize very small metal nanoparticles,[36-38] as well as to study their photochemistry[39] and reactivity.[40-43] They also synthesized larger metal particles of Au[44,45] and Ag[50] with narrow size distributions by enlarging Au and Ag seeds, respectively, which in the case of Au constituted a useful material to study fast electronic processes.[46] Additionally, the same group applied radiation chemistry techniques to produce core-shell bimetallic particles. In the presence of suitable stabilizers, shells of different metals and various thickness could be grown around chemically performed seeds by varying the type and concentration of the metal ions in solution. Thus, high quality metal particles such as, Au@Pb,[47,48] Au@Pb@Cd,[47] Ag@Au,[49] Pt@Ag,[50] Au@Hg,[51] Pt@Au[52,53] and Pd@Au[48,54] were produced to study their optical and chemical properties.

Other groups like those of Belloni[55-57] or Doudna[58,59] also applied γ-radiolysis to reduce metals. Most of the systems reported by these groups were metallic alloys with catalytic applications. Recently, Doudna et al.[60] reported a method based on radiolytic reduction to generate filament-like nanostructures of Ag@Pt.

### 2.1.4. Growth on Preformed Nanoparticles

There are many synthetic methods to produce metal nanoparticles with sizes typically below 20 nm, but only a few methods produce particles of uniform size. To obtain larger particles, the difficulty in controlling the nucleation and growth steps occurring at intermediate stages results in a broad particle size distribution.[61] A common approach for controlling the size of nanoparticles is seeding growth.[45,62-65] In seeding growth methods, small metal particles are prepared first and later used as seeds (nucleation centers) for the preparation of particles with larger size. Providing a controlled number of preformed seeds and growth conditions that inhibit any secondary nucleation, the particle size can be controlled simply by varying the ratio between seed and metal salt concentrations. The main difficulty is to find suitable growth conditions to avoid that a new nucleation process takes place during the growth stages, which strongly limits the application of these methods.[66] Henglein employed radiolytic reduction for seeded growth of gold and silver particles, since the flux of $\gamma$-rays can be conveniently controlled to vary the amount of metal atoms produced.[45,67] Natan and co-workers used a reducing agent (hydroxylamine) that is too weak to reduce the gold metal salt without the presence of seeds, since these act as catalysts.[30] Similarly, Murphy and co-workers used ascorbic acid as the reducing agent and added cetyltrimethyl ammonium bromide in the growth solution to complex the gold salt and prevent its reduction in the absence of seeds.[65] The growth mechanism consists of an electron transfer from the reducing agent to the metal particles, followed by the reduction of the gold salt on the surface of the seed, so that the growth process follows electrochemical mixed current theory.[68,69] Schematically,

$$M_n^{(x+1)-} + AuCl_4^- \rightarrow Au_{(n+1)}^{x-} + 4Cl^- \tag{1}$$

where $M_n^{(x+1)-}$ is a metal particle that has already accumulated x electrons.

To avoid additional nucleation, a step-by-step particle enlargement method is more useful, allowing a large metal to salt ratio to be maintained throughout successive growth steps.[66] A limitation of the seeded growth method is the lack of availability of smaller seeds with a narrow size distribution. The seeding growth method has been also used for the preparation of different metallic core-shell particles; Au@Ag,[70] Ag@Au,[71] Pt@Au and Au@Pt.[52]

### 2.1.5. Growth of Silica Shells on Metal Nanoparticles

Among the various surface modification processes developed for metal (and other) colloids, the growth of silica shells with tailored thickness has recently gathered great attention. One of the major reasons for silica coating is the anomalously high stability of silica colloids, especially in aqueous media, but other reasons are the easy control on the deposition process, its processibility, chemical inertness, controllable porosity, and optical transparency. All these properties make of silica an ideal material to modulate surface properties, while maintaining the physical properties of the underlying cores.

Silica coating of metal colloids presents the difficulty of a chemical mismatch between the core and the shell materials. Liz-Marzán and Philipse[72] first approached the synthesis of Au@SiO$_2$ particles through borohydride reduction of AuCl$_4^-$ in the presence of small silica spheres, followed by extensive silica growth in ethanol of the resulting heteroaggregates. Although coated gold nanoparticles were obtained, these were mixed with pure silica spheres, which were difficult to separate.

A method that has provided substantially better results was later designed by Liz-Marzán et al.[73,74] using silane coupling agents as primers to deposit a first monolayer of silanol groups on the surface, which would then be receptive toward deposition of thin silica layers from sodium silicate solution. Subsequent shell growth could be carried out in ethanol using the so-called Stöber method.[75] Monodisperse colloids can be obtained using this procedure with nearly perfect core-shell structures. Examples of such particles are shown in Figure 5 below. This system has proved extremely useful for instance for systematic studies of optical properties, both in solution[74,76,77] and within nanostructures.[11,78,79] A similar strategy can be applied to the coating of silver colloids,[29] with the additional difficulty that dissolution of the Ag cores was observed when concentrated ammonia was added to increase the thickness of the silica shell, leading to formation of hollow silica shells.[80]

Other methods have been devised for the preparation of Ag@SiO$_2$, such as consecutive reactions within microemulsion droplets,[81] or the reduction of Ag$^+$ by *N,N*-dimethylformamide (DMF) in the presence of an amino-silane.[82] However, none of these methods have been used for systematic studies until now.

## 2.2. Spherical Nanoparticles in Organic Solvents

### *2.2.1. Two-Phase Reduction*

One of the currently most used methods for the preparation of gold nanoparticles has been developed by Brust et al.[83,84] in 1994. This method combines borohydride reduction with the use of thio- or amino-derivatives as steric stabilizers, by means of a two-phase reaction. Basically, HAuCl$_4$ is dissolved in water and subsequently transported into toluene by means of tetraoctylammonium bromide, which acts as a phase transfer agent. The toluene solution is then mixed and thoroughly stirred together with an aqueous solution of sodium borohydride, in the presence of thio-alkanes or amino-alkanes, which readily bind to the formed Au nanoparticles. Depending on the ratio between Au salt and capping agent (thiol/amine), the particle size can be tuned between ca. 1 and 10 nm. Several refinements of the preparative procedure including the development of analogous methods for the preparation of silver particles have been reported.[85,86] Murray and co-workers also contributed to enhance the method's popularity by offering an interesting and elegant alternative to the two-phase reduction method which has opened a new field of preparative chemistry. They explored routes to functionalized monolayer protected clusters by ligand place exchange reactions.[87-91] Simple alkane thiol ligands can be completely or partially exchanged by a variety of functional groups, which can be for instance electrochemically active[87,89,90] or photoluminiscent,[90] allowing the chemical modification of their ligand shell.

### *2.2.2. Reduction by the Solvent*

Several examples exist on the reduction of metallic salts by organic solvents. Probably the most popular one has been ethanol, which was long used by Toshima and co-workers for the preparation of metal nanoparticles such as Pt, Pd, Au or Rh (suitable for catalytic applications) in the presence of a protecting polymer.[92-94] Another interesting example is found in Figlarz's polyol method, which was initially developed for the formation of larger colloidal particles,[95,96] though later was also adapted and improved for the production of silver nanoparticles.[97] Propanol was also used by Mills and coworkers for the preparation of silver colloids in basic conditions.[98] Related to these processes we can also mention the reduction of noble metal salts by non-ionic surfactants, and more specifically by those with a large number of ethoxy groups, to which the reducing ability has been assigned.[99,100]

N,N-dimethylformamide (DMF) is also one of the usual organic solvents for numerous processes, and it has been shown to behave as an active reducing agent under suitable conditions. Pastoriza-Santos et al. reported the ability of DMF to reduce $Ag^+$ ions to the zero-valent metal, even at room temperature and in the absence of any external reducing agent.[101,102] Stable spherical silver nanoparticles were synthesized in DMF, using poly(vinyl pyrrolidone) as a stabilizer,[103] but additionally silica-[82] and titania-[104] coated nanoparticles were produced by the same method, in the presence of aminopropyl-trimethoxysilane and titanium tetrabutoxide, respectively. It was also shown that formamide[105] can act as a powerful reductant for silver or gold salts in the absence of oxygen. More recently, Diaz and coworkers reported the preparation of silver nanoparticles by the spontaneous reduction of silver 2-ethylhexanoate in dimethyl sulfoxide (DMSO) at room temperature.[106]

*2.2.3. Reduction within Microemulsions*

Microemulsions have been used as confined reaction media during the past two decades, since, due to the very small size of the droplets, they can act as microreactors capable to control the size of the particles and at the same time to inhibit the aggregation by adsorption of the surfactants on the particle surface when the particle size approaches that of the microreactor droplet. The synthesis of nanoparticles using reactions in microemulsions was first described by Boutonnet and coworkers[107] They synthesized monodispersed metal particles of Pt, Pd, Rh and Ir by reduction of metal salts with hydrogen or hydrazine in water in oil (w/o) microemulsions. Since then, many different types of materials have been prepared using microemulsions, including metal carbonates,[108] metal oxides,[109,110] metal chalcogenides,[111-113] polymers,[114] etc.

In the case of noble metals, several protocols have been described to produce metal nanosized particles by chemical reduction in microemulsions. Pileni and co-workers synthesized metallic particles of Ag,[115] Cu,[116] Co[117] (and other materials) by reduction of the corresponding salts in w/o microemulsions using borohydride or hydrazine as reducing agents. They also showed a correlation between the structure of the mesophase in the surfactant system with the size and shape of the formation of metallic particle.

The wide range of applications (in catalysis, as biological stains, as condensers for electron storage in artificial photosynthesis or as ferrofluids) has also motivated studies by other groups on the formation of colloidal Ag,[118] Au,[119] Ni,[120] Cu,[121] Pd[122] and Co[123] particles. However, there is still a strong controversy on whether the microemulsion droplet size actually controls the final size of the obtained nanoparticles.

In recent years, the use of supercritical solvents has offered a new alternative in the synthesis of nanoparticles in microemulsions which presents several advantages over conventional organic solvents, because their solvation properties can be easily controlled by varying pressure. Colloidal silver[124-126] and copper[126] have been synthesized using microemulsions of water in supercritical $CO_2$ with fluorinated surfactants.

## 2.3. Nanorods and Nanoprisms in Water

### *2.3.1. Synthesis of Nanorods within Porous Membranes*

The first procedure reported for the production of metal nanorods is called template synthesis.[127,128] This method entails the preparation or deposition of the desired material within the cylindrical and monodisperse pores of a nanopore membrane. Martin and co-workers used polycarbonate filters, prepared by the track-etch method,[129] and nanopore aluminas prepared electrochemically from Al foil,[130] as template materials. This method allows the preparation of cylindrical nanostructures with monodisperse diameters and lengths, and depending on the nature of the membrane and the synthetic method used, these may be solid nanowires or hollow nanotubes.[131]

The earliest applications of the template method consisted of the preparation of microscopic and macroscopic electrodes prepared by depositing Au within the pores of policarbonate membranes using electrochemical[132] or electroless plating[133] methods. In the electrochemical method, the Au deposition is more uniform and starts at the pore walls creating, at short deposition times, hollow Au nanotubes within the pores.[134] Further deposition results in the formation of Au nanowires. The electroless method comprises the application of a catalyst to all the surfaces of the templating membrane. The membrane is then immersed into the electroless plating bath which contains $Au^I$ and a chemical reducing agent. Considering that the reduction of $Au^I$ to metallic Au only occurs in the presence of the catalyst, Au nanotubes are obtained along the pore walls. The thickness of the nanotube walls increases and the inner diameter of the tubes decreases with electroless plating time. At longer plating times, membranes containing nanowires are obtained.

A sequential electrodeposition within a porous template can be used to prepare striped nanowires with tailorable dimensions and composition.[135] Variation in composition along the length of the wire can be used to incorporate electrical functionalities,[136] optical contrast,[137] and the desired surface chemistry.[138]

Although very uniform thickness and controlled length were achieved, the amounts that could be made following this procedure were always very small, but still considerable basic work on the optical properties and alignment of these rods could be achieved.[139-141]

### *2.3.2. Nanorods from Wet Synthesis in Solution*

Wang and co-workers have developed an electrochemical solution phase synthesis of gold nanorods.[142,143] Their synthetic approach is to control the growth by introducing into the electrochemical system appropriate "shape-inducing" cationic surfactants and other additives that were found empirically to favor rod formation and act as both the supporting electrolyte and the stabilizer for the resulting cylindrical Au nanoparticles. In the electrochemical method for Au nanorod formation the micellar system consists of two cationic surfactants: cetyltrimethyl ammonium bromide (CTAB) and the much more hydrophobic surfactant tetradecyl ammonium bromide (TDAB) or tetraoctyl ammonium bromide (TOAB). The ratio between the surfactants controls the average aspect ratio of the gold nanorods. The synthesis works in a small scale and is very difficult to carry out on a large scale, but represents a landmark in terms of shape control. The groups of El-Sayed and Hartland have shown that such rods possess quite unique and interesting optical properties. The particles may be selectively melted using laser pulses and show exciting coupling of the electron gas to bending modes.[144,145] Additionally, very strong optical polarization effects have been observed in nanorod polymer films.[146]

Subsequently, Murphy and co-workers discovered reduction conditions that enable the entire synthesis of gold and silver nanorods to be carried out directly in solution,[32,147,148] and their protocols enabled excellent aspect ratio control. In this seed mediated method a preformed silver or gold seed was used to promote gold or silver growth in solution in the presence of CTAB to promote nanorod formation. A mild reducing agent, such as ascorbic acid was used to avoid any extra nucleation, while the addition of different volumes of the seed solution produced gold nanorods with different aspect ratios. In the case of gold nanorods, the reproducibility depends on the presence of adventitious ions such as $Ag^+$ and stringent seed preparation conditions. A mechanism for rod formation could not be provided.

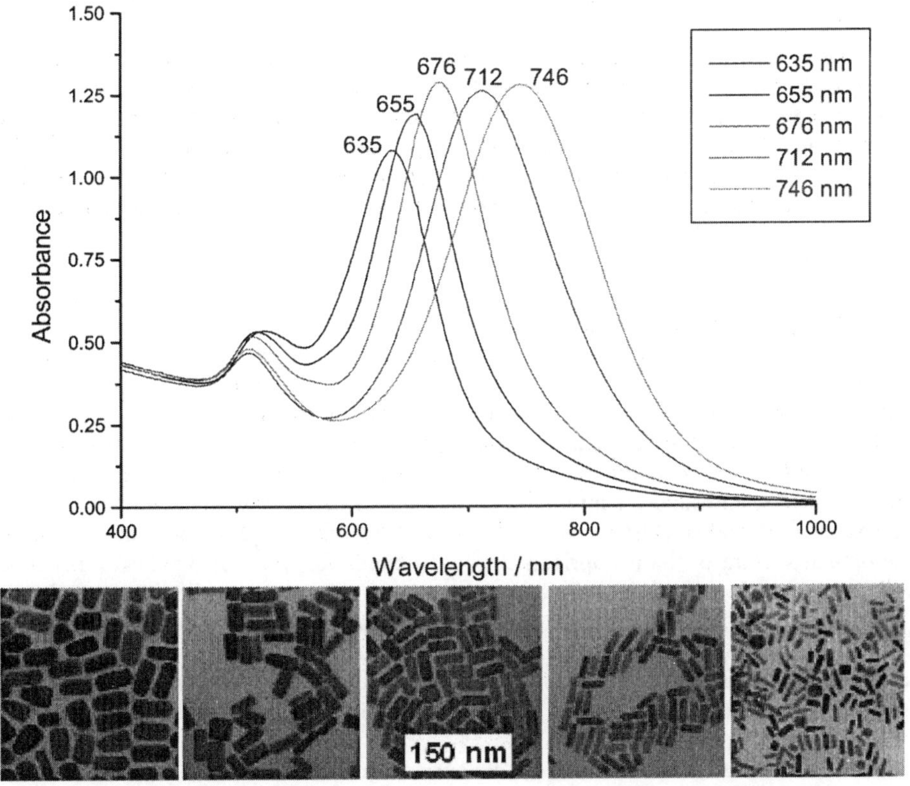


**Figure 2.** UV-visible spectra and TEM micrographs of Au nanorods prepared by chemical growth on small seeds (varying seed concentration), in the presence of CTAB and $AgNO_3$.

The seed mediated method has been recently improved by El-Sayed and co-workers,[149] resulting in a spectacular increase in the yield of rod formation, similar to the photochemical method but easier to produce. Two modifications were applied to the seed-mediated growth method: (a) replacement of citrate with CTAB molecules in the seed formation step and (b) adjustment of the silver content of the growth solution to grow nanorods with controlled aspect ratios, resulting in the formation of only 0-1% of spheres. Examples of rods formed using this procedure are shown in Figure 2. In the same figure, the corresponding UV-visible spectra were plotted, to show how an increase in aspect ratio leads to a red-shift of the longitudinal plasmon band, while the transversal resonance only slightly varies due to differences in particle thickness.

Other methods that were recently reported include a seedless and surfactantless wet chemical method by Murphy and co-workers, to obtain silver nanowires in the presence of sodium citrate and sodium hydroxide,[150] as well as a photochemical method developed by Yang et al. for the synthesis of gold nanorods.[151] In contrast to the seed mediated mechanism proposed by Murphy, in the photochemical method the addition of a preformed gold seed is not needed. However, different amounts of silver nitrate are added to the growth solution containing CTAB and the gold salt before irradiating the solution with UV light. Again, the role of the $Ag^+$ in controlling the aspect ratio of the rods could not be determined.

Some work has also been recently carried out on the solution synthesis of bimetallic nanorods, employing gold nanorods as preformed cores. Jang and co-workers synthesized Au@Ag core-shell nanorods using the electrochemical method to obtain the gold nanorods and the seed-mediated mechanism to get the desired core-shell structure.[152] The combination of the different optical resonances from the core and the shell metals leads to a quite spectacular optical spectrum from these composite, anisotropic nanoparticles.

### 2.3.3. Synthesis of Nanoprisms in Water

Although some triangular and polygonal flat nanoprisms are often obtained during every metal colloid preparation, the synthesis of metal flat nanoprisms with relatively high yield is quite recent. Only in November 2001,[153] Jin et al. reported the conversion of citrate-stabilized silver nanospheres into (truncated) triangular nanoprisms driven by irradiation with a 40-W fluorescent lamp, in the presence of bis(p-sulfonatophenyl) phenylphosphine. Apparently, the initial silver nanoparticles are first fragmented into smaller culsters, and subsequently there is a growth of particles with nanoprism shape, at the cost of the smaller clusters, which is induced by the phosphine present in solution. The optical spectra of the nanoprisms displayed bands for in-plane dipole resonance, as well as for in-plane and out-of-plane quadrupole resonances, while the out-of-plane dipole was only reflected as a small shoulder. Truncated triangles were also obtained by Chen and Carrol[154] using a procedure quite similar to the seeded growth method described for Au nanorod formation. In this case, the reduction was performed using ascorbic acid to reduce silver ions on silver seeds in a basic solution of concentrated CTAB.

Malikova et al.[155] synthesized Au nanoprisms in aqueous solution by reduction of neutralized $HAuCl_4$ with salicylic acid at 80 °C. Although the yield of nanoprisms was not outstanding, the formation of thin films showed that strong optical coupling occurs when the nanoprisms are close enough to each other (see Section 3.3 below).

## 2.4. Nanorods and Nanoprisms in Organic Solvents

### 2.4.1. Reduction within Microemulsions

While the microemulsion method has been widely applied to the production and stabilization of spherical metal particles with various sizes and compositions, shape control of noble metallic particles using this procedure has only been demonstrated in a handful of studies to date. Pileni and co-workers demonstrated that it is possible to control nanocrystal shape to some extent within microemulsions.[156] Although the shape of the templates plays a role during the growth of the nanocrystals, these authors showed that the particle shape can be controlled even if the microscopic structure of the self-assembled surfactant system used as a template remains unchanged and that addition of salt to the templates can induce drastic changes in the particle shape. Recently, the same group also reported the synthesis of silver nanodisks in reverse micellar solution by reduction of Ag(AOT) with hydrazine, with various sizes that depended on the relative amount of hydrazine, but with constant aspect ratio.[157]

Ni nanorods have been synthesized by reduction of nickel chloride with hydrazine hydrate in w/o microemulsion.[158] Another use of interfaces to drive the synthesis of nanomaterials of variable morphology was introduced very recently by Sanyal and coworkers,[159] who demonstrated that the reduction of aqueous chloroaurate ions by anthracene anions bound to a liquid-liquid interface produced thin nanosheets of gold.

### 2.4.2. Reduction by the Solvent

Xia and coworkers recently demonstrated that the polyol method[95] can be applied for the production of silver nanowires by reducing silver nitrate with ethylene glycol in the presence of poly(vinyl pyrrolidone)(PVP).[160-162] Silver nanoparticles with high aspect ratios were only formed in the presence of (Pt) seeds formed *in situ* prior to the addition of the silver salt. In the course of refluxing, These authors claim that Ostwald ripening takes place at the high temperatures used, forming nanoparticles with larger sizes at the expense of smaller ones. It is likely that PVP controls the growth rates of various faces of silver by coordinating to the surface. As a result, silver nanowires with diameters in the range of 30-60nm and lengths up to ~50μm were formed. They could control the dimensions of the silver nanowires by varying the experimental conditions (temperature, seed concentration, ratio between silver salt and PVP, etc...). In a later report,[163] the same group showed that, again through control of the concentrations of silver salt and PVP, silver nanocubes can be produced, which are single crystals, but with slightly truncated corners and edges. Additionally, all these silver particles with different shapes can be converted into hollow nanostructures (nanotubes, nanocubes, etc.) by reacting with salts of Au, Pt or Pd in water at reflux,[164] which opens a new way to exciting morphologies with interesting properties.

### 2.4.3. Shape Control Using DMF

As it is previously shown, N,N-dimethylformamide (DMF) can reduce $Ag^+$ ions to the zero-valent metal, even at room temperature and in the absence of any external reducing agent,[82] which takes place through the following reaction:

$$HCONMe_2 + 2Ag^+ + H_2O \rightarrow 2Ag^0 + Me_2NCOOH + 2H^+ \tag{2}$$

A basic difference between this reaction and the reduction by other organic compounds is that it proceeds at a meaningful rate even when performed at room temperature and in the dark. This is readily observed through yellow coloration of the solution, which deepens with time. This reducing ability under mild conditions points toward a larger tendency of this solvent for the reduction of $Ag^+$ as compared to ethanol or other organic solvents.

Interestingly, the shape (and size) of the obtained nanoparticles depends on several parameters, such as silver salt and stabilizer concentrations, temperature and reaction time. Specifically, when poly(vinyl pyrrolidone) (PVP) is used as a protecting agent, spherical nanoparticles form at low $AgNO_3$ concentration (0.76 mM), while increasing silver concentration (up to 0.02 M) the formation of anisotropic particles is largely favored, though again the concentration of PVP and the reaction temperature strongly influence the shape of the final particles.

Figure 3 shows examples of nanospheres, nanorods and nanoprisms obtained by reduction of $AgNO_3$ in DMF, in the presence of PVP. It can be clearly observed that for high Ag concentration, at low PVP content nanorods are mainly formed, but when higher PVP concentrations are used, the formation of nanorods is suppressed, and mainly trianglular and other polygonal nanoplates predominate, which indicates that at low PVP:Ag ratios some crystallographic facets can become more active with respect to nanoparticle growth. In fact, TEM observation shows that initially, small spheres are formed which then assemble into certain shapes, which are determined by the crystallographic structure of the initial nanocrystals, and ultimately a restructuration takes place, which leads to fcc single crystals with well-defined shapes.[165] Additionally, in the case of nanoprisms, it was also observed that they become larger with time, and a wider variety of shapes are found for longer boiling times.

Atomic force microscopy (AFM) measurements demonstrated the flat geometry of the nanoprisms, as shown in Figure 3, indicating an average thickness of ca. 35 nm for lateral dimensions of the order of 200 nm. It can also be observed in this image that most of the nanoprisms are truncated triangles, which sometimes leads to other polygonal shapes.

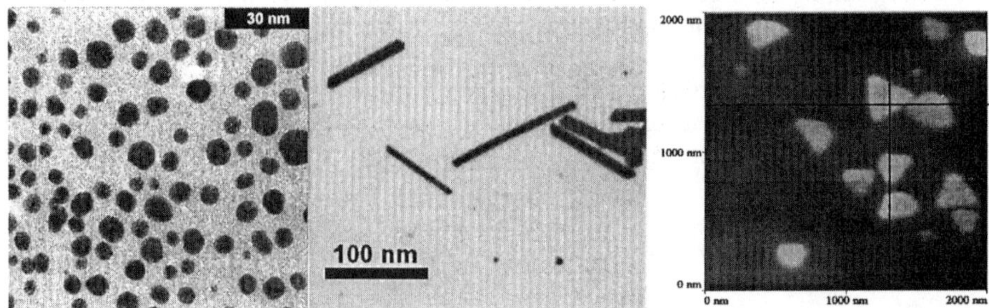


**Figure 3.** TEM micrographs showing Ag nanoparticles obtained by reduction in DMF in the presence of PVP: (a) [PVP]=0.76 mM, [$Ag^+$]=0.76 mM; (b) [PVP]=0.06 mM, [$Ag^+$]=0.02 M. (c) AFM image of Ag nanosprisms obtained by reduction in DMF ([PVP]=0.4 mM, [$Ag^+$]=0.02 M).

# 3. METAL COLLOID STRUCTURES THROUGH LAYER-BY-LAYER ASSEMBLY

## 3.1. Layer-by-Layer Assembly

The use of polyelectrolytes for the assembly of nanoparticles was developed only recently,[166] but has experienced a tremendous advance during the last decade. The basis of this, so-called layer-by-layer (LBL) technique is the electrostatic attraction between oppositely charged species.[167] Briefly, a charged substrate is immersed within a solution of an oppositely charged polyelectrolyte, so that the surface charge is reversed, which is known as overcompensation effect. A subsequent immersion in a solution of nanoparticles with the appropriate surface charge leads to the deposition of a homogeneous, compact nanoparticle monolayer on the surface. The process can be repeated in successive cycles to deposit more monolayers and build up a hybrid film. If we compare this process to traditional deposition techniques, such as sputtering and vacuum evaporation, layer-by-layer assembly possesses features that make it specially valuable for the formation of complex nanostructures. Importantly, LBL can be carried out at room temperature and does not require subsequent annealing of the deposited film, so that a wide variety of substrates can be used. Furthermore, the shape and size of the constituting nanoparticles can be selected prior to assembly, which makes possible a careful design of the properties of the film by means of the proper choice of the colloid. With respect to other standard film formation techniques, such as Langmuir-Blodgett deposition, LBL assembly is much simpler to carry out and does not require special equipment such as Langmuir troughs. However, it is also true that Langmuir-Blodgett films show a higher in-plane order.

All these features make LBL an ideal technique for the assembly of core-shell and anisotropic nanoparticles (among others), since the nanoparticles can be completely synthesized prior to the assembly process, and the desired properties can be designed in the first synthetic stage. We only describe here two examples of the versatility of this technique for the assembly of metal nanoparticles, which allows to obtain structures with special optical features.

## 3.2. Assembly of Au@SiO$_2$

The interest of the assembly of silica-coated gold nanoparticles relies on the possibility of controlling the particle volume fraction by means of the variation of the silica shell thickness, provided that the assembled films are close-packed. In close-packing conditions, the separation between metal cores is just twice the thickness of the coating shell, which can be controlled during the synthesis of the colloids. This system has been recently studied in detail by Ung et al.,[11,168] and described here are the main results on the influence of metal nanoparticles volume fraction (through interparticle interactions) on the optical properties of the films.

In refs. 11 and 168, experimental and calculated absorbance and reflectance spectra are compared for thin films with different metal contents (through variation of silica shell thickness). This comparison shows that, as the gold volume fraction increases (the separation between nanoparticles decreases), there is a red-shift of the plasmon resonance, as well as a broadening of the band. This effect originates in the dipole-dipole interactions between neighboring nanoparticles, and a separation of just 15 nm is sufficient to screen such interactions, so that the thin films display basically the same properties as a dilute dispersion of the same nanoparticles in water. The agreement between the experimental results and the calculated spectra demonstrates that Maxwell-Garnett effective medium theory effectively accounts for such dipole-dipole interactions.

Thus, both the transmission and reflection properties of thin gold films can be easily controlled by means of an adjustment of the thickness of the silica shell surrounding each nanoparticle, so that dipole-dipole interparticle interactions are effectively screened. Detailed studies using different metal core sizes have not been carried out yet, but it is expected that the distance at which interactions are effectively screened will scale up with particle size, being of the order of one particle diameter.

The LBL method can also be used for the assembly of nanoparticles on suitable colloid templates. This procedure was recently developed by Caruso et al.[169,170] and has found application in a great variety of materials.[171] In this system, the optical effects observed for the macroscopic thin films are obtained for colloidal particles, which possess themselves a core-shell geometry and can be used in turn for the construction of more complex, nanoscale materials. The resulting spheres are essentially different to continuous metal shells grown on colloid templates, which have been recently reported.[31,172] Such continuous shells display optical properties associated to resonances along the whole shell, and are therefore extremely sensitive to both core size and shell thickness, while in the system presented here the optical properties only depend on the nature and dimensions of the constituting units and the number of deposited monolayers.[173,174]

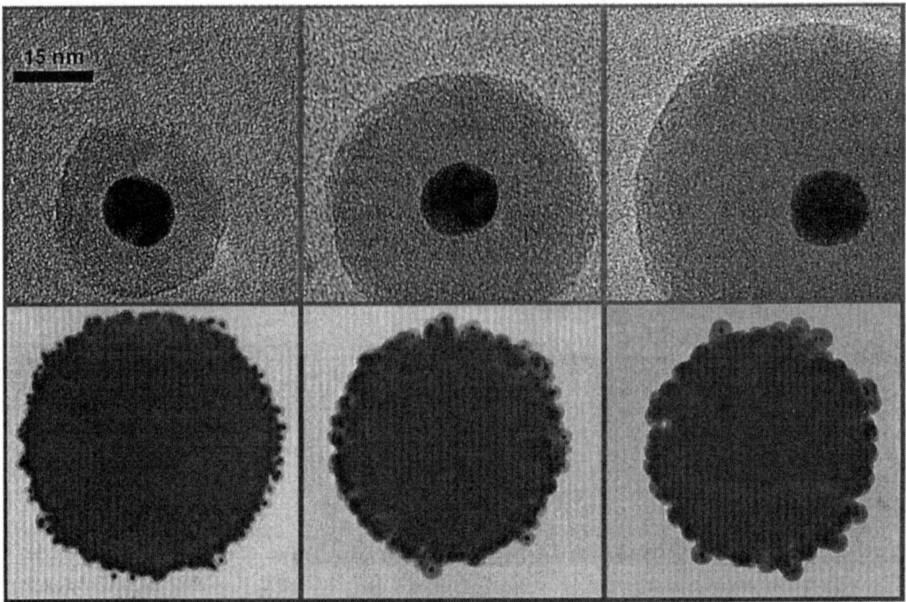


**Figure 4.** Top row: TEM micrographs of Au@SiO$_2$ nanoparticles with 13 nm Au cores and varying shell thicknesses (8, 18, 28 nm). Bottom row: TEM micrographs of 640 nm latex spheres coated with five monolayers of the Au@SiO$_2$ nanoparticles shown on the top row.

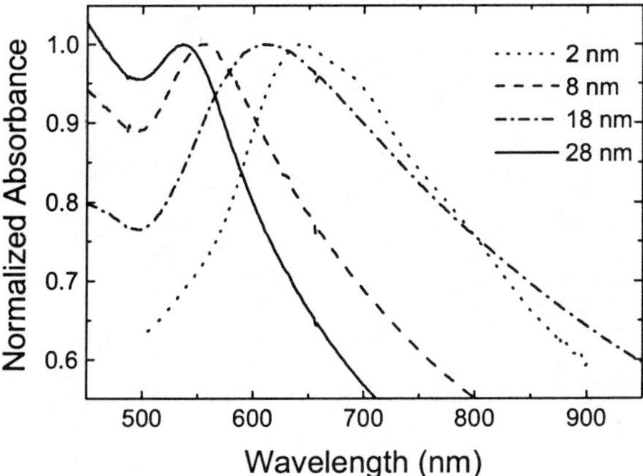

**Figure 5.** Normalized UV-visible spectra of dilute dispersions of 640 nm latex spheres coated with five monolayers of Au@SiO₂ nanoparticles. The thickness of the corresponding silica shells is indicated.

Examples of nanostructured colloids formed by assembly on 640 nm ploystyrene spheres, of Au@SiO₂ with identical cores but silica shells with various thickness are shown in Figure 4. It is clear that, although the outer surface is rough, the obtained coating layers are very compact and uniform, so that, again, the interparticle separation will be determined by the corresponding thickness of the silica shell. UV-visible spectra of 640 nm polystyrene colloids coated with five monolayers of Au@SiO₂ nanoparticles of various shell thickness are shown in Figure 5. As expected, for thicker silica shells (increased separation between gold nanoparticles), the plasmon resonance is close to that of the initial, isolated Au nanoparticles. Due to the increase in particle size, there is also an increased scattering contribution as more monolayers are deposited, which is reflected in the dramatic increase of the low-wavelength tail of the spectra. These scattering effects also slightly affect the actual position of the plasmon band, as demonstrated by a larger shift as the size of the polystyrene cores is increased.[174] Similar effects have also been observed using Ag cores, thus showing that by proper choice of the nanoparticle units, composite colloids with tailored optical features can be synthesized.

### 3.3. Assembly of Au Nanoprisms

Gold nanoplates with polygonal shapes can be synthesized by reduction with salicylic acid in water at 80 °C.[155] An example of a triangular nanoplate is shown in the AFM image of Figure 6, which clearly demonstrates the flat geometry. However, these nanoprisms are accompanied during the synthesis by a relatively large amount of spheres of similar size, which basically impedes a good separation, and therefore, the optical spectrum of the colloid shows two distinct plasmon bands, one at 535 nm due to the dipole resonance of the spheres and another at 840 nm corresponding to the in-plane resonance of the nanoprisms. Due to the polydispersity, the band at high wavelength is quite broad. The stabilizing layer of salicylic acid renders the formed particles negatively charged, and can thus be assembled by LBL using a positively charged polyelectrolyte, such as poly(diallyldimethylammonium chloride) (PDDA).

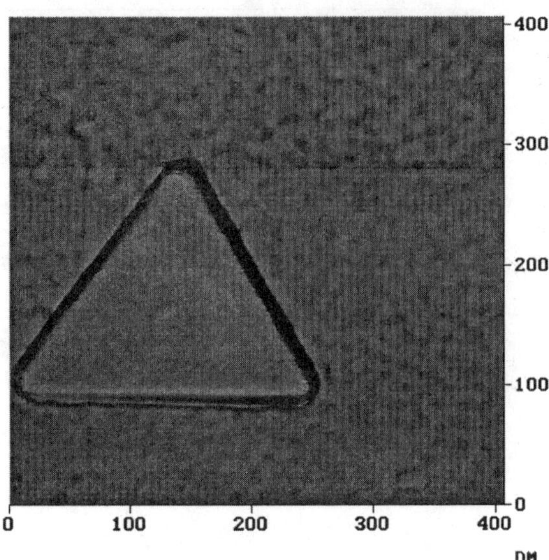

**Figure 6.** AFM image showing the flat nature of the Au triangular nanosprisms prepared by reduction with salicylic acid.

AFM analysis of the deposited monolayers shows that the polygons lie parallel to the substrate, and are embedded in a compact monolayer of the spherical nanoparticles. Since this is repeated for each deposited monolayer, it results in a homogeneous increase in the absorbance spectrum as successive monolayers are deposited (see Figure 7a). However, although the first deposited gold monolayer displays absorption bands very similar to those of the starting nanoparticle dispersion, a new optical feature develops as the number of layers increases. An intense band is formed between the two previous ones becoming clearly visible in the fourth layer and becoming dominant by the seventh bilayer. This new band at 650 nm arises from interparticle interactions between neighboring monolayers, in a similar way to what was discussed in the previous section.

This new band is very likely due to interactions between spherical nanoparticles within neighboring layers, but also with the nanoprisms. In the latter case, the coupling strength should be higher than in former because the geometry of sphere on a plane results in greater integral electrostatic attraction than that between two spheres. The variety of different geometrical arrangements causes band broadening, which can also be seen in Figure 7a. The origin of this new band was confirmed[155] by means of further LBL experiments in which insulting layers were used to separate the Au nanoparticle bilayers. This was accomplished by depositing a nanoscale spacer made from montmorillonite clay platelets, which are better electrical insulators than polyelectrolytes and prevent interdigitation between adjacent layers due to their sheet-like morphology.[175] The resulting optical spectra are plotted in Figure 7b, and show that in this case the position of both bands remains basically unaltered, which is due to insulation of the adjacent gold nanoparticle layers. There is still a slight red-shift of the maxima in comparison to the UV-visible spectrum of the dispersion because of the refractive index increase after the deposition of the montmorillonite sheets.

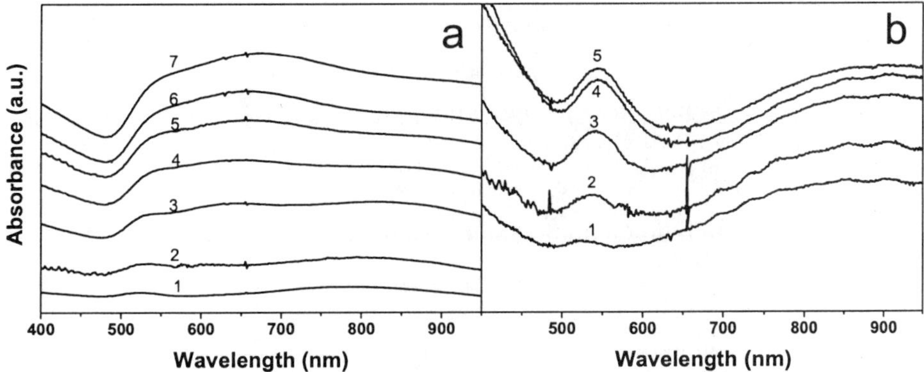

**Figure 7.** UV-visible spectra of sequentially deposited (PDDA/Au)$_n$, $n=1$-7  (a); and (PDDA/ Au/PDDA/clay)$_n$, $n = 1$-5 films (b).

## 4. CONCLUSIONS

A large variety of synthetic procedures have been developed for the preparation of noble metal nanoparticles with various sizes and shapes, and which are stable in a wide range of solvents. The optical properties of these nanomaterials are currently quite well understood, since the quality of the current preparation methods allows to obtain a high monodispersity, so that simplified theories can be used.

Since it was not within the scope of this chapter, we have only presented a couple of examples of structures that can be fabricated using noble metal colloids as building units. These examples showed that the coupling between particles in close contact leads to modifications in the optical properties, which ultimately allows to engineer the distribution of nanoparticles and thereby obtain the desired optical properties. Since very little has been done in this field by taking profit of the wealth of properties arising from the newly devised anisotropic metal nanoparticles, we should expect huge developments in this area within the coming years.

## 6. ACKNOWLEDGEMENTS

The authors are grateful to all the collaborators who contributed to carry out part of the results described in this chapter. In particular, Benito Rodríguez-González and Verónica Salgueiriño-Maceira are thanked for providing some of the figures. Financial support from Ministerio de Ciencia y Tecnología and Xunta de Galicia is acknowledged.

## 7. REFERENCES

1.    http://www.thebritishmuseum.ac.uk/science/text/lycurgus/sr-lycurgus-p1-t.html.
2.    http://www.magicalcolloids.com/
3.    M. Faraday, Experimental relations of gold (and other metals) to light, *Philos. Trans. Roy. Soc. Lon.* **147**, 145-181 (1857).

4.   U. Kreibig and M. Vollmer, *Optical Properties of Metal Clusters* (Springer-Verlag, Berlin, 1996).
5.   J. A. A. Perenboom, P. Wyder, and F. Meier, Electronic properties of small metallic particles, *Phys. Rep.* **78**, 173-292 (1981).
6.   C. F. Bohren and D. F. Huffman, *Absorption and Scattering of Light by Small Particles* (Wiley, New York, 1983).
7.   J. A. Creighton and D. G. Eadon, Ultraviolet-visible absorption spectra of the colloidal metallic elements, *J. Chem. Soc. Faraday Trans.* **87**, 3881-3891 (1991).
8.   S. Link and M. A. El-Sayed, Spectral properties and relaxation dynamics of surface plasmon electronic oscillations in gold and silver nanodots and nanorods, *J. Phys. Chem. B* **103**, 8410-8426 (1999).
9.   P. Mulvaney, Surface plasmon spectroscopy of nanosized metal particles, *Langmuir* **12**, 788-800 (1996).
10.  S. Underwood and P. Mulvaney, Effect of the solution refractive index on the color of gold colloids, *Langmuir*, 3427-3430 (1994).
11.  T. Ung, L. M. Liz-Marzán, and P. Mulvaney, Optical properties of thin films of Au@SiO2 particles, *J. Phys. Chem. B* **105**, 3441-3452 (2001).
12.  R. Elghanian, J. J. Storhoff, R. C. Mucic, R. L. Letsinger, and C. A. Mirkin, Selective colorimetric detection of polynucleotides based on the distance-dependent optical properties of gold nanoparticles, *Science* **227**, 1078-1080 (1997).
13.  J. S. Bradley, The chemistry of transition metal colloids. In: *Clusters and Colloids. From Theory to Application*, edited by G. Schmid (VCH, Wienhiem, 1994), pp. 459-544.
14.  N. Toshima, Metal nanoparticles for catalysis. In: *Nanoscale Materials*, edited by L. M. Liz-Marzán and P. V. Kamat (Kluwer Academic Publishers, Boston, 2003), pp. 79-96.
15.  J. Turkevich, P. C. Stevenson, and J. Hillier, A study of the nucleation and growth processes in the synthesis of colloidal gold, *Discuss. Faraday Soc.* **11**, 55-75 (1951).
16.  B. V. Enüstun and J. Turkevich, Coagulation of colloidal gold, *J. Am. Chem. Soc.* **85**, 3317-3328 (1963).
17.  G. Frens, Controlled nucleation for the regulation of the particle size in monodisperse gold suspensions, *Nature Phys. Science* **241**, 20-22 (1973).
18.  M. K. Chow, C. F. Zukoski, and F. Grieser, The role of colloidal stability in the formation of gold sols, *J. Colloid Interface Sci.* **160**, 511-513 (1993).
19.  M. K. Chow and C. F. Zukoski, Gold sol formation mechanisms: role of colloidal stability, *J. Colloid Interface Sci.* **165**, 97-109 (1994).
20.  S. Link, Z. L. Wang, and M. A. El-Sayed, Alloy formation of gold-silver nanoparticles and the dependence of the plasmon absorption on their composition, *J. Phys. Chem. B* **103**, 3529-3533 (1999).
21.  B. Rodríguez-González, A. Sánchez-Iglesias, M. Giersig, and L. M. Liz-Marzán, AuAg Bimetallic nanoparticles: formation, silica-coating and selective etching, *Faraday Discuss.*, in press.
22.  D. N. Furlong, A. Launikonis, W. H. F. Sasse, and J. V. Sanders, Colloidal platinum sols. Preparation, characterization, and stability towards salt, *J. Chem. Soc., Faraday Trans. 1* **80**, 571-88 (1984).
23.  J. Turkevich, R. S. Miner, Jr., and L. Babenkova, Further studies on the synthesis of finely divided platinum, *J. Phys. Chem.* **90**, 4765-4767 (1986).
24.  L. Yiping, G. C. Hadjipanayis, C. M. Sorensen, and K. J. Klabunde, Magnetic properties of fine iron-cobalt-boron particles prepared by chemical reduction, *J. Appl. Phys.* **69**, 5141-5143 (1991).
25.  G. N. Glavee, K. J. Klabunde, C. M. Sorensen, and G. C. Hadjapanayis, Borohydride reductions of metal ions. A new understanding of the chemistry leading to nanoscale particles of metals, borides, and metal borates, Langmuir **8**, 771-773 (1992).
26.  H. C. Brown and C. A. Brown, A simple preparation of highly active platinum metal catalysts for catalytic hydrogenation, *J. Am. Chem. Soc.* **84**, 1494-1495 (1962).
27.  P. R. Van Rheenen, M. J. McKelvy, and W. S. Glausinger, Synthesis and characterization of small platinum particles formed by the chemical reduction of chloroplatinic acid, *J. Solid State Chem.* **67**, 151-169 (1987).
28.  L. M. Liz-Marzán and A. P. Philipse, Stable hydrosols of metallic and bimetallic nanoparticles immobilized on imogolite fibers, *J. Phys. Chem.* **99**, 15120-15128 (1995).
29.  T. Ung, L. M. Liz-Marzán, and P. Mulvaney, Controlled method for silica coating of silver colloids. Influence of coating on the rate of chemical reactions, *Langmuir* **14**, 3740-3748 (1998).
30.  K. R. Brown and M. J. Natan, Hydroxylamine seeding of colloidal Au nanoparticles in solution and on surfaces, *Langmuir* **14**, 726-728 (1998).
31.  S. J. Oldenburg, R. D. Averitt, S. L. Westcott, and N. J. Halas, Nanoengineering of optical resonances, *Chem. Phys. Lett.* **288**, 243-247 (1998).
32.  N. R. Jana, L. Gearheart, and C. J. Murphy, Wet chemical synthesis of silver nanorods and nanowires of controllable aspect ratio, *Chem. Comm.* 617-618 (2001).

33. A. Henglein, Physicochemical properties of small metal particles in solution: "microelectrode" reactions, chemisorption, composite metal particles, and the atom-to-metal transition, *J. Phys. Chem.* **97**, 5457-5471 (1993).

34. A. Henglein, Small-particle research: physicochemical properties of extremely small colloidal metal and semiconductor particles, *Chem. Rev.* **89**, 1861-1873 (1989).

35. Arnim Henglein's Festschrift, special issue in *J. Phys. Chem. B* (2003).

36. A. Henglein, Nonmetallic silver clusters in aqueous solution: stabilization and chemical reactions, *Chem. Phys. Lett.* **154**, 473-476 (1989).

37. S. Mosseri, A. Henglein, and E. Janata, Reduction of dicyanoaurate(I) in aqueous solution: formation of nonmetallic clusters and colloidal gold, *J. Phys. Chem.* **93**, 6791-6795 (1989).

38. P. Mulvaney and A. Henglein, Long-lived nonmetallic silver clusters in aqueous solution: a pulse radiolysis study of their formation, *J. Phys. Chem.* **94**, 4182-4188 (1990).

39. T. Linnert, P. Mulvaney, and A. Henglein, Photochemistry of colloidal silver particles: the effects of nitrous oxide and adsorbed cyanide ion, *Ber. Bunsen-Ges. Phys. Chem.* **95**, 838-841 (1991).

40. P. Mulvaney, T. Linnert, and A. Henglein, Surface chemistry of colloidal silver in aqueous solution: observations on chemisorption and reactivity, *J. Phys. Chem.* **95**, 7843-7846 (1991).

41. A. Henglein, P. Mulvaney, T. Linnert, and A. Holzwarth, Surface chemistry of colloidal silver: reduction of adsorbed cadmium(2+) ions and accompanying optical effects, *J. Phys. Chem.* **96**, 2411-2414 (1992).

42. A. Henglein, P. Mulvaney, and T. Linnert, Chemistry of silver $Ag_n$ aggregates in aqueous solution: nonmetallic oligomeric clusters and metallic particles, *Faraday Discuss.* **92**, 31-44 (1992).

43. F. Strelow and A. Henglein, Time resolved chemisorption of $I^-$ and $SH^-$ on colloidal silver particles (a stopped-flow study), *J. Phys. Chem.* **99**, 11834-11838 (1995).

44. A. Henglein, Radiolytic preparation of ultrafine colloidal gold particles in aqueous solution: optical spectrum, controlled growth, and some chemical reactions, *Langmuir* **15**, 6738-6744 (1999).

45. A. Henglein and D. Meisel, Radiolytic control of the size of colloidal gold nanoparticles, *Langmuir* **14**, 7392-7396 (1998).

46. J. H. Hodak, A. Henglein, and G. V. Hartland, Photophysics of nanometer sized metal particles: electron-phonon coupling and coherent excitation of breathing vibrational modes, *J. Phys. Chem. B* **104**, 9954-9965 (2000).

47. F. Henglein, A. Henglein, and P. Mulvaney, Surface chemistry of colloidal gold: deposition and reoxidation of Pb, Cd, and Tl, *Ber. Bunsen-Ges. Phys. Chem.* **98**, 180-189 (1994).

48. P. Mulvaney, M. Giersig, and A. Henglein, Surface chemistry of colloidal gold: deposition of lead and accompanying optical effects, *J. Phys. Chem.* **96**, 10419-10424 (1992).

49. P. Mulvaney, M. Giersig, and A. Henglein, Electrochemistry of multilayer colloids: preparation and absorption spectrum of gold-coated silver particles, *J. Phys. Chem.* **97**, 7061-7064 (1993).

50. A. Henglein, Reduction of $Ag(CN)_2^-$ on silver and platinum colloidal nanoparticles, *Langmuir* **17**, 2329-2333 (2001).

51. A. Henglein and M. Giersig, Optical and chemical observations on gold-mercury nanoparticles in aqueous solution, *J. Phys. Chem. B* **104**, 5056-5060 (2000).

52. A. Henglein, Preparation and optical absorption spectra of $Au_{core}Pt_{shell}$ and $Pt_{core}Au_{shell}$ colloidal nanoparticles in aqueous solution, *J. Phys. Chem. B* **104**, 2201-2203 (2000).

53. J. H. Hodak, A. Henglein, and G. V. Hartland, Tuning the spectral and temporal response in PtAu core-shell nanoparticles, *J. Chem. Phys.* **114**, 2760-2765 (2001).

54. A. Henglein, Colloidal palladium nanoparticles. Reduction of Pd(II) by H2; PdCoreAuShellAgShell particles, *J. Phys. Chem. B* **104**, 6683-6685 (2000).

55. J. Belloni, Contribution of radiation chemistry to the study of metal clusters, *Radiat. Res.* **150**, S9-S20 (1998).

56. M. Treguer, C. de Cointet, H. Remita, J. Khatouri, M. Mostafavi, J. Amblard, J. Belloni, and R. de Keyzer, Dose rate effects on radiolytic synthesis of gold-silver bimetallic clusters in solution, *J. Phys. Chem. B* **102**, 4310-4321 (1998).

57. J. L. Marignier, J. Belloni, M. O. Delcourt, and J. P. Chevalier, Microaggregates of non-noble metals and bimetallic alloys prepared by radiation-induced reduction, *Nature* **317**, 344-345 (1985).

58. C. M. Doudna, M. F. Bertino, and A. T. Tokuhiro, Structural investigation of Ag-Pd clusters synthesized with the radiolysis method, *Langmuir* **18**, 2434-2435 (2002).

59. C. M. Doudna, J. F. Hund, and M. F. Bertino, Synthesis of nanometer-sized (bi)metallic clusters with a nuclear reactor, *International Journal Mod. Phys. B* **15**, 3302-3307 (2001).

60. C. M. Doudna, M. F. Bertino, F. D. Blum, A. T. Tokuhiro, D. Lahiri-Dey, S. Chattopadhyay, and J. Terry, Radiolytic synthesis of bimetallic Ag-Pt nanoparticles with a high aspect ratio, *J. Phys. Chem. B* **107**, 2966-2970 (2003).

61. J. Belloni, Metal nanocolloids, *Curr. Op. Colloid Interface Sci.* **1**, 184-196 (1996).

62. J. T. G. Overbeek, Monodisperse colloidal systems, fascinating and useful, *Adv. Colloid Interface Sci.* **15**, 251-277 (1982).
63. T. Teranishi, M. Hosoe, T. Tanaka, and M. Miyake, Size control of monodispersed Pt nanoparticles and their 2D organization by electrophoretic deposition, *J. Phys. Chem. B* **103**, 3818-3827 (1999).
64. K. R. Brown, D. G. Walter, and M. J. Natan, Seeding of colloidal Au nanoparticle solutions. 2. Improved control of particle size and shape, *Chem. Mater.* **12**, 306-313 (2000).
65. N. R. Jana, L. Gearheart, and C. J. Murphy, Seeding growth for size control of 5-40 nm diameter gold nanoparticles, *Langmuir* **17**, 6782-6786 (2001).
66. N. R. Jana, L. Gearheart, and C. J. Murphy, Evidence for seed-mediated nucleation in the chemical reduction of gold salts to gold nanoparticles, *Chem. Mater.* **13**, 2313-2322 (2001).
67. A. Henglein and M. Giersig, Formation of colloidal silver nanoparticles. Capping action of citrate, *J. Phys. Chem. B* **103**, 9533-9539 (1999).
68. D. S. Miller, A. J. Bard, G. McLendon, and J. Ferguson, Catalytic water reduction at colloidal metal "microelectrodes". 2. Theory and experiment, *J. Am. Chem. Soc.* **103**, 5336-5341 (1981).
69. P. L. Freund and M. Spiro, Catalysis by colloidal gold of the reaction between ferricyanide and thiosulfate ions, *J. Chem. Soc., Faraday Trans. 1* **82**, 2277-2282 (1986).
70. L. Lu, H. Wang, Y. Zhou, S. Xi, H. Zhang, J. Hu, and B. Zhao, Seed-mediated growth of large, monodisperse core-shell gold-silver nanoparticles with Ag-like optical properties, *Chem. Comm.* 144-145 (2002).
71. I. Srnova-Sloufova, F. Lednicky, A. Gemperle, and J. Gemperlova, Core-shell (Ag)Au bimetallic nanoparticles: analysis of transmission electron microscopy images, *Langmuir* **16**, 9928-9935 (2000).
72. L. M. Liz-Marzán and A. P. Philipse, Synthesis and optical properties of gold-labeled silica particles, *J. Colloid Interface Sci.* **176**, 459-466 (1995).
73. L. M. Liz-Marzán, M. Giersig, and P. Mulvaney, Silica coating of vitreophobic colloids, *Chem. Comm.* 731-732 (1996).
74. L. M. Liz-Marzán, M. Giersig, and P. Mulvaney, Synthesis of nanosized gold-silica core-shell particles, *Langmuir* **12**, 4329-4335 (1996).
75. W. Stöber, A. Fink, and E. Bohn, Controlled growth of monodisperse silica spheres in the micron size range, *J. Colloid Interface Sci.* **26**, 62-69 (1968).
76. L. M. Liz-Marzán and P. Mulvaney, Au@SiO2 colloids: effect of temperature on the surface plasmon absorption, *N. J. Chem.* **22**, 1285-1288 (1998).
77. M. Hu, G. V. Hartland, V. Salgueiriño-Maceira, and L. M. Liz-Marzán, *Chem. Phys. Lett.* **372**, 767-772 (2003).
78. F. García-Santamaría, V. Salgueiriño-Maceira, C. López, and L. M. Liz-Marzán, Synthetic opals based on silica-coated gold nanoparticles, *Langmuir* **18**, 4519-4522 (2002).
79. D. Wang, V. Salgueiriño-Maceira, L.M. Liz-Marzán, and F. Caruso, Gold-silica inverse opals by colloidal crystal templating, *Adv. Mater.* **14**, 908-912 (2002).
80. M. Giersig, T. Ung, L. M. Liz-Marzán, and P. Mulvaney, Direct observation of chemical reactions in silica-coated gold and silver nanoparticles, *Adv. Mater.* **9**, 570-575 (1997).
81. T. Li, J. Moon, A. A. Morrone, J. J. Mecholsky, D. R. Talhman, and J. H. Adair, Preparation of Ag/SiO$_2$ Nanosize Composites by a Reverse Micelle and Sol-Gel Technique, *Langmuir* **15**, 4328-4334 (1999).
82. I. Pastoriza-Santos and L. M. Liz-Marzán, Formation and stabilization of silver nanoparticles through reduction by N,N-Dimethylformamide, *Langmuir* **15**, 948-951 (1999).
83. M. Brust, M. Walker, D. Bethell, D. J. Schiffrin, and R. Whyman, Synthesis of thiol-derivatized gold nanoparticles in a two-phase liquid-liquid system, *Chem. Comm.* 801-802 (1994).
84. M. Brust, J. Fink, D. Bethell, D. J. Schiffrin, and C. Kiely, Synthesis and reactions of functionalized gold nanoparticles, *Chem. Comm.* 1655-1656 (1995).
85. J. R. Heath, C. M. Knobler, and D. V. Leff, Pressure/temperature phase diagrams and superlattices of organically functionalized metal nanocrystal monolayers: the influence of particle size, size distribution, and surface passivant, *J. Phys. Chem. B* **101**, 189-197 (1997).
86. B. A. Korgel, S. Fullam, S. Connolly, and D. Fitzmaurice, Assembly and self-organization of silver nanocrystal superlattices: Ordered "soft spheres", *J. Phys. Chem. B* **102**, 8379-8388 (1998).
87. M. J. Hostetler, S. J. Green, J. J. Stokes, and R. W. Murray, Monolayers in three dimensions: synthesis and electrochemistry of ω-functionalized alkanethiolate-stabilized gold cluster compounds, *J. Am. Chem. Soc.* **118**, 4212-4213 (1996).
88. R. S. Ingram, M. J. Hostetler, and R. W. Murray, Poly-hetero-ω-functionalized alkanethiolate-stabilized gold cluster compounds, *J. Am. Chem. Soc.* **119**, 9175-9178 (1997).

89.  A. C. Templeton, M. J. Hostetler, E. K. Warmoth, S. Chen, C. M. Hartshorn, V. M. Krishnamurthy, M. D. E. Forbes, and R. W. Murray, Gateway reactions to diverse, polyfunctional monolayer-protected gold clusters, *J. Am. Chem. Soc.* **120**, 4845-4849 (1998).

90.  A. C. Templeton, D. E. Cliffel, and R. W. Murray, Redox and fluorophore functionalization of water-soluble, tiopronin-protected gold clusters, *J. Am. Chem. Soc.* **121**, 7081-7089 (1999).

91.  M. J. Hostetler, A. C. Templeton, and R. W. Murray, Dynamics of place-exchange reactions on monolayer-protected gold cluster molecules, *Langmuir* **15**, 3782-3789 (1999).

92.  H. Hirai, Y. Nakao, and N. Toshima, Preparation of colloidal transition metals in polymers by reduction with alcohols or ethers, *J. Macromol. Sci., Chem.* **13**, 727-750 (1979).

93.  H. Hirai, Y. Nakao, N. Toshima, and K. Adachi, Colloidal rhodium in poly(vinyl alcohol) as a hydrogenation catalyst of olefins, *Chem. Lett.* 905-910 (1976).

94.  Y. Wang and N. Toshima, Preparation of Pd-Pt bimetallic colloids with controllable core/shell structures, *J. Phys. Chem. B* **101**, 5301-5306 (1997).

95.  F. Fievet, J. P. Lagier, B. Blin, B. Beaudoin, and M. Figlarz, Homogeneous and heterogeneous nucleations in the polyol process for the preparation of micron and submicron size metal particles, *Solid State Ionics* **32-33**, 198-205 (1989).

96.  C. Ducamp-Sanguesa, R. Herrera-Urbina, and M. Figlarz, Synthesis and characterization of fine and monodisperse silver particles of uniform shape, *J. Solid State Chem.* **100**, 272-280 (1992).

97.  P.-Y. Silvert, R. Herrera-Urbina, N. Duvauchelle, V. Vijayakrishnan, and K. T. Elhsissen, Preparation of colloidal silver dispersions by the polyol process. Part 1. Synthesis and characterization, *J. Mater. Chem.* **6**, 573-577 (1996).

98.  Z. Y. Huang, G. Mills, and B. Hajek, Spontaneous formation of silver particles in basic 2-propanol, *J. Phys. Chem.* **97**, 11542-11550 (1993).

99.  P. P. Barnickel and A. Wokaun, Synthesis of metal colloids in inverse microemulsions, *Mol. Phys.* **69**, 1-9 (1990).

100. L. M. Liz-Marzán and I. Lado-Touriño, Reduction and stabilization of silver nanoparticles in ethanol by nonionic surfactants, Langmuir **12**, 3585-3589 (1996).

101. I. Pastoriza-Santos and L. M. Liz-Marzán, Reduction of silver nanoparticles in DMF. Formation of monolayers and stable colloids, *Pure Appl. Chem.* **72**, 83-90 (2000).

102. I. Pastoriza-Santos, C. Serra-Rodríguez, and L. M. Liz-Marzán, Self-assembly of silver particle monolayers on glass from $Ag^+$ solutions in DMF, *J. Colloid Interface Sci.* **221**, 236-241 (2000).

103. I. Pastoriza-Santos and L. M. Liz-Marzán, Preparation of PVP-protected metal nanoparticles in DMF, *Langmuir* **18**, 2888-2894 (2002).

104. I. Pastoriza-Santos, D.S. Koktysh, A.A. Mamedov, M. Giersig, N.A. Kotov, and L.M. Liz-Marzán, One-pot synthesis of Ag@TiO$_2$ core-shell nanoparticles and their layer-by-layer assembly, *Langmuir* **16**, 2731-2735 (2000).

105. M. Y. Han, C. H. Quek, W. Huang, C. H. Chew, and L. M. Gan, A simple and effective chemical route for the preparation of uniform nonaqueous gold colloids, *Chem. Mater.* **11**, 1144-1147 (1999).

106. G. Rodríguez-Gattorno, D. Díaz, L. Rendón, and G. O. Hernández-Segura, Metallic nanoparticles from spontaneous reduction of silver(I) in DMSO. Interaction between nitric oxide and silver nanoparticles, *J. Phys. Chem. B* 106, 2482-2487 (2002).

107. M. Boutonnet, J. Kizling, P. Stenius, and G. Maire, The preparation of monodisperse colloidal metal particles from microemulsions, *Colloids Surf.* **5**, 209-225 (1982).

108. K. Kandori, K. Konno, and A. Kitahara, Dispersion stability of nonaqueous calcium carbonate dispersion prepared in water core of W/O microemulsion, *J. Colloid Interface Sci.* **115**, 579-572 (1987).

109. E. Stathatos, P. Lianos, F. Del Monte, D. Levy, and D. Tsiourvas, Formation of TiO$_2$ nanoparticles in reverse micelles and their deposition as thin films on glass substrates, *Langmuir* **13**, 4295-4300 (1997).

110. F. J. Arriagada, and K. Osseo-Assare, Synthesis of nanosize silica in aerosol OT reverse microemulsions, J. Colloid Interface Sci., **170**, 8-17 (1995).

111. P. Lianos and J. K. Thomas, Small cadmium sulfide particles in inverted micelles, *J. Colloid Interface Sci.* **117**, 505-512 (1987).

112. A. R. Kortan, R. Hull, R. L. Opila, M. G. Bawendi, M. L. Steigerwald, P. J. Carroll, and L. E. Brus, Nucleation and growth of cadmium selendie on zinc sulfide quantum crystallite seeds, and vice versa, in inverse micelle media, *J. Am. Chem. Soc.* 112, 1327-1332 (1990).

113. S. K. Haram, A. R. Mahadeshwar, and S. G. Dixit, Synthesis and characterization of copper sulfide nanoparticles in triton-X 100 water-in-oil microemulsions, *J. Phys. Chem.* **100**, 5868-5873 (1996).

114. M. Antonietti, W. Bremser, D. Mueschenborn, C. Rosenauer, B. Schupp, and M. Schmidt, Synthesis and size control of polystyrene latices via polymerization in microemulsion, *Macromolecules* **24**, 6636-6643 (1991).

115. C. Petit, P. Lixon, and M. P. Pileni, In situ synthesis of silver nanocluster in AOT reverse micelles, *J. Phys. Chem.* **97**, 12974-12983 (1993).

116. I. Lisiecki and M. P. Pileni, Synthesis of copper metallic clusters using reverse micelles as microreactors, *J. Am. Chem. Soc.* **115**, 3887-3896 (1993).

117. J. Tanori, N. Duxin, C. Petit, I. Lisiecki, P. Veillet, and M. P. Pileni, Synthesis of nanosize metallic and alloyed particles in ordered phases, *Colloid Polym. Sci.* **273**, 886-892 (1995).

118. P. Barnickel, A. Wokaun, W. Sager, and H. F. Eicke, Size tailoring of silver colloids by reduction in w/o microemulsions, *J. Colloid Interface Sci.* **148**, 80-90 (1992).

119. J. P. Wilcoxon, R. L. Williamson, and R. Baughman, Optical properties of gold colloids formed in inverse micelles, *J. Chem. Phys.* **98**, 9933-9950 (1993).

120. D.-H. Chen and S.-H. Wu, Synthesis of nickel nanoparticles in water-in-oil microemulsions, *Chem. Mater.* **12**, 1354-1360 (2000).

121. Q. Limin, M. Jiming, and S. Julin, Synthesis of copper nanoparticles in nonionic water-in-oil microemulsions, *J. Colloid Interface Sci.* **186**, 498-500 (1997).

122. C. C. Wang, D. H. Chen, and T. C. Huang, Synthesis of palladium nanoparticles in water-in-oil microemulsions, *Colloid Surf., A* **189**, 145-154 (2001).

123. J. P. Chen, K. M. Lee, C. M. Sorensen, K. J. Klabunde, and G. C. Hadjipanayis, Magnetic properties of microemulsion synthesized cobalt fine particles, *J.Appl. Phys.* **75**, 5876-5878 (1994).

124. Y.-P. Sun, P. Atorngitjawat, and M. J. Meziani, Preparation of silver nanoparticles via rapid expansion of water in carbon dioxide microemulsion into reductant solution, *Langmuir* **17**, 5707-5710 (2001).

125. M. Ji, X. Chen, C. M. Wai, and J. L. Fulton, Synthesizing and dispersing silver nanoparticles in a water-in-supercritical carbon dioxide microemulsion, *J. Am. Chem. Soc.* **121**, 2631-2632 (1999).

126. H. Ohde, F. Hunt, and C. M. Wai, Synthesis of silver and copper nanoparticles in a water-in-supercritical-carbon dioxide microemulsion, *Chem. Mater.* **13**, 4130-4135 (2001).

127. C. R. Martin, Nanomaterials: a membrane-based synthetic approach, *Science* **266**, 1961-1966 (1994).

128. J. C. Hulteen and C. R. Martin, A general template-based method for the preparation of nanomaterials, *J. Mater. Chem.* **7**, 1075-1087 (1997).

129. R. L. Fleischer, P. B. Price, and R. M. Walker, *Nuclear Tracks in Solids* (University of California Press, Berkeley, 1975).

130. G. L. Hornyak, C. J. Patrissi, and C. R. Martin, Fabrication, characterization, and optical properties of gold nanoparticle/porous alumina composites: The nonscattering Maxwell-Garnett limit, *J. Phys. Chem. B* **101**, 1548-1555 (1997).

131. M. Wirtz and C. R. Martin, Template-fabricated gold nanowires and nanotubes, *Adv. Mater.* **15**, 455-458 (2003).

132. R. M. Penner and C. R. Martin, Preparation and electrochemical characterization of ultramicroelectrode ensembles, *Anal. Chem.* **59**, 2625-2630 (1987).

133. V. P. Menon and C. R. Martin, Fabrication and evaluation of nanoelectrode ensembles, *Anal. Chem.* **67**, 1920-1928 (1995).

134. K. B. Jirage, J. C. Hulteen, and C. R. Martin, Effect of thiol chemisorption on the transport properties of gold nanotubule membranes, *Anal. Chem.* **71**, 4913-4918 (1999).

135. C. D. Keating and M. J. Natan, Striped metal nanowires as building blocks and optical tags, *Adv. Mater.* **15**, 451-454 (2003).

136. L. Piraux, J. M. George, J. F. Despres, C. Leroy, E. Ferain, R. Legras, K. Ounadjela, and A. Fert, Giant magnetoresistance in magnetic multilayered nanowires, *Appl. Phys. Lett.* **65**, 2484-2486 (1994).

137. B. D. Reiss, R. G. Freeman, I. D. Walton, S. M. Norton, P. C. Smith, W. G. Stonas, C. D. Keating, and M. J. Natan, Electrochemical synthesis and optical readout of striped metal rods with submicron features, *J. Electroanal. Chem.* **522**, 95-103 (2002).

138. B. R. Martin, D. J. Dermody, B. D. Reiss, M. Fang, L. A. Lyon, M. J. Natan, and T. E. Mallouk, Orthogonal self-assembly on colloidal gold-platinum nanorods, *Adv. Mater.* **11**, 1021-1025 (1999).

139. C. Schoenenberger, B. M. I. van der Zande, L. G. J. Fokkink, M. Henny, C. Schmid, M. Krueger, A. Bachtold, R. Huber, H. Birk, and U. Staufer, Template synthesis of nanowires in porous polycarbonate membranes: electrochemistry and morphology, *J. Phys. Chem. B* **101**, 5497-5505 (1997).

140. V. M. Cepak and C. R. Martin, Preparation and stability of template-synthesized metal nanorod sols in organic solvents, *J. Phys. Chem. B* **102**, 9985-9990 (1998).

141. N. Al-Rawashdeh and C. A. Foss, Jr., UV/visible and infrared spectra of polyethylene/nanoscopic gold rod composite films: effects of gold particle size, shape and orientation, *Nanostruct. Mater.* **9**, 383-386 (1997).

142. Y.-Y. Yu, S.-S. Chang, C.-L. Lee, and C. R. C. Wang, Gold nanorods: electrochemical synthesis and optical properties, *J. Phys. Chem. B* **101**, 6661-6664 (1997).

143. S.-S. Chang, C.-W. Shih, C.-D. Chen, W.-C. Lai, and C. R. C. Wang, The shape transition of gold nanorods, *Langmuir* **15**, 701-709 (1999).

144. S. Link, C. Burda, B. Nikoobakht, and M. A. El-Sayed, Laser-induced shape changes of colloidal gold nanorods using femtosecond and nanosecond laser pulses, *J. Phys. Chem. B* **104**, 6152-6163 (2000).
145. G. V. Hartland, M. Hu, O. Wilson, P. Mulvaney, and J. E. Sader, Coherent excitation of vibrational modes in gold nanorods, *J. Phys. Chem. B* **106**, 743-747 (2002).
146. O. Wilson, G. J. Wilson, and P. Mulvaney, Laser writing in polarized silver nanorod films, *Adv. Mater.* **14**, 1000-1004 (2002).
147. N. R. Jana, L. Gearheart, and C. J. Murphy, Wet chemical synthesis of high aspect ratio cylindrical gold nanorods, *J. Phys. Chem. B* **105**, 4065-4067 (2001).
148. C. J. Murphy and N. R. Jana, Controlling the aspect ratio of inorganic nanorods and nanowires, *Adv. Mater.* **14**, 80-82 (2002).
149. B. Nikoobakht and M. A. El-Sayed, Preparation and growth mechanism of gold nanorods (NRs) using seed-mediated growth method, *Chem. Mater.* **15**, 1957-1962 (2003).
150. K. K. Caswell, C. M. Bender, and C. J. Murphy, Seedless, surfactantless wet chemical synthesis of silver nanowires, *Nano Lett.* **3**, 667-669 (2003).
151. F. Kim, J. H. Song, and P. Yang, Photochemical synthesis of gold nanorods, *J. Am. Chem. Soc.* **124**, 14316-14317 (2002).
152. C. S. Ah, S. D. Hong, and D.-J. Jang, Preparation of $Au_{core}Ag_{shell}$ nanorods and characterization of their surface plasmon resonances, *J. Phys. Chem. B* **105**, 7871-7873 (2001).
153. R. Jin, Y. Cao, C. A. Mirkin, K. L. Kelly, G. C. Schatz, and J. G. Zheng, Photoinduced conversion of silver nanospheres to nanoprisms, *Science* **294**, 1901-1903 (2001).
154. S. Chen and D. L. Carroll, Synthesis and characterization of truncated triangular silver nanoplates, *Nano Lett.* **2**, 1003-1007 (2002).
155. N. Malikova, I. Pastoriza-Santos, M. Schierhorn, N.A. Kotov, and L.M. Liz-Marzán, Layer-by-layer assembled mixed spherical and planar gold nanoparticles: control of interparticle interactions, *Langmuir* **18**, 3694-3697 (2002).
156. A. Filankembo and M. P. Pileni, Is the template of self-colloidal assemblies the only factor that controls nanocrystal shapes?, *J. Phys. Chem. B* **104**, 5865-5868 (2000).
157. M. Maillard, S. Giorgio, and M.-P. Pileni, Tuning the size of silver nanodisks with similar aspect ratios: Synthesis and optical properties, *J. Phys. Chem. B* **107**, 2466-2470 (2003).
158. X.-M. Ni, X.-B. Su, Z.-P. Yang, and H.-G. Zheng, The preparation of nickel nanorods in water-in-oil microemulsion, *J. Cryst. Growth* **252**, 612-617 (2003).
159. A. Sanyal and M. Sastry, Gold nanosheets via reduction of aqueous chlroaurate ions by anthracene anions bound to a liquid-liquid interface, *Chem. Comm.* 1236-1237 (2003).
160. Y. Sun, B. Gates, B. Mayers, and Y. Xia, Crystalline silver nanowires by soft solution processing, *Nano Lett.* **2**, 165-168 (2002).
161. Y. Sun and Y. Xia, Large-scale synthesis of uniform silver nanowires through a soft, self-seeding, polyol process, *Adv. Mater.* **14**, 833-837 (2002).
162. Y. Sun, Y. Yin, B. T. Mayers, T. Herricks, and Y. Xia, Uniform silver nanowires synthesis by reducing $AgNO_3$ with ethylene glycol in the presence of seeds and poly(vinyl pyrrolidone), *Chem. Mater.* **14**, 4736-4745 (2002).
163. Y. Sun and Y. Xia, Shape-controlled synthesis of gold and silver nanoparticles, *Science* **298**, 2176-2179 (2003).
164. Y. Sun, B. Mayers, and Y. Xia, Metal nanostructures with hollow interiors, *Adv. Mater.* **15**, 641-646 (2003).
165. M. Giersig, I. Pastoriza-Santos, and L.M. Liz-Marzán, manuscript submitted (2003).
166. N. A. Kotov, I. Dekàny, and J. H. Fendler, Layer-by-layer self-assembly of polyelectrolyte-semiconductor nanoparticle composite films, *J. Phys. Chem.* **99**, 13065-13069 (1995).
167. G. Decher, Fuzzy nanoassemblies: toward layered polymeric multicomposites, *Science* **277**, 1232-1237 (1997).
168. T. Ung, L. M. Liz-Marzán, and P. Mulvaney, Gold nanoparticle thin films, *Colloid Surf. A* **202**, 119-126 (2002).
169. F. Caruso, R. A. Caruso, and H. Möhwald, Nanoengineering of inorganic and hybrid hollow spheres by colloidal templating, *Science* **282**, 1111-1114 (1998).
170. F. Caruso, R. A. Caruso, and H. Möhwald, Production of hollow microspheres from nanostructured composite particles, *Chem. Mater.* **11**, 3309-3314 (1999).
171. F. Caruso, Nanoengineering of particle surfaces, *Adv. Mater.* **13**, 11-22 (2001).
172. C. Graf and A. Van Blaaderen, Metallodielectric colloidal core-shell particles for photonic applications, *Langmuir* **18**, 524-534 (2002).
173. F. Caruso, M. Spasova, V. Salgueiriño-Maceira, and L.M. Liz-Marzán, Multilayer assemblies of silica-encapsulated gold nanoparticles on decomposable colloid templates, *Adv. Mater.* **13**, 1090-1095 (2001).

174. V. Salgueiriño-Maceira, F. Caruso, and L.M. Liz-Marzán, submitted, 2002.
175. N. A. Kotov, T. Haraszti, L. Turi, G. Zavala, R. E. Geer, I. Dékány, and J. H. Fendler, Mechanism of and defect formation in the self assembly of polymeric polycation montmorillonite ultrathin films, *J. Am. Chem. Soc.* **119**, 6821-6832 (1997).

# 2

# NEAR-FIELD SCANNING OPTICAL MICROSCOPY: ALTERNATIVE MODES OF USE FOR NSOM PROBES

David S. Moore-Nichols and Robert C. Dunn[*]

## 1. INTRODUCTION

Near-field scanning optical microscopy (NSOM) is a scanning probe technique with a potential for revealing novel insights into the natural world at the sub-microscopic level. The technique circumvents the classical diffraction limit that constrains the spatial resolution of conventional light microscopy, unlocking new opportunities for probing sample optical properties at the mesoscopic dimension.

NSOM relies on the formation of a sub-wavelength sized light source that is scanned in close proximity to a sample surface. The resulting spatial resolution in such a configuration is limited only by the size of the light source and its position near the sample. (Betzig, Trautman et al. 1991; Pohl 1991) Resolution is not limited by the wavelength of illumination as it is in microscopy techniques utilizing conventional optical lenses. For example, Figure 1 shows a NSOM fluorescence image of single molecules in a lipid film. Each bright spot represents the fluorescence from a single fluorescent molecule in the film, illustrating the low detection limits possible with this technique. Moreover, the full-width-at-half-maximum of each feature is approximately 28 nm, demonstrating the high spatial resolution possible with NSOM. (Hollars and Dunn, unpublished data.) To further illustrate this point, the image size presented in Figure 1 is 514 nm; the same dimension as the excitation wavelength used in this measurement. Thus, NSOM is clearly capable of greatly exceeding the resolution limits of traditional methods of light microscopy.

The various methods used in forming sub-wavelength apertures and implementing NSOM techniques have been the subject of several recent reviews and will not be recounted in this chapter. (de Lange, Cambi et al. 2001; Dunn 1999; Edidin 2001; Higgins 2002; Lewis, Radko et al. 1999; Shiku and Dunn 1999) Moreover, readers are

---

[*]David S. Moore-Nichols and Robert C. Dunn, Department of Chemistry, University of Kansas, Lawrence, Kansas, 66045

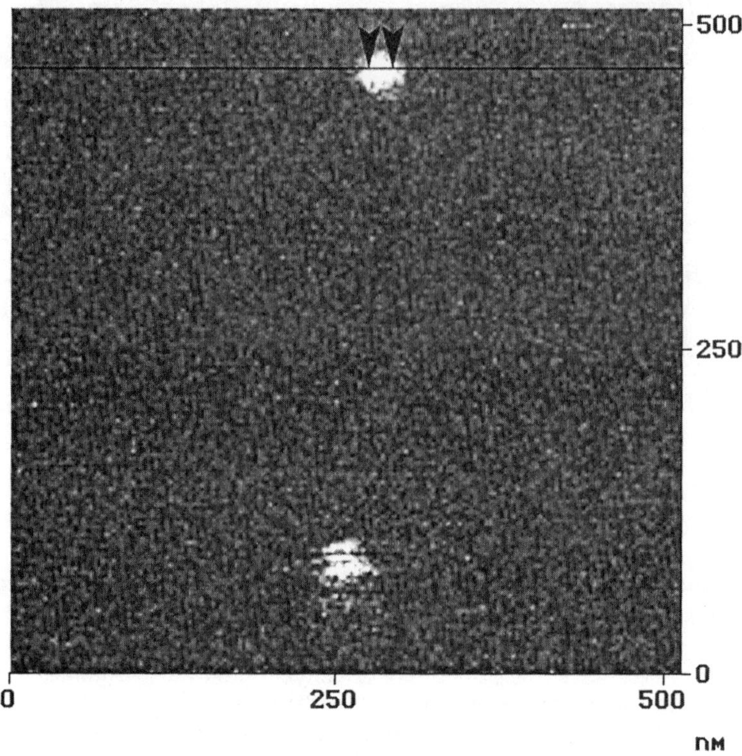


**Figure 1.** NSOM image of single fluorescent molecules in a DPPC lipid film created using the Langmuir-Blodgett technique. Each bright spot represents the fluorescence from a single fluorescent molecule of diIC$_{18}$ in the film, illustrating the low detection limits possible with this technique. The full-width-at-half-maximum of each feature is ~28 nm, (black arrows) demonstrating the high spatial resolution that is possible with NSOM. The image size presented in Figure 1 is 514 nm, the same as the excitation wavelength, illustrating the high resolution achieved with NSOM.

referred to an excellent chapter within this edition for a discussion of NSOM fluorescence measurements which, to date, have proven to be the most popular and informative of the NSOM contrast mechanisms. This chapter will instead provide a selective overview of some non-traditional uses of NSOM probes in scanning probe microscopy and molecular detection. Descriptions of the NSOM methods found here are not meant to reiterate the expanding spectroscopic applications of NSOM, but are rather meant to introduce NSOM variants permitting sample characterization not typically accessible by traditional scanning probe techniques. While approaches covered in this chapter continue to take advantage of the sub-wavelength light source of traditional NSOM probes, these modified techniques offer enhanced or unique contrast mechanisms that expand the information available to this form of scanning probe microscopy.

## 2. SCANNING NEAR-FIELD FRET MICROSCOPY

Fluorescence resonance energy transfer (FRET) is an increasingly useful technique in the biological sciences for measuring small distances at the Ångstrom level.(Emptage 2001; Gaits and Hahn 2003; Sekar and Periasamy 2003) The technique takes advantage of the strong distance-dependence of Förster energy transfer between molecules possessing the appropriate spectral properties.(Lakowicz 1983; Van Der Meer, Coker et al. 1994) Experimentally, two dye molecules are chosen so that the emission of the donor dye spectrally overlaps the absorption of the acceptor dye. Upon excitation of donor dye, non-radiative energy transfer to the acceptor dye may occur when the dyes move toward each other in the appropriate orientation. Thus, emission from the acceptor dye reports relative donor-acceptor positioning, and acceptor emission can be used to probe distances between distinct molecules or between structures within a given macromolecule or complex.

Acceptor emission is dependent on the rate of energy transfer, $k_T$, between donor and acceptor dyes which is given by

$$k_T = \tau_d^{-1}(R_o/r)^6$$

where $\tau_d$ is the unperturbed donor lifetime, r is the distance between the donor-acceptor pair, and $R_o$, the Förster distance, is the distance at which transfer efficiency is 50% of its maximum for the particular donor-acceptor pair.(Lakowicz 1983; Van Der Meer et al. 1994; Vickery and Dunn 1999) Förster distances depend on the selected donor-acceptor pair and usually range from 20 to 50 Å. $R_o$ in Ångstroms is given by

$$R_o = [8.79 \times 10^{-5} J(\lambda) \, \phi_D \, n^{-4} \, \kappa^2]^{1/6}$$

where $J(\lambda)$ is a function representing the spectral overlap between donor emission and acceptor absorption ($nm^4$ $M^{-1}$ $cm^{-1}$), $\phi_D$ is the quantum yield of the donor, $n$ is the refractive index of the medium in which energy transfer occurs, and the $\kappa^2$ term is an orientation factor that takes into account the spatial orientation of the transition dipoles during energy transfer. The inverse sixth power dependence of the rate upon r is what leads to the exquisite, 20 to 80 Å, distance sensitivity of FRET measurements. (Bardo, Collinson et al. 2001; Deniz, Dahan et al. 1999; Ha, Enderle et al. 1996; Lakowicz 1983; Sekatskii, Shubeita et al. 2000; Shubeita, Sekatskii et al. 1999; Van Der Meer et al. 1994)

Several groups have explored combining FRET with NSOM to exploit the distance dependence of FRET and enhance NSOM imaging capabilities.(Kirsch, Subramaniam et al. 1999; Sekatskii et al. 2000; Shubeita, Sekatskii et al. 2002; Vickery et al. 1999) As a result, scanning near–field fluorescence energy transfer microscopy combines both the imaging capabilities of NSOM, and the intermolecular distance mapping capacity of FRET.

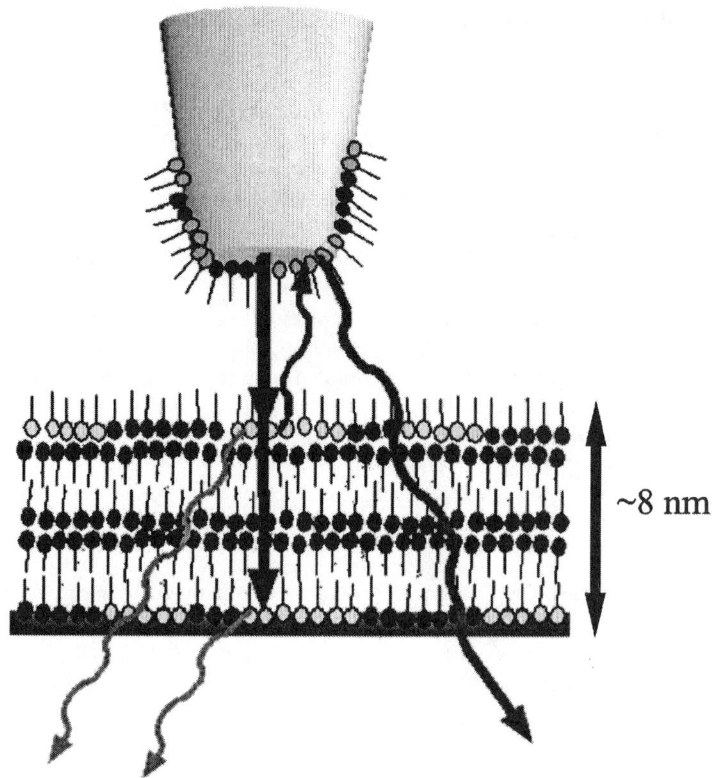

**Figure 2.** See Fig. 2.2 in the color insert at the end of this volume. The above illustration represents the configuration of tip and sample used in FRET/NSOM. The acceptor dye of the FRET pair is rhodamine. Rhodamine was incorporated into a DPPC monolayer at 0.5 mol % and used to coat an NSOM probe lacking any metal. Fluorescein was used as the donor dye in the sample, and was incorporated into two DPPC/0.5 mol % fluorescein layers separated by a spacer of three arachidic acid layers. Light exiting the tip, (*blue arrow*) excites the donor dye in the sample but does not directly excite the acceptor dye on the tip. When the modified NSOM probe is near the sample, however, energy transfer from the excited donor in the monolayer to the rhodamine acceptor (*dark green arrow*) on the tip leads to a new emission, shifted to the red (*red arrow*) of the donor emission (*light green arrows*). By monitoring rhodamine fluorescence, it is possible to optically probe only those structures within nanometers of the NSOM tip. Reproduced with permission from (Vickery et al. 1999) . Copyright 1999 Biophysical Society.

One approach for combining FRET with NSOM is shown schematically in Figure 2.(Vickery et al. 1999) An acceptor dye of a FRET pair is attached to an NSOM probe while a donor dye is dispersed in the sample. Light exciting the NSOM probe is resonant with the donor dye in both monolayers of the sample, but not with the acceptor dye attached to the tip. Specifically monitoring only the emission from the donor dye provides information about its spatial distribution in the sample, much as in conventional NSOM measurements. However, as the tip nears the sample surface, non-radiative energy transfer from excited donor molecules in the sample to tip-bound acceptor molecules leads to emission from the acceptor dye. With the proper use of spectral filters, emission farther to the red than that of the donor can be selectively monitored. This provides information on only those donor molecules in closest proximity to the modified NSOM tip.

In this example, an uncoated, heat-pulled fiber optic NSOM probe was coated with a lipid monolayer containing the acceptor dye rhodamine using the Langmuir-Blodgett (LB) deposition technique. The sample consisted of a multi-layer film consisting of two DPPC monolayers, each containing the donor dye fluorescein, separated by a spacer group of three arachidic acid layers (Figure 2). The multilayer film was also created using the LB technique. Fluorescein and rhodamine are a well-characterized FRET pair with a $R_o$ of ~ 55 Å.

Figure 3 shows 50 μm x 50 μm NSOM fluorescence images of the multi-layer film following excitation at 458 nm, which preferentially excites the donor dye in the sample. The emission signal in the left image (A) was collected through a 548 nm band-pass filter that selectively passes the fluorescein fluorescence. This arrangement, therefore, is sensitive to the donor emission from the sample. This image can be compared with the fluorescence image shown on the right in Figure 3B. This image was collected over the same sample area using a 590 nm band-pass filter, so that the rhodamine fluorescence from the NSOM tip was monitored. Since 458 nm excitation was used in this image, which is non-resonant with rhodamine, the fluorescence reflects energy transfer from the excited fluorescein in the sample to the rhodamine attached to the end of the NSOM tip. (Vickery et al. 1999)

Striking differences exist between the fluorescence features observed in the two images, examples of which are denoted by the circled region and the features marked by

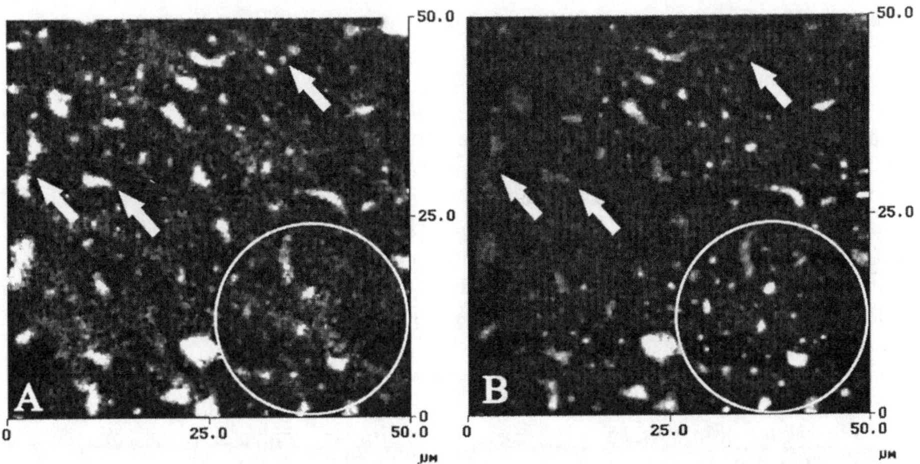


**Figure 3.** Fluorescein (A) and rhodamine (B) fluorescence images of the same 50 x 50 μm region of the lipid multi-layer using the FRET/NSOM scheme shown in Figure 2. Fluorescein fluorescence was collected at 548 nm following excitation at 458 nm. Rhodamine fluorescence was collected at 590 nm while exciting fluorescein at 458 nm. The arrows in the donor emission image (A) denote features that are reduced in intensity in the energy transfer image (B) These areas, therefore, distinguish regions where the donor dye is located in the bottom monolayer of the multi-layer film. Other regions such as those encircled, become more intense and are better resolved in (B) indicating that donor is present in the top monolayer. Reproduced with permission from (Vickery et al. 1999). Copyright 1999 Biophysical Society.

arrows. In the donor emission image (left), dye fluorescence from both layers of the multi-layer film contributes to the signal. In the image on the right, however, only those donor dye molecules in close proximity to the tip-bound acceptor dye contribute to the energy transfer signal. This signal, therefore, is less sensitive to donor dye located in the bottom layer of the film and most sensitive to that residing in the top layer, closest to the modified NSOM probe. The arrows denote regions where these changes are particularly apparent. It should also be noted that the dye-containing monolayers in these films were only separated by approximately 8 nm. Comparing contributions from both signals, therefore, permits one to optically discriminate fluorescence from each of the labeled monolayers. Essentially, this permits the optical sectioning of fluorescently labeled features at different depths beneath a sample surface, but with nanometer resolution! As this ability to optically section samples at different depths is not limited to flat samples, this could be important, for example, in probing the location and distribution of proteins in and across biological membranes with more complicated topographies.

There is also a noticeable change in spatial resolution between the two images shown in Figure 3. For example, the circled region shows that the resolution increases in going to the energy transfer mode of imaging. This again is a manifestation of the energy transfer process. The strong distance dependence in the FRET signal leads to energy transfer from only those areas of the sample where the donor is in closest proximity to the tip. This effectively reduces the excitation volume and thus increases the spatial resolution. As in this demonstration, this effectively relaxes the constraints on NSOM aperture formation since the metal coatings are no longer needed to form a well-defined aperture. In fact, an aperture may be altogether omitted, since far field excitation may serve to excite donor fluorescence without affecting the resolution of detected acceptor emission. This was recently demonstrated using dye modified atomic force microscopy (AFM) tips for high-resolution fluorescence measurements.(Vickery and Dunn 2001)

Although combining FRET with NSOM endows scanning probe microscopy with several unique imaging capabilities, these measurements are very difficult and several obstacles will need to be overcome before it becomes a routine and robust tool. Photobleaching of tip-attached dye results in a shortening of the useful lifetime of each tip. This is somewhat remedied by attaching the acceptor dye to the tip instead of the donor; reducing its duty cycle for excitation and extending its useful life. Yet, even the most robust organic dyes will eventually succumb to low probability events that lead to photobleaching. One avenue worth exploring would be the incorporation of solid-state materials, such as quantum dots, into probe designs. Such materials would be more robust and resistant to photobleaching.

As illustrated in Fig. 3 and reviewed above, combining the distance sensitivity of FRET with the high resolution of NSOM yields a method with unparalleled ability to optically section thin surface membranes while similarly enhancing optical resolution. Although the distance sensitivity obtained in the FRET geometry shown in Figure 3 is somewhat less than the inverse sixth power found in bulk applications, the nanometer-scale depth discrimination of this hybrid technique has obvious applications in the biological sciences. This technique should prove useful for probing constituents in and across biological membranes or in single point dynamic measurements of conformational motions. The reduced constraints on the actual tip geometry and properties should also open new avenues for high-resolution optical imaging as has already been demonstrated by combining FRET with AFM.(Vickery et al. 2001)

## 3. NANOMETRIC BIOSENSORS AND BIOPROBES

A rapidly expanding area of use for NSOM probes is as nanometric optical sensors and probes.(Cullum and Vo-Dinh 2000; Marchese-Ragona and Haydon 1997; Meixner and Kneppe 1998; Subramaniam, Kirsch et al. 1998; Wolfbeis 2002) Generally, these types of detectors are said to fall into one of two different categories. In the first, the NSOM tip is used to monitor the optical properties of an intrinsic, environmentally sensitive indicator within biological solutions or, in certain applications, inside the plasma membrane of a living cell. In the second, an optically active indicator is immobilized at the end of an NSOM tip and is used to probe the tip's local environment. In this section, only applications utilizing the latter experimental arrangement will be discussed.

Typically, agents used to modify probes are selected so that their optical properties, i.e. fluorescence, are modified in the presence of a relevant ligand. Although within the fiber optic sensor field a distinction is sometimes made between sensors, which are said to detect events continuously, and probes, which are thought of as single use instruments, for clarity and simplicity, those distinctions will not be made here and the terms will be used interchangeably.(Vo-Dinh, Alarie et al. 2000)

The general approach for combining sensing elements with NSOM is to functionalize the very end of the NSOM tip with an environmentally sensitive molecule that responds optically to specific changes in its surroundings. The small area of the NSOM probe over which these molecules are immobilized creates sub-micron surfaces that permit detection of substances in highly localized regions. For example, the pioneering work of Kopelman and coworkers used this approach to create small pH- and nitric oxide-sensitive indicators confined to the distal end of NSOM tips. The small size of the tips was used to impale embryos and track pH changes through the developmental cycle while minimally disrupting cell structure.(Tan, Shi et al. 1992) Additionally, the same group used chloride nanoprobes and nitrite sensors to sample ion concentrations and nitric oxide formation during organogenesis in rat fetuses in a preparation mimicking *in utero* conditions.(Barker, Thorsrud et al. 1998) The small size of such probes proved it was possible monitor otherwise difficult to observe biological processes in a minimally invasive manner.

An interesting recent development in the use of NSOM tips as biological probes involves the use of antibodies as localized antigenic probes. Immobilized antibodies at the submicron apertures of NSOM probes are used to create nanometric fluorescence-based assays of antigen binding. Although still in the early stages of development, the concept is straightforward and may provide a unique new tool for intracellular diagnostics. Current approaches mainly involve attachment of an antibody to a NSOM aperture that recognizes a fluorescent ligand. Upon exposure of the functionalized tip to a recognized ligand, the fluorescence signal at the tip grows in with time as recognition and binding occur. The fluorescence, presumably, can then be quantified relative to comparable exposures to standard solutions and used to quantify ligand binding.

Following this approach, Vo-Dinh et al. have recently developed an antibody-based NSOM probe for the detection of benzopyrene tetrol (BPT) within single living cells.(Vo-Dinh et al. 2000) BPT, a carcinogen in humans, is an aromatic hydrocarbon that fluoresces at 439 nm following excitation at 374 nm. In their experiments, an optical fiber was pulled to sub-micron diameters and coated with silver using conventional means. Silver was used in lieu of the more common aluminum coating to avoid toxic effects within the cells where measurements were to be taken. Anti-BPT antibodies were then attached to the tip apex by covalent immobilization. Exposure to BPT resulted in increased fluorescence.

It was reported that the high binding affinity between BPT and the immobilized antibody meant that once bound ligand saturated the antibodies, each tip was no longer active and had to be replaced. Practically, this allowed only one run for each tip fabricated. Because of this, calibration of the sensor response had to be done with many sensors prepared under similar conditions and not with the actual sensor used in the experiment.

Thus, these initial proof-of-principle experiments demonstrate the ability to use the epitope specificity of antibodies to specifically detect a molecule of interest within the complex intracellular milieu of the cytoplasm, albeit with limitations. However, the small size of these probes did allow for membrane penetration within specific and distinct sub-cellular regions, which perhaps will permit differential localization of target molecules within a cell in the future.

Ion sensing is an additional area of interest in development of novel NSOM techniques, where the reduced size of the NSOM probe may provide new means for localizing second messengers within living cells or dynamic, noninvasive measurements of events such as ion channel gating.(Cullum et al. 2000; Dunn 1999; Lee, Talley et al. 1999; Vo-Dinh 2002) For the latter, one could imagine attaching an ion sensitive dye to the end of a NSOM aperture and positioning the small aperture above the plasma membrane of an excitable cell. As the cells are activated, ion channels in the cellular

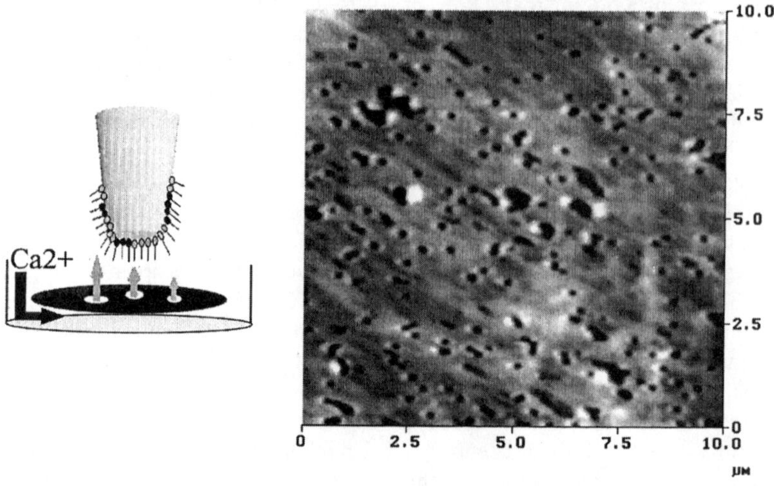


**Figure 4.** A schematic of the experimental cell used for the calcium sensor experiments. A reservoir, into which a calcium-containing solution could be added, was covered by a nitrocellulose membrane with pores of ~ 200nm. A 10 X 10 µ m AFM image of a representative nitrocellulose membrane used in those experiments is shown on the right.

membrane would open and ion flux would modify the fluorescent properties of the nearby, tip-attached reporter dye. Much like patch-clamp electrical measurements, these experiments might permit the gating of ions through ion channels to be followed in real time. However, unlike patch clamping, formation of a high-resistance seal with the membrane would not be required, since such measurements would not be based upon detecting small ion currents through a physically isolated membrane patch. Such an approach might avoid many of the perturbative effects that patch clamping produces in whole-cell recordings where minimal disruptions of cellular processes are desired.

Preliminary results using this type of approach appear promising. Figure 4 presents a schematic of a test chamber fabricated to simulate ion channels in the cellular membrane. In this configuration, a thin nitrocellulose membrane possessing 200 nm pores separated two chambers and a calcium gradient was formed across the membrane. The small pores provided spatially confined regions of increased calcium concentrations. In order to form a calcium-sensitive probe, a Langmuir-Blodgett film containing the calcium indicator, Calcium Green $C_{18}$ (Molecular Probes, C-6804), was transferred onto

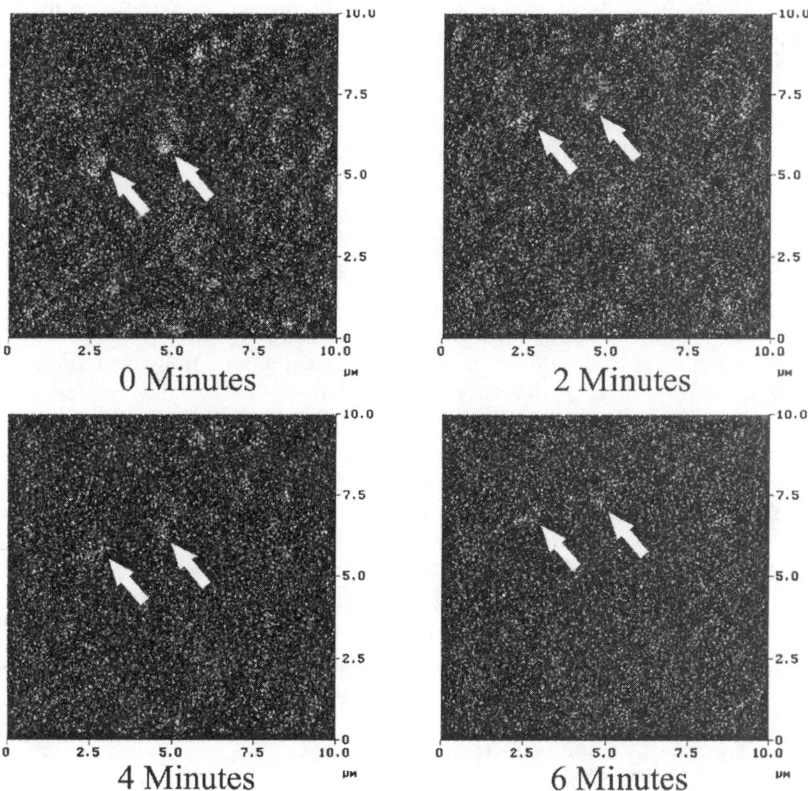


**Figure 5.** Above images represent preliminary results of the calcium sensor experiments. Arrows denote regions of increased fluorescence detected with the calcium green-modified NSOM tips as calcium diffused through the pores. Over time, the fluorescence diminishes due to equilibration of calcium levels across the membrane and photobleaching of the tip-bound dye.

the end of a NSOM tip. Figure 5 shows a series of four images taken consecutively over the same region of the membrane while monitoring the fluorescence from the tip-bound detector dye. Regions of calcium flux through the small pores can be detected as indicated by the arrows in the Figure 5. While the detected flux is seen to diminish as a function of time as the calcium gradient between the two chambers equalizes, it is also likely that photobleaching of the tip-bound reporter dye also contributed to such decreases. Clearly, the sensitivity of this technique will need to be improved before single channel recordings on small ion channels is possible, but these encouraging initial results do illustrate the feasibility of such studies.


## 4. APPLIED VOLTAGE COMBINED WITH NSOM FOR STRUCTURE/ DYNAMIC MEASUREMENTS

In NSOM, heat-pulled or chemically etched fiber optic probes are typically coated with aluminum to confine the light to a small aperture left uncoated at the apex of the NSOM tip. Normally, aperture formation is the primary purpose for the metal coating. However, the conductive properties of the aluminum also permit additional uses. By applying a potential to a metal-coated NSOM tip, the field produced at the tip can be used to modify the sample or alter the orientation of molecules within certain types of samples. Changes in the orientation of such molecules, in turn, may be monitored optically using light exiting the NSOM probe. Such an approach, therefore, provides an additional means by which NSOM probes may be used to perturb sample properties while dynamically following changes in the optical properties of samples.(Higgins 2002)

In order to study the structure and dynamics of polymer-dispersed liquid crystal films, Higgins and coworkers exploited this hybrid NSOM technique in a particularly useful and interesting manner.(Higgins, Liao et al. 2001; Higgins, Mei et al. 1999; Mei and Higgins 1998; Mei and Higgins 1998; Mei and Higgins 2000) Typically, polymer-dispersed liquid crystal (PDLC) films are comprised of small, micron sized, liquid crystal droplets dispersed in an optically transparent polymer film. Liquid crystal films are considered a useful class of optical materials based on their ability to change from a translucent to a transparent state upon application of a potential. The structuring and dynamics of liquid crystal within micron-sized droplets in response to the application of an external field is of particular interest in PDLC films. How the alignment of the liquid crystal within the droplet is altered by potential, the manner in which this is effected by the walls of the cavity forming the droplet, and the influences that this has on dynamics are all interesting questions that are difficult to address with far-field techniques. This approach, therefore, would seem to be an ideal system for NSOM studies. The small size of the NSOM aperture provides sub-droplet spatial information that is ideally suited for polarized light experiments. The metal-coated NSOM probe acts as a small electrode with which an external field can be applied while simultaneous topography measurements yield information concerning droplet size.

Before discussing the results, it is instructive to consider what the fields look like at the NSOM tip when it is biased. Obviously, an understanding of the shape and magnitude of the fields at, and surrounding, the metal-coated NSOM tips is necessary before their effects on samples can be interpreted. The fields near an idealized NSOM tip were visualized by solving Laplace's equation in two dimensions.(Mei et al. 2000) As expected, the fields were found to be concentrated annularly around the tip

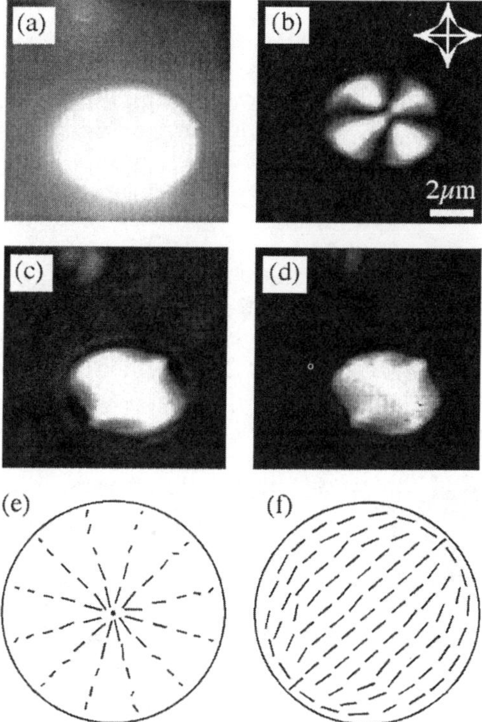

**Figure 6.** Images above represent: (a) NSOM topography of a spherical PDLC droplet, (b) and (c) static NSOM transmitted light images of PDLC droplets, illustrating that there is no preferred alignment of the optical axis in any specific direction. (d) A dynamic optical image collected while a potential was sinusoidally oscillated at a frequency of 100 Hz (4.2 V peak-to-peak). The polarization of the incident and detected light are indicated by arrows in the upper right of panel (b). Panels (e) and (f), respectively, represent models for the liquid crystal orientation found in (b) and (c). Reproduced with permission from (Mei et al. 2000). Copyright 2000 American Institute of Physics.

aperture and to fall off rapidly with increasing distance from the tip. More sophisticated simulations which included details of the PDLC beneath the tip quantized the penetration and attenuation of the fields into the sample and provided guidance into the actual volume probed as a function of applied potential.

An emulsion of E7, a common nematic liquid crystal, in an aqueous solution of poly (vinyl alcohol) (PVA) was deposited onto the surface of an indium-tin-oxide (ITO) coated substrate. The ITO coating provided a transparent conducting surface that formed the counter-electrode to the biased NSOM tip. Polarized light NSOM experiments were performed on both pure E7 droplets and those doped with the fluorescent probe molecule BODIPY.(Mei et al. 2000)

Figure 6 shows an example of both the static and dynamic NSOM measurements on PDLC droplets.(Mei et al. 2000) Fig. 6A displays the topography of an encapsulated droplet, measured using the shear-force feedback method for NSOM tip regulation.

Panels B and C show transmitted light NSOM measurements through typical droplets using a crossed-polarizer configuration. The difference in the two images reflects differences in the orientation of the optical axis within the PDLC. Models of the bipolar configuration seen in panels B and C are shown schematically in panels E and F, respectively. These static NSOM images provide sub-droplet structural information while the image shown in Figure 6D provides insight into the dynamic response of these droplets. This image, taken on the same droplet as that shown in panel C, was collected while applying a sine wave bias to the NSOM tip. The bias was modulated at a frequency of 100 Hz with a peak potential of 4.2 V. The gray scale image reflects the amplitude of the modulated transmitted optical signal.

In a series of papers, these and similar measurements were used by Higgins and coworkers to characterize the response of PDLCs to changes in applied voltage. The sub-diffraction limited resolution in these measurements found a spatial variation in the liquid crystal reorganization threshold that had contributions from several sources including interactions with the polymer encapsulation interface.

In the context of the current review, these measurements demonstrate the added dimension contributed to NSOM measurements by combining a tip-applied voltage with the more standard NSOM contrast mechanisms of transmitted light, fluorescence, and force. These measurements demonstrate that a bias can be applied to the complex geometry of an NSOM probe in an understandable and predictive manner. One can easily imagine similar measurements providing new insights into function of biological molecules where small voltage changes at the nanometric dimension may influence a number of important processes. A potential application of this technique, for instance, could be the use of similar approaches to locally perturb the gating properties of voltage-sensitive ion channels while simultaneously following conformational dynamics using fluorescent toxins or structurally specific probes.


## 5. INTERFEROMETRIC NSOM MEASURMENTS

NSOM is unique among other scanning probe techniques such as AFM and STM in that light is exiting the tip. This rather obvious statement is what leads to NSOM's unique niche among these other very powerful techniques. This enables specific and sensitive contrast mechanisms based on spectroscopic properties, lifetime measurements, or sample interactions with polarized light to all be incorporated into the measurements. Indeed, there is a growing NSOM literature in which all of these properties have been exploited to reveal new insights into sample properties at the nanometric dimension and single molecule limit (Reviewed in Dunn 1999). The fact that light is exiting a nanometric aperture also allows one to take advantage of the interference properties of light, which is another powerful attribute of NSOM.

Several groups have shown that the light exiting the NSOM can be used to create a small interferometer.(Guttroff, Keto et al. 1996; Kramer, Hartmann et al. 1998; Kramer, Hartmann et al. 1995; Shiku and Dunn 1999; Shiku, Krogmeier et al. 1999) As the NSOM tip nears a sample surface, light exiting the tip and that reflecting off the sample surface can interfere with one another. Monitoring this signal as the tip approaches the sample surface; amplitude oscillations are observed as the signal sweeps through constructive and destructive interferences. To date, this has mainly been devoted to creating new feedback methods for tip-sample height regulation.

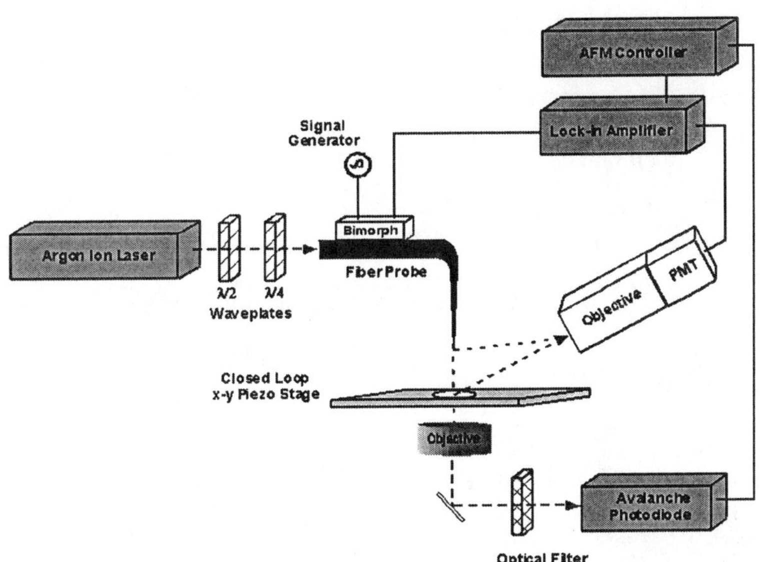


**Figure 7.** Schematic of NSOM adapted for both tapping–mode and interferometric optical feedback imaging. For both feedback methods, the cantilevered NSOM tip is oscillated at its resonant frequency using a piezo bimorph. In tapping-mode feedback, the amplitude of the oscillation is monitored by reflecting a laser off the bend of the NSOM tip and onto a split photodiode (not shown). In optical feedback mode, a long working distance microscope collects both light exiting the NSOM tip and that reflecting off of the sample surface and focuses it onto a PMT. The signal generated by the PMT is fed to a lock-in amplifier that is referenced to the oscillation frequency of the signal controlling the bimorph. Reproduced with permission from (Shiku et al. 1999). Copyright 1999 American Chemical Society.

As an example, Shiku et. al. reported an adaptation of the interferometric optical feedback method for use with cantilevered NSOM probes oscillated normal to the sample surface.(Shiku et al. 1999; Talley, Cooksey et al. 1996) In contrast to straight fiber optic NSOM probes used in shear-force feedback, cantilevered NSOM probes incorporate a near 90° bend near the aperture. These tips can then be operated in a tapping-mode arrangement for tip-sample gap regulation, much tapping-mode AFM.(Talley et al. 1996)

Using the arrangement shown in Figure 7, light exiting a cantilevered NSOM probe was detected from the side with a long working distance microscope coupled to a photomultiplier tube.(Shiku et al. 1999) The tip was oscillated vertically to the sample surface using a bimorph and both the tapping-mode force signal, detected through conventional means, and the signal received at the PMT were collected simultaneously.

Figure 8 shows the approach curves for both signals. The top panel displays the conventional tapping-mode force signal that remains constant until interactions between the tip and the sample cause a dampening in the signal.(Shiku et al. 1999) The bottom panel displays the optical interferometric signal recorded simultaneously from the side. Unlike the force signal, the optical signal sweeps through constructive and destructive interferences as the tip approaches the surface (right to left), until the signal finally drops to zero at the sample surface.

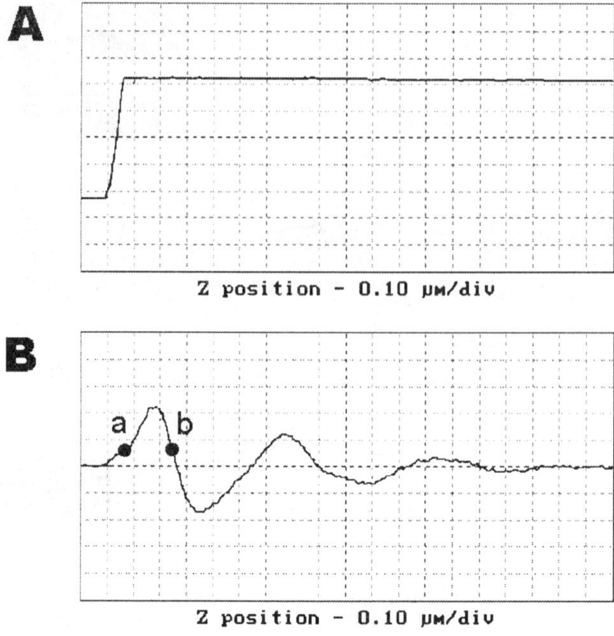



**Figure 8.** Panel (A) displays the approach curve of the NSOM tip using tapping-mode feedback. Panel (B) displays the approach curve collected using the optical feedback mode. Comparison of the two simultaneously collected approach curves, indicates that the optical feedback approach curve begins to oscillate within micrometers of the surface, while tapping mode feedback remains unaltered until the tip is very close to the sample surface Reproduced with permission from (Shiku et al. 1999). Copyright 1999 American Chemical Society.

Comparison of the two panels shown in Figure 8 clearly demonstrates that, unlike the force signal, the optical signal is capable of detecting the presence of the sample surface long before actual contact. This provides a means of implementing a truly non-contact feedback signal for tip-sample gap regulation. In fact, this method is capable of tip-sample gap regulation at the air/liquid interface, as has been demonstrated.(Shiku et al. 1999)

Perhaps of more interest for the current chapter on non-traditional uses of NSOM, this technique also allows one to make single point dynamic measurements. Positioning the tip over a region of interest and holding the tip stationary allows monitoring of optical interferometric signals in order to follow small, Ångstrom level, height level changes in the sample. To demonstrate this capability, a NSOM tip was positioned above a piezoelectric bimorph to which a sine wave was applied. The applied sine wave caused the bimorph to oscillate vertically towards the stationary NSOM tip with a controlled amplitude and frequency. The interferometric optical signal was monitored from the side using the same experimental arrangement as that shown in Figure 7. Figures 9A and 9B plot both the sinusoidal potential applied to the bimorph and

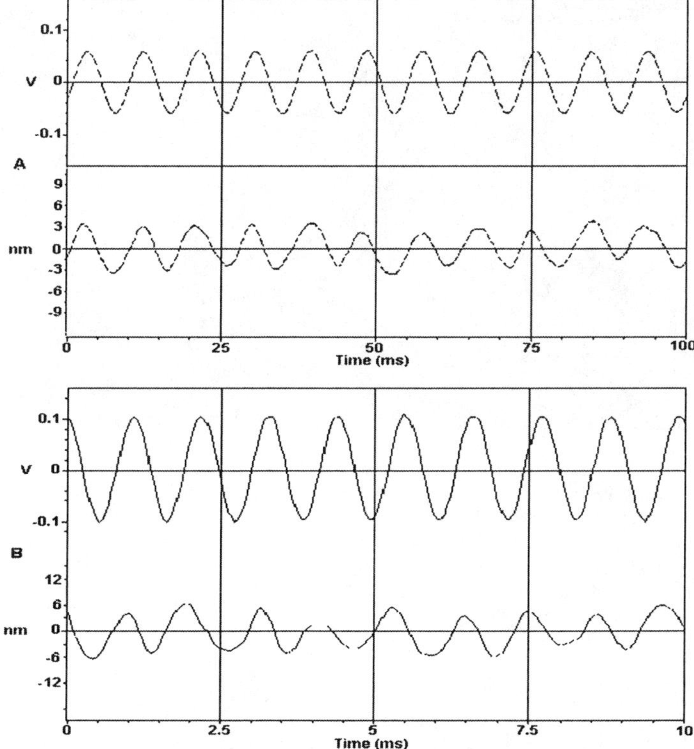

**Figure 9.** While holding the NSOM tip stationary, above an oscillating bimorph, small height changes are detected using the interferometric technique shown in Figure 8. In (A) and (B) the top trace displays the oscillating voltage driving the piezo bimorph, while the bottom trace tracks the change in height detected by the interferometric signal. In (A), the bimorph was oscillated at a frequency of 100Hz, at peak-to- peak amplitudes of 6 nm. In (B), the bimorph was oscillated at a frequency of 1000Hz, at peak amplitudes of ~12nm. Reproduced with permission from (Shiku et al. 1999). Copyright 1999 American Chemical Society.

the resulting optical signal recorded from the interference between the light exiting the NSOM and that reflecting off the bimorph.(Shiku et al. 1999)

In Figure 9A, the bimorph is oscillated with a peak-to-peak amplitude of 6 nm at a frequency of 100 Hz. The optical signal plotted below can clearly follow this motion with good signal-to-noise. In Figure 9B, the bimorph oscillation is increased to 1000 Hz with an amplitude of approximately 12 nm. Again the optical signal shown below is able to track the bimorph motion. These measurements demonstrate that this technique is capable of following very small height level changes on a fast time scale, in a region defined by the small aperture of the NSOM probe. This offers several intriguing avenues for use, especially in the biological sciences. For instance, all that is needed to make these measurements is a refractive index difference between the sample and surrounding media and the measurements are completely non-invasive. Fluorescent probes or other extrinsic tags, therefore, are not needed to conduct these measurements. This reduces sample manipulations and also accounts for the temporal resolution, since the signal scales as the

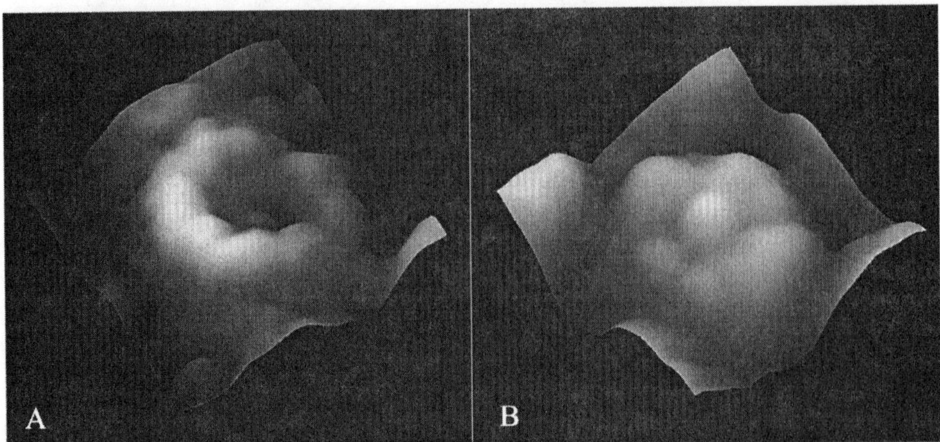


**Figure 10.** AFM measurements of nuclear pore complex (NPC) conformation in the "open" (A) and "closed" (B) configuration. Nuclear pores are ~ 125 MDa protein assemblies that control transport across the nuclear envelope. The outer diameter of each NPC is ~ 120 nm. Several groups have found evidence for a calcium dependent displacement of the central granule in the pore region from recessed (A), to extended (B), following calcium release. (Fahrenkrog, Stoffler et al. 2001; Moore-Nichols, Arnott et al. 2002; Perez-Terzic, Pyle et al. 1996; Stoffler, Feja et al. 2003; Wang and Clapham 1999) Reproduced with permission from (Vickery et al. 1999). Copyright 2000 Biophysical Society.

output power of the NSOM tip and is not limited by the photophysical or photochemical properties of the sample.(Shiku et al. 1999)

There are many applications one can envision that would benefit from the nanometer height sensitivity and millisecond time resolution, combined with the high lateral resolution inherent in NSOM. For instance, nuclear pore complexes (NPC) are large macromolecular protein channels spanning the nuclear envelope of eukaryotic cells that apparently undergo large conformation changes in order to gate transport. (Lee, Dunn et al. 1998; Perez-Terzic et al. 1996) Figure 10A shows a single nuclear pore complex measured using AFM.(Moore-Nichols et al. 2002) Several groups have characterized a calcium-dependent conformational change that leads to the displacement of a "central granule" by approximately 5 nm, similar to that shown in Figure 10B.(Fahrenkrog et al. 2001; Moore-Nichols et al. 2002; Perez-Terzic et al. 1996; Wang et al. 1999) For systems like this, it may be possible to use measurements such as those shown in Figure 9 to non-invasively follow the dynamics of the central granule in real time, without the addition of external tags.

## 6. FLUORESCENCE, TOPOGRAPHY AND COMPLIANCE MEASUREMENTS USING TAPPING-MODE NSOM

As mentioned in the previous section, there are various ways in which feedback may be implemented in order to hold the tip-sample constant during scanning. The shear-force mechanism, where the tip is dithered laterally with respect to the sample surface, is commonly implemented when using straight NSOM probes. For cantilevered NSOM

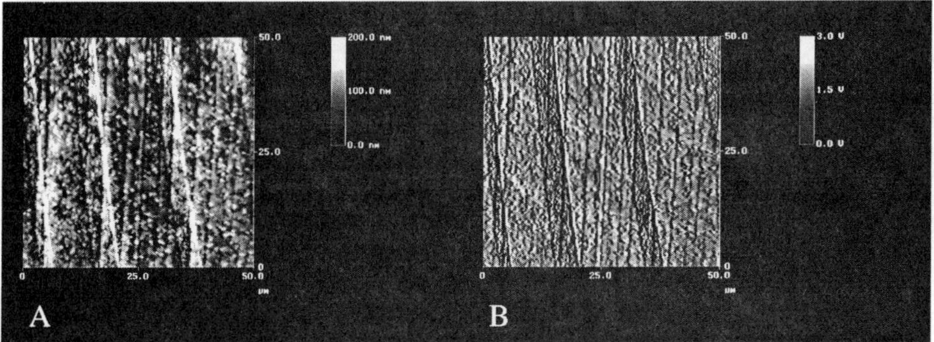


**Figure 11.** TM-NSOM was applied to a test sample of layered polyethylene (PE). The image in (A) represents the topography image of a 50 X 50 μm area of layered PE, showing a small difference in height associated with the alternating PE layers in the plain of the microtome slice. The image in (B) represents the compliance measured over the same region of sample, more clearly revealing the alternating layers of low- and high-density PE. High-density PE regions are revealed as the thinner bands of dark contrast, while the low-density regions are the broader bands of lighter contrast. Reproduced with permission from (Hollars and Dunn 1997). Copyright 1997 American Chemical Society.

probes, a tapping-mode arrangement where the tip is oscillated normal to the surface provides the mechanism for sensing the sample surface. The latter is analogous to tapping mode AFM feedback and can thus take advantage of the extensive literature dealing with tip-sample interactions in both understanding the nature and magnitude of the forces involved and also in the development of new contrast mechanisms.(Talley et al. 1996; Talley, Lee et al. 1998)

In the other sections we have mainly dealt with the enhanced capabilities of NSOM that take advantage by the light exiting the NSOM tip. It should also be mentioned that expanding the contrast mechanisms associated with the force feedback offers a powerful approach for extending the information content in NSOM measurements. For example, it has been shown in tapping-mode AFM measurements, that additional information about sample properties can be obtained by monitoring the phase of the tip resonance. These measurements provide information on the rigidity or compliance of the sample. Similarly, phase measurements on sample compliance can also be incorporated into tapping-mode NSOM measurements, leading to simultaneous fluorescence, height, and compliance information at the nanometric level.

The ability to simultaneously measure near-field fluorescence, topography and compliance using TM-NSOM was demonstrated in 1997. In these experiments, the feedback signal was directed through a lock-in amplifier referenced to the drive frequency of the tip resonance. Both amplitude (height) and phase (compliance) from the phase-sensitive detector were recorded simultaneously, along with the near-field fluorescence signal. To demonstrate the compliance sensitivity of TM-NSOM, height and compliance measurements (Figure 11) were conducted on samples of micro-layered polyethylene (PE) containing layers of differing density. Slicing PE samples perpendicular to the layered planes using an ultra-microtome knife created a test sample surface comprised of alternating low- and high-density PE micro-layers. (Hollars et al. 1997)

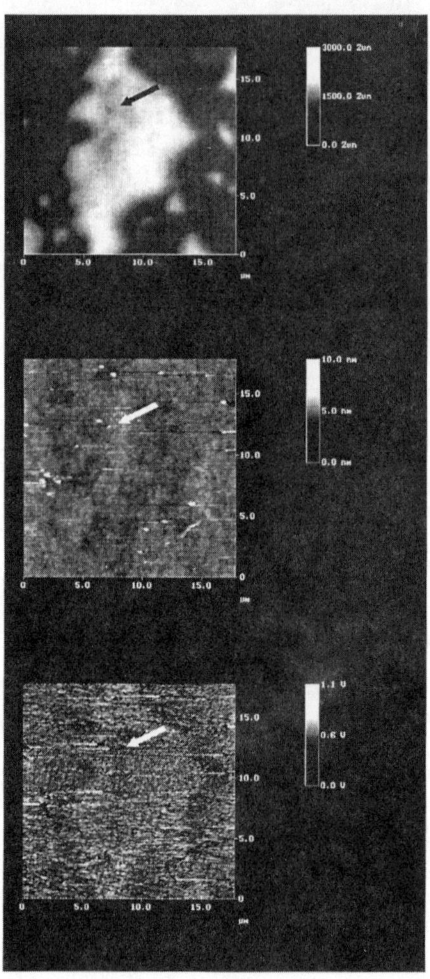

**Figure 12.** Fluorescence, height, and compliance images of a DPPC monolayer doped with 0.25 mol % diIC$_{18}$. Image (A) is the NSOM fluorescence image of diIC$_{18}$ distribution within the monolayer. Fluorescence is distributed primarily within a network comprised of one of two different lipid phases. Note that even within the more intensely fluorescing regions that some variability in dye distribution can be observed. (B) Topography image of the same region of monolayer. Small differences in height (~5 Å) are apparent that correspond to the coexisting lipid phases defined by the fluorescence distributions. (C) Phase image illustrating compliance across different regions of the monolayer. Arrows indicate a region within the large, fluorescent, domain in which fluorescence is attenuated by 30%, the height is increased, and compliance is decreased. Such a region is consistent with the existence of a condensed phase region inside the larger, expanded phase domain. Reproduced with permission from (Hollars et al. 1997). Copyright 1997 American Chemical Society.

The topography image presented in of Figure 11A shows the height differential between the low-density and high density PE-Layers (60 –70 nm), and cutting marks from the slicing process. Figure 11B shows the simultaneously collected compliance image, measured by monitoring the change in phase of the TM-NSOM tip oscillation. Clearly visible in Figure 11B are contrast features reflecting compliance differences between the softer, low density PE sections (light contrast) and the more rigid, high density PE layers (dark contrast). Previous AFM studies performed on similar samples showed comparable patterns of contrast across low- and high- density PE micro-layers.

To further demonstrate the utility of this technique, experiments were conducted on artificial lipid monolayers that included measurements of fluorescence marker distributions. Figure 12 displays the simultaneously measured TM-NSOM fluorescence, topography, and compliance of a L-$\alpha$-dipalmitoylphosphatidylcholine (DPPC) monolayer.(Hollars et al. 1997) The lipid monolayer was formed using the Langmuir-Blodgett technique and was transferred onto a substrate in a region of the pressure isotherm where coexisting expanded and condensed phases of the film exist. A small amount, 0.25mol%, of the fluorescent dye diIC$_{18}$ was doped into the monolayer. Previous studies have shown that this marker preferentially partitions in the expanded phase of the film thus providing a marker of the phase distribution in the film.

The near-field fluorescence image shown in Fig. 12A can therefore be used to map the distribution of the two phases in the lipid monolayer. The bright, fluorescent regions of the film mark the expanded phase of the film while the dark areas denote condensed regions in the film. This can be compared to the TM-NSOM force image shown in Fig. 12B. Small, 4-7Å, height level changes are observed which correlate with the fluorescence image. Regions of lower topography correspond to bright regions in the fluorescence image.

Finally, the bottom panel displays the phase image. Features in the phase image correspond to features observed in both fluorescence and topographical images. The condensed regions in the film induce less of a phase shift than do the more expanded regions. This is consistent with the expanded regions being more compliant than the condensed regions. It should also be emphasized that small, but measurable phase changes are observed even though the change in topography is negligible. This suggests that the two contrast mechanisms are decoupled and provide independent and complementary measures of the sample properties.

These experiments demonstrate the utility and feasibility of simultaneously measuring near-field fluorescence, topography, and compliance using TM-NSOM. As in the previous sections, the addition of compliance contrast permits yet another simultaneous and independent measure of sample properties.


## 7. CONCLUSIONS

The continued development of NSOM is expanding its capabilities as an analytical tool and here we have examined only a few of those techniques currently in development. However, a general trend is clear; the means by which NSOM may be used to characterize molecular properties at the nanometric scale is rapidly diversifying, and is not merely limited to traditional fluorescence measurements. Whether through modification of the NSOM probe or the experimental approach, these new NSOM capabilities will undoubtedly find applications across disciplines. For the biological sciences in particular, NSOM seems poised to make a significant impact as we have tried to point out throughout this survey. Progress in this area has been slowed by problems associated with imaging soft and fragile samples. However, recent advances in tip and feedback designs are rapidly advancing the technique and it seems hard to imagine that this final hurdle will not be overcome before the publication of this edition.


## 8. ACKNOWLEDGEMENTS

Much of the work described in this chapter was carried out in collaboration with a very talented group of former colleagues including Christopher W. Hollars, Jeffery R. Krogmeier, M. Annie Lee, Hitoshi Shiku, Chad E. Talley, and Sarah A. Vickery. Support for this work was generously provided by NIH GM065964 and the Madison and Lila Self Foundation.


## 9. REFERENCES

Bardo, A. M., M. M. Collinson and D. A. Higgins (2001). "Nanoscale properties and matrix-dopant interactions in dye-doped organically modified silicate thin films." *Chemistry of Materials* **13**(8): 2713-2721.

Barker, S. L., B. A. Thorsrud and R. Kopelman (1998). "Nitrite- and chloride-selective fluorescent nano-optodes and in vitro application to rat conceptuses." *Anal Chem* **70**(1): 100-4.

Betzig, E., J. K. Trautman, T. D. Harris, J. S. Weiner and R. L. Kostelak (1991). "Breaking the diffraction barrier: Optical microscopy on a nanometric scale." *Science* **251**: 1468-1470.

Cullum, B. M. and T. Vo-Dinh (2000). "The development of optical nanosensors for biological measurements." *Trends in Biotechnology* **18**(9): 388-393.

de Lange, F., A. Cambi, R. Huijbens, B. de Bakker, W. Rensen, M. Garcia-Parajo, N. van Hulst and C. G. Figdor (2001). "Cell biology beyond the diffraction limit: Near-field scanning optical microscopy." *J Cell Sci* **114**(Pt 23): 4153-60.

Deniz, A. A., M. Dahan, J. R. Grunwell, T. Ha, A. E. Faulhaber, D. S. Chemla, S. Weiss and P. G. Schultz (1999). "Single-pair fluorescence resonance energy transfer on freely diffusing molecules: Observation of forster distance dependence and subpopulations." *Proc Natl Acad Sci U S A* **96**(7): 3670-5.

Dunn, R. C. (1999). "Near-field scanning optical microscopy." *Chem Rev* **99**(10): 2891-928.

Edidin, M. (2001). "Near-field scanning optical microscopy, a siren call to biology." *Traffic* **2**(11): 797-803.

Emptage, N. J. (2001). "Fluorescent imaging in living systems." *Curr Opin Pharmacol* **1**(5): 521-5.

Fahrenkrog, B., D. Stoffler and U. Aebi (2001). "Nuclear pore complex architecture and functional dynamics." *Curr Top Microbiol Immunol* **259**: 95-117.

Gaits, F. and K. Hahn (2003). "Shedding light on cell signaling: Interpretation of fret biosensors." *Sci STKE* **2003**(165): PE3.

Guttroff, G., J. M. Keto, C. K. Shih, A. Anselm and B. G. Streetman (1996). "A design of reflection scanning near-field optical microscope and its application to algaas/gaas heterostructures." *Applied Physics Letters* **68**(25): 3620-3622.

Ha, T., T. Enderle, D. F. Ogletree, D. S. Chemla, P. R. Selvin and S. Weiss (1996). "Probing the interaction between two single molecules: Fluorescence resonance energy transfer between a single donor and a single acceptor." *Proc Natl Acad Sci U S A* **93**(13): 6264-8.

Higgins, D. A. (2002). "Materials characterization: Optics up close and personal." *Nat Mater* **1**(2): 83-5.

Higgins, D. A., X. Liao, J. E. Hall and E. Mei (2001). "Simultaneous near-field optical birefringence and fluorescence contrast applied to the study of dye-doped polymer-dispersed liquid crystals." *Journal of Physical Chemistry B* **105**(25): 5874-5882.

Higgins, D. A., E. Mei and X. Liao (1999). "Electric-field-induced molecular reorientation dynamics by near-field scanning optical microscopy." *Proceedings of SPIE-The International Society for Optical Engineering* **3607**(Scanning and Force Microscopies for Biomedical Applications): 26-35.

Hollars, C. W. and R. C. Dunn (1997). "Submicron fluorescence, topology, and compliance measurements of phase-separated lipid monolayers using tapping-mode near-field scanning optical microscopy." *Journal of Physical Chemistry B* **101**(33): 6313-6317.

Kirsch, A. K., V. Subramaniam, A. Jenei and T. M. Jovin (1999). "Fluorescence resonance energy transfer detected by scanning near-field optical microscopy." *Journal Of Microscopy* **194**(2-3): 448-54.

Kramer, A., T. Hartmann, R. Eschrich and R. Guckenberger (1998). "Scanning near-field fluorescence microscopy of thin organic films at the water/air interface." *Ultramicroscopy* **71**(1-4): 123-132.

Kramer, A., T. Hartmann, S. M. Stadler and R. Guckenberger (1995). "An optical tip-sample distance control for a scanning near-field optical microscope." *Ultramicroscopy* **61**(1-4, Selected Papers from the 3rd International Conference on Near-Field Optics and Related Techniques, 1995): 191-195.

Lakowicz, J. R. (1983). *Principles of fluorescence spectroscopy*. New York, Plenum Press.

Lee, M. A., R. C. Dunn, D. E. Clapham and L. Stehno-Bittel (1998). "Calcium regulation of nuclear pore permeability." *Cell Calcium* **23**(2-3): 91-101.

Lee, M. A., C. E. Talley, S. A. Vickery, J. R. Krogmeier, C. W. Hollars, H. Shiku and R. C. Dunn (1999). "Progress toward imaging biological samples with nsom." *Proceedings of SPIE-The International Society for Optical Engineering* **3607**(Scanning and Force Microscopies for Biomedical Applications): 60-66.

Lewis, A., A. Radko, N. Ben Ami, D. Palanker and K. Lieberman (1999). "Near-field scanning optical microscopy in cell biology." *Trends Cell Biol* **9**(2): 70-3.

Marchese-Ragona, S. P. and P. G. Haydon (1997). "Near-field scanning optical microscopy and near-field confocal optical spectroscopy: Emerging techniques in biology." *Ann N Y Acad Sci* **820**: 196-206; discussion 206-7.

Mei, E. and D. A. Higgins (1998). "Near-field scanning optical microscopy studies of electric-field-induced molecular reorientation dynamics." *Journal of Physical Chemistry A* **102**(39): 7558-7563.

Mei, E. and D. A. Higgins (1998). "Polymer-dispersed liquid crystal films studied by near-field scanning optical microscopy." *Langmuir* **14**(8): 1945-1950.

Mei, E. and D. A. Higgins (2000). "Nanometer-scale resolution and depth discrimination in near-field optical microscopy studies of electric-field-induced molecular reorientation dynamics." *Journal of Chemical Physics* **112**(18): 7839-7847.

Meixner, A. J. and H. Kneppe (1998). "Scanning near-field optical microscopy in cell biology and microbiology." *Cell Mol Biol (Noisy-le-grand)* **44**(5): 673-88.

Moore-Nichols, D., A. Arnott and R. C. Dunn (2002). "Regulation of nuclear pore complex conformation by ip(3) receptor activation." *Biophys J* **83**(3): 1421-8.

Perez-Terzic, C., J. Pyle, M. Jaconi, L. Stehno-Bittel and D. E. Clapham (1996). "Conformational states of the nuclear pore complex induced by depletion of nuclear ca2+ stores." *Science* **273**(5283): 1875-7.

Pohl, D. W. (1991). Scanning near-field optical microscopy (snom). *Advances in optical and electron microscopy*. T. Mulvey and C. J. R. Sheppard. London, Academic Press. **12**: 243-312.

Sekar, R. B. and A. Periasamy (2003). "Fluorescence resonance energy transfer (fret) microscopy imaging of live cell protein localizations." *J Cell Biol* **160**(5): 629-33.

Sekatskii, S. K., G. T. Shubeita, M. Chergui, G. Dietler, B. N. Mironov, D. A. Lapshin and V. S. Letokhov (2000). "Towards the fluorescence resonance energy transfer (fret) scanning near-field optical microscopy: Investigation of nanolocal fret processes and fret probe microscope." *Journal of Experimental and Theoretical Physics (Translation of Zhurnal Eksperimental'noi i Teoreticheskoi Fiziki)* **90**(5): 769-777.

Shiku, H. and R. C. Dunn (1999). "Near-field scanning optical microscopy." *Anal Chem* **71**(1): 23A-29A.

Shiku, H. and R. C. Dunn (1999). "Near-field scanning optical microscopy studies of l-alpha-dipalmitoylphosphatidylcholine monolayers at the air-liquid interface." *J Microsc* **194**(Pts 2-3): 461-6.

Shiku, H., J. R. Krogmeier and R. C. Dunn (1999). "Noncontact near-field scanning optical microscopy imaging using an interferometric optical feedback mechanism." *Langmuir* **15**(6): 2162-2168.

Shubeita, G. T., S. K. Sekatskii, M. Chergui, G. Dietler and V. S. Letokhov (1999). "Investigation of nanolocal fluorescence resonance energy transfer for scanning probe microscopy." *Applied Physics Letters* **74**(23): 3453-3455.

Shubeita, G. T., S. K. Sekatskii, G. Dietler and V. S. Letokhov (2002). "Local fluorescent probes for the fluorescence resonance energy transfer scanning near-field optical microscopy." *Applied Physics Letters* **80**(15): 2625-2627.

Stoffler, D., B. Feja, B. Fahrenkrog, J. Walz, D. Typke and U. Aebi (2003). "Cryo-electron tomography provides novel insights into nuclear pore architecture: Implications for nucleocytoplasmic transport." *J Mol Biol* **328**(1): 119-30.

Subramaniam, V., A. K. Kirsch and T. M. Jovin (1998). "Cell biological applications of scanning near-field optical microscopy (snom)." *Cell Mol Biol (Noisy-le-grand)* **44**(5): 689-700.

Talley, C. E., G. A. Cooksey and R. C. Dunn (1996). "High resolution fluorescence imaging with cantilevered near-field fiber optic probes." *Applied Physics Letters* **69**(25): 3809-3811.

Talley, C. E., M. A. Lee and R. C. Dunn (1998). "Single molecule detection and underwater fluorescence imaging with cantilevered near-field fiber optic probes." *Applied Physics Letters* **72**(23): 2954-2956.

Tan, W., Z. Y. Shi, S. Smith, D. Birnbaum and R. Kopelman (1992). "Submicrometer intracellular chemical optical fiber sensors." *Science* **258**(5083): 778-81.

Van Der Meer, B. W., G. Coker and S. Simon Chen (1994). *Resonance energy transfer theory and data*. New York, VCH Publishers, Inc.

Vickery, S. A. and R. C. Dunn (1999). "Scanning near-field fluorescence resonance energy transfer microscopy." *Biophys J* **76**(4): 1812-8.

Vickery, S. A. and R. C. Dunn (2001). "Combining afm and fret for high resolution fluorescence microscopy." *J Microsc* **202**(Pt 2): 408-12.

Vo-Dinh, T. (2002). "Nanobiosensors: Probing the sanctuary of individual living cells." *J Cell Biochem Suppl* **39**: 154-61.

Vo-Dinh, T., J. P. Alarie, B. M. Cullum and G. D. Griffin (2000). "Antibody-based nanoprobe for measurement of a fluorescent analyte in a single cell." *Nature Biotechnology* **18**: 764-767.

Wang, H. and D. E. Clapham (1999). "Conformational changes of the in situ nuclear pore complex." *Biophys J* **77**(1): 241-7.

Wolfbeis, O. S. (2002). "Fiber-optic chemical sensors and biosensors." *Anal Chem* **74**(12): 2663-77.

# 3

# NANOPARTICLES WITH TUNABLE LOCALIZED SURFACE PLASMON RESONANCES
## TOPICS IN FLUORESCENCE SPECTROSCOPY

Christy L. Haynes, Amanda J. Haes, Adam D. McFarland, and Richard P. Van Duyne[*]

## 1. INTRODUCTION

### 1.1. General Overview

The last decade has brought enormous progress in the area of tunable noble metal nanoparticle optical properties. An understanding of the optical properties of noble metal nanoparticles holds both fundamental and practical significance. Fundamentally, it is important to systematically explore nanostructure characteristics that cause optical property variation as well as provide access to regimes of predictable behavior. Practically, the tunable optical properties of nanostructures can be applied as materials for surface-enhanced spectroscopy, optical filters, waveguides, and sensors. In all of these cases, the optical property of interest is the localized surface plasmon resonance (LSPR). The LSPR condition is fulfilled when a specific wavelength of light impinges on the nanostructure, generating enhanced electromagnetic fields that are localized at the surface of the nanostructure. The wavelength of light that excites the collective oscillation of conduction electrons is absorbed and scattered by the nanostructure. Accordingly, UV-visible-NIR-IR extinction spectroscopy can be used to monitor the nanostructure LSPR.

---

[*] Northwestern University, Department of Chemistry, 2145 Sheridan Road, Evanston, Illinois 60208-3113

*Topics in Fluorescence Spectroscopy, Volume 8: Radiative Decay Engineering*
Edited by Geddes and Lakowicz, Springer Science+Business Media, Inc., New York, 2005

                                                    

The simplest theoretical approach available for modeling the optical properties of nanoparticles is the Mie theory estimation of the extinction of a metallic sphere in the long wavelength, electrostatic dipole limit.  In the following equation:[1]

$$E(\lambda) = \frac{24\pi^2 N_A a^3 \varepsilon_m^{\frac{3}{2}}}{\lambda \cdot \ln(10)} \left[ \frac{\varepsilon_i}{\left(\varepsilon_r + 2\varepsilon_m\right)^2 + \varepsilon_i^2} \right] \qquad (1.1)$$

$E(\lambda)$ is the extinction which is, in turn, equal to the sum of absorption and Rayleigh scattering, $N_A$ is the areal density of nanoparticles, "a" is the radius of the metallic nanosphere, $\varepsilon_m$ is the dielectric constant of the medium surrounding the metallic nanosphere (assumed to be a positive, real number and wavelength independent), $\lambda$ is the wavelength of the absorbing radiation, $\varepsilon_i$ is the imaginary portion of the metallic nanosphere's dielectric function, and $\varepsilon_r$ is the real portion of the metallic nanosphere's dielectric function. Even in this most primitive model, it is abundantly clear that the LSPR spectrum of an isolated metallic nanosphere embedded in an external dielectric medium will depend on the nanoparticle radius "a", the nanoparticle material ($\varepsilon_i$ and $\varepsilon_r$), and the nanoenvironment's dielectric constant ($\varepsilon_m$). Furthermore, when the nanoparticles are not spherical, as is always the case in real samples, the extinction spectrum will depend on the nanoparticle's in-plane diameter, out-of-plane height, and shape. The dependence of the extinction spectrum on these nanoparticle structural parameters has been recognized in the form of simplified model calculations such as those for ellipsoidal nanoparticle geometries.[2]  In this case the resonance term from the denominator of equation 1.1 is replaced with:

$$\left(\varepsilon_r + \chi\varepsilon_m\right)^2 \qquad (1.2)$$

where $\chi$ is a term that describes the nanoparticle aspect ratio.  The values for $\chi$ increase from 2 (for a sphere) up to, and beyond, values of 17 for a 5:1 aspect ratio nanoparticle. In addition, many of the samples considered in this work contain an ensemble of nanoparticles that are supported on a substrate.  Thus, the LSPR will also depend on interparticle spacing and substrate dielectric constant.


## 1.2. Fabrication of Nanostructures with Tunable Optical Properties

In recent years a world-wide effort has been undertaken using three main nanofabrication techniques to systematically probe each of the characteristics affecting the LSPR.  These techniques are: (1) chemical/electrochemical synthesis, (2) electron beam lithography, and (3) nanosphere lithography.  A large portion of nanoscale fabrication is accomplished by chemical reduction of metal salts in the presence of geometry-determining templates or surface-modifying surfactants.  Research groups have advanced significantly from the traditional citrate-reduction of Ag or Au salts used to generate spherical nanoparticles.[3]  Halas and coworkers have successfully coated latex nanospheres and microspheres both completely and partially to create Ag and Au nanoshells[4] and nanocups.[5] Small polystyrene nanospheres (100 nm diameter) have also been used as templates by Hupp and coworkers to fabricate size-monodisperse Ag nanodisks.[6] Triangular platelets have been produced by Mirkin and coworkers in a photochemical reaction between Ag salt and a bis(p-sulfonatophenyl) phenylphosphine surfactant.[7] Schultz and coworkers take advantage of a diversity of nanoparticle geometries, including spheres, triangles, and pentagons, resulting from the gelatin-stabilized reduction of Ag salt onto small nucleating Au nanospheres.[8] As will be discussed in Section 1.3, the LSPR characteristics of the nanoparticles fabricated using chemical reduction have been interrogated to better understand and apply size-dependent optical properties.

Nanostructures fabricated by electrochemical and electroless plating represent another class of nanoparticles produced by the reduction of noble metal compounds. In all cases, control of the resultant nanostructure is achieved by plating the metal into a well-defined nanoscale template. Electrochemical plating into micelle templates[9] has been implemented by El-Sayed and coworkers to fabricate homogeneous high aspect ratio Au nanorods.[10] Martin and coworkers pioneered an approach where Au, among other materials, is electrochemically plated into an anodic Al template in order to create parallel arrays of Au cylinders.[11-13] After dissolution of the Al template in a strong base, the homogeneous Au nanoparticles are released. Foss and coworkers have further controlled the plating conditions into anodic Al templates to fabricate non-centrosymmetric Au cylinder/sphere pairs.[14] Polycarbonate membranes have been employed by Schultz and coworkers in concert with electroless and electrochemical plating of Ag, Au, and/or Ni metal.[15] The resulting striped nanocylinders can also be released by dissolving the polycarbonate membrane.

In the aforementioned examples, structural homogeneity of noble metal nanostructures was achieved. The control of LSPR-defining factors has been further manipulated by embedding nanoparticles created by chemical or electrochemical reduction into dielectric matrices. For example, Feldman and coworkers embedded monodisperse Au colloids in a matrix of titanium dioxide before performing near-field scanning optical microscopy (NSOM) measurements.[16] Foss and coworkers exploited the highly-oriented nature of polymer chains in a thin film of poly(tetrafluoroethylene) to create a homogeneous dielectric environment around thiol-coated Au colloids[17] and templated Au cylinders.[13] A polymer matrix was also employed by Dirix and coworkers; in this case, Ag nanoparticles were dispersed in a poly(ethylene) matrix that was then drawn to create ordered arrays of nanoparticles.[18] Chumanov and coworkers implanted Ag nanoparticles into a matrix of polydimethylsiloxane elastomer so that interparticle distances could be easily varied.[19] In all of these cases, strict control of nanoparticle structure and environment simplified the interpretation of both fundamental and applied results.

All nanoparticle fabrication methods discussed thus far are used to create suspensions of nanoparticles either in solution or in a dielectric matrix. Another class of techniques address substrate-bound nanostructure fabrication. The standard approach for making substrate-bound nanostructures in the area of nanoelectronics is electron beam lithography (EBL). In EBL, the desired pattern is serially produced by exposing a thin layer of photoresist to high-energy electrons followed by chemical development and deposition of the noble metal (Figure 1). While EBL is an expensive and time-consuming nanofabrication technique, the resultant flexibility in nanostructure design is a powerful asset. Aussenegg and coworkers have used EBL extensively.[20-32] To take only a single

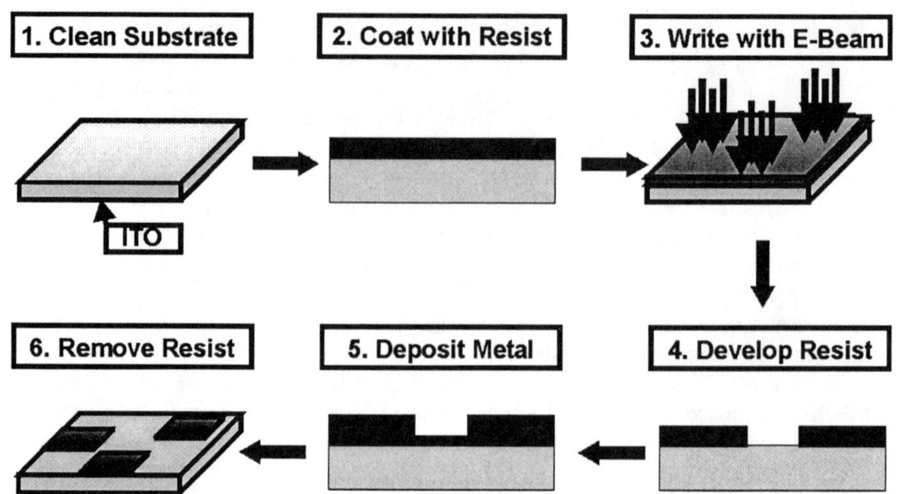



**Figure 1.** Schematic illustration of the electron beam lithography fabrication technique. The clean substrate is coated with a thin film of an electron sensitive resist, and then the electron beam exposes the resist as desired. After exposure, the patterns are developed in xylene, creating a patterned resist film. The desired noble metal is then deposited in a high vacuum thin film vapor deposition system. In the last step of the sample preparation, the lift-off step, the resist is dissolved in a strong solvent.

example from their work, one-dimensional chains of Ag disks and rods are fabricated to study nanoparticle-nanoparticle communication.[22] Atwater and coworkers also fabricate one-dimensional chains of Ag nanorods to study the fundamental principles of waveguides.[33] Käll and coworkers used EBL to fabricate homogeneous two-dimensional arrays of Au and Ag disks and triangular platelets in both hexagonal and square-packed arrangements.[34] Intricate patterns of Au disks were created by Bozhevolnyi and coworkers to define explicit gaps for waveguiding applications.[35]

While noble metal deposition through an EBL-defined mask is common, there are other thermal deposition techniques that address the limitations of EBL. Thermal deposition of Ag or Au onto a smooth substrate under high or ultrahigh vacuum conditions has been used for many years to create well-ordered nanoparticles or metal island films.[36, 37] Quasi-periodic nanoparticle arrays can be fabricated, without a template, by exploiting lattice strain between the substrate and nanoparticle material.[38] Meanwhile, metal island films are a collection of non-confluent, heterogeneous noble metal islands. The island film nanostructure has been exploited in recent years by Weimer and coworkers; their work utilizes stringent control of the substrate temperature, deposition rate, and noble metal film thickness to create a diverse set of nanostructures.[39] Siiman and coworkers have used Ag and Au island films deposited onto nanospheres and microspheres as optical labels in applied experiments.[40] Van Duyne and coworkers fabricate homogeneous substrate-bound nanoparticles by performing thermal deposition onto and through a two-dimensional colloidal crystal mask consisting of either a single or double layer of polystyrene or silica nanospheres. After nanosphere lift-off, an ordered array of nanoparticles remains on the substrate.[41-63] Due to the specific relevance

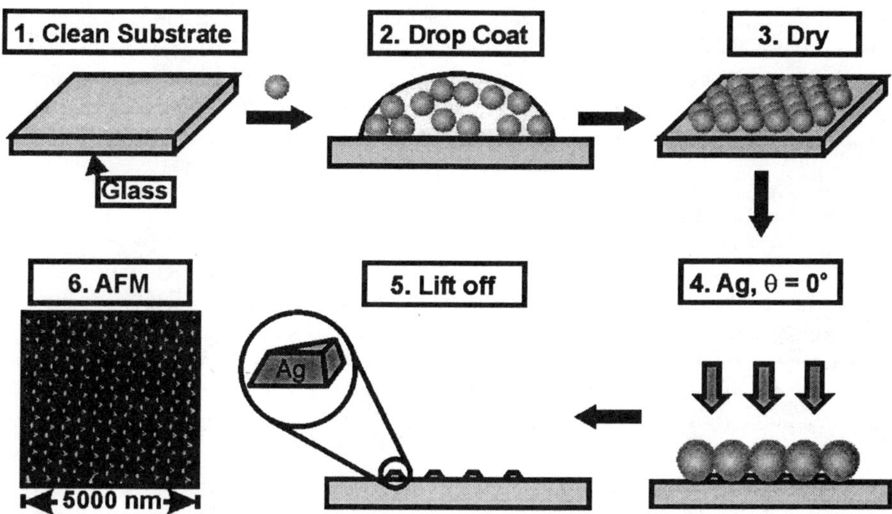


**Figure 2.** Schematic illustration of the nanosphere lithography fabrication technique. A small volume of nanosphere solution is drop-coated onto the clean substrate. As the solvent evaporates, the nanospheres assemble into a two-dimensional colloidal crystal mask. The desired noble metal is then deposited in a high vacuum thin film vapor deposition system. In the last step of the sample preparation, the lift-off step, the nanospheres are removed by sonication in absolute ethanol.

throughout the remainder of the presented work, this technique, known as nanosphere lithography, will be explained in detail.

Nanosphere lithography (NSL) is a powerful fabrication technique to inexpensively produce nanoparticle arrays with controlled shape, size, and interparticle spacing (Figure 2).[52] The need for monodisperse, reproducible, and materials general nanoparticles has driven the development and refinement of the most basic NSL architecture as well as many new nanostructure derivatives. Every NSL structure begins with the self-assembly of size-monodisperse nanospheres of diameter D to form a two-dimensional colloidal crystal deposition mask. Methods for deposition of a nanosphere solution onto the desired substrate include spin coating,[52] drop coating,[51] and thermoelectrically-cooled angle coating.[64] All of these deposition methods require that the nanospheres be able to freely diffuse across the substrate seeking their lowest energy configuration. This is often achieved by chemically modifying the nanosphere surface with a negatively charged functional group such as carboxylate or sulfate that is electrostatically repelled by the negatively charged surface of a substrate such as mica or glass. As the solvent (water) evaporates, capillary forces draw the nanospheres together, and the nanospheres crystallize in a hexagonally close-packed pattern on the substrate. As in all naturally occurring crystals, nanosphere masks include a variety of defects that arise as a result of nanosphere polydispersity, site randomness, point defects (vacancies), line defects (slip dislocations), and polycrystalline domains. Typical defect-free domain sizes are in the 10 - 100 μm range. Following self-assembly of the nanosphere mask, a metal or other material is then deposited by thermal evaporation, electron beam deposition, or pulsed laser deposition from a collimated source normal to the substrate through the nanosphere mask to a controlled mass thickness, $d_m$. After metal deposition, the nanosphere mask is removed, typically by sonicating the entire sample in a solvent, leaving behind the material deposited through the nanosphere mask and onto the substrate. It is possible to vary the shape and interparticle spacing between nanoparticles by mounting the nanosphere mask at a non-normal angle relative to the deposition beam.[48]

## 1.3. Fundamental Studies of Tunable Optical Properties

Section 1.2 outlined three major classes of nanofabrication techniques that are currently in use in laboratories around the world to produce noble metal nanoparticles. As previously discussed, noble metal nanoparticles are of particular interest due to their extraordinary optical properties. This section will review a number of fundamental studies carried out to understand and exploit both the tunability of the resonance condition and the resultant amplification of the local electromagnetic fields.

### 1.3.1. Defining the Fundamental Characteristics of the Localized Surface Plasmon Resonance

The recent increase in single molecule fluorescence and scattering studies indicates the significant general interest in exploring the chemical and physical properties in non-ensemble-averaged systems. In the last decade, scientists have begun to exploit the very large extinction coefficient of noble metal nanostructures in order to study the optical properties of single nanoparticles. Matsuo and coworkers implemented a novel technique to capture transient extinction spectra of single Ag nanoparticles.[65] Differential interference contrast microscopy with pulsed laser excitation generates information about the amplitude and phase of the absorbed and scattered photons. Feldman and coworkers measured dark-field scattering spectra from single Au nanospheres and nanorods.[66] Because single nanoparticle spectra do not suffer from inhomogeneous broadening, it is easy to determine the LSPR dephasing times. In this case, Feldman and coworkers found that increased nanoparticle aspect ratio systematically yields decreased dephasing rates (dephasing times are 1.4 – 5.0 femtoseconds for nanospheres and 6.0 – 18.0 femtoseconds for nanorods). Aussenegg and coworkers approached the question of LSPR dephasing times using evanescent excitation of single EBL-fabricated Au nanoparticles and found similar results (4.0 – 7.3 femtoseconds).[25]

While the previously discussed experiments used the homogeneous linewidth spectrum of single nanoparticle LSPR spectra to understand dephasing times, an entirely different class of experiments has been performed to directly image LSPR dynamics and electromagnetic field propagation. The most basic work has been done by Aussenegg and coworkers using NSOM.[23] Under the assumption that the detected extinction intensity from EBL-fabricated nanoparticles is proportional to the square of the electric field, this group created images of LSPR interference patterns with varied nanoparticle aspect ratio and lattice spacing. In later experiments, Aussenegg and coworkers exploited fluorescence imaging to measure LSPR dynamics.[30, 31] In this work, a partial monolayer of a fluorescent dye was applied over a surface containing silica-coated Ag nanoparticles. The dye absorption band and nanoparticle LSPR were well-matched so that excitation of the LSPR generates a spatial map of the resultant electromagnetic fields. The results provide high fidelity images of nanosphere, nanorod, and nanoparticle pair electromagnetic fields as well as quantitative information regarding electromagnetic field strength through measured fluorescence intensity and photobleaching rate.

### 1.3.2. Controlling the Localized Surface Plasmon Resonance

Section 1.3.1 demonstrates a firm empirical understanding of the single nanoparticle resonance condition, LSPR dephasing time, and spatial maps of the resulting electromagnetic fields. In order to exploit nanoparticle optics, it is necessary to methodically control all parameters that determine the LSPR characteristics. With a sufficiently flexible nanofabrication method, it is possible to manipulate all factors affecting the LSPR, varying one parameter systematically to examine the outcome. For example, Mulvaney wrote a comprehensive review of the tunable optical properties of Au nanoparticles with changing size and dielectric coating.[67] Weimer and coworkers exploited very fine control of substrate temperature and deposition rate when producing Ag and Au island films in order to systematically tune the LSPR.[39] They were able to empirically create a three-parameter plot whereby knowledge of the chosen metal, substrate temperature and deposition rate allows prediction of the resulting LSPR. Two-dimensional EBL arrays of Ag disks were fabricated by Aussenegg and coworkers in order to probe the LSPR as a function of nanoparticle aspect ratio and the refractive index of the local environment.[21] Both increase in aspect ratio and increase in local refractive index caused systematic shifts of the LSPR to lower energies. Extensive studies have been completed by Van Duyne and coworkers using NSL-fabricated substrates whereby it is possible to systematically vary nanoparticle aspect ratio,[58] shape,[54] substrate,[42] dielectric environment,[55] and effective thickness of a chemisorbed monolayer.[43] In all cases, the experiments revealed systematic shifts in the LSPR: increased aspect ratio shifts the LSPR to lower energies, retraction of sharp tetrahedral tips shifts the LSPR to higher energies, increased refractive index of the substrate or solvent environment shifts the LSPR to lower energies, and increased thickness of chemisorbed molecules shifts the LSPR to lower energies within the limit of the electromagnetic field decay length.

The spatial maps of electromagnetic fields measured using NSOM and fluorescence imaging discussed in Section 1.3.1 are measured from EBL samples. EBL has a distinct advantage over chemical nanoparticle preparations in that the nanoparticle structure is well defined. Schultz and coworkers performed a correlated dark-field scattering/transmission electron microscopy study on chemically-produced Ag spherical, triangular, and pentagonal nanoparticles ranging in size from 40–140 nm.[8] In a plot of LSPR extinction maximum versus overall nanoparticle size, the three different shapes are largely segregated from one another. This result is consistent with the extensive body of theoretical studies addressing the effects of nanoparticle size and shape on the LSPR.[56, 57, 59, 60, 68] Multiple research groups have invested significant effort in the theoretical modeling of LSPR behavior in noble metal nanoparticles. Schatz and coworkers used the discrete dipole approximation in electrodynamic theory to successfully identify the dipole and quadrupole contributions to the LSPR spectrum of Ag nanodisks that were chemically synthesized by Hupp and coworkers[6] and Ag nanoprisms that were chemically synthesized by Mirkin and coworkers.[7] Exploration of higher order multipoles is a relatively new area, accessible only with advanced nanofabrication techniques that are capable of producing high aspect ratio nanoparticles. Aussenegg and coworkers used EBL to fabricate Ag nanorods with lengths varying from 100 – 1000 nm.[29] In the highest aspect ratio nanorod, LSPR spectra demonstrate bands corresponding to the multipolar order of six. Martin and coworkers have used a numerical discretization theory, similar to the discrete dipole approximation, to calculate the LSPR, and resulting electromagnetic fields, for infinite nanowires with arbitrarily-shaped cross sections.[69, 70] Non-equilateral triangle cross sections generate very intense electromagnetic fields at a single vertex with predicted electromagnetic field enhancements of up to $10^3$. Pairs of interacting nanowires demonstrate extreme distance-dependent behavior with predicted electromagnetic field enhancements greater than $10^4$ in the area between the nanowires.[71]

Fundamental studies of electromagnetic coupling between nanoparticles are driven by the fact that the design of plasmonic nanodevices relies heavily on the nature of the electromagnetic interactions between nanoparticles in the devices. These interactions can be evaluated by measuring the LSPR wavelength because an explicit LSPR peak shift occurs as the electromagnetic coupling changes, i.e., when nanoparticles come closer together or further apart. Deki and coworkers explored the effects of electromagnetic coupling on the LSPR by varying the distribution of embedded Ag nanoparticles in a nylon film.[72] Upon heating the nylon film, the interparticle distance increased, while nanoparticle size/shape as well as the dielectric function of the nylon remained unchanged. The result was an explicit shift in the LSPR to higher energies. This shift to higher energies was generally understood to be a result of the effective decrease in nanoparticle aspect ratio as the coupling character decreased. However, Chumanov and coworkers measured a shift to higher energies when a polymer film moved embedded Ag nanoparticles within one diameter of one another. The authors attribute the narrow, blue-shifted extinction spectrum to a new optical phenomenon wherein the LSPRs of the nanoparticles couple coherently. Schatz and coworkers observed a similar trend in LSPR shift when monitoring the optical properties of two-dimensional arrays of EBL-fabricated nanoparticles.[34] Upon increasing the lattice spacing, while holding all other parameters constant, the LSPR shifted to lower energies. Extensive theoretical work demonstrated that the contribution of radiative dipole effects could cause shifts to lower or higher energies depending on the nanoparticle size and regime of lattice spacing. Aussenegg and coworkers utilize NSOM and a simple arrangement of Au EBL-fabricated nanospheres and nanorods to demonstrate the nature of electromagnetic coupling.[27] The samples included an isolated row of nanospheres, an isolated row of nanorods, and a row of nanosphere/nanorod pairs; the nanosphere and nanorods were designed to have vastly different LSPR behavior. When the resonance condition of the nanospheres is satisfied, the optical micrograph reveals that the electromagnetic fields generated from the nanosphere excite the nearby nanorod. The efficiency of the electromagnetic coupling is highly dependent on the relative positions of the nanosphere and nanorod. Similar circumstances were simulated by Martin and coworkers using a numerical discretization method.[71] In this case, the electromagnetic coupling character, and resulting fields, of infinitely long interacting nanowires were modeled as a function of interparticle distance. The results demonstrated an optimal distance between nanowires with a given cross section to maximize electromagnetic coupling.

Clearly, many research groups are working to understand and tune the plasmonic properties of noble metal nanostructures. The relationship between the LSPR and parameters such as nanoparticle size or refractive index of the local environment is relatively straightforward. On the other hand, parameters such as nanoparticle shape or lattice spacing present more complex behavior. The theoretical work done with both electrostatic and electrodynamic models has greatly increased the general understanding of the tunable LSPR phenomenon.

### 1.3.3. Implications for Related Phenomena

One of the major implications of the tunable LSPR in noble metal nanostructures lies in the effects of the generated electromagnetic fields. The serendipitous, and previously misunderstood, enhanced Raman spectrum of pyridine on a roughened Ag electrode[73] represents the first example in the literature of a large class of techniques now known as surface-enhanced spectroscopies. Surface-enhanced Raman scattering (SERS) was first recognized by Jeanmaire and Van Duyne in 1977[74] and named in 1979.[75] In the years since this seminal 1977 manuscript, surface-enhanced analogs of hyper-Raman scattering, infrared absorption, fluorescence, second harmonic generation, and many other photon driven processes have been examined. Only the recent advances in each of these areas will be detailed below.

Of the many surface-enhanced spectroscopies, SERS is by far the most widely used, especially by the surface science, electrochemistry, and catalysis communities. SERS occurs when a Raman-active molecule is placed onto or near a noble metal nanostructured surface where the LSPR has been excited. The intense electromagnetic fields generate an increased induced dipole in the molecule, resulting in Raman scattering that is enhanced by up to fourteen orders of magnitude when compared to normal Raman scattering (NRS). A small portion of the enhancement factor (EF) is attributed to a change in the electronic band structure of the molecule upon adsorption to the metal surface. There have been many years of contentious discussion in the SERS community on the topic of the responsible enhancement mechanism(s).[76-79] While the scientific community generally agrees that both electromagnetic and chemical enhancement mechanisms contribute to the measured enhancement factors, the magnitude of contribution is still under dispute. Recent work by El-Sayed and coworkers demonstrates that Raman excitation far from the LSPR of Au nanorods yields SERS EFs for pyridine and 2-aminothiophenol of $10^4 - 10^5$.[80] El-Sayed and coworkers argue that the chemical mechanism is completely responsible for these large EFs. However, recent work by Van Duyne and coworkers demonstrates a novel technique to optimize enhancement factors by systematically varying the LSPR of NSL substrates.[63] This work concludes that SERS EFs as large as $10^8$ can be almost completely attributed to the electromagnetic enhancement mechanism.

The understanding of SERS enhancement mechanisms has been further complicated in recent years by the discovery of the single molecule SERS phenomenon, yielding enhancement factors as high as $10^{14}$.[81,82] Many research groups have observed the temporal fluctuations in SER spectra that are characteristic of the single molecule phenomenon when very low analyte concentrations are used.[83-86] Current hypotheses suggest that a junction between two noble metal nanostructures acts as an optical trap, confining single molecules for seconds or more within very large electromagnetic fields.[87] Single molecule SERS will have extensive applications once the enhancement mechanism and the required nanostructure geometry is determined.

Hyper-Raman scattering (HRS) is a non-linear process whereby scattered photons are Raman-shifted relative to the second harmonic of the excitation frequency. Because HRS is a three-photon process, the selection rules are different than those for NRS. In fact, the NR and HR spectra of centrosymmetric molecules should not have any common bands. However, the SER and surface-enhanced hyper-Raman (SEHR) spectra of centrosymmetric molecules often look very similar, causing speculation about whether the SEHRS effect was occurring at all. Van Duyne and coworkers demonstrated, using SERS, SEHRS, and *ab initio* calculations, that the comparable bands are due to similar vibrational modes but different symmetries.[88] In fact, SEHRS follows the three photon selection rules as expected. The SEHRS technique is often paired with SERS to measure complementary information on surface-bound molecules. Li and coworkers performed spectroelectrochemical studies of pyridine, pyrazine, and phenazine, in each case verifying that the expected selection rules are obeyed.[89, 90] Further, the spectra of certain intermediate species, such as protonated semiquinone radicals, could only be measured by SEHRS.

Surface-enhanced infrared absorption (SEIRA) has also been employed to garner vibrational information from surface-bound molecules. Wandlowski and coworkers used electrochemical SEIRA to study the specific molecular orientation of 4,4'-bipyridine, a bifunctional ligand often participating in electron transfer events in macromolecules.[91] Three distinct orientations of 4,4'-bipyridine were found under varied temperature and potential conditions. As in SERS, significant effort is devoted to optimizing the nanostructured substrate in order to achieve the largest SEIRA signals possible. Van Duyne and coworkers compared the relative EFs for para-nitrobenzoic acid adsorbed onto standard Ag island films and Ag NSL-fabricated substrates.[58] Though the LSPR spectra of NSL-fabricated nanoparticles are significantly narrower than those of Ag island films, the resulting EFs were comparable. Brown and coworkers attempted to increase the SEIRA signal by assembling a capture layer of organic thiols onto Au island film substrates before exposure to the analyte of interest.[92] A thiophenol capture layer yielded improved spectra of trinitrotoluene analogs, with a total EF ~ 20.

Surface-enhanced fluorescence (SEF) is a counter-intuitive concept because the required noble metal substrate naturally quenches the fluorescence of contacted molecules. However, the electromagnetic fields extend beyond the length of most small molecule adsorbates, generating SEF in molecules that are not in direct contact with the metal substrate. Different SEF substrates (glass, Ag island films, and silica-coated Ag island films) were investigated by Tarcha and coworkers.[93] Upon adsorbing a perylene dye onto each substrate, the authors found that the silica-coated Ag island film was significantly more effective than the other two substrates, EF ~ 11. The distance-dependence of SEF was probed by Lakowicz and coworkers using Ag island films with alternating layers of biotinylated bovine serum albumin and avidin to create spacer layers of varied lengths.[94] After assembling the desired number of layers, the surface was exposed to a fluorophore-labeled, biotinylated oligonucleotide. The fluorescence EF varied between 2 and 12, depending on the spacer layer thickness. Strekal and coworkers employed the SEF technique to study the intermolecular interactions of doxorubicin, an anti-tumor agent.[95] Correlation of SEF spectra with surface-enhanced resonance Raman scattering spectra demonstrated extensive intermolecular hydrogen bonding. SEF enhancement factors as high as 50 have been measured by Aroca and coworkers when studying aggregation schemes of perylene tetracarboxylic derivatives in Langmuir-Blodgett films.[96, 97] Based on the SEF spectra, the authors determined that the perylene molecules are organized with parallel and overlapping aromatic rings and that the chromophore is oriented perpendicular to the Ag island film substrate. Recently, Bawendi and coworkers demonstrated five-fold enhancement of excitonic fluorescence upon spin-coating semiconductor quantum dots onto a roughened Au surface.[98] Single quantum dots studies also showed decreased blinking of the fluorescence signal and shorter exciton lifetimes.

Second harmonic generation (SHG) is the process by which photons of double the incident frequency are generated after interacting with a surface. Surface-enhanced second harmonic generation (SESHG) is used because it is very sensitive to changes in the electronic structure of an interface. SESHG experiments have been performed on both smooth and roughened noble metal surfaces. Van Duyne and coworkers found that the EFs measured from nanostructured biperiodic gratings were $10^4$ times larger than those measured from smooth Ag films.[99] Smolyaninov and coworkers further investigated the properties of smooth films and metal-coated gratings by measuring NSOM images from these surfaces during SESHG experiments.[100] They concluded that the most intense SHG results from crystalline defects in both cases. While the features necessary for optimized enhancement are clearly not yet well-understood, Brolo and coworkers have used SESHG, in conjunction with SERS, to examine the orientation of L-cysteine on a Ag electrode.[101] If the orientation of L-cysteine can be controlled, it may find application as a template for the patterning of proteins on surfaces. The SESHG study reveals a change in orientation of the L-cysteine at the surface potential of –650 mV. Correlated SERS data revealed that a protonated amino group points toward the Ag electrode at potentials more positive than –650 mV while a carboxylate group points toward the Ag electrode at potentials more negative than –650 mV.

## 1.4. Applications of Tunable Optical Properties

In closing this review of recent work in the area of tunable optical properties, it is fitting to discuss the practical uses of noble metal nanoparticles. The most obvious application of high aspect ratio nanoparticles lies in the area of dichroic filters. Foss and coworkers distributed both spherical and cylindrical Au nanoparticles into oriented polyethylene films, generating stripes of nanoparticles.[13, 17] The dichroic behavior is apparent upon varying the nanoparticle assembly orientation relative to a polarized excitation source. The LSPR spectra differ drastically in extinction maximum and width when exciting parallel and perpendicular to the long axis of the oriented nanoparticles. In a similar experiment, Dirix and coworkers show color images of the oriented nanoparticle films in each polarization matching condition.[18] When the polarized excitation is parallel to the oriented nanoparticle stripes, the film appears blue in color; accordingly, polarized excitation perpendicular to the oriented nanoparticle stripes yields a red film.

Just as nanoparticle assemblies can act to block light of particular wavelengths, they can also be applied to guide light along a prescribed path. Nanoparticle waveguides hold enormous potential in the general field of optical communication. Bozhevolnyi and coworkers used EBL to design intricate substrates consisting of networks of Au cylinders.[35] The ability to guide light of a particular wavelength was defined by the intentional absence of Au cylinders in a specific geometry. NSOM images showed strong reflection of excited surface plasmon polaritons in the areas of high-density cylinders and efficient waveguiding along the line defects created by the intentional cavities. One-dimensional chains of EBL-fabricated Au nanoparticles were used by Aussenegg and coworkers to investigate the waveguiding process.[102] Upon excitation with light polarized parallel to the nanoparticle chain, NSOM images reveal periodic bright spots, indicating that light is propagating from one nanoparticle to the next. Theoretical simulations suggest that these bright spots lie between nanoparticles.

Atwater and coworkers excite the end of an EBL-fabricated array of Ag nanorods with an NSOM tip and then probe energy transport by monitoring the fluorescence of nanospheres placed in close proximity to the waveguide structure.[33] The resultant fluorescence images demonstrate that noble metal nanoparticle waveguides have energy attenuation lengths of several hundred nanometers. While guiding light along a predetermined path will be necessary in future devices, it will also be essential to confine certain wavelengths of light in a particular location. Toward this idea, Girard and coworkers have simulated the optical analog of the quantum corral.[103] Based on their calculations, nanoparticles, spaced regularly around the perimeter of a circle, should sustain particular wavelengths of light within that circle. Upon calculating the probability of finding a photon at each location within the circle, the authors found that interference patterns form in the evanescent fields. Understanding these patterns may have significant impact on the interpretation of NSOM images.

Noble metal nanoparticles also hold the potential to act as optical data storage elements. Aussenegg and coworkers suggest mixing various sizes, shapes, and arrangements of nanoparticles to create dense data storage. With single nanoparticles or small nanoparticle assemblies, the width of the LSPR spectra will be significantly less than 100 nm. When working only in the visible region of the spectrum, it would be possible to distinguish at least five separate resonances, yielding at least five times current data storage density. In a proof-of-concept experiment, Aussenegg and coworkers created three different classes Ag nanoparticle assemblies using EBL, each having a different LSPR. The different nanoparticle assemblies, or bits, are easily distinguished as optical images are created of scattered light. Dickson and coworkers took advantage of the photoactivated emission of Ag nanoparticles on $Ag_2O$ films to demonstrate high-density optical data storage.[104] By "writing" with blue light and "reading" with red light, data were readily stored for many hours.

While the application of noble metal nanostructures as optical data storage elements is still under development, companies have already formed to exploit noble metal nanostructures as biological labels. Nanoscale cylinders with stripes of different metals have been used as immunoassay tags; readout mechanisms have been based on both the extinction[15] and reflectivity.[105] In another example, Siiman and coworkers have applied thin coatings of Ag or Au to latex microspheres that then act as flow cytometry labels.[40] The noble metal island films make it easy to identify different subpopulations of white blood cells without significantly changing the bead's flow characteristics. Mirkin and coworkers have taken advantage of the SERS effect by attaching Raman-active labels to Au nanoparticles that are labeled with single stranded DNA.[106] When the labeled strand of DNA is exposed to a complimentary single strand of DNA immobilized on a glass substrate, the Au nanoparticle is bound to the surface. A flatbed scanner can be used to identify DNA matches, and SERS readout from the Raman-active label distinguishes the DNA sequence. In a proof-of-concept experiment, Mirkin and coworkers were able to simultaneously screen for DNA sequences relevant to hepatitis A, hepatitis B, HIV, Ebola, smallpox, and anthrax without any false negatives or false positives.

Noble metal nanoparticles can go beyond acting as labels; recent advances show that changes in nanoparticle optical properties can act as the signal transduction mechanism in chemosensing and biosensing events. Van Duyne and coworkers have exploited the extreme LSPR sensitivity of NSL-fabricated nanoparticles to changes in local refractive index in order to sense small molecules, amino acids, proteins, and antibodies. The LSPR shifts systematically to lower energies as the local dielectric constant increases;

accordingly, chemisorption of alkanethiols of increasing length cause a systematic red-shift in the LSPR (3 nm/methylene unit).[43, 46] Electrostatic adsorption of polylysine to a carboxylic acid-terminated alkanethiol monolayer on a NSL-fabricated substrate produces further red-shift of the LSPR.[43]   As a consequence of its enormous affinity constant, $K_a \sim 10^{11}$ M$^{-1}$, binding of the protein streptavidin to nanoparticle-bound biotin can be detected down to low picomolar concentrations through the LSPR shift signal transduction mechanism.[44]   Detection of anti-biotin was also possible; the limit of detection in this case, as governed by the affinity constant of $K_a \sim 10^8$ M$^{-1}$, was less than 700 picomolar.[61]   These nanoparticle sensors take advantage of tunable noble metal nanostructure optical properties in order to sensitively and selectively detect trace levels of the target molecule. These experiments are further detailed in Section 3.1.

## 1.5. Goals and Organization

The remainder of this chapter is organized as follows.  Section two describes recent results using the three major fabrication methods to produce nanostructures with tunable optical properties: chemical reduction, electron beam lithography, and nanosphere lithography.  Section three details recent achievements in the areas of nanoparticle array sensing and single nanoparticle sensing using the LSPR as the signal transduction mechanism.  The use of the tunable LSPR to optimize the surface-enhanced Raman spectroscopy phenomenon is also described in Section three.  Finally, Section four suggests future directions and applications of tunable nanoparticle optics.

## 2. TUNABLE LOCALIZED SURFACE PLASMON RESONANCE

### 2.1. Introduction to Colloidal Nanoparticles

Metal nanoparticles have been used for hundreds of years to produce color in stained glass, but artists had little control over the resulting tints.  Though scientists understood that different metal dispersions gave diverse characteristic hues, it wasn't until the advent of nanoscale analysis that they began to understand how material properties such as metal, size, shape, and local dielectric affect the apparent color of a metal suspension. Previous research results have demonstrated that the localized surface plasmon resonance (LSPR) of surface-confined noble metal nanoparticles can be tuned throughout the visible region of the spectrum and that the extinction coefficient is very large.[54] These characteristics indicate that noble metal nanoparticles may find more general application as coloring agents – these intense, tunable nanoparticles could even replace molecular organic dyes.  The use of organic dyes in food, drugs, and cosmetics is widespread because the coloring agents give products an attractive, uniform appearance.  There is strong evidence, however, that some of these dyes pose serious health hazards ranging from skin irritations to cancer.[107-111]  Noble metal nanoparticles present a safe, non-carcinogenic dye replacement system that is completely tunable across the color spectrum.  While all previous work in the area of LSPR tunability in the Van Duyne group has focused on the nanosphere lithography (NSL) technique, the goal of creating organic dye replacements has inspired LSPR studies of chemically-synthesized metal nanoparticles. One of the major limitations of the NSL nanofabrication technique is that, while millions of nanoparticles are made simultaneously, it is not a technique with high enough throughput to satisfy the requirement of large-scale manufacturing. In section 2.2, details of the chemical fabrication methods used to create nanoparticles of varied shape, size, and material as colloidal suspensions are given. Section 2.3 discusses the structural and optical characterization of these materials.

## 2.2. Colloidal Nanoparticle Experimental Section

### 2.2.1. Fabrication of Surfactant-Modified Silver Nanoparticles

The chemical fabrication scheme for Ag nanotriangles was adapted from work done by Jin and coworkers.[7] The following chemicals were used: silver nitrate, sodium citrate dihydrate, sodium borohydride, and bis(p-sulfonatophenyl) phenylphosphine dihydrate dipotassium salt (BSPP). Before beginning nanotriangle synthesis, all glassware was treated with aqua regia (3:1 $HCl:HNO_3$), rinsed thoroughly with ultrapure water (18.2 $M\Omega$ cm$^{-1}$), and dried under reduced pressure at 80 °C. The following solutions were each made in 1 mL of ultrapure water: 52.8 mM sodium borohydride (2.0 mg), 30.3 mM sodium citrate dihydrate (8.9 mg), 2.2 mM BSPP (1.1 mg), and 8.8 mM silver nitrate (1.5 mg). The solutions were added in the order listed above to 96 mL of cold ultrapure water while stirring. Immediately after adding silver nitrate, the solution was irradiated with a conventional 13-watt fluorescent light while stirring was continued. The color of the solution changed with varied time exposure to the fluorescent light source.

The fabrication scheme for Ag nanorods was adapted from work done by Yin and coworkers.[112] Silver nitrate (99.998%), platinum chloride (99.99+%), anhydrous ethylene glycol (99.5+%), and poly(vinylpyrrolidone) (MW ~ 40,000) were used. Before beginning nanorod synthesis, a 3-neck roundbottom flask was treated with aqua regia (3:1 $HCl:HNO_3$), rinsed thoroughly with ultrapure water (18.2 $M\Omega$ cm$^{-1}$), and dried under reduced pressure at 80 °C. The roundbottom flask was placed in an oil bath heated to 160 °C, and then 5 mL of anhydrous ethylene glycol was placed in the flask. One mg of platinum chloride was suspended in 1 mL of ultrapure water by sonication, and then 20 μL of the suspension was added to the heated ethylene glycol, while stirring, to form 5 nm Pt nanoparticle catalysts. After 4 minutes, 118 mM silver nitrate (50 mg in 2.5 mL ethylene glycol) and 0.125-0.5 mM poly(vinylpyrrolidone) (25-100 mg in 5 mL ethylene glycol) were added dropwise, simultaneously over 5 minutes. The solution was allowed to heat and stir for 1 hour before being stored in brown glass bottle.

### 2.2.2. Fabrication of Core-Shell Nanoparticles

The fabrication scheme for core-shell nanoparticles was adapted from work done by Lee and Meisel[3] and Ung et al.[113] Silver nitrate (99.998%), hydrogen tetrachloro-aurate(III) (99.999%), sodium citrate dihydrate (99+%), (3-aminopropyl)trimethoxysilane (97%), and sodium trisilicate solution (27%) were used. Absolute ethanol was used for quenching the $SiO_2$ shell growth. In all cases, the metallic core of the core-shell nanoparticle was synthesized before growing the dielectric shell.

The Ag nanotriangle synthesis outlined in section 2.2.1 was one of the two ways that the metallic Ag cores were fabricated. The second Ag core fabrication method began with the treatment of all glassware with aqua regia (3:1 $HCl:HNO_3$) followed by a thorough rinse with ultrapure water (18.2 $M\Omega$ cm$^{-1}$). Ninety mg of silver nitrate was dissolved in 0.5 L of ultrapure water and brought to a boil while stirring. Once boiling, 34 mM sodium citrate (100 mg in 10 mL) was added, and the solution was allowed to continue boiling and stirring for 30 minutes. This synthesis is known to yield polydisperse Ag colloids (diameter = 25-65 nm). The solution was stored in a brown glass bottle to avoid photodegredation.

Synthesis of the metallic Au cores began with the treatment of all glassware with aqua regia (3:1 $HCl:HNO_3$) followed by a thorough rinse with ultrapure water (18.2 M$\Omega$ cm$^{-1}$) from a Millipore academic system. A 1 mM aqueous solution of hydrogen tetrachloroaurate(III) (0.2 g in 0.5 L water) was placed in a roundbottom flask equipped with a reflux condenser. After the solution was brought to a boil, 39 mM sodium citrate dihydrate (1.15 g in 100 mL water) was added through the top of the condenser and the solution was allowed to boil for 10 minutes.

The colloid solutions described above (Ag nanotriangles, citrate-reduced Ag, and citrate-reduced Au) were dialyzed against pH=5 ultrapure water for 48 hours before further modification. Based on the known size and density of the metallic cores, 1 mM aqueous (3-aminopropyl)trimethoxysilane (390 µL) was added to 10 mL of the dialyzed colloid solution. After 15 minutes, the pH of the colloid/(3-aminopropyl)trimethoxy-silane solution was raised to 11.5 by adding aqueous potassium hydroxide. Finally, the silica shell began to form after 27% sodium silicate (2.1 µL) was added to the basic solution. Quenching of the shell growth was accomplished by adding 40 mL of absolute ethanol to the nanoparticle solution.

### 2.2.3. Transmission Electron Microscopy Characterization

In all cases, electron microscopy samples were prepared by placing a drop of the desired colloidal nanoparticle sample onto a 100 mesh copper grid and allowing the solvent to evaporate. A 200 kV accelerating voltage on a tungsten filament served as the electron source in all cases. Both images (up to 500,000× magnification) and diffraction data were collected on photographic film.

## 2.3. Structural and Optical Properties of Colloidal Nanoparticles

The structural characteristics of the nanotriangle solution were monitored by performing TEM of aliquots removed after the reaction had proceeded for varied time periods. Initially, the solution is yellow in color; TEM images show that the yellow solution consists of inhomogeneous Ag nanoparticles including 15% trigonal prisms (perpendicular bisector = 70 ± 40 nm) and 85% polygon platelet nanoparticles (diameter = 40 ± 15 nm). After approximately 2 days, the yellow solution takes on a green tint, eventually appearing bright green after 4 days. The bright green solution is made up of 50% trigonal prisms (perpendicular bisectors = 120 ± 62 nm) and 50% polygon platelet nanoparticles (diameter = 35 ± 13 nm). If the green solution is continually exposed to the fluorescent light source for more than one month, the nanoparticle solution turns dark blue in color. This dark blue solution is made up of 25% trigonal prisms with rounded tips (perpendicular bisector = 50 ± 12 nm) and 75% polygon platelet nanoparticles (diameter = 60 ± 12 nm). If further fluorescent light exposure is allowed, the dark blue solution transitions through a violet color to a gray solution. There are no trigonal prism nanoparticles in the violet solution; in fact, the nanoparticles are largely oblong in nature (major axis = 63 ± 18 nm), appearing as though multiple small nanoparticles have fused. TEM images demonstrate that the gray solution is made up exclusively of spheroidal nanoparticles (diameter = 100 ± 25 nm).

As the initial polygon platelet nanoparticles are converted into trigonal prisms and finally into large spheroidal nanoparticles, the LSPR shifts due to the changing size and shape of the nanoparticles. This characteristic, along with the fact that these nanoparticles can be made in high yield, suggests that these Ag nanoparticles are candidates to replace molecular organic dyes. A series of studies were conducted altering the concentrations of silver nitrate and surfactant present in solution, as well as changing the temperature and light exposure during the course of the reaction, to determine how to best control the chemical synthesis of Ag nanoparticles. This nanoparticle synthesis is very sensitive to small procedural changes, and it was often impossible to repeat previously attained results. Figure 3 shows the appearance of a range of nanoparticle solutions and the correlated TEM images. Current and future work on this nanoparticle dye research requires an increase in nanoparticle homogeneity at each stage of the synthesis as well as coating of the nanoparticles with an inert substance so that the nanoparticles are stable to varied environmental parameters.

In hopes of achieving further LSPR tunability, perhaps into the infrared region of the spectrum, a surfactant-mediated Ag nanorod synthesis was also pursued. Variation of the overall poly(vinylpyrrolidone) concentration from 0.05 – 0.2 mM yielded significant differences in the resultant nanoparticle shapes. Representative TEM images of Ag nanorods are shown in Figure 5A and B. The initial protocol, using 0.2 mM surfactant, yielded 5% nanorods with an average length of 58 ± 24 nm. When the surfactant concentration was halved, to 0.1 mM, the nanorod yield increased to 14%, and the average length was 117 ± 39 nm. When the surfactant concentration was halved again, to 0.05 mM, the nanorod yield increased further to 30%. In this case, the average nanorod length was 273 ± 203 nm. While decreased surfactant concentration results in higher aspect ratio Ag nanorods, the size distribution grows as well. In all cases, the nanorod solutions were gray-brown in appearance; measured extinction spectra were broad and featureless. This nanorod preparation needs to be further refined before the nanoparticles can be considered for use as coloring or contrast agents.

In addition to being able to create nanoparticles in a vast range of sizes, shapes, and materials to produce an almost continuous spectrum of colors, it is also possible to encapsulate the nanoparticles to prevent aggregation and to protect them from undesired interactions during use. This work adapts a procedure established by Ung and coworkers[113] for creating $Ag@SiO_2$ and $Au@SiO_2$ core-shell nanoparticles. In this process, a silane coupling agent (3-aminopropyltrimethoxysilane) is used to make the metal core receptive towards silica monomers or oligomers. In the preliminary core-shell work presented here, attempts were made to form silica shells on citrate-reduced Au colloids, citrate-reduced Ag colloids, and the Ag nanotriangle colloids. TEM investigation demonstrated that the silica shells do not localize around the citrate-reduced Au colloid core. Instead, web-like networks of silica form, and the Au colloids are segregated from this network. Silica shells were successfully grown on both citrate-reduced Ag colloids and Ag nanotriangle colloids (Figures 5C and D). The silica shells formed on the citrate-reduced Ag colloids were approximately 15 nm thick and largely homogeneous in appearance. There was more variance in the silica shells grown on the

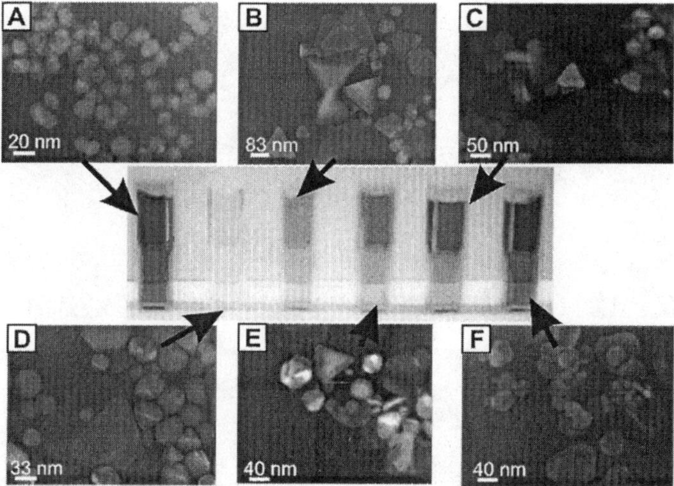

**Figure 3.** See Fig. 3.3 in the color insert at the end of this volume. Solutions of nanoparticle dyes with corresponding transmission electron micrographs. (A) The red solution is made up of homogeneous Au nanospheres with D=13 nm. (B) The yellow solution is made up of inhomogeneous Ag nanoparticles: 15% trigonal prisms and 85% polygon platelets. (C) The green solution is made up of 50% trigonal prisms and 50% polygon platelets. (D) The light blue solution is made up of Ag nanoparticles including 30% trigonal prisms and 70% polygon platelets. (E) The dark blue solution is made up of 25% trigonal prisms with rounded tips and 75% polygon platelets. (F) The purple solution is made up of inhomogeneous oblong Ag nanoparticles.

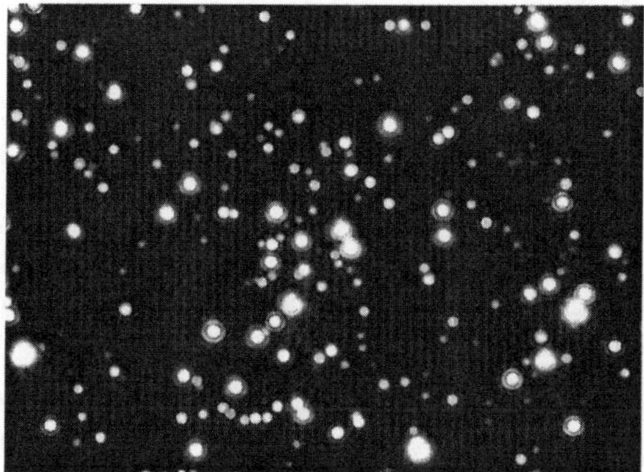

**Figure 4.** See Fig. 3.4 in the color insert at the end of this volume. A dark-field optical image of a field of Ag nanoparticles. The field of view is approximately 130 μm x 170 μm. The nanoparticles were fabricated by citrate reduction of silver ions in aqueous solution and drop-coated onto a glass coverslip.

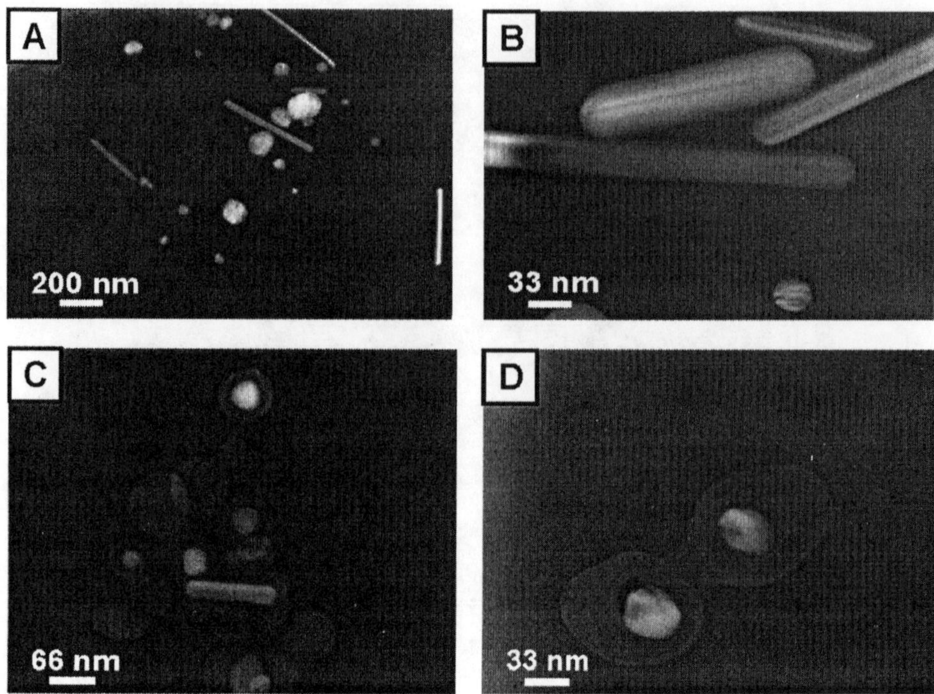

**Figure 5.** Ag nanorod and Ag@SiO$_2$ core-shell nanoparticle transmission micrographs. (A, B) TEM images of Ag nanorods synthesized with 0.05 mM poly(vinylpyrrolidone) surfactant. (C, D) TEM images of citrate-reduced Ag@SiO$_2$ core-shell nanoparticles. The SiO$_2$ shells are approximately 15 nm thick.

Ag nanotriangle colloids; the average shell thickness was 25 ± 7 nm. In order to assess the effect of the silica shell on colloid stability, a simple aggregation study was executed. Bare citrate-reduced Ag colloids aggregate quickly when an aqueous solution of strong base is added dropwise. The aggregation is apparent because the color of the solution darkens and, eventually, the large nanoparticle aggregates crash out of the colloidal solution. When the same aggregation experiment is performed on a solution of Ag@SiO$_2$ core-shell nanoparticles, the colloidal nanoparticles do not respond. The spectral characteristics of the metal nanoparticles are retained when encapsulated. Surprisingly, no significant LSPR shift is apparent upon adding the silica shells to the metal cores. While further refinement of this core-shell technique is needed, it is clear that these nanoparticles hold great promise as stable, inert dyes for introduction into a variety of systems, including food, drugs, and cosmetics.

## 2.4. Study of Electromagnetic Coupling using Electron Beam Lithography Substrates

In order to satisfy the technological demand for ever-decreasing device feature sizes and ever-increasing device performance, it is imperative to explore size-dependent chemical and physical properties. The optical,[114, 115] magnetic,[116] catalytic,[117] thermodynamic,[118] electrical transport,[119] and electrochemical[120] properties of nanoscale materials differ significantly from the properties found in their bulk, molecular, or atomic counterparts. Studies of nanoscale noble metal materials are especially important because these materials have potential as optical filters,[18] plasmonic waveguides,[121] bio/chemosensors,[46, 122, 123] and substrates for surface-enhanced spectroscopies.[124, 125] Due to interest in nanoparticle-based devices, many research groups are currently fabricating and studying nanoparticle arrays. Chumanov et al. have studied nanoparticle coupling effects by dispersing nanoparticles onto polymer-coated quartz substrates.[126] The varying lattice spacing between the randomly dispersed nanoparticles prohibits a thorough understanding of the relation between nanoparticle arrangement and the optical properties of the array. The extinction spectra of Ag and Au nanowire gratings of various dimensions and square arrays of Au cylinders have been studied by Aussenegg et. al.[28, 127] Atwater et. al. have studied one-dimensional arrays of Au spheroidal nanoparticles for application as optical waveguides.[128, 129] Other research groups have focused on the optical properties of two-dimensional arrays in order to utilize these nanoparticle assemblies as surface-enhanced spectroscopy substrates.[130-132]

The work presented herein utilizes two-dimensional noble metal nanoparticle arrays fabricated using electron beam lithography (EBL) to systematically probe the effect of electromagnetic coupling on the localized surface plasmon resonance (LSPR). The LSPR is a collective oscillation of the nanoparticle conduction electrons.[79] This oscillation can be localized on a single nanoparticle, or it may involve many coupled nanoparticles. The electron oscillation frequency is highly dependent on the effective size and shape of the noble metal nanostructure, and it also depends on nanoparticle arrangement and separation. The LSPR for noble metal nanoparticles in the 20 to a few hundred nanometer size regime occurs in the visible and IR regions of the spectrum and can be measured by UV-visible-IR extinction spectroscopy.[49]

Nanoscale devices are likely to require assemblies of nanoparticles for functionalities that include sensing and waveguide applications. The design of plasmonic nanodevices relies heavily on the nature of the electromagnetic interactions between nanoparticles in the devices. These interactions can be evaluated by measuring the LSPR wavelength because an explicit LSPR peak shift occurs as the electromagnetic coupling changes, i.e., when nanoparticles come closer together or further apart. Some aspects of these electromagnetic interactions are well understood. For example, the LSPR wavelength of a pair of nanoparticles exhibits a red shift as the nanoparticles approach and the polarization of the incident light is parallel to the interparticle axis, while blue-shifting occurs when the polarization is perpendicular to the axis.[57] This can easily be explained in terms of static dipolar coupling between the nanoparticles. Parallel polarization leads to collinear induced dipoles where fields constructively add to the applied field, thus achieving resonance at lower frequencies. Alternately, perpendicular polarization leads to destructive interaction with the applied field, and therefore, higher frequency resonances. The consequences of static dipolar interactions for planar arrays is known from the work by Murray and Bodoff where red shifts were observed for coupled vibrational dipoles in planar arrays with in-plane polarization.[133] These interactions have also been considered in three dimensions; Lazarides and Schatz demonstrated that red-shifts occur for three dimensional cubic arrays of 13 nm gold nanoparticles with separations of 5-15 nm (separations small enough that electrostatic interactions should dominate).[134] However, for other nanoparticle configurations, and for nanoparticle separations comparable with a wavelength of light, electromagnetic coupling between nanoparticles can lead to more complex behavior. Computational modeling is employed to explore this behavior in a companion theoretical study.[135] This complimentary work illustrates the complex behavior of a planar array of spherical nanoparticles as a function of lattice spacing. With in-plane polarization, a decrease in lattice spacing yields a LSPR blue shift for large separations (>100 nm), but a LSPR red shift for smaller lattice spacing. Whereas red shifts are the expected result for static dipolar interactions (which have a $1/d^3$ dependence on nanoparticle separation), the blue shifts demonstrate the importance of radiative dipolar coupling (which has a $1/d$ dependence on nanoparticle separation) and retardation (which multiplies the dipole field by $e^{ikd}$) for large interparticle separations.

This work provides experimental evidence for this long-range coupling mechanism using a variety of two-dimensional nanoparticle arrays fabricated with EBL. EBL readily lends itself to precision control of nanoparticle features and arbitrary array geometry,[136] which is ideal for studying electromagnetic coupling. Because the nanoparticles produced by EBL for this work are cylinders or trigonal prisms, the work done in the companion study[135] has been extended to allow for computational modeling of dipolar coupling of spheroidally-shaped nanoparticles. This work presents the most complete, systematic study of interparticle coupling by pairing experimental measurements with coupled dipole simulations.

## 2.5. Experimental Methods

### 2.5.1. Sample Fabrication

The samples were prepared by EBL on soda glass substrates. First, the clean glass substrate was spin coated with a 70 nm thin film of an electron sensitive resist, ZEP 520 diluted 1:2 in methoxybenzene. Before exposing the pattern, the resist film was coated with a 10 nm thin film of Au to make the surface conductive. During patterning, the electron beam passes straight through this thin metal coating and exposes the resist as desired. The resolution of the EBL system used is approximately 20 nm employing an accelerating voltage of 50 kV. After exposure, the Au film was removed by etching in an aqueous solution of 4 g KI and 1 g $I_2$ in 150 mL of deionized water.[137] The patterns were developed in xylene, creating a patterned resist film on top of which metal (Ag or Au) was deposited in a high vacuum thin film vapor deposition system (AVAC HVC 600). The deposited thickness and deposition rate were measured by a quartz crystal microbalance. The deposition rate was maintained at ~1 Å/s in order to create a smooth film. In the last step of the sample preparation, the lift-off step, the resist was dissolved in a strong solvent, which also removes the material deposited on top of the resist. To ensure that the metal film on top of the resist does not have any physical contact with the metal deposited directly on the substrate, samples were prepared with an "undercut" in the resist film. This is accomplished by overdeveloping the resist slightly.

In this work, 60 μm x 60 μm arrays of Ag or Au nanoparticles were prepared with varied shapes and arrangements. Keeping all the other parameters in the array constant, the lattice spacing of the array has been varied from near contact to a few hundred nanometer separation.

## 2.5.2. Optical Characterization of Nanoparticle Arrays

The extinction measurements reported here were recorded with an Ocean Optics spectrometer over the range 350 nm to 850 nm. The spectrometer was fiber optically coupled to an inverted microscope. White light from the microscope lamp was collimated before being passed through the sample. The transmitted light was collected with a 20x (NA = 0.5) objective. The light was then spatially filtered at the image plane on the side port of the microscope before being focused into the 400 μm core diameter fiber coupled to the spectrometer. Because of the spatial filtering, the probed spot size was approximately 50 μm in diameter at the sample. Data processing included boxcar smoothing and identification of the extinction maximum ($\lambda_{max}$) using a derivative routine.

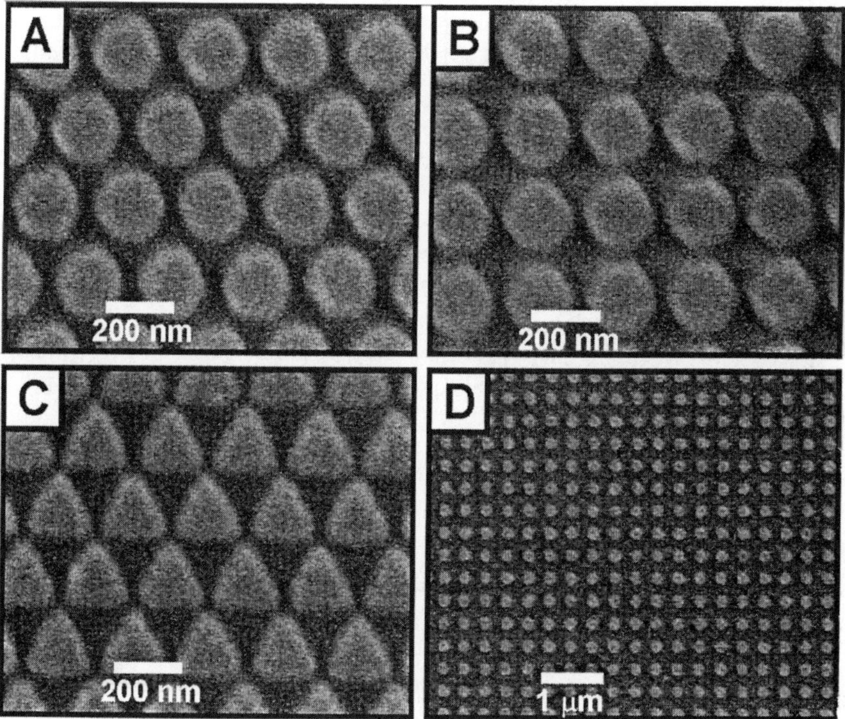



**Figure 6.** Scanning electron micrograph images of electron beam lithography arrays. In all cases, nanoparticles were 40 nm in height. (A) Cylindrical nanoparticles with a diameter of 200 nm in hexagonal arrangement with a lattice spacing of 260 nm. (B) Cylindrical nanoparticles with a diameter of 200 nm in square arrangement with a lattice spacing of 250 nm. (C) Trigonal prismatic nanoparticles with a perpendicular bisector of 173 nm in hexagonal arrangement with a lattice spacing of 230 nm. (D) Cylindrical nanoparticles with a diameter of 200 nm in square arrangement with a lattice spacing of 350 nm. Reproduced with permission from *J. Phys. Chem. B* 2003, ASAP, 04/30/2003. Copyright 2003 Am. Chem. Soc.

### 2.5.3. Structural Characterization of Nanoparticle Arrays

The EBL patterns were characterized using scanning electron microscopy, and the height of the nanoparticles were determined by atomic force microscope (AFM) measurements. In Figure 6, images of different shapes and arrangements are shown.

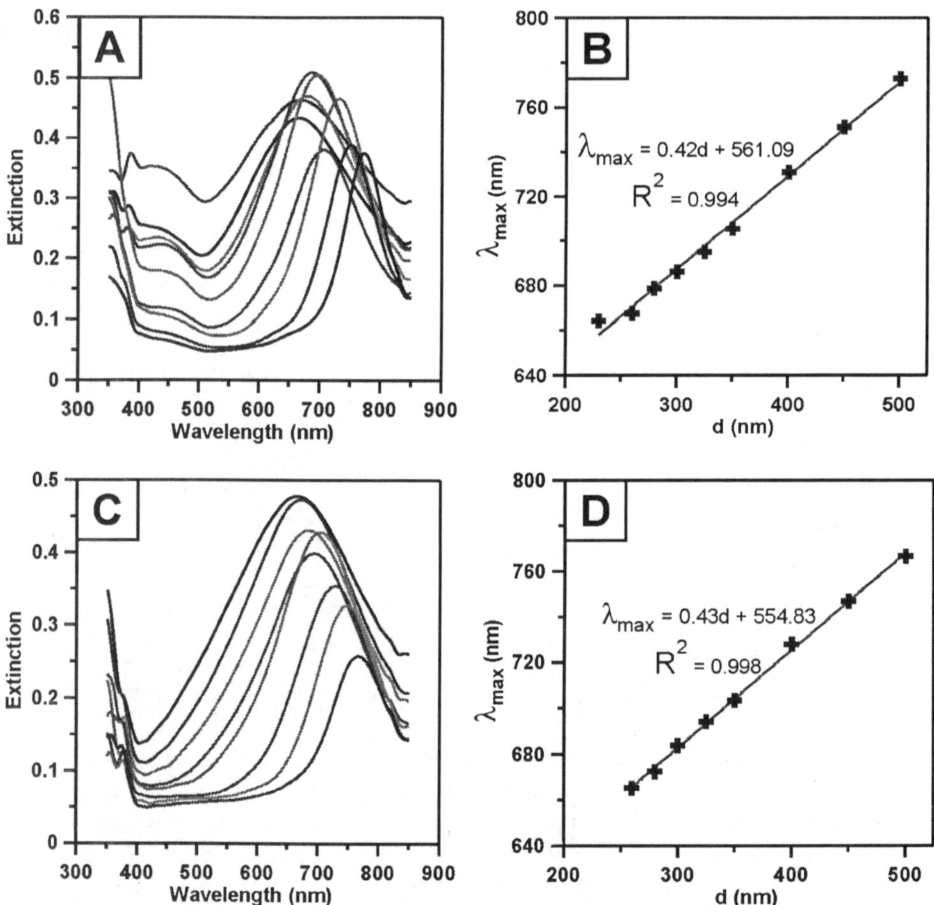

**Figure 7.** Extinction spectra of Au and Ag cylinders in hexagonal arrays. In this case, nanoparticle diameter was 200 nm and the height was 40 nm with varied nanoparticle spacing. (A) Au cylinders, lattice spacing = 230-500 nm; (B) Corresponding plot of lattice spacing (d) versus LSPR $\lambda_{max}$; (C) Ag cylinders, lattice spacing = 260-500 nm; (D) Corresponding plot of lattice spacing (d) versus LSPR $\lambda_{max}$. Reproduced with permission from *J. Phys. Chem. B* 2003, ASAP, 04/30/2003. Copyright 2003 Am. Chem. Soc.

## 2.6. Optical Properties of Electron Beam Lithography-Fabricated Nanoparticle Arrays

Figure 7 shows the extinction spectra for hexagonal arrays of Au and Ag cylinders (diameter (D) = 200 nm, height (h) = 35 nm, and lattice spacing (d) = 230-500 nm). This shows the expected LSPR extinction maximum ($\lambda_{max}$) in the 700-800 nm range; the measured $\lambda_{max}$ blue shifts as the nanoparticle separation decreases for both Au and Ag nanoparticle arrays. The wavelength of the $\lambda_{max}$ shifts 42 and 43 optical nm for a 100 nm change in nanoparticle lattice spacing for the Au and Ag arrays, respectively.

In the case of square arrays of Au and Ag cylinders (D = 200 nm, h = 35 nm, and d = 250-500 nm), the extinction $\lambda_{max}$ shifts 62 and 55 optical nm for a 100 nm change in nanoparticle lattice spacing for the Au and Ag arrays, respectively. Based on the linear regressions, the coupling effects appear to be slightly more significant in the square arrays.

In the case of hexagonal arrays of Au and Ag trigonal prisms (a = 170 nm, h = 35 nm, and d = 280-500 nm) the extinction spectra still show a maximum for wavelengths in the 700-800 nm range, and the $\lambda_{max}$ shifts 36 and 35 optical nm for a 100 nm change in nanoparticle lattice spacing for the Au and Ag arrays, respectively. The $\lambda_{max}$ of the array of trigonal prisms shows a slightly smaller dependence on lattice spacing than the array of cylinders.   Note also that there are well-defined LSPR features near 500 nm for Ag and 570 nm for Au. By analogy with what has previously been observed for trigonal prisms in solution,[7] these features are assigned to in-plane quadrupole resonances. These resonances have a weaker dependence on lattice spacing than the corresponding dipole resonances at 700-800 nm, but blue shifts are also observed as lattice spacing decreases.

This work has demonstrated that the LSPR of Au and Ag nanoparticle arrays exhibit pronounced blue shifts as lattice spacing is decreased.  The blue shifts were found to occur for both hexagonal and square lattices, with slight differences between lattices. These differences correlate with lattice density such that denser lattices give bluer $\lambda_{max}$. In addition, the Au and Ag results are very similar, which demonstrates that these two metals have similar dielectric properties for LSPR in the far-red and near-infrared in spite of the fact that they have significant differences in behavior below 700 nm.

The blue shifts are observed for nanoparticle arrays with relatively large lattice spacings (approximately half the LSPR wavelength) where radiative dipole interactions between the nanoparticles play an important role and retardation effects are large.  For smaller lattice spacings (<100 nm), one expects red shifts (for the case where the electric polarization is in the plane of the array) and indeed such red shifts have been observed for three dimensional aggregates of Au nanoparticles.[138] For the highly oblate nanoparticles considered in this study, the size of the blue shift is very large, corresponding to a 40 nm shift in $\lambda_{max}$ for each 100 nm change in the lattice spacing.

## 2.7. Tunable Localized Surface Plasmon Resonance using Nanosphere Lithography

### 2.7.1. Effect of Nanoparticle Material on the LSPR

Within the last 10 years, there have been examples of NSL structures fabricated from the inorganic materials: Ag, Al, Au, Co, Cr, Cu, Fe, In, KBr, $MgF_2$, NaCl, Ni, Pd, Pt, Rh, $SiO_x$, and Zn. The first nanoparticles fabricated from an organic material, cobalt phthalocyanine (CoPc), were also fabricated using the NSL technique.[77] Experiments are currently in progress to systematically study the nanoparticle optics of the many material systems capable of supporting a LSPR. However, in the following sections attention is focused exclusively on the remarkable nanoparticle optics of silver.

### 2.7.2. Effect of Nanoparticle Size on the Ag LSPR

The relationship between nanoparticle size and the LSPR extinction maximum, $\lambda_{max}$, has been recognized, though not fully understood, for many years. As atomic force microscopy, scanning electron microscopy, and tunneling electron microscopy have

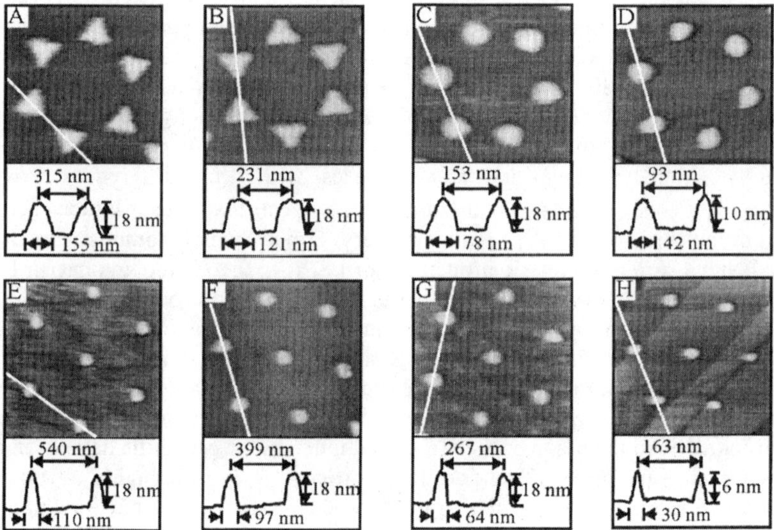


**Figure 8.** AFM images and line scans of representative Ag nanoparticle arrays on mica substrates. The line scan values reported here have not been corrected for tip broadening effects. (A) 870 nm x 870 nm image, D = 542 nm, dm = 18 nm; (B) 610 nm x 610 nm image, D = 401 nm, dm = 18 nm; (C) 420 nm x 420 nm image, D = 264 nm, dm = 18 nm; (D) 260 nm x 260 nm image, D = 165 nm, dm = 14 nm; (E) 1200 nm x 1200 nm image, D = 542 nm, dm = 18 nm; (F) 1000 nm x 1000 nm image, D = 401 nm, dm = 18 nm; (G) 670 nm x 670 nm image, D = 264 nm, dm = 18 nm; (H) 410 nm x 410 nm image, D = 165 nm, dm = 0.5 nm. Reproduced with permission from *J. Phys. Chem. B* 2003, 105, 5599-5611. Copyright 2003 Am. Chem. Soc.

become standard laboratory techniques, nanoparticle dimensions have been determined, and electrodynamic theory is helping to explain the experimental light scattering and absorption trends. NSL is particularly useful in this regard because nanoparticle size is easily varied by changing the nanosphere mask diameter, D, and the deposited mass thickness, $d_m$. The dimensions of each nanoparticle are defined by two parameters: (1) the length of the triangle's perpendicular bisector, "a", and (2) the measured out-of-plane height, "b". As seen earlier, NSL is particularly useful because both large and small size extremes are accessible with the NSL technique (Figure 8). Very large nanoparticles (for example, with a = 830 nm and b = 50 nm) absorb and scatter light in the near-infrared and mid-infrared region of the spectrum.[98,99] The smallest DL PPA nanoparticles, with a = 21 nm and b = 4 nm, contain only 4 ×10$^4$ atoms, and allow investigation of surface cluster properties.[78]

### 2.7.3. Effect of Nanoparticle Shape on the Ag LSPR.

Nanoparticle shape also has a significant effect on the $\lambda_{max}$ of the LSPR. When the

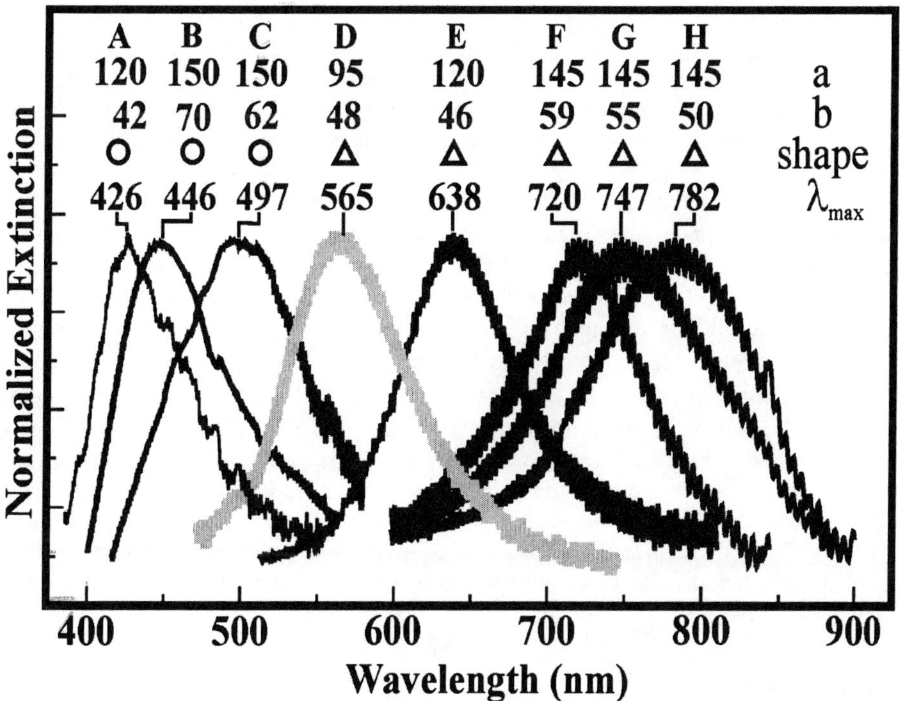


**Figure 9.** UV-visible extinction spectra of Ag NSL-fabricated nanoparticles on mica substrates. Reported spectra are raw, unfiltered data. The oscillatory signal superimposed on the LSPR spectrum seen in the data is due to interference of the probe beam between the front and back faces of the mica. Reproduced with permission from *J. Phys. Chem. B* 2003, 105, 5599-5611. Copyright 2003 Am. Chem. Soc.

standard triangular nanoparticles are thermally annealed at 300 °C under vacuum, the nanoparticle shape is modified, increasing in out-of-plane height and becoming ellipsoidal. This structural transition results in a blue shift of ~200 nm in the $\lambda_{max}$ of the LSPR. It is often difficult to decouple size and shape effects on the LSPR wavelength, and so they are considered together as the nanoparticle aspect ratio (a/b). Large aspect ratio values represent oblate nanoparticles and aspect ratios with a value of unity represent spheroidal nanoparticles.

Figure 9 shows a series of extinction spectra collected from Ag NSL-fabricated nanoparticles of varied shapes and aspect ratios on mica substrates. These extinction spectra were recorded in standard transmission geometry. All macroextinction measurements were recorded using unpolarized light with a probe beam size of approximately 2-4 mm$^2$. All parameters, except those listed explicitly in Figure 9, were held constant throughout this series of measurements. To identify each parameter's effect on the LSPR, one must examine three separate cases. First, extinction peaks F, G, and H all have the same "a" value (signifying that the same diameter nanosphere mask was used for each sample), but the "b" value is varied as the shape is held constant. Note that the LSPR $\lambda_{max}$ shifts to the red as the out-of-plane nanoparticle height is decreased, i.e. the aspect ratio is increased. Secondly, extinction peaks D, E, and H have varying "a" values, very similar "b" values, and constant shape. In this case, the LSPR $\lambda_{max}$ shifts to the red with increased nanosphere diameter (a larger "a" value). Again, as the aspect ratio increases, the LSPR shifts to longer wavelengths. Finally, extinction peaks C and F were measured from the same sample before and after thermal annealing. Note the slight increase in nanoparticle height ("b") as the annealed nanoparticle dewets the mica substrate. The 223 nm blue shift upon annealing is in accordance with the decreasing nanoparticle aspect ratio as the nanoparticles transition from oblate to ellipsoidal geometries.

The range of possible LSPR $\lambda_{max}$ values extends beyond those shown in Figure 9. In fact, $\lambda_{max}$ can be tuned continuously from ~400 nm to 6000 nm by choosing the appropriate nanoparticle aspect ratio and geometry.[99] Recent experiments exploring the sensitivity of the LSPR $\lambda_{max}$ to changes in "a" and "b" values support the assertion that it is not always possible to decouple the in-plane width and out-of-plane height from one another. Figure 9 demonstrates an in-plane width sensitivity of $\Delta\lambda_{max}/\Delta a = 4$ and an out-of-plane height sensitivity of $\Delta\lambda_{max}/\Delta b = 7$. In order to further investigate the out-of-plane height sensitivity, a larger data set was collected in the $\lambda_{max} = 500\text{-}600$ nm region using nanosphere lithography masks made from D = 310±7 nm nanospheres. In this case, both the in-plane width and the shape were held constant as the nanoparticle height was varied. From this larger data set, the calculated out-of-plane height sensitivity was $\Delta\lambda_{max}/\Delta b = 2$. The variance in the two $\Delta\lambda_{max}/\Delta b$ values suggests that nanoparticles with smaller in-plane widths (a = 90±6 nm versus a = 145±6 nm) are less sensitive to changes in nanoparticle height. This conclusion supports the relationship between nanoparticle aspect ratio and LSPR shift susceptibility noted above.

### 2.7.4. Effect of the External Dielectric Medium on the Ag LSPR

Next, the role of the external dielectric medium on the optical properties of these surface-confined nanoparticles is considered.  Just as it is difficult to decouple the effects of size and shape from one another, the dielectric effects of the substrate and external dielectric medium (i.e., bulk solvent) are inextricably coupled because together they describe the entire dielectric environment surrounding the nanoparticles.  The

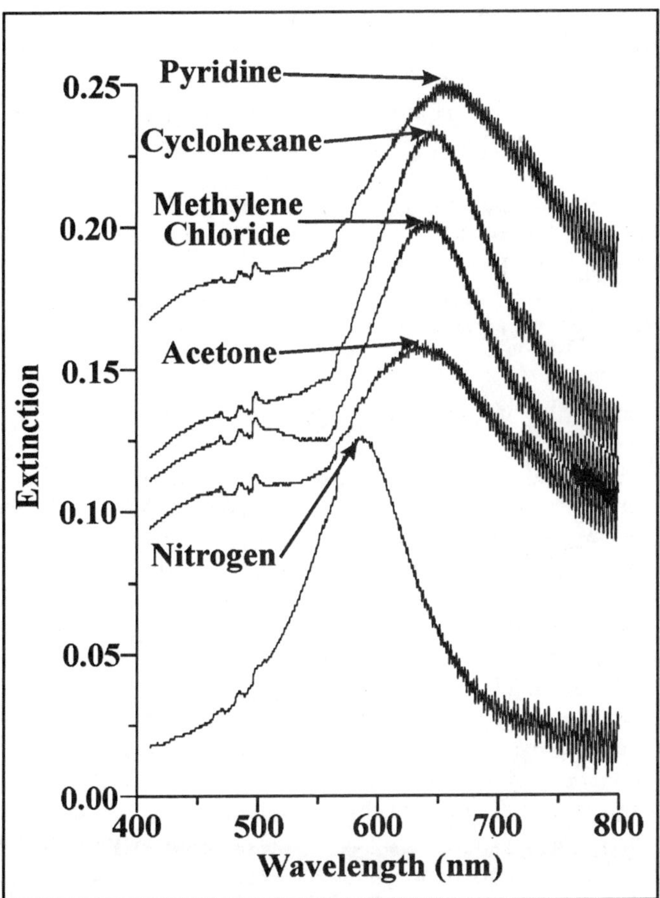


**Figure 10.**  Extinction spectra of Ag  NSL-fabricated nanoparticles having an aspect ratio of 2.17 measured in various solvents.  Spectra have been arbitrarily spaced along the extinction axis for clarity.  Reproduced with permission from *J. Phys. Chem. B* 2003, 105, 5599-5611.  Copyright 2003 Am. Chem. Soc.

nanoparticles are surrounded on one side by the substrate and on the other four sides by a chosen environment. A systematic study of the relationship between the LSPR $\lambda_{max}$ and the external dielectric constant on the four non-substrate-bound faces of the nanoparticles was done by immersing a series of varied aspect ratio nanoparticle samples (aspect ratios 3.32, 2.17, and 1.64) in a variety of solvents.[101] These solvents represent a progression of refractive indices: nitrogen (1.0), acetone (1.36), methylene chloride (1.42), cyclohexane (1.43), and pyridine (1.51). Each sample was equilibrated in a $N_2$ environment before solvent treatment. Extinction measurements were made before, during, and after each solvent cycle. With the exception of pyridine, the LSPR peak always shifted back to the $N_2$ value when the solvent was purged from the sample cell. The measured LSPR $\lambda_{max}$ values progress toward longer wavelengths as the solvent refractive index increases (Figure 10). A plot of solvent refractive index versus the LSPR $\lambda_{max}$ shift is linear for all three aspect ratio samples. The highest aspect ratio nanoparticles (most oblate) demonstrate the greatest sensitivity to external dielectric environment. In fact, the LSPR $\lambda_{max}$ of the 3.32 aspect ratio nanoparticles shifts 200 nm per refractive index unit (RIU).

After completing the aspect ratio/solvent sensitivity experiment, a duplicate of the 2.17 aspect ratio sample was thermally annealed. This annealed sample, with an aspect ratio of 2.14, was then subjected to the same series of solvent treatments and extinction measurements. The resultant solvent refractive index versus LSPR $\lambda_{max}$ shift still shows a linear trend, but a significantly decreased sensitivity of 100 nm shift per RIU.

### 2.7.5. Effect of Thin Film Dielectric Overlayers on the LSPR

The dielectric environment of the nanoparticles in the aforementioned experiments had two components: substrate on one face and solvent or $N_2$ on the other four faces. Clearly, the LSPR $\lambda_{max}$ can be tuned by choosing the refractive index of an encapsulating solvent. Another means of controlling the nanoenvironment is to encapsulate the Ag nanoparticle in a thin film dielectric shell. Such a dielectric overlayer could function to provide chemical protection for the nanoparticle surface or be a molecularly selective permeation layer to control molecular access to the surface of the nanoparticle. A study of the effect of nanoparticle encapsulation made use of varying thicknesses of $SiO_x$ overlayers. The $SiO_x$ overlayer system presents a more complex dielectric environment including the substrate, the porous $SiO_x$, and the environment above the $SiO_x$ layer. $SiO_x$ overlayer thicknesses of 0, 15, 26, and 36 nm were vapor deposited onto otherwise identical samples. The LSPR $\lambda_{max}$ red shifted as the ratio of $SiO_x$ to air increased, again showing that increased refractive index decreases the frequency of electron oscillation. By plotting the $SiO_x$ thickness versus the LSPR $\lambda_{max}$ peak shift, one concludes that the LSPR peak linearly shifts 4 nm to the red for every 1 nm of deposited $SiO_x$. It was demonstrated in section 3.3 that the LSPR $\lambda_{max}$ can be tuned from ~400-6000 nm by changing the size and shape of the nanoparticle. Now, it is obvious that dielectric tuning with substrate, solvent, and overlayers will also provide a similar range of LSPR $\lambda_{max}$ values.

## 2.7.6. Effect of the Substrate Dielectric Constant on the LSPR

A systematic study was also done with Ag nanoparticles holding all parameters constant with the exception of substrate.[102] Theoretical predictions suggest that the LSPR $\lambda_{max}$ cannot be correctly modeled without accounting for the effect of the substrate dielectric constant.[103] The refractive index of the substrate was systematically varied by using fused silica (1.46), borosilicate optical glass (1.52), mica (1.6), and specialty SF-10 glass (1.73). AFM investigation of the Ag nanoparticles on the four substrates showed that the nanoparticle shape is independent of the substrate choice, i.e. the Ag metal wets each substrate to approximately the same extent. The importance of geometric similarity is apparent from previous discussions of LSPR $\lambda_{max}$ sensitivity to nanoparticle size and shape. The LSPR $\lambda_{max}$ was measured for each sample in a controlled environment, and results demonstrated a red shift for increasing substrate refractive index. When the LSPR $\lambda_{max}$ for each substrate was plotted versus the refractive index, a straight line could be fitted to the experimental points indicating a sensitivity of 87 nm per substrate RIU.

Earlier, it was demonstrated that high aspect ratio nanoparticles are most sensitive to changes in environmental refractive index. It was hypothesized that high aspect ratio nanoparticles would exhibit similarly enhanced sensitivity to the refractive index of the substrate. This was tested by exposing identical nanoparticles deposited on the four aforementioned substrates (fused silica, borosilicate glass, mica, and SF-10 specialty glass) to a variety of refractive index environments.[102] Plots of solvent refractive index verses the LSPR $\lambda_{max}$ peak shift establish sensitivities of 238 nm/RIU, 229 nm/RIU, 206 nm/RIU, and 258 nm/RIU, respectively. Interestingly, there appears to be no correlation between substrate refractive index and the contribution to solvent sensitivity. Also, the sensitivity gained by using the substrate with the highest sensitivity factor (SF-10) versus the substrate with the lowest sensitivity factor (mica) is only a 25% advantage. One concludes that, although there is slightly improved sensitivity of the LSPR to the external dielectric environment when the Ag nanoparticle is deposited on a high refractive index substrate, the advantage is not sufficient to justify the use of the expensive SF-10 substrate in place of other inexpensive substrates such as borosilicate glass.

# 3. RECENT APPLICATIONS OF THE TUNABLE LOCALIZED SURFACE PLASMON RESONANCE

## 3.1. Sensing with Nanoparticle Arrays

The development of biosensors for the diagnosis and monitoring of diseases, drug discovery, proteomics, and the environmental detection of pollutants and/or biological agents is an extremely significant problem.[139] Fundamentally, a biosensor is derived from the coupling of a ligand-receptor binding reaction to a signal transducer. Much biosensor research has been devoted to the evaluation of the relative merits of various signal transduction methods including optical, radioactive, electrochemical, piezoelectric, magnetic, micromechanical, and mass spectrometric. The development of large scale biosensor arrays composed of highly miniaturized signal transducer elements that enable the real-time, parallel monitoring of multiple species is an important driving force in biosensor research. Recently, several research groups have begun to explore alternative strategies for the development of optical biosensors and chemosensors based on the extraordinary optical properties of noble metal nanoparticles. Recently, we demonstrated that nanoscale chemosensors and biosensors can be realized through shifts in the LSPR $\lambda_{max}$ of triangular silver nanoparticles. These wavelength shifts are caused by adsorbate-induced local refractive index changes in competition with charge-transfer interactions at the nanoparticle surface. In this review, a detailed study is presented demonstrating that triangular silver nanoparticles fabricated by nanosphere lithography (NSL) function as extremely sensitive and selective nanoscale affinity chemosensors and biosensors. It will be shown that these nanoscale biosensors based on LSPR spectroscopy operate in a manner totally analogous to propagating surface plasmon resonance (SPR) sensors by transducing small changes in refractive index near the noble metal surface into a measurable wavelength shift response.

### 3.1.1. Experimental Procedure

11-Mercaptoundecanoic acid (11-MUA), 1-octanethiol (1-OT), anti-biotin, 1-ethyl-3-[3-dimethylaminopropyl]carbodiimide hydrochloride (EDC), streptavidin, 10 mM and 20 mM phosphate buffered saline (PBS), pH = 7.4, (+)-biotinyl-3,6-dioxaoctanediamine (biotin), hexanes, methanol, ethanol, Ag wire (99.99%, 0.5 mm diameter), borosilicate glass substrates, tungsten vapor deposition boats, polystyrene nanospheres with diameters of 400±7 nm were received as a suspension in water and were used without further treatment. All materials were used without further purification.

NSL was used to fabricate monodisperse, surface-confined Ag nanoparticles. For these experiments, single layer colloidal crystal nanosphere masks were prepared and 50.0 nm Ag films were deposited onto the samples. Following Ag deposition, the nanosphere mask was removed by sonicating the sample in ethanol for 3 minutes. Macroscale UV-vis extinction measurements were collected using an Ocean Optics spectrometer. All spectra collected are macroscopic measurements performed in standard transmission geometry with unpolarized light. The probe beam diameter was approximately 4 mm. A home built flow cell was used to control the external environment of the Ag nanoparticle substrates. Prior to modification, the Ag nanoparticles were solvent annealed with hexanes and methanol. Dry $N_2$ gas and solvent were cycled through the flow cell until the $\lambda_{max}$ of the sample stabilized.

### 3.1.2. Effect of the Alkanethiol Chain Length on the LSPR

Triangular silver nanoparticles have been shown to unexpectedly sensitive to alkanethiol adsorbates.[43, 46] For this study, the LSPR extinction maximum was compared before and after incubation in a given alkanethiol. For nanoparticles with in-plane widths of 100 nm and out-of-plane heights of 50.0 nm Ag, it has been shown that the LSPR extinction wavelength shifts 3.06 nm for every carbon atom in an adsorbed

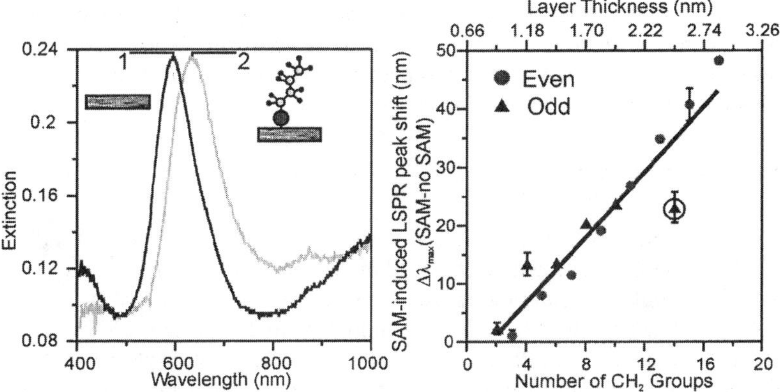

**Figure 11.** LSPR spectra of nanoparticles in a $N_2$ environment. (A) Ag nanoparticles before chemical modification, $\lambda_{max}$ = 594.8 nm, and after modification with 1 mM hexadecanethiol, $\lambda_{max}$ = 634.8 nm. (B) Alkanethiol chain length and layer thickness dependence on the LSPR spectral peak shifts. Linear fit: y = 3.06(x) - 5.5. The circled data point was not used when performing the regression analysis.

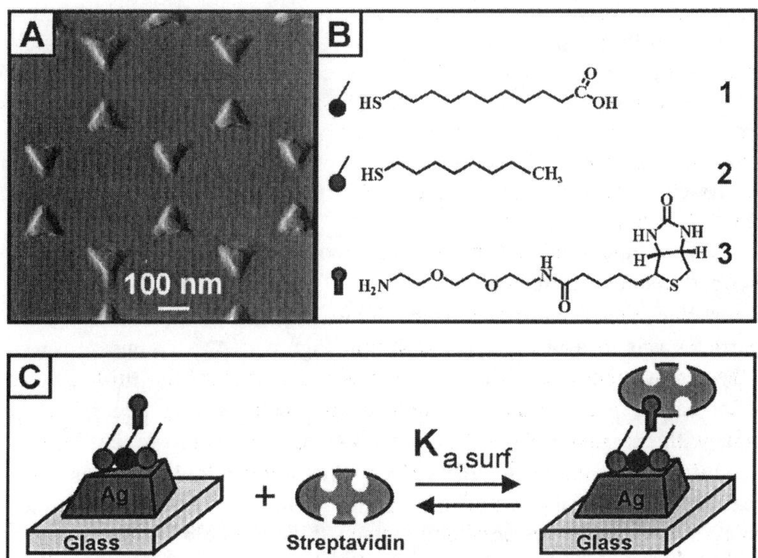



**Figure 12.** LSPR Nanobiosensor. (A) AFM image of the Ag nanoparticles. (B) Surface chemistry of the Ag nanobiosensor. A mixed monolayer of 11-mercaptoundecanoic acid and 1-octanethiol is formed on the exposed surfaces of the Ag nanoparticles followed by the covalent linking of biotin to the carboxyl groups. (C) Schematic representation of streptavidin binding to a biotinylated Ag nanobiosensor. Reproduced with permission from *J. Am. Chem. Soc.* 2002, 124, 10596-10604. Copyright 2002 Am. Chem. Soc.

$CH_3(CH_3)_xSH$ monolayer for x=2-11, 13-15, and 17 (Figure 11). Additionally, the y-intercept, which is related to the Ag-S charge transfer interaction, is -5.5 nm. Furthermore, it should be noted that these LSPR wavelength shifts are caused by only 60,000 alkanethiol molecules per nanoparticle.

### 3.1.3. Streptavidin Sensing using LSPR Spectroscopy

The well-studied biotin-streptavidin system with its extremely high binding affinity ($K_a \sim 10^{13}$ $M^{-1}$) is chosen to illustrate the attributes of these LSPR-based nanoscale affinity biosensors. The biotin-streptavidin system has been studied in great detail by SPR spectroscopy and serves as an excellent model system for the LSPR nanosensor.[44] Streptavidin, a tetrameric protein, can bind up to four biotinylated molecules (i.e. antibodies, inhibitors, nucleic acids, etc.) with minimal impact on its biological activity and, therefore, will provide a ready pathway for extending the analyte accessibility of the LSPR nanobiosensor.

NSL was used to create surface-confined triangular Ag nanoparticles supported on a glass substrate (Figure 12A). The Ag nanotriangles have in-plane widths of ~ 100 nm and out-of-plane heights of ~ 51 nm as determined by AFM. To prepare the LSPR nanosensor for biosensing events, the Ag nanotriangles are first functionalized with a self-assembled monolayer (SAM) composed of 3:1 1-OT (Figure 12B-2):11-MUA (Figure 12B-1) resulting in a surface coverage corresponding to 0.1 monolayer of carboxylate binding sites. Since the maximum number of alkanethiol molecules per nanoparticle is 60 000, this is equivalent to ~ 6000 carboxylate binding sites per nanoparticle. Next, biotin (Figure 12B-3) was covalently attached to the carboxylate groups using a zero-length coupling reagent. The number of resulting biotin sites will be determined by the yield of this coupling reaction. Since this is likely to be ~1-5% efficient one expects there to be only 60-300 biotin sites per nanoparticle at maximum coverages. A schematic illustration of the LSPR nanobiosensor depicting its exposure to streptavidin is shown in Figure 12C.

Before surface functionalization, the Ag nanoparticles were exposed to solvent and $N_2$ as described above. In this study, the $\lambda_{max}$ of the Ag nanoparticles were monitored during each surface functionalization step (Figure 13). First, the LSPR $\lambda_{max}$ of the bare Ag nanoparticles was measured to be 561.4 nm (Figure 13A). To ensure a well-ordered SAM on the Ag nanoparticles, the sample was incubated in the thiol solution for 24 hours. After careful rinsing and thorough drying with $N_2$ gas, the LSPR $\lambda_{max}$ after modification with the mixed SAM (Figure 13B) was measured to be 598.6 nm. The LSPR $\lambda_{max}$ shift corresponding to this surface functionalization step was a 38 nm red-shift, hereafter + will signify a red-shift and - a blue-shift, with respect to bare Ag nanoparticles. Next, biotin was covalently attached via amide bond formation with a two unit polyethylene glycol linker to carboxylated surface sites (Figure 13C). The LSPR $\lambda_{max}$ after biotin attachment (Figure 13C) was measured to be 609.6 nm corresponding to an additional + 11 nm shift. The LSPR nanosensor has now been prepared for exposure to the target analyte. Exposure to 100 nM streptavidin, resulted in LSPR $\lambda_{max}$ = 636.6 nm (Figure 13D) corresponding to an additional +27 nm shift. It

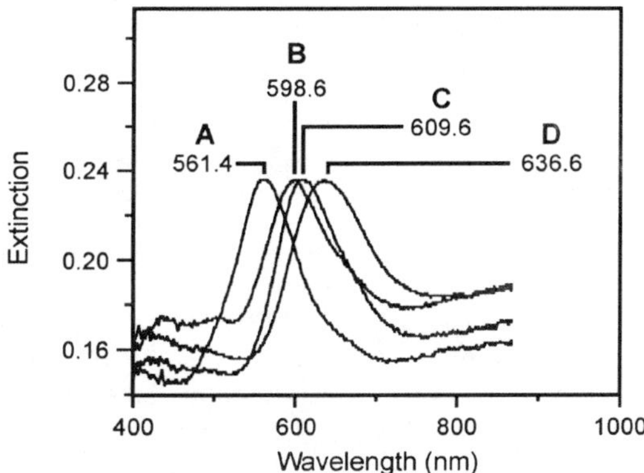

**Figure 13.** LSPR spectra in $N_2$ of each step in the surface modification for the specific binding of streptavidin to Ag nanoparticles (A) before chemical modification, $\lambda_{max}$ = 561.4 nm, (B) after modification with the mixed monolayer, $\lambda_{max}$ = 598.6 nm, (C) after modification with 1 mM biotin, $\lambda_{max}$ = 609.6 nm, and (D) after modification with 100 nM streptavidin, $\lambda_{max}$ = 636.6 nm. Reproduced with permission from *J. Am. Chem. Soc.* 2002, 124, 10596-10604. Copyright 2002 Am. Chem. Soc.

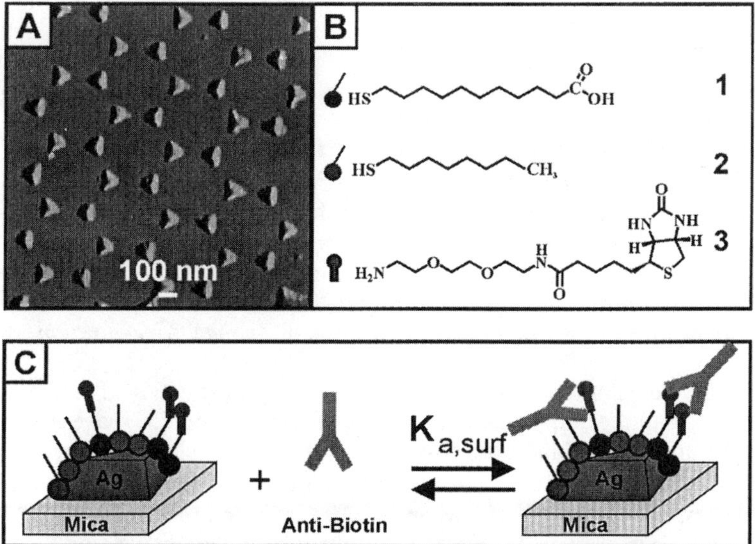

**Figure 14.** LSPR Nanobiosensor. (A) AFM image of the Ag nanoparticles. (B) Surface chemistry of the Ag nanobiosensor. A mixed monolayer is formed on the exposed surfaces of the Ag nanoparticles followed by the covalent linking of biotin to the carboxyl groups. (C) Schematic representation of anti-biotin sensing. Reproduced with permission from *J. Phys. Chem. B* 2003 107, 1772-1780. Copyright 2003 Am. Chem. Soc.

should be noted that the signal transduction mechanism in this nanosensor is a reliably measured wavelength shift rather than an intensity change as in many previously reported nanoparticle-based sensors.

### 3.1.4. Anti-biotin Sensing using LSPR Spectroscopy

A field of particular interest is the study of the interaction between antigens and antibodies. For these reasons we have chosen to focus the present LSPR nanobiosensor study on the prototypical immunoassay involving biotin and anti-biotin, an IgG antibody. In this study, we report the use of Ag nanotriangles synthesized using NSL as a LSPR biosensor that monitors the interaction between a biotinylated surface and free anti-biotin in solution.[61] The importance of this study is that it demonstrates the feasibility of LSPR biosensing with a biological couple whose binding affinity is significantly lower ($1.9 \times 10^6$ - $4.98 \times 10^8$ $M^{-1}$) than in the biotin/streptavidin model.

NSL was used to create massively parallel arrays of Ag nanotriangles on a mica substrate, as shown in Figure 14A. AFM studies indicate that these nanotriangles have perpendicular bisectors of 90 nm and out of plane heights of 50 nm. A SAM of 1:3 1-MUA (Figure 14B-1) :1-OT (Figure 14B-2) was formed on the surface by incubation for 24 hours. A zero length coupling agent was then used to covalently link biotin (Figure 14B-3) to the carboxylate groups. Figure 14C provides a schematic representation of the binding of anti-biotin to biotinylated Ag nanoparticles.

Each step of the functionalization of the samples was monitored using UV-vis spectroscopy, as shown in Figure 15. After a 24 hour incubation in SAM, the LSPR extinction wavelength of the Ag nanoparticles was measured to be 670.3 nm (Figure 15A). Samples were then incubated for 3 hours in biotin to ensure that the amide bond between the amine and carboxyl groups had been formed. The LSPR wavelength shift due to this binding event was measured to be +12.7 nm, resulting in a LSPR extinction wavelength of 683.0 nm (Figure 15B). At this stage, the nanosensor was ready to detect the specific binding of anti-biotin. Incubation in 700 nM anti-biotin for three hours resulted in a LSPR wavelength shift of + 42.6 nm, giving a $\lambda_{max}$ of 725.6 nm (Figure 15C).

### 3.1.5. Monitoring the Specific Binding of Streptavidin to Biotin and Anti-biotin to Biotin and the LSPR Response as a Function of Analyte Concentration

The well-studied biotin/streptavidin[44] system with its extremely high binding affinity ($K_a \sim 10^{13}$ $M^{-1}$) and the antigen-antibody couple, biotin/anti-biotin[61] have been chosen to illustrate the attributes of these LSPR-based nanoscale affinity biosensors. The LSPR $\lambda_{max}$ shift, $\Delta R$, vs. [analyte] response curve was measured over the concentration range $1 \times 10^{-15}$ M < [streptavidin] < $1 \times 10^{-6}$ M and $7 \times 10^{-10}$ M < [anti-biotin] < $7 \times 10^{-6}$ M (Figure 16).[44, 61] Each data point is an averaged resulted from the analysis of three different sample at identical concentrations. The lines are not a fit to the data. Instead, the line was computed from a response model. It was found that this response could be interpreted quantitatively in terms of a model involving: (1) 1:1 binding of a ligand to a multivalent receptor with different sites but invariant affinities and (2) the assumption

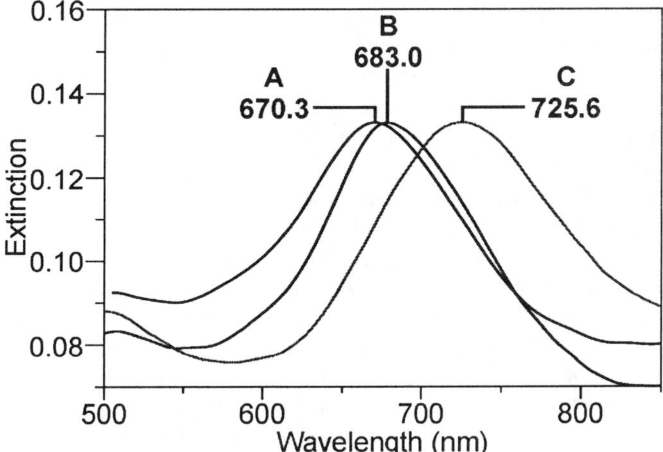

**Figure 15.** Smoothed LSPR spectra for each step of the preparation of the Ag nanobiosensor and the specific binding of anti-biotin to biotin. (A) Ag nanoparticles after modification with the mixed monolayer, $\lambda_{max}$ = 670.3 nm, (B) Ag nanoparticles after modification with 1 mM biotin, $\lambda_{max}$ = 683.0 nm, and (C) Ag nanoparticles after modification with 700 nM anti-biotin, $\lambda_{max}$ = 725.6 nm. All spectra were collected in a $N_2$ environment. Reproduced with permission from *J. Phys. Chem. B* 2003 107, 1772-1780. Copyright 2003 Am. Chem. Soc.

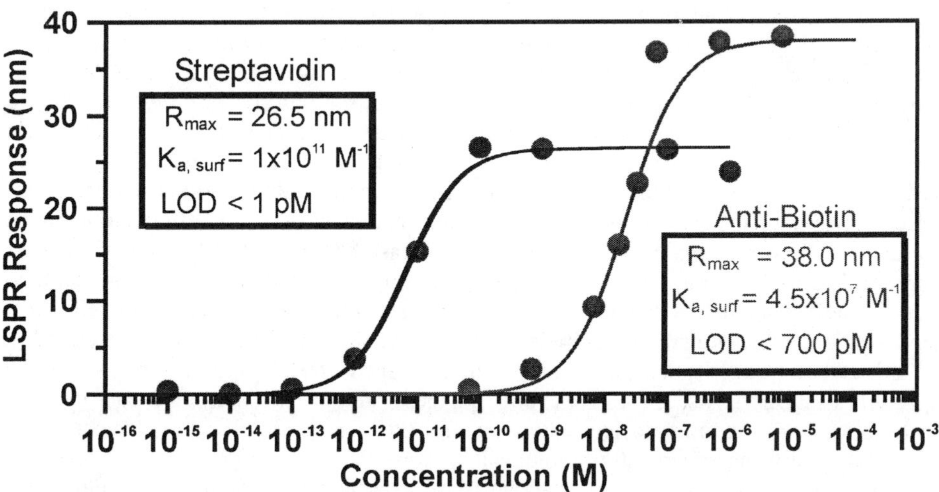


**Figure 16.** The specific binding of streptavidin (left) and anti-biotin (right) to a biotinylated Ag nanobiosensor is shown in the response curves. All measurements were collected in a $N_2$ environment. The solid line is the calculated value of the nanosensor's response.

that only adsorbate-induced local refractive index changes were responsible for the operation of the LSPR nanosensor.

The binding curve provides three important characteristics regarding the system being studied. First, the mass and dimensions of the molecules affect the magnitude of the LSPR shift response. Comparison of the data with theoretical expectations yielded a saturation response, $\Delta R_{max}$ = 26.5 nm for streptavidin, a 60 kDa molecule, and 38.0 nm for anti-biotin, a 150 kDa molecule. Clearly, a larger mass density at the surface of the nanoparticle results in a larger LSPR response. Next, the surface-confined thermodynamic binding constant $K_{a,surf}$ can be calculated from the binding curve and is estimated to be $1 \times 10^{11}$ $M^{-1}$ for streptavidin and $4.5 \times 10^{7}$ $M^{-1}$ for anti-biotin. These numbers are directly correlated to the third important characteristic of the system, the limit of detection (LOD). The LOD is less than 1 pM for streptavidin and 100 pM for anti-biotin. As predicted, the LOD of the nanobiosensor studied is lower for systems with higher binding affinities such as for the well studied biotin-streptavidin couple and higher for systems with lower binding affinities as seen in the anti-biotin system.


## 3.2. Sensing with Single Nanoparticles

The extension of LSPR sensing technique to the single nanoparticle limit provides several improvements over existing array- or cluster-based techniques. First, absolute detection limits are dramatically reduced. The surface area of chemically prepared Ag nanoparticles is typically less than 20,000 $nm^2$, which requires that a complete monolayer of adsorbate must constitute fewer than approximately 100 zeptomole. As demonstrated above, the formation of alkanethiol monolayers Ag nanoparticles can result in a LSPR $\lambda_{max}$ shift of greater that 40 nm, a change that is over 100 times larger than the resolution of convention UV-visible spectrometers. This suggests that the limit of detection for single nanoparticle-based LSPR sensing will be well below 1,000 molecules for small molecule adsorbates. For larger molecules such as antibodies and proteins that result a greater change in the local dielectric environment upon surface adsorption, the single molecule detection limit may be achieved. Second, the extreme sensitivity of single nanoparticle sensors dictates that only very small sample volumes (viz., attoliters) are necessary to induce a measurable response. This characteristic would eliminate the need for analyte amplification techniques (e.g., polymerase chain reaction) required by other analytical methods. Third, single nanoparticle sensing platforms are readily applicable to multiplexed detection schemes. By controlling the size, shape, and chemical modification of individual nanoparticles, multiple sensing platforms can be generated in which each unique nanoparticle can be distinguished from the others based on the spectral location of its LSPR. Several of these unique nanoparticles can then be incorporated into a one device, allowing for the rapid, simultaneous detection of thousands of different chemical or biological species.

The key to exploiting single nanoparticles as sensing platforms is developing a technique to monitor the LSPR of individual nanoparticles with a reasonable signal-to-noise ratio. UV-visible absorption spectroscopy does not provide a practical means of accomplishing this task. Even under the most favorable experimental conditions, the

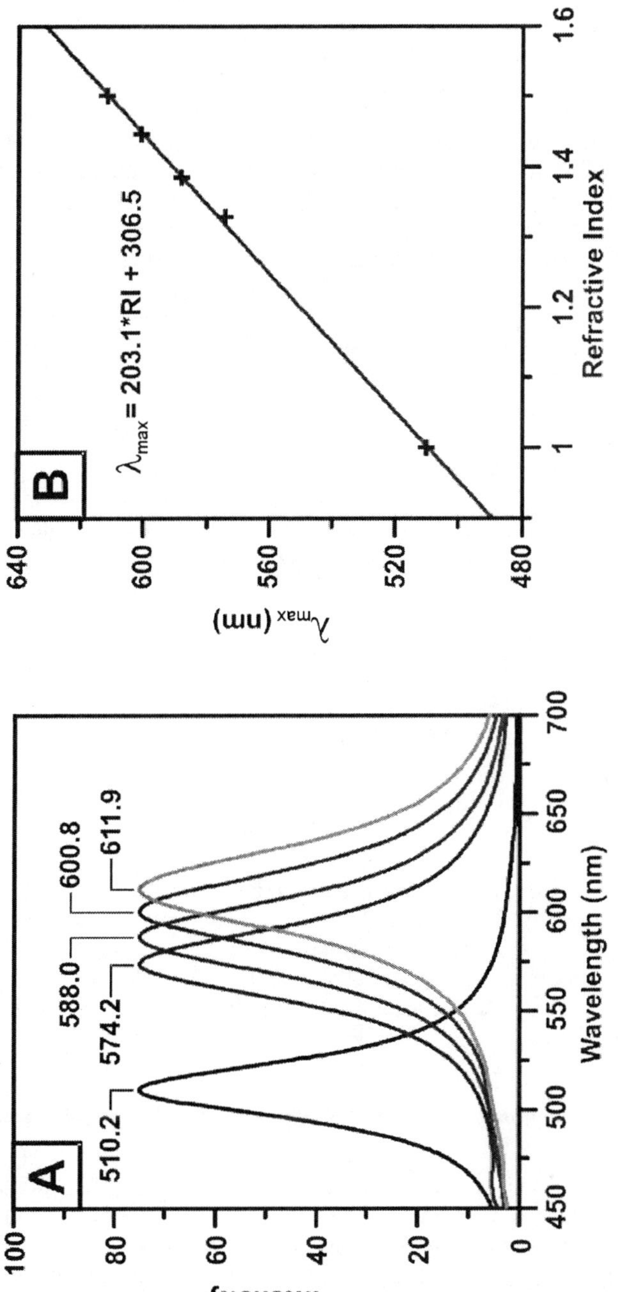


**Figure 17.** (A) Resonant Rayleigh scattering spectra of an individual silver nanoparticle in various dielectric environments ($N_2$, methanol, 1-propanol, chloroform, and benzene). (B) A plot depicting the linear relationship between the solvent refractive index and the nanoparticle's $\lambda_{max}$.

absorbance of a single nanoparticle is very close to the shot noise-governed limit of detection. Instead, resonant Rayleigh scattering spectroscopy is the most straightforward means of characterizing the optical properties of individual metallic nanoparticles. Similar to fluorescence spectroscopy, the advantage of scattering spectroscopy lies in the fact that the scattering signal is being detected in the presence of a very low background. The instrumental approach for performing these experiments generally involves using high magnification microscopy coupled with oblique or evanescent illumination of the nanoparticles. Klar et al. utilized a near-field scanning optical microscope coupled to a tunable laser source to measure the scattering spectra of individual gold nanoparticles embedded in a $TiO_2$ film.[140] Sonnichsen et al. were able to measure the scattering spectra of individual EBL-fabricated nanoparticles using conventional light microscopy.[141] Their technique involved illuminating the nanoparticles with the evanescent field produced by total internal reflection of light in a glass prism. The light scattered by the nanoparticles was collected with a microscope objective and coupled into a spectrometer for analysis. Matsuo and Sasaki employed differential interference contrast microscopy to perform time-resolved laser scattering spectroscopy of single silver nanoparticles.[65] Mock et al. correlated conventional dark-field microscopy and TEM in order to investigate the relationship between the structure of individual metallic nanoparticles and their scattering spectra.[8] They have also used the same light microscopy techniques to study the response of the scattering spectrum to the particle's local dielectric environment by immersing the nanoparticle in oils of various refractive indexes.[142]

### 3.2.1. Experimental Procedure

Colloidal Ag nanoparticles were prepared by reducing silver nitrate with sodium citrate in aqueous solution according to the procedure developed by Lee and Meisel.[143] These nanoparticles were immobilized by drop coating approximately 5 µL of the colloidal solution onto a cover slip and allowing the water to evaporate. The cover slip was then inserted into a flow cell that would allow for continuous optical monitoring of the nanoparticles as they were exposed to various dielectric environments or molecular adsorbates. Prior to all experiments, the nanoparticles in the flow cell were repeatedly rinsed with methanol and dried under nitrogen. All optical measurements were performed using an inverted microscope equipped with an imaging spectrograph. The samples were illuminated using a dark-field condenser and scattered light was collected with a 100X objective. An image of a field of nanoparticles acquired with the apparatus is shown in Figure 4.

### 3.2.2. Single Nanoparticle Refractive Index Sensitivity

The local refractive index sensitivity of the LSPR of a single Ag nanoparticle was measured by recording the resonant Rayleigh scattering spectrum of the nanoparticle as it was exposed to various solvent environments inside the flow cell. As illustrated in Figure 17, the LSPR $\lambda_{max}$ systematically shifts to longer wavelength as the solvent refractive index is increased. Linear regression analysis for this nanoparticle yielded a refractive index sensitivity of 203.1 nm RIU$^{-1}$. The refractive index sensitivity of several individual Ag nanoparticles was measured and typical values were determined to be 170-235 nm RIU$^{-1}$. These are similar to the values obtained for arrays of triangular nanoparticles.

### 3.2.3. Single Nanoparticle Response to Adsorbates

In order to investigate the application of single nanoparticles as chemical sensing platforms, the LSPR response to molecular adsorbates was measured.   First, the scattering spectrum of an individual Ag nanoparticle in a nitrogen environment was recorded. A 1.0 mM solution of the adsorbate in ethanol was then injected into the flow cell.   The particles were allowed to incubate in the solution for at least 1 hour, although real-time monitoring of the scattering spectrum indicates that the majority of the response occurs in the first few minutes after injection. After incubation, the flow cell was flushed several times with ethanol, methanol, and hexane in order to ensure that no more than one monolayer of adsorbate had accumulated on the surface. The nanoparticle was then dried under nitrogen and a final scattering spectrum was collected.  Figure 18 displays the scattering spectrum of an individual Ag nanoparticle in nitrogen before and after modification with thiol monolayer. Based on the average surface area of the nanoparticles prepared and the monolayer packing density of HDT and BZT on Ag, these responses correspond to the detection of fewer than 60,000 and 20,000 molecules of HDT and BZT, respectively.

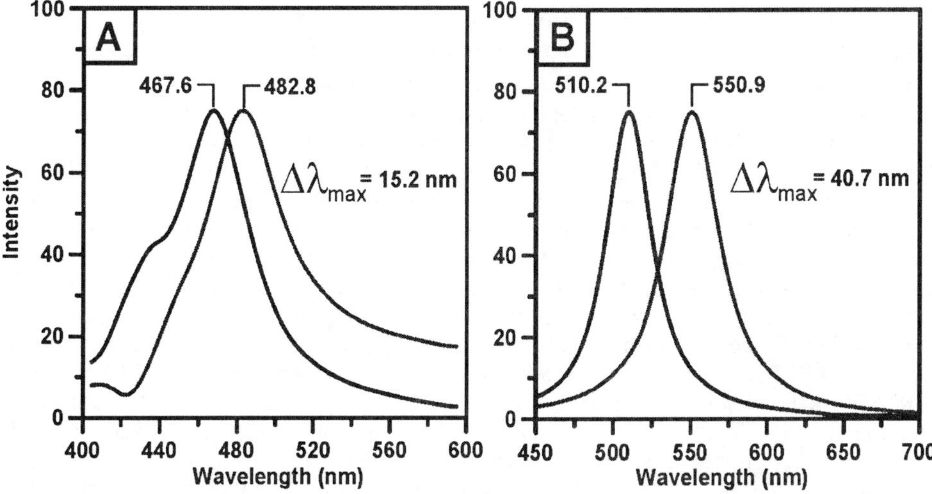

**Figure 18.** Resonant Rayleigh scattering spectra of individual silver nanoparticles before and after exposure to (A) 1 mM benzenethiol and (B) 1 mM 1-hexadecanethiol.

### 3.3. Plasmon-Sampled Surface-Enhanced Raman Excitation Spectroscopy

The current widespread interest in size-dependent optical properties of nanomaterials[45] is a consequence of their many applications in research areas such as chemosensors,[144] biosensors,[46] nanoparticle optics,[54] and surface-enhanced spectroscopy.[6, 78] Recently, interest in surface-enhanced Raman spectroscopy (SERS) has been rekindled by the observation of single molecule SERS.[145, 146] Single molecule SER spectra are characterized by enormous EFs ($\sim 10^{14}$-$10^{16}$) and temporal fluctuation in the vibrational spectrum. The mechanism is not understood, though many research groups are actively pursuing this topic. It is not yet clear if single molecule SERS occurs at a single nanoparticle or if a nanoparticle assembly (viz., dimer, aggregate, etc.) is required. The work presented here focuses not on the single molecule SERS phenomenon, but normal SERS. In fact, this work reports the first SERS data measured from nanosphere lithography (NSL) samples.

Surprisingly, in the 25 years since the discovery of the SERS effect,[147] there has never been a detailed comparison of the localized surface plasmon resonance (LSPR) spectra of highly ordered, structurally well-defined, nanoparticle surfaces and the surface-enhanced Raman excitation spectra (SERES) even though they are intimately linked through the electromagnetic (EM) enhancement mechanism.[6] There are more than 4000 papers concerning SERS and its many applications; however, less than 30 of these address SERES. The paucity of SERES data in the literature is a consequence of the difficulty of SERES experiments, its typically low data point density, and, therefore, its low information content.

The traditional wavelength scanned approach to SERES (WS-SERES)[148-150] involves the measurement of SER spectra from a single substrate with many laser excitation wavelengths, $\lambda_{ex}$. WS-SERE spectra are plots of Raman intensity for a particular vibrational band, $\lambda_{vib}$, versus $\lambda_{ex}$. These plots report the optimized $\lambda_{ex}$ for that specific substrate, adsorbate, and local dielectric environment. The inherent difficulty in WS-SERES lies in the fact that the number of data points is limited either by excitation and/or detection tunability. WS-SERES studies with low data point density have reported EF = $5 \times 10^6$ and $2 \times 10^6$ for microlithographically fabricated Ag nanoparticle arrays[148, 151] and Ag island films,[152] respectively. Broadly tunable WS-SERES experiments are now technically possible using CW mode-locked Ti:Sapphire lasers, their harmonics, and optical parametric amplifiers in combination with a triple spectrograph/CCD detector; however, such equipment is not widely available. Consequently, to our knowledge, only one such WS-SERES experiment has ever been reported.[149] In contrast, the innovative approach to SERES detailed herein describes a method, plasmon sampled-SERES (PS-SERES), that yields information rich SERE spectra using readily available instrumentation. The PS-SERE spectra reported here have been measured under ambient conditions using irreversibly bound adsorbates and nanoparticle surfaces fabricated by nanosphere lithography (NSL).[49] Previous work has demonstrated that these nanoparticle surfaces have LSPRs that are extremely sensitive to their size and shape and are spectrally narrow (viz., < 100 nm).[54] In the PS-SERES experiment, the LSPR and SERS measurements are both spatially resolved and spatially correlated using far-field optical microscopy. By interchanging a white light source with UV-vis extinction detection and a laser light source with Raman detection, it is possible to measure the LSPR ($\sim$25 µm diameter spot size) and SER (~4 $\mu$m diameter spot size) spectra from the same domain on each sample. Thus, PS-SERE spectra plot the SERS intensity or EF for a particular adsorbate vibrational mode versus the LSPR $\lambda_{max}$. Empirically, the SERS signal intensity does not vary from spot to spot on the 25 $\mu$m scale. The results garnered using the PS-SERES technique are information rich when compared to results gained using the WS-SERES technique. Instead of determining the best $\lambda_{ex}$ to use with a particular substrate, adsorbate, and environment, the PS-SERES gives generalized optimization conditions. Accordingly, the results of this work increase the practical utility of the SERS method.

An entire PS-SERES experiment (viz., ~100 data points) is done with a single $\lambda_{ex}$. Consequently, readily available fixed wavelength lasers and notch filter/single monochromator/CCD detection systems may be used. The number of data points obtainable in PS-SERES is limited only by the number of sample regions with different LSPR $\lambda_{max}$ characteristics and the number of samples measured. In this work, we exploit the naturally occurring variation in the LSPR $\lambda_{max}$ on each sample caused by: (1) the small, random structural variations inherent in the NSL fabrication technique;[51] (2) inhomogeneities in the local dielectric environment due to adsorbed water;[153-155] and (3) any residual electromagnetic coupling effects.[57] Previous work has firmly established the sensitivity of the LSPR $\lambda_{max}$ of NSL-fabricated nanoparticles to all the details of their local dielectric environment.[43, 46, 56] An adsorbed water layer is impossible to eliminate under ambient conditions due to the high number density of oxide groups at the surface of the glass substrate.[156] Because PS-SERES benefits from the distribution of LSPR $\lambda_{max}$ values, all extinction and SER spectra presented herein are measured from NSL-fabricated Ag nanoparticles on glass substrates where the standard deviation of a Gaussian fit to the LSPR $\lambda_{max}$ distribution is ~10 nm.


### 3.3.1. Experimental Procedure

Pretreatment of glass substrates required $H_2SO_4$, $H_2O_2$, and $NH_4OH$. Surfactant-free, white, carboxyl-substituted polystyrene latex nanospheres with diameters, D, of 280 ± 4 nm, 310 ± 8 nm, 400 ± 8 nm were used to fabricate nanosphere masks. Benzenethiol, 1,4-benzenedithiol, and 3,4-dichlorobenzenethiol were used as received.

Spatially-resolved extinction and SER spectra were measured using a modified confocal microscope with a 20x objective in back-scattering geometry. When recording an extinction spectrum, the tungsten-halogen microscope lamp provided white light excitation (the laser was blocked from entering the input fiber optic) and the output fiber optic was coupled to an Ocean Optics spectrometer. When recording a Raman spectrum, excitation was provided by lasers of the following wavelengths: $\lambda_{ex}$ = 514.5 nm, $\lambda_{ex}$ = 532.0 nm, or $\lambda_{ex}$ = 632.8 nm. In each case, the laser light was coupled into a 200 $\mu$m core diameter fiber using a fiber launch, and appropriate interference filters and holographic notch filters were placed in the beam path. For SER spectra, the output fiber optic was coupled to a single monochromator with the entrance slit set at 250 $\mu$m with a liquid $N_2$-cooled CCD detector.

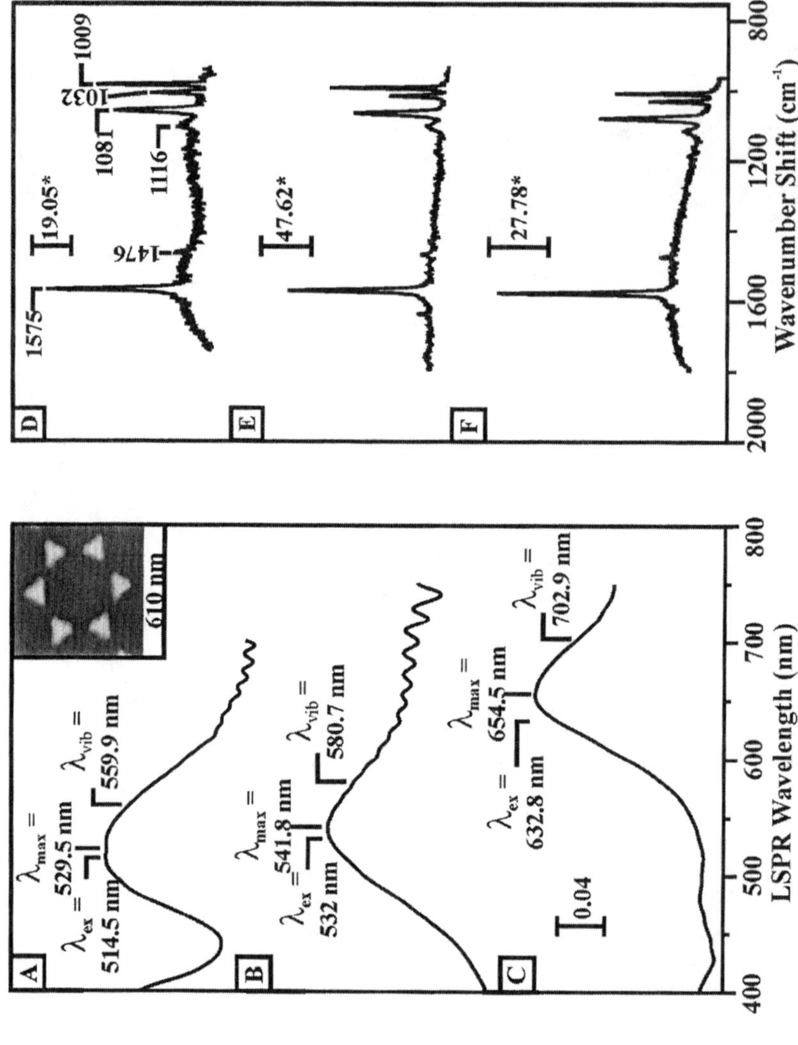



**Figure 19.** Correlated, spatially-resolved LSPR and SER spectra of benzenethiol-dosed NSL substrates with maximized enhancement factors. (A) and (D) were measured from D = 280 nm, $d_m$ = 36 nm, (B) and (E) were measured from D = 280 nm, $d_m$ = 36 nm Ag nanoparticles with 0.7 mW $\lambda_{ex}$ = 514.5 nm, (B) and (E) were measured from D = 280 nm, $d_m$ = 36 nm Ag nanoparticles with 0.7 mW $\lambda_{ex}$ = 532.0 nm, and (C) and (F) were measured from D = 400 nm, $d_m$ = 56 nm Ag nanoparticles with 1.2 mW $\lambda_{ex}$ = 632.8 nm. * represents units of adu mW⁻¹ sec⁻¹. Reproduced with permission from *J. Phys. Chem. B* 2003, ASAP, 03/29/2003. Copyright 2003 Am. Chem. Soc.

### 3.3.2. Varying the Excitation Wavelength in PS-SERES

Figure 19 shows the correlated LSPR and SER spectra corresponding to the maxima of the PS-SERES experiments performed on NSL-fabricated samples with adsorbed benzenethiol using $\lambda_{ex}$ = 514.5 nm, 532.0 nm, and 632.8 nm. These three examples show that the highest intensity SER spectra are achieved when the LSPR $\lambda_{max}$ is located between $\lambda_{ex}$ and $\lambda_{vib}$. The EFs for the $\nu_{8a}$ (symmetric C-C ring stretch) mode[157] at 1575 cm$^{-1}$ shift in spectra D, E, and F are 7.6 x 10$^7$, 6.3 x 10$^7$, and 9.0 x 10$^7$, respectively.

Figures 20A-C show PS-SERE spectra for the $\nu_{8a}$ mode of benzenethiol excited at $\lambda_{ex}$ = 514.5 nm, 532.0 nm, and 632.8 nm respectively. In each case, the paired extinction and SER spectra seen in Figure 19 generate the information necessary to define the optimized point for the PS-SERES plot.  The PS-SERE spectra clearly show that the maximum SERS EFs occur when the LSPR $\lambda_{max}$ lies between $\lambda_{ex}$ and $\lambda_{vib}$.  However, the EF does not vary by more than a factor of 10 when the SERS signal is measurable above background. Attempts to capture SERS spectra from substrates with a LSPR $\lambda_{max}$ outside the x-axis LSPR ranges were unsuccessful.  The solid line overlaid on the PS-SERES data points represents the binned average values of the LSPR $\lambda_{max}$ and EF.  The bin widths were determined by dividing the LSPR $\lambda_{max}$ data range by the square root of the number of data points.  Note that in all cases, there are a small number of data points representing EFs two to four times the average EF.  In order to quantify the conditions for maximizing the EF within this window, we have calculated the LSPR $\lambda_{max}$ range that yields the top 20% of the EF values. In plots 20A-C, these ranges are: (20A) 529-583 nm (18,904–17,153 cm$^{-1}$), (20B) 539-615 nm (18,553-16,260 cm$^{-1}$), and (20C) 638-688 nm (15,674-14,535 cm$^{-1}$).  All of the other vibrational modes of benzenethiol behave similarly.

It is important to note that the spectra depicted in this study are significantly different than those measured in single molecule SERS experiments.  This is most evident when one considers that the measured SER spectrum does not vary over the 25 µm diameter spot size probed by the white light source during extinction measurements and that there are no evident temporal spectral fluctuations.  This work investigates fundamental electromagnetic effects without being hindered by issues of site-to-site adsorbate motion and non-ensemble averaged spectral characteristics.


### 3.3.3. Varying the Molecular Adsorbate in PS-SERES

In order to verify that these optimization conditions are molecularly general, PS-SERES investigations were performed using 3,4-dichlorobenzenethiol, 1,4-benzenedithiol, and Fe(bpy)$_3^{2+}$ on Ag NSL-fabricated substrates. Each of these molecules is known to adsorb irreversibly to the Ag substrate.  Figure 21 displays correlated extinction and SER spectra from the maximum of PS-SERES experiments for each of the aforementioned adsorbates.  In all cases, high signal-to-noise ratio spectra are observed. The correlated extinction and SER spectra for 3,4-dichlorobenzenethiol are shown in Figures 21A and 21D.  In this case, EFs were calculated using neat, liquid 3,4-

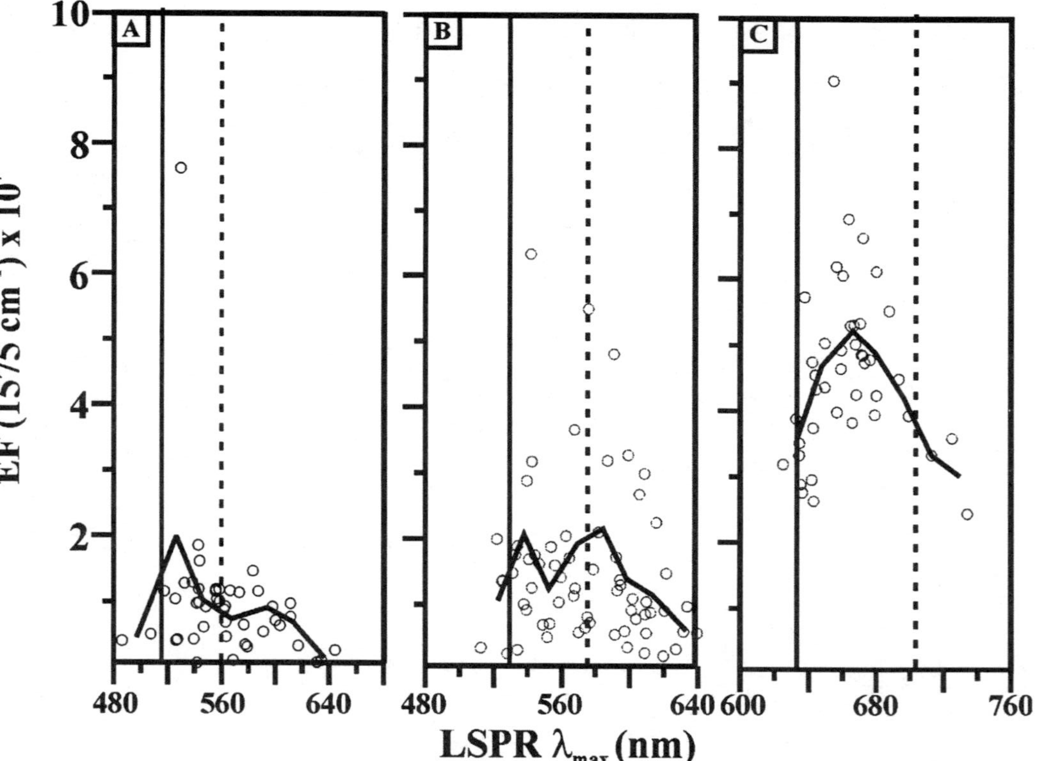

**Figure 20.** Plasmon-sampled surface-enhanced Raman spectra for the $v_{8a}$ (1575 cm$^{-1}$) band of benzenethiol with three different excitation wavelengths: A) $\lambda_{ex}$ = 514.5 nm, B) $\lambda_{ex}$ = 532.0 nm, and C) $\lambda_{ex}$ = 632.8 nm. For each $\lambda_{ex}$, both the wavelength location of the excitation (solid line) and the scattering (dashed line) are marked. The overlaid line represents the bin-averaged values of the LSPR $\lambda_{max}$ and EF. Bin widths are 24 nm (6A), 16 nm (6B), and 16 nm (6C). Reproduced with permission from *J. Phys. Chem. B 2003*, ASAP, 03/29/2003. Copyright 2003 Am. Chem. Soc.

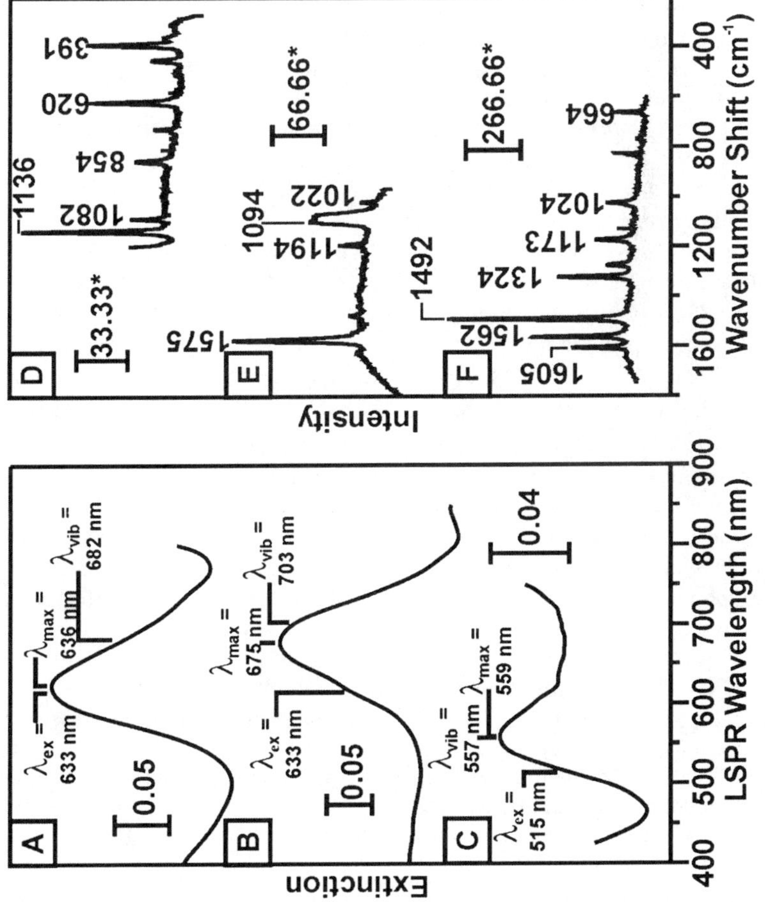

**Figure 21.** Correlated, spatially-resolved LSPR and SER spectra of two other non-resonant adsorbates and one resonant adsorbate on NSL substrates with maximized enhancement factors. (A) and (D) 3,4-dichlorobenzenethiol adsorbed onto D = 400 nm, $d_m$ = 56 nm Ag samples, $\lambda_{ex}$ = 632.8 nm, and P = 1 mW. (B) and (E) 1,4-benzenedithiol adsorbed onto D = 400 nm, $d_m$ = 52 nm Ag samples, $\lambda_{ex}$ = 632.8 nm, and P = 1 mW. (C) and (F) Fe(bpy)$_3$(PF$_6$)$_2$ adsorbed onto D = 310 nm, $d_m$ = 52 nm Ag samples, $\lambda_{ex}$ = 514.5 nm, and P = 0.5 mW. The solid lines in (A), (B), and (C) indicate $\lambda_{ex}$. * represents units of adu mW$^{-1}$ sec$^{-1}$. Reproduced with permission from *J. Phys. Chem. B* 2003, ASAP, 03/29/2003. Copyright 2003 Am. Chem. Soc.

dichlorobenzenethiol as the standard and the packing density of benzenethiol because no coverage measurements for 3,4-dichlorobenzenethiol have been reported in the literature. This assumption likely overestimates the packing density, and accordingly, underestimates the EF of $2.3 \times 10^7$ for the 1136 cm$^{-1}$ shift in-plane ring deformation mode of 3,4-benzenedithiol. The correlated extinction and SER spectra for 1,4-benzenedithiol are shown in Figures 21B and 21E. In this case, EFs were calculated using neat, liquid benzenethiol as the standard because 1,4-benzenedithiol is a solid and the packing density of benzenethiol since no coverage measurements for 1,4-benzenedithiol have been reported. Because benzenethiol and 1,4-benzenedithiol are structurally similar, neither assumption is likely to significantly perturb the calculated EF of $1.4 \times 10^8$ for the 1094 cm$^{-1}$ $\nu_2$ symmetric C-C ring stretching mode of 1,4-benzenedithiol. The correlated extinction and surface-enhanced resonance Raman (SERR) spectra for $Fe(bpy)_3^{2+}$ are shown in Figures 21C and 21F. In this case, EFs were calculated using the normal Raman spectrum of $Fe(bpy)_3^{2+}$ solution excited outside the molecular resonance band with $\lambda_{ex} = 632.8$ nm. The experimentally measured packing density for $Ru(bpy)_3^{2+}$ on Ag[158] of $0.96 \times 10^{14}$ molecules/cm$^2$ was used to calculate the SERRS EF of $7.1 \times 10^9$ for the 1492 cm$^{-1}$ shift $\nu_7$ in-plane bipyridine ring vibration mode[159] of $Fe(bpy)_3^{2+}$.

It is clear from the data presented in Figure 21 that NSL-fabricated surfaces are effective SERS substrates for irreversibly bound adsorbates. Other data made a direct comparison between PS-SERE spectra for the same vibrational mode (the $\nu_2$ symmetric C-C ring stretching mode)[157] in three related adsorbates: benzenethiol, 1,4-benzenedithiol, and 3,4-dichlorobenzenethiol. In all cases, $\lambda_{ex} = 632.8$ nm. These spectra clearly show the same pattern seen in Figure 20 when the $\lambda_{ex}$ was varied. The EFs vary by less than a factor of 10 when the LSPR $\lambda_{max}$ is within approximately a 120 nm window. As before, SERS spectra were not measurable above background from substrates with an LSPR $\lambda_{max}$ outside the x-axis range shown. Overall, the optimized EFs are $1.0 \times 10^8$, $1.3 \times 10^8$, and $2.1 \times 10^6$, respectively. Generally, the EF is relatively constant over the demonstrated LSPR $\lambda_{max}$ window, yielding relatively lenient matching conditions. However, in each case, there are data points representing EF values that exceed the average by a factor of two to three.

In each of the cases presented, the PS-SERES follow a pattern where correlation of the substrate LSPR $\lambda_{max}$ allows achievement of high S/N ratio SER spectra. It is important to note that in all PS-SERE spectra collected to date, including those presented herein, a small percentage of measured EFs fall significantly above (viz., two to four times) the average EF values. While this occurrence is repeatable from sample to sample and for different excitation wavelengths, adsorbates, vibrational modes, and electronic resonance conditions, its cause is currently unknown and under active investigation.


## 4. CONCLUSIONS

This chapter highlights representative research accomplishments in the area of fundamental and applied studies of the tunable LSPR. Specific applications in the Van Duyne laboratory include exploitation of the LSPR as a signal transduction mechanism for sensing applications and optimization of SERS signals. The optical properties of metallic nanostructures will find future application in the areas of dichroic filters, plasmonic waveguides, data storage, biological labels and sensors. Commercialization of nanoparticle devices relies on better understanding of the electromagnetic interaction between nanoparticles as well as the development of techniques that will preserve nanoparticle optical activity in adverse environments.

## 5. ACKNOWLEDGMENTS

The authors wish to acknowledge research collaboration with Traci R. Jensen, Michelle Duval Malinsky, Karen Shafer-Peltier, George C. Schatz, Linlin Zhao, Mikael Käll, Linda Gunnarsson, Juris Prikulis, and Bengt Kasemo. The authors also acknowledge the support of the Nanoscale Science and Engineering Initiative of the National Science Foundation under NSF Award Number EEC-0118025. Any opinions, finding, and conclusions or recommendations expressed in this material are those of the authors and do not necessarily reflect those of the National Science Foundation. Support was also provided by the Air Force Office of Scientific Research MURI program (Grant F49620-02-1-0381) and the Materials Research Science and Engineering Center under NSF Award Number DMR-0076097.

## 6. REFERENCES

1. U. Kreibig and M. Vollmer, *Optical Properties of Metal Clusters*, (Springer-Verlag, Heidelberg, Germany, 1995).
2. S. Link and M. A. El-Sayed, Spectral properties and relaxation dynamics of surface plasmon electronic oscillations in gold and silver nanodots and nanorods, *J. Phys. Chem. B* **103**, 8410-8426 (1999).
3. P. C. Lee and D. Meisel, Surface-enhanced Raman scattering of colloid-stabilizer systems, *Chem. Phys. Lett.* **99**, 262-265 (1983).
4. S. J. Oldenburg, R. D. Averitt, S. L. Westcott and N. J. Halas, Nanoengineering of optical resonances, *Chem. Phys. Lett.* **288**(2,3,4), 243-247 (1998).
5. N. J. Halas and R. K. Bradley, *Partial coverage metal nanoshells and method of making same.* Patent application, 2002.
6. E. Hao, K. L. Kelly, J. T. Hupp and G. C. Schatz, Synthesis of silver nanodisks using polystyrene mesospheres as templates, *J. Am. Chem. Soc.* **124**, 15182-15183 (2002).
7. R. Jin, Y. Cao, C. A. Mirkin, K. L. Kelly, G. C. Schatz and J. G. Zheng, Photoinduced conversion of silver nanospheres to nanoprisms, *Science* **294**(5548), 1901-1903 (2001).
8. J. J. Mock, M. Barbic, D. R. Smith, D. A. Schultz and S. Schultz, Shape effects in plasmon resonance of individual colloidal silver nanoparticles, *J. Chem. Phys.* **116**(15), 6755-6759 (2002).
9. Y.-Y. Yu, S.-S. Chang, C.-L. Lee and C. R. C. Wang, Gold nanorods: Electrochemical synthesis and optical properties, *J. Phys. Chem. B* **101**(34), 6661-6664 (1997).
10. S. Link, M. B. Mohamed and M. A. El-Sayed, Simulation of the optical absorption spectra of gold nanorods as a function of their aspect ratio and the effect of the medium dielectric constant, *J. Phys. Chem. B* **103**(16), 3073-3077 (1999).
11. C. R. Martin, Nanomaterials - a membrane-based synthetic approach, *Science* **266**, 1961-1966 (1994).
12. J. C. Hulteen and C. R. Martin, A general template-based method for the preparation of nanomaterials, *J. Mater. Chem.* **7**(7), 1075 - 1087 (1997).
13. N. A. F. Al-Rawashdeh, M. L. Sandrock, C. J. Seugling and C. A. Foss, Visible region polarization spectroscopic studies of template-synthesized gold nanoparticles oriented in polyethylene, *J. Phys. Chem. B* **102**, 361-371 (1998).
14. M. L. Sandrock, C. D. Pibel, F. M. Geiger and C. A. Foss, Synthesis and second-harmonic generation of noncentrosymmetric gold nanostructures, *J. Phys. Chem. B* **103**, 2668-2673 (1999).

15. J. J. Mock, S. J. Oldenburg, D. R. Smith, D. A. Schultz and S. Schultz, Composite plasmon resonant nanowires, *Nano Lett.* **2**(5), 465-469 (2002).
16. T. Klar, M. Perner, S. Grosse, G. von Plessen, W. Spirkl and J. Feldmann, Surface-plasmon resonances in single metallic nanoparticles, *Phys. Rev. Lett.* **80**(19), 4249-4252 (1998).
17. A. H. Lu, G. H. Lu, A. M. Kessinger and C. A. Foss, Dichroic thin layer films prepared from alkanethiol-coated gold nanoparticles, *J. Phys. Chem. B* **101**(45), 9139-9142 (1997).
18. Y. Dirix, C. Bastiaansen, W. Caseri and P. Smith, Oriented pearl-necklace arrays of metallic nanoparticles in polymers: A new route toward polarization-dependent color filters, *Adv. Mater.* **11**, 223-227 (1999).
19. S. Malynych and G. Chumanov, Light-induced coherent interactions between silver nanoparticles in two-dimensional arrays, *J. Am. Chem. Soc.* **125**, 2896-2898 (2003).
20. W. Gotschy, K. Vonmetz, A. Leitner and F. R. Aussenegg, Optical dichroism of lithographically designed silver nanoparticle films, *Opt. Lett.* **21**(15), 1099-1101 (1996).
21. W. Gotschy, K. Vonmetz, A. Leitner and F. R. Aussenegg, Thin films by regular patterns of metal nanoparticles: Tailoring the optical properties by nanodesign, *App. Phys. B* **63**, 381-384 (1996).
22. J. R. Krenn, W. Gotschy, D. Somitsch, A. Leitner and F. R. Aussenegg, Investigation of localized surface plasmons with the photon scanning tunneling microscope, *Appl. Phys. A* **61**, 541-545 (1995).
23. J. R. Krenn, R. Wolf, A. Leitner and F. R. Aussenegg, Near-field optical imaging the surface plasmon fields of lithographically designed nanostructures, *Opt. Comm.* **137**, 46-50 (1997).
24. M. Quinten, A. Leitner, J. R. Krenn and F. R. Aussenegg, Electromagnetic energy transport via linear chains of silver nanoparticles, *Opt. Lett.* **23**(17), 1331-1333 (1998).
25. C. Sonnichsen, S. Geier, N. E. Hecker, G. von Plessen, J. Feldmann, H. Ditlbacher, B. Lamprecht, J. R. Krenn, F. R. Aussenegg, V. Z.-H. Chan, J. P. Spatz and M. Moller, Spectroscopy of single metallic nanoparticles using total internal reflection microscopy, *Appl. Phys. Lett.* **77**(19), 2949-2951 (2000).
26. H. Ditlbacher, J. R. Krenn, B. Lamprecht, A. Leitner and F. R. Aussenegg, Spectrally coded optical data storage by metal nanoparticles, *Opt. Lett.* **25**(8), 563-565 (2000).
27. J. R. Krenn, J.-C. Weeber, A. Dereux, E. Bourillot, J. P. Goudonnet, G. Schider, A. Leitner, F. R. Aussenegg and C. Girard, Direct observation of localized surface plasmon coupling, *Phys. Rev. B* **60**(7), 5029-5033 (1999).
28. B. Lamprecht, G. Schider, R. T. Lechner, H. Ditlbacher, J. R. Krenn, A. Leitner and F. R. Aussenegg, Metal nanoparticle gratings: Influence of dipolar particle interaction on the plasmon resonance, *Phys. Rev. Lett.* **84**(20), 4721-4724 (2000).
29. J. R. Krenn, G. Schider, W. Rechberger, B. Lamprecht, A. Leitner, F. R. Aussenegg and J.-C. Weeber, Design of multipolar plasmon excitations in silver nanoparticles, *Appl. Phys. Lett.* **77**(21), 3379-3381 (2000).
30. H. Ditlbacher, N. Felidj, J. R. Krenn, B. Lamprecht, A. Leitner and F. R. Aussenegg, Electromagnetic interaction of fluorophores with designed two-dimensional silver nanoparticle arrays, *App. Phys. B* **73**, 373-377 (2001).
31. H. Ditlbacher, J. R. Krenn, N. Felidj, B. Lamprecht, G. Schider, M. Salerno, A. Leitner and F. R. Aussenegg, Fluorescence imaging of surface plasmon fields, *Appl. Phys. Lett.* **80**(3), 404-406 (2002).
32. M. Scharte, R. Porath, T. Ohms, M. Aeschlimann, B. Lamprecht, H. Ditlbacher and F. R. Aussenegg, Lifetime and dephasing of plasmons in Ag nanoparticles, *Proc. SPIE* **4456**, 14-21 (2001).
33. S. A. Maier, P. G. Kik, H. A. Atwater, S. Meltzer, E. Harel, B. E. Koel and A. A. G. Reuicha, Local detection of electromagnetic energy transport below the diffraction limit in metal nanoparticle plasmon waveguides, *Nat. Mater.* **2**, 229 - 232 (2003).
34. C. L. Haynes, A. D. McFarland, L. Zhao, R. P. Van Duyne, G. C. Schatz, L. Gunnarsson, J. Prikulis, B. Kasemo and M. Käll, Nanoparticle optics: The importance of radiative dipole coupling in two-dimensional nanoparticle arrays, *J. Phys. Chem. B* **107**, in press (2003).
35. S. I. Bozhevolnyi, J. Erland, K. Leosson, P. M. W. Skovgaard and J. M. Hvam, Waveguiding in surface plasmon polariton band gap structures, *Phys. Rev. Lett.* **86**(14), 3008-3011 (2001).
36. D. D. Chambliss, R. J. Wilson and S. Chiang, Nucleation of ordered nickel island arrays on gold(111) by surface-lattice dislocations, *Phys. Rev. Lett.* **66**(13), 1721-1724 (1991).
37. R. S. Sennett and G. D. Scott, The structure of evaporated metal films and their optical properties, *J. Opt. Soc. Am.* **40**, 203-211 (1950).
38. S. Aggarwal, A. P. Monga, S. R. Perusse, R. Ramesh, V. Ballarotto, E. D. Williams, B. R. Chalamala, Y. Wei and R. H. Reuss, Spontaneous ordering of oxide nanostructures, *Science* **287**, 2235 - 2237 (2000).
39. W. A. Weimer and M. J. Dyer, Tunable surface plasmon resonance silver films, *Appl. Phys. Lett.* **79**(19), 3164-3166 (2001).
40. O. Siiman and A. Burshteyn, Preparation, microscopy, and flow cytometry with excitation into surface plasmon resonance bands of gold or silver nanoparticles on aminodextran-coated polystyrene beads, *J. Phys. Chem. B* **104**, 9795-9810 (2000).

41. M. Duval Malinsky Ph.D. Thesis, Northwestern University, Evanston, IL, 2000.
42. M. Duval Malinsky, L. Kelly, G. C. Schatz and R. P. Van Duyne, Nanosphere lithography: Effect of the substrate on the localized surface plasmon resonance spectrum of silver nanoparticles, *J. Phys. Chem. B*, 2343-2350 (2001).
43. M. Duval Malinsky, L. Kelly, G. C. Schatz and R. P. Van Duyne, Chain length dependence and sensing capabilities of the localized surface plasmon resonance of silver nanoparticles chemically modified with alkanethiol self-assembled monolayers, *J. Am. Chem. Soc.* **123**, 1471-1482 (2001).
44. A. J. Haes and R. P. Van Duyne, A nanoscale optical biosensor: Sensitivity and selectivity of an approach based on the localized surface plasmon resonance spectroscopy of triangular silver nanoparticles, *J. Am. Chem. Soc.* **124**(35), 10596 - 10604 (2002).
45. A. J. Haes, C. L. Haynes and R. P. Van Duyne, Nanosphere lithography: Self-assembled photonic and magnetic materials, *MRS Symp. Proc.* **636**, D4.8/1-D4.8/6 (2001).
46. A. J. Haes and R. P. Van Duyne, A nanoscale optical biosensor: The short range distance dependence of the localized surface plasmon resonance of noble metal nanoparticles, *J. Phys. Chem. B*, in preparation.
47. C. L. Haynes, A. J. Haes and R. P. Van Duyne, Nanosphere lithography: Synthesis and application of nanoparticles with inherently anisotropic structures and surface chemistry, *MRS Symp. Proc.*, C.6.3/1-C6.3/6 (2001).
48. C. L. Haynes, A. D. McFarland, M. T. Smith, J. C. Hulteen and R. P. Van Duyne, Angle resolved nanosphere lithography: Manipulation of nanoparticle size, shape, and interparticle spacing, *J. Phys. Chem. B* **106**, 1898-1902 (2002).
49. C. L. Haynes and R. P. Van Duyne, Nanosphere lithography: A versatile nanofabrication tool for studies of size-dependent nanoparticle optics, *J. Phys. Chem. B* **105**, 5599-5611 (2001).
50. J. C. Hulteen Ph.D. Thesis, Northwestern University, Evanston, IL, 1995.
51. J. C. Hulteen, D. A. Treichel, M. T. Smith, M. L. Duval, T. R. Jensen and R. P. Van Duyne, Nanosphere lithography: Size-tunable silver nanoparticle and surface cluster arrays, *J. Phys. Chem. B* **103**, 3854-3863 (1999).
52. J. C. Hulteen and R. P. Van Duyne, Nanosphere lithography: A materials general fabrication process for periodic particle array surfaces, *J. Vac. Sci. Technol. A* **13**, 1553-1558 (1995).
53. T. R. Jensen Ph. D., Northwestern University, Evanston, IL, 1999.
54. T. R. Jensen, M. Duval Malinsky, C. L. Haynes and R. P. Van Duyne, Nanosphere lithography: Tunable localized surface plasmon resonance spectra of silver nanoparticles, *J. Phys. Chem. B* **104**, 10549-10556 (2000).
55. T. R. Jensen, M. L. Duval, L. Kelly, A. Lazarides, G. C. Schatz and R. P. Van Duyne, Nanosphere lithography: Effect of the external dielectric medium on the surface plasmon resonance spectrum of a periodic array of silver nanoparticles, *J. Phys. Chem. B* **103**, 9846-9853 (1999).
56. T. R. Jensen, K. L. Kelly, A. A. Lazarides and G. C. Schatz, Electrodynamics of noble metal nanoparticles and nanoparticle clusters, *Clust. Sci.* **10**(2), 295-317 (1999).
57. T. R. Jensen, G. C. Schatz and R. P. Van Duyne, Nanosphere lithography: Surface plasmon resonance spectrum of a periodic array of silver nanoparticles by ultraviolet-visible extinction spectroscopy and electrodynamic modeling, *J. Phys. Chem. B* **103**(13), 2394-2401 (1999).
58. T. R. Jensen, R. P. Van Duyne, S. A. Johnson and V. A. Maroni, Surface-enhanced infrared spectroscopy: A comparison of metal island films with discrete and non-discrete surface plasmons, *Appl. Spectrosc.* **54**, 371-377 (2000).
59. K. L. Kelly, E. Coronado, L. L. Zhao and G. C. Schatz, The optical properties of metal nanoparticles: The influence of size, shape and dielectric environment, *J. Phys. Chem. B* **107**(3), 668-677 (2003).
60. K. L. Kelly, T. R. Jensen, A. A. Lazarides and G. C. Schatz, Modeling metal nanoparticle optical properties, in: *Metal Nanoparticles: Synthesis, Characterization, and Applications,* edited by D. L. Feldheim and C. A. Foss, (Marcel Dekker, Inc., New York, 2001).
61. J. C. Riboh, A. J. Haes, A. D. McFarland, C. Ranjit and R. P. Van Duyne, A nanoscale optical biosensor: Real-time immunoassay in physiological buffer enabled by improved nanoparticle adhesion, *J. Phys. Chem. B* **107**, 1772-1780 (2003).
62. M. T. Smith Ph.D. Thesis, Northwestern University, Evanston, IL, 1999.
63. C. L. Haynes and R. P. Van Duyne, Plasmon-sampled surface-enhanced Raman excitation spectroscopy, *J. Phys. Chem. B* **107**, in press, (2003).
64. R. Micheletto, H. Fukuda and M. Ohtsu, A simple method for the production of a two-dimensional, ordered array of small latex particles, *Langmuir* **11**, 3333-3336 (1995).
65. Y. Matsuo and K. Sasaki, Time-resolved laser scattering spectroscopy of a single metallic nanoparticle, *Jpn. J. Appl. Phys.* **40**, 6143-6147 (2001).
66. C. Sonnichsen, T. Franzl, T. Wilk, G. von Plessen, J. Feldmann, O. Wilson and P. Mulvaney, Drastic reduction of plasmon damping in gold nanorods, *Phys. Rev. Lett.* **88**(7), 077402-1-077402-4 (2002).

67. P. Mulvaney, Not all that's gold does glitter, *MRS Bull.* **26**(12), 1009 - 1014 (2001).
68. W. H. Yang, G. C. Schatz and R. P. Van Duyne, Discrete dipole approximation for calculating optical absorption spectra and surface-enhanced Raman intensities for adsorbates on metal nanoparticles with arbitrary shapes, *J. Chem. Phys.* **103**, 869-875 (1995).
69. J. P. Kottmann, O. J. F. Martin, D. R. Smith and S. Schultz, Non-regularly shaped plasmon resonant nanoparticle as localized light source for near-field microscopy, *J. Microscopy* **202**(1), 60-65 (2001).
70. J. P. Kottmann, O. J. F. Martin, D. R. Smith and S. Schultz, Dramatic localized electromagnetic enhancement in plasmon resonant nanowires, *Chem. Phys. Lett.* **341**, 1-6 (2001).
71. J. P. Kottmann and O. J. F. Martin, Plasmon resonant coupling in metallic nanowires, *Opt. Express* **8**(12), 655-663 (2001).
72. K. Akamatsu, N. Tsuboi, Y. Hatakenaka and S. Deki, In situ spectroscopic and microscopic study on dispersion of Ag nanoparticles in polymer thin films, *J. Phys. Chem. B* **104**(44), 10168-10173 (2000).
73. M. Fleishman, P. J. Hendra and A. J. McQuillan, Raman spectra of pyridine adsorbed at a silver electrode, *Chem. Phys. Lett.* **26**, 163-166 (1974).
74. D. L. Jeanmaire and R. P. Van Duyne, Surface Raman spectroelectrochemistry part I. Heterocyclic, aromatic, and aliphatic amines adsorbed on the anodized silver electrode, *J. Electroanal. Chem.* **84**, 1-20 (1977).
75. R. P. Van Duyne, Laser excitation of Raman scattering from adsorbed molecules on electrode surfaces, in: *Chemical and Biochemical Applications of Lasers,* edited by C. B. Moore, (Academic Press, New York, 1979).
76. M. Moskovits, Surface-enhanced spectroscopy, *Rev. Mod. Phys.* **57**, 783-826 (1985).
77. A. Otto, I. Mrozek, H. Grabhorn and W. Akemann, Surface-enhanced Raman scattering, *J. Phys. Cond. Matt.* **4**, 1143-1212 (1992).
78. A. Campion and P. Kambhampati, Surface-enhanced Raman scattering, *Chem. Soc. Rev.* **27**, 241-250 (1998).
79. G. C. Schatz and R. P. Van Duyne, Electromagnetic mechanism of surface-enhanced spectroscopy, in: *Handbook of Vibrational Spectroscopy,* edited by J. M. Chalmers and P. R. Griffiths, (Wiley, New York, 2002).
80. B. Nikoobakht, J. Wang and M. A. El-Sayed, Surface-enhanced Raman scattering of molecules adsorbed on gold nanorods: Off-surface plasmon resonance condition, *Chem. Phys. Lett.* **366**, 17-23 (2002).
81. S. Nie and S. R. Emory, Probing single molecules and single nanoparticles by surface-enhanced Raman scattering, *Science* **275**, 1102-1106 (1997).
82. K. Kneipp, Y. Wang, H. Kneipp, L. T. Perelman, I. Itzkan, R. R. Dasari and M. S. Feld, Single molecule detection using surface-enhanced Raman scattering (SERS), *Phys. Rev. Lett.* **78**, 1667-1670 (1997).
83. H. Xu, E. J. Bjerneld, M. Käll and L. Börjesson, Spectroscopy of single hemoglobin molecules by surface enhanced Raman scattering, *Phys. Rev. Lett.* **83**, 4357-4360 (1999).
84. E. J. Bjerneld, P. Johansson and M. Käll, Single molecule vibrational fine-structure of tyrosine adsorbed on Ag nano-crystals, *Single Molecules* **1**, 239-248 (2000).
85. A. M. Michaels, M. Nirmal and L. E. Brus, Surface enhanced Raman spectroscopy of individual rhodamine 6G molecules on large Ag nanocrystals, *J. Am. Chem. Soc.* **121**, 9932-9939 (1999).
86. C. J. L. Constantino, T. Lemma, P. A. Antunes and R. Aroca, Single-molecule detection using surface-enhanced resonance Raman scattering and Langmuir-Blodgett monolayers, *Anal. Chem.* **73**, 3674-3678 (2001).
87. A. M. Michaels, J. Jiang and L. Brus, Ag nanocrystal junctions as the site for surface-enhanced Raman scattering of single rhodamine 6G molecules, *J. Phys. Chem. B* **104**, 11965 - 11971 (2000).
88. W. H. Yang, J. C. Hulteen, G. C. Schatz and R. P. Van Duyne, A surface-enhanced hyper-Raman and surface-enhanced Raman scattering study of trans-1,2-bis(4-pyridyl)ethylene adsorbed onto silver film over nanospheres electrodes. Vibrational assignments: Experiment and theory, *J. Chem. Phys.* **104**(11), 4313-4323 (1996).
89. W. H. Li, X. Y. Li and N. T. Yu, Surface-enhanced hyper-Raman spectroscopy (SEHRS) and surface-enhanced Raman spectroscopy (SERS) studies of pyrazine and pyridine adsorbed on silver electrodes, *Chem. Phys. Lett.* **305**(3-4), 303-310 (1999).
90. W. H. Li, X. Y. Li and N. T. Yu, Surface-enhanced hyper-Raman scattering and surface-enhanced Raman scattering studies of electroreduction of phenazine on silver electrode, *Chem. Phys. Lett.* **327**(3-4), 153-161 (2000).
91. T. Wandlowski, K. Ataka and D. Mayer, In situ infrared study of 4,4'-bypyridine adsorption on thin gold films, *Langmuir* **18**, 4331-4341 (2002).
92. J. A. Seelenbinder and C. W. Brown, Comparison of organic self-assembled monolayers as modified substrates for surface-enhanced infrared absorption spectroscopy, *Appl. Spectrosc.* **56**(3), 295-299 (2002).

93. P. J. Tarcha, J. Desaja-Gonzalez, S. Rodriguel-Llorente and R. F. Aroca, Surface-enhanced fluorescence on $SiO_2$-coated silver island films, *Appl. Spectrosc.* **53**(1), 43-48 (1999).
94. J. Malicka, I. Gryczynski, Z. Gryczynski and J. R. Lakowicz, Effects of fluorophore-to-silver distance on the emission of cyanine-dye-labeled oligonucleotides, *Anal. Biochem.* **315**, 57-66 (2003).
95. N. Strekal, A. German, G. Gachko, A. Maskevich and S. Maskevich, The study of the doxorubicin adsorbed onto chemically modified silver films by surface-enhanced spectroscopy, *J. Mol. Struct.* **563-564**, 183-191 (2001).
96. C. J. L. Constantino and R. F. Aroca, Surface-enhanced resonance Raman scattering imaging of Langmuir-Blodgett monolayers of bis(benzimidazo)perylene on silver island films, *J. Raman Spec.* **31**, 887-890 (2000).
97. P. A. Antunes, C. J. L. Constantino, R. F. Aroca and J. Duff, Langmuir and Langmuir-Blodgett films of perylene tetracarboxylic derivatives with varying alkyl chain length: Film packing and surface-enhanced fluoresence studies, *Langmuir* **17**, 2958-2964 (2001).
98. K. T. Shimizu, W. K. Woo, B. R. Fisher, H. J. Eisler and M. G. Bawendi, Surface-enhanced emission from single semiconductor nanocrystals, *Phys. Rev. Lett.* **89**(11), 117401-1 - 117401-4 (2002).
99. A. C. R. Pipino, G. C. Schatz and R. P. Van Duyne, Surface-enhanced second-harmonic diffraction: Experimental investigation of selective enhancement, *Phys. Rev. B* **53**, 4162-4169 (1996).
100. I. I. Smolyaninov, C. H. Lee, C. C. Davis and S. Rudin, Near-field imaging of surface-enhanced second harmonic generation, *J. Microscopy* **194**(2-3), 532-536 (1999).
101. A. G. Brolo, P. Germain and G. Hager, Investigation of the adsorption of L-cysteine on a polycrystalline silver electrode by surface-enhanced Raman scattering (SERS) and surface-enhanced second harmonic generation (SESHG), *J. Phys. Chem. B* **106**, 5982-5987 (2002).
102. J. R. Krenn, A. Dereux, J.-C. Weeber, E. Bourillot, Y. Lacroute, J. P. Goudonnet, G. Schider, W. Gotschy, A. Leitner, F. R. Aussenegg and C. Girard, Squeezing the optical near-field zone by plasmon coupling of metallic nanoparticles, *Phys. Rev. Lett.* **82**(12), 2590-2593 (1999).
103. G. C. des Francs, C. Girard, J.-C. Weeber, C. Chicane, T. David, A. Dereux and D. Peyrade, Optical analogy to electronic quantum corrals, *Phys. Rev. Lett.* **86**(21), 4950-4953 (2001).
104. L. A. Peyser, A. E. Vinson, A. P. Bartko and R. M. Dickson, Photoactivated fluorescence from individual silver nanoclusters, *Science* **291**, 103 -106 (2001).
105. S. R. Nicewarner-Peña, R. G. Freeman, B. D. Reiss, L. He, D. J. Peña, I. D. Walton, R. Cromer, C. D. Keating and M. J. Natan, Submicrometer metallic barcodes, *Science* **294**, 137-141 (2001).
106. Y. C. Cao, R. Jin and C. A. Mirkin, Nanoparticles with Raman spectroscopic fingerprints for DNA and RNA detection, *Science* **297**, 1536-1540 (2002).
107. A. H. A. Aziz, S. A. Shouman, A. S. Attia and S. F. Saad, A study on the reproductive toxicity of erythrosine in male mice, *Pharm. Res.* **35**(5), 457-462 (1997).
108. M. S. Bhatia, Allergy to tartrazine in psychotropic drugs, *J. Clin. Psych.* **61**(7), 473-476 (2000).
109. C. Dees, M. Askari, S. Garett, K. Gehrs, D. Henley and C. M. Ardies, Estrogenic and DNA-damaging activity of red No 3 in human breast cancer cells, *Environ. Health Perspec.* **105**, 625-632 (1997).
110. S. W. Huang and P. R. Borum, Study of skin rashes after antibiotic use in young children, *Clin. Ped* **37**(10), 601-607 (1998).
111. M. D. Lowry, C. F. Hudson and J. P. Callen, Leukocytoclastic vasculitis caused by drug additives, *J. Am. Acad. Derm.* **30**(5), 854-855 (1994).
112. Y. Yin, Y. Lu, Y. Sun and Y. Xia, Silver nanowires can be directly coated with amorphous silica to generate well-controlled coaxial nanocables of silver/silica, *Nano Lett.* **2**(4), (2002).
113. T. Ung, L. M. Liz-Marzan and P. Mulvaney, Controlled method for silica coating of silver colloids. Influence of coating on the rate of chemical reactions, *Langmuir* **14**, 3740-3748 (1998).
114. P. Mulvaney, Surface plasmon spectroscopy of nanosized metal particles, *Langmuir* **12**(3), 788-800 (1996).
115. M. J. Feldstein, C. D. Keating, Y. H. Liau, M. J. Natan and N. F. Scherer, Electronic relaxation dynamics in coupled metal nanoparticles, *J. Am. Chem. Soc.* **119**(28), 6638-6647 (1997).
116. J. Shi, S. Gider, K. Babcock and D. D. Awschalom, Magnetic clusters in molecular beams, metals, and semiconductors, *Science* **271**, 937-941 (1996).
117. S. C. Street, C. Xu and D. W. Goodman, The physical and chemical properties of ultrathin oxide films, *Ann. Rev. Phys. Chem.* **48**, 43-68 (1997).
118. Z. L. Wang, J. M. Petroski, T. C. Green and M. A. El-Sayed, Shape transformation and surface melting of cubic and tetrahedral platinum nanocrystals, *J. Phys. Chem. B* **102**, 6145-6151 (1998).
119. R. P. Andres, J. D. Bielefeld, J. I. Henderson, D. B. Janes, V. R. Kolagunta, C. P. Kubiak, W. J. Mahoney and R. G. Osifchin, Self-assembly of a two-dimensional superlattice of molecularly linked metal clusters, *Science* **273**(5282), 1690-1693 (1996).

120. S. Gorer, J. A. Ganske, J. C. Hemminger and R. M. Penner, Size-selective and epitaxial electrochemical/chemical synthesis of sulfur-passivated cadimum sulfide nanocrystals on graphite, *J. Am. Chem. Soc.* **130**, 9584-9593 (1998).

121. W. Knoll, Interfaces and thin films as seen by bound electromagnetic waves, *Ann. Rev. Phys. Chem.* **49**, 569-638 (1998).

122. R. C. Mucic, J. J. Storhoff, C. A. Mirkin and R. L. Letsinger, DNA-directed synthesis of binary nanoparticle network materials, *J. Am. Chem. Soc.* **120**, 12674-12675 (1998).

123. G. Pan, R. Kesavamoorthy and S. A. Asher, Nanosecond switchable polymerized crystalline colloidal array Bragg diffracting materials, *J. Am. Chem. Soc.* **120**, 6525-6530 (1998).

124. M. Kahl, E. Voges, S. Kostrewa, C. Viets and W. Hill, Periodically structured metallic substrates for SERS, *Sens. Actuator B-Chem.* **51**(1-3), 285-291 (1998).

125. R. G. Freeman, K. C. Grabar, K. J. Allison, R. M. Bright, J. A. Davis, A. P. Guthrie, M. B. Hommer, M. A. Jackson, P. C. Smith, D. G. Walter and M. J. Natan, Self-assembled metal colloid monolayers: An approach to SERS substrates, *Science* **267**, 1629-1632 (1995).

126. S. Malynych, I. Luzinov and G. Chumanov, Poly(vinyl pyridine) as a universal surface modifier for immobilization of nanoparticles, *J. Phys. Chem. B* **106**, 1280-1285 (2002).

127. G. Schider, J. Krenn, W. Gotschy, B. Lamprecht, H. Ditlbacher, A. Leitner and F. Aussenegg, Optical properties of Ag and Au nanowire gratings, *J. Appl. Phys.* **90**, 3825-3830 (2001).

128. S. A. Maier, M. L. Brongersma, P. G. Kik and H. A. Atwater, Observation of near-field coupling in metal nanoparticle chains using far-field polarization spectroscopy, *Phys. Rev. B* **65**, 193408-1 - 193408-4 (2002).

129. S. A. Maier, P. G. Kik and H. A. Atwater, Observation of coupled plasmon-polariton modes in Au nanoparticle chain waveguides of different lengths: Estimation of waveguide loss, *Appl. Phys. Lett.* **81**(9), 1714-1716 (2002).

130. L. Gunnarsson, E. J. Bjerneld, H. Xu, S. Petronis, B. Kasemo and M. Käll, Interparticle coupling effects in nanofabricated substrates for surface-enhanced Raman scattering, *Appl. Phys. Lett.* **78**(6), 802-804 (2001).

131. T. R. Jensen, G. C. Schatz and R. P. Van Duyne, Nanosphere lithography: Surface plasmon resonance spectrum of a periodic array of silver nanoparticles by UV-vis extinction spectroscopy and electrodynamic modeling, *J. Phys. Chem. B* **103**, 2394-2401 (1999).

132. N. Felidj, J. Aubard, G. Levi, J. R. Krenn, M. Salerno, G. Schider, B. Lamprecht, A. Leitner and F. R. Aussenegg, Controlling the optical response of regular arrays of gold particles for surface-enhanced Raman scattering, *Phys. Rev. B* **65**, 075419-1 - 075419-9 (2002).

133. C. A. Murray and S. Bodoff, Depolarization effects in Raman scattering from monolayers on surfaces: The classical microscopic local field, *Phys. Rev. Lett.* **52**(25), 2273-6 (1984).

134. A. A. Lazarides and G. C. Schatz, DNA-linked metal nanosphere materials: Structural basis for the optical properties, *J. Phys. Chem. B* **104**(3), 460-467 (2000).

135. L. Zhao, K. L. Kelly and G. C. Schatz, The extinction spectra of silver nanoparticle arrays: Influence of array structure on plasmon resonance wavelength and widths, *J. Phys. Chem. B* **107**, in press, (2003).

136. T. Ito and S. Okazaki, Pushing the limits of lithography, *Nature* **406**, 1027-1031 (2000).

137. W. Eidelloth and R. L. Sandstrom, Wet etching of gold films compatible with high $T_c$ superconducting thin films, *Appl. Phys. Lett.* **59**(13), 1632-1634 (1991).

138. J. J. Storhoff, A. A. Lazarides, R. C. Mucic, C. A. Mirkin, R. L. Letsinger and G. C. Schatz, What controls the optical properties of DNA-linked gold nanoparticle assemblies?, *J. Am. Chem. Soc.* **122**, 4640-50 (2000).

139. A. P. F. Turner, Biosensors - sense and sensitivity, *Science* **290**, 1315-1317 (2000).

140. T. Klar, M. Perner, S. Grosse, G. von Plessen, W. Spirkl and J. Feldmann, Surface-plasmon resonances in single metallic nanoparticles, *Phys. Rev. Lett.* **80**, 4249-4252 (1998).

141. C. Sönnichsen, S. Geier, N. E. Hecker, G. von Plessen, J. Feldmann, H. Ditlbacher, B. Lamprecht, J. R. Krenn, F. R. Aussenegg, V. Z.-H. Chan, J. P. Spatz and M. Möller, Spectroscopy of single metallic nanoparticles using total internal reflection microscopy, *Appl. Phys. Lett.* **77**, 2949-2951 (2000).

142. J. J. Mock, D. R. Smith and S. Schultz, Local refractive index dependence of plasmon resonance spectra from individual nanoparticles, *Nano Lett.* **3**(4), 485-491 (2003).

143. P. C. Lee and D. Meisel, Adsorption and surface-enhanced Raman of dyes on silver and gold sols, *J. Phys. Chem.* **86**(17), 3391-5 (1982).

144. R. I. Altkorn, I. Koev, R. P. Van Duyne and M. Duval Malinsky, Intensity considerations in liquid core optical fiber Raman spectroscopy, *Appl. Spectrosc.*, **55**, 373-381 (2001).

145. E. Danielson, J. H. Golden, E. W. McFarland, C. M. Reaves, W. H. Weinberg and X. D. Wu, A combinatorial approach to the discovery and optimization of luminescent materials, *Nature* **389**, 944-948 (1997).

146. K. Kneipp, H. Kneipp, I. Itzkan, R. R. Dasari and M. S. Feld, Surface-enhanced Raman scattering: A new tool for biomedical spectroscopy, *Curr. Sci.* **77**, 915-924 (1999).
147. D. L. Jeanmaire Dissertation, Northwestern University, Evanston, IL, 1977.
148. J. G. Bergman, D. S. Chemla, P. F. Liao, A. M. Glass, A. Pinczuk, R. M. Hart and D. H. Olson, Relationship between surface-enhanced Raman scattering and the dielectric properties of aggregated silver films, *Opt. Lett.* **6**, 33-35 (1981).
149. B. W. Gregory, B. K. Clark, J. M. Standard and A. Avila, Localization of image state electrons in the sulfur headgroup region of alkanethiol self-assembled films, *J. Phys. Chem. B* **105**, 4684-4689 (2001).
150. W. Akemann and A. Otto, Roughness induced reactions of $N_2$ and $CO_2$ on noble and alkali metals, *Surf. Sci.* **272**, 211-219 (1992).
151. P. F. Liao, Silver structures produced by microlithography, in: *Surface Enhanced Raman Scattering*, edited by R. K. Chang and T. E. Furtak, (Plenum Press, New York, 1982).
152. D. A. Weitz, S. Garoff and T. J. Gramila, Excitation spectra of surface-enhanced Raman scattering on silver-island films, *Opt. Lett.* **7**, 168-170 (1982).
153. R. D. Piner, J. Zhu, F. Xu, S. H. Hong and C. A. Mirkin, "Dip-pen" nanolithography, *Science* **283**(5402), 661-663 (1999).
154. M. Salmeron and B. Hendrick, Structure and properties of ice and water film interfaces in equilibrium with vapor, *Surf. Rev. Lett.* **6**, 1275-1281 (1999).
155. P. B. Miranda, L. Xu, Y. R. Shen and M. Salmeron, Icelike water monolayer adsorbed on mica at room temperature, *Phys. Rev. Lett.* **81**, 5876-5879 (1998).
156. C. Noguera, *Physics and Chemistry at Oxide Surfaces*, (Cambridge University Press, Cambridge, 1996).
157. G. Varsanyi, *Vibrational Spectra of Benzene Derivatives*, (Academic Press, New York, 1969).
158. A. M. Stacy and R. P. Van Duyne, Surface-enhanced Raman and resonance Raman spectroscopy in a non-aqueous electrochemical environment: Tris (2,2'-bipyridine) ruthenium (II) adsorbed on silver from acetonitrile, *Chem. Phys. Lett.* **102**, 365-370 (1983).
159. K. Maruszewski, K. Bajdor, D. P. Strommen and J. R. Kincaid, Position-dependent deuteration effects on the nonradiative decay of the 3MLCT state of tris(bipyridine)ruthenium (II). An experimental evaluation of radiationless transition theory, *J. Phys. Chem.* **99**, 6286-6293 (1995).

# 4

# COLLOID SURFACE CHEMISTRY

Arnim Henglein

## 1. Introduction

The investigation of the chemistry that occurs on the surface of colloidal metal particles is important for our understanding of many properties of nano-particles, such as their catalytic effects, their growth, and their electrochemical and photochemical behaviour. The electronics of nano-sized metal particles are changed when chemical events occur on the surface, and, consequently, certain physical properties such as their optical absorption are affected. On the other hand, optical measurements can often be used to trace the chemical modifications on the surface.

In this chapter, special attention is given to the use of radiation for the synthesis of colloidal nano-particles and for the investigation of their chemical and electronic properties. Radiolytic preparation under strictly anaerobic conditions has been shown to be an especially useful method, which allows one to control particle formation in a reproducible manner. Most studies in the literature are concerned with particles of the noble metals, i.e. of Au, Pt, Ag and Cu. The radiolysis method enables one to prepare also colloids of the more electronegative metals, such as of cadmium and thallium. In addition, this method has also been used to initiate chemical reactions on the surface of nano-particles in a controlled manner.

Time resolved chemical experiments on colloidal metals have also been reported. Processes that occur in the millisecond to minute range can be studied using stopped-flow techniques. Faster processes are investigated using pulse radiolysis.

An interesting question is how the chemical and electronic behaviour of particles depend on their size, especially in the <1 nm size range where the transition from non-metallic to metallic properties occurs. In the reduction of metal ions by radiolysis it has sometimes been possible to stabilize intermediate clusters in this size range and to study their chemical and physical properties.

In this chapter, a selective overview of the aspects just mentioned is given without aspiring to present an exhaustive review.

## 2. Radiolytic Methods

Gamma radiation from commercial $^{60}$Co-sources and electron radiation from accelerators are available for studies in colloid science. These radiations deeply penetrate through matter. Solutions can therefore be exposed in ordinary glass vessels, no windows being necessary as in photochemical experiments, and the solutions are uniformly irradiated. As most experiments are carried out with rather dilute solutions, the direct absorption of radiation by the solutes can be neglected as compared to the absorption by the solvent. In aqueous solutions, for example, free OH radicals, H-atoms, and hydrated electrons are formed in the ionization and dissociation of water molecules. These radicals subsequently attack the solutes, the OH radicals having strongly oxidizing and the H-atoms and electrons strongly reducing properties. The radiation chemical yields of these primary radiolysis products are well known. They are expressed as "G-values", i.e. the number of species formed per 100 eV of absorbed radiation energy: $G(OH) = 2.7$, $G(H) = 0.5$, $G(e_{aq}^-) = 2.7$ (besides these free radical products, small amounts of "molecular" products, $H_2O_2$ and $H_2$, are formed). Since the absorbed dose can easily be measured by chemical dosimetry (for example by the oxidation of ferrous ions) one knows within a few percent the amount of radicals formed. This allows one to study the reactions of the radicals with metal ions or metal particles in a quantitative manner.

When the reduction of a solute is desired, the oxidizing hydroxyl radicals have to be removed before they are able to attack the solute. This is achieved by adding an alcohol to the solution at a concentration much higher than that of the solute. Alcohols do not react with the hydrated electron, but scavenge the OH radical. Propanol-2, for example, reacts according to:

$$OH \cdot + (CH_3)_2CH(OH) \rightarrow H_2O + (CH_3)_2C \cdot (OH) \tag{1}$$

The 1-hydroxyalkyl radical formed is a strong reducing agent and reduces the solute in addition to the hydrated electron. When the oxidation of a solute is desired, the hydrated electron has to be removed. This is achieved by irradiating the solution under an atmosphere of nitrous oxide. This gas dissolves at a concentration of $2.5 \times 10^{-2}$ M, i.e. at a concentration much higher than that of the solute. $N_2O$ is known to react rapidly with the hydrated electron according to

$$N_2O + e_{aq}^- + H_2O \rightarrow N_2 + OH^- + OH \cdot \tag{2}$$

If only an organic radical is to attack the solute, both $e_{aq}^-$ and OH have to be scavenged. This can be done by adding both propanol-2 and acetone. The latter is relatively little reactive towards OH, but rapidly reacts with $e_{aq}^-$:

$$e_{aq}^- + (CH_3)_2CO + H_2O \rightarrow (CH_3)_2C \cdot (OH) + OH^- \tag{3}$$

The rate constants of the reactions of the primary radicals and of the alcohol radicals with many inorganic and organic solutes are tabulated in the literature.[1] In addition, the redox potentials of these radicals are known.[2] One can thus "tailor" a system to achieve a desired reaction upon irradiation.

Pulse radiolysis is used to measure the rate of fast reactions of radicals. The solution is exposed to a nanosecond pulse of high energy electrons, which produces free radicals at a concentration of several micromoles per liter. The light absorption of the solution is recorded by suitable electronics before, during and after the pulse. The disappearance of the radicals after the pulse and appearance of the products of their reactions can thus be monitored optically. Many reactions of free radicals are accompanied by changes in the conductivity of the solution. These changes can also be used to trace the intermediates.

## 3. Silver Colloid Preparation

The reduction of silver ions in aqueous solution under the influence of ionizing radiation has often been studied. The elementary processes and intermediate structures are known in many details, due to the fact that silver nano-particles and certain oligomeric precursors have strong optical absorptions by which they can be traced. Regard the case, where a $10^{-4}$ M $AgClO_4$ solution, containing also 0.1 M propanol-2 and $3 \cdot 10^{-4}$ M sodium citrate is $\gamma$-irradiated under an atmosphere of nitrous oxide. During the irradiation, the 380 nm absorption of colloidal silver growths in. The complete reduction of $Ag^+$ is reached when this band does no longer increase. The organic radicals, which are first formed via reactions 1 to 3, attack the silver ions:

$$(CH_3)_2C^{\cdot}(OH) + Ag^+ \rightarrow (CH_3)_2CO + H^+ + Ag^0 \tag{4}$$

The citrate anion, which is present in relatively low concentration, is practically not attacked. It is, however, needed to stabilize the final colloidal nano-particles. Reaction 4 is relatively slow. However, it is fast enough to make sure that all radicals undergo reaction with silver ions, i.e. no radicals are lost by mutual deactivation. The silver atoms formed build up various oligomeric clusters which mutually condense to form larger and larger particles. At a certain cluster size (about 1 nm) the clusters are stabilized by adsorbed citrate anions, i.e. they start to behave like real colloidal particles by building up a stabilizing electrical double layer. Further growth can now occur by reduction of $Ag^+$ ions on the surface of these "nuclei" by the free radicals. This reduction is a fast process which practically occurs at the first encounter of a radical with a silver particle. Thus, a situation exists where initial nuclei formation and growth of the nuclei are well separated in time; under these circumstances, the final nano-particles have a narrow size distribution. **Figure 1** shows high-resolution electron micrographs of the particles obtained this way. They are nicely facetted polyhedra with no noticeable defects in the crystal lattice. Their absorption peaks at 381 nm, and the width of the absorption band at half height amounts to 40 nm.[3]

Particles with a much narrower absorption band of only 23 nm have been made by "push-pull" reduction of silver ions. The solution contained $2 \cdot 10^{-4}$ M sodium poly-phosphate as stabilizer and hydrogen gas as OH radical scavenger: $H_2 + OH^{\cdot} \rightarrow H_2O + H^{\cdot}$. However, only part of the OH radicals were scavenged, i.e. a situation was created where both reducing and oxidizing radicals acted on the silver species. With a small surplus of reducing radicals, the reduction very slowly proceeded during irradiation. Spherical nanoparticles with a narrow size distribution resulted, which had the most narrow and intense plasmon absorption band ever reported. This absorption band is shown in **Figure 2**. The spectrum also contains - well separated from the plasmon band - a weaker band around 240 nm which is caused by interband transitions. The electron microscope investigation revealed a crystal structure without defects.[4]

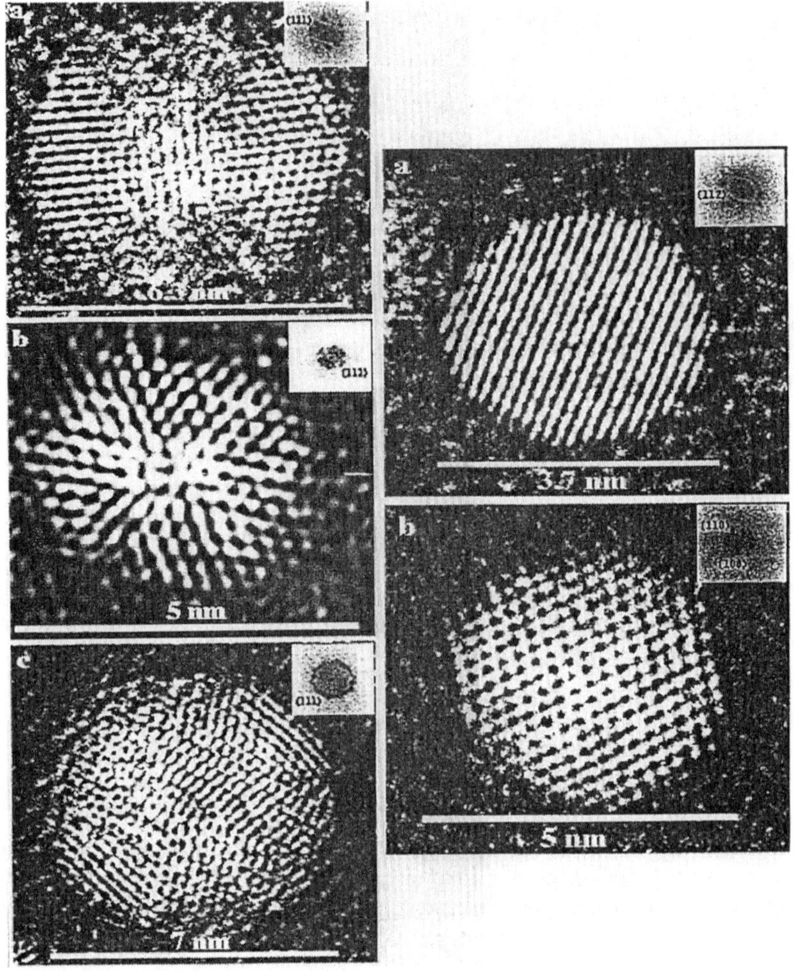

**Figure 1.** High resolution electron micrographs of silver particles produced radiolytically. Left: icosahedra particles viewed along the 2-fold (a), 3-fold (b). and 5-fold symmetry axis. Right: Projections of cuboctahedra particles, oriented along the 112 (a) and 110 (b) direction.[3]

The width of the plasmon absorption band of silver particles has often been discussed in the literature, most of the observations having been made on particles in glasses.[5] In the case of aqueous silver colloids, the height and the width of the band strongly depend on the kind of preparation. This is due to the adsorption of substances which change the electronics of the nanoparticles (see below) and to defects in the crystal lattice of the particles. Another property that has not yet been considered is the outer shape of the particles: In the two preparations described in the experiments of Figures 1

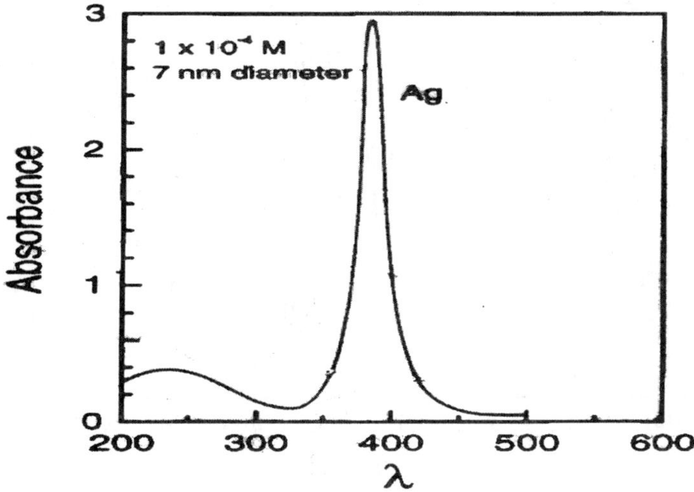


**Figure 2**. Absorption spectrum of silver particles produced by "push-pull" reduction, where each silver atom undergoes several reduction-reoxidation cycles before it is finally reduced. The initiation of reduction occurs by gamma irradiation. The solution contained $1.0 \cdot 10^{-4}$ M $AgClO_4$ and $1.0 \cdot 10^{-4}$ M sodium polyphosphate.[4]

particles of about equal size and of perfect lattices were obtained. Nevertheless, the widths of their optical absorption bands are quite different. One may thus suspect that the different widths are brought about by differences in the outer shape of the particles, the spherical particles in Figure 2 exhibiting a narrower absorption band than the facetted ones in Figure 1.

## 4. Pulsed Particle Formation

Particle growth can be initiated by applying a single pulse of high energy radiation to a $10^{-4}$ M $AgClO_4$ solution. Some $10^{-6}$ M free silver atoms are formed in the reaction of the hydrated electrons with $Ag^+$ ions, i.e. only a small fraction of the silver ions is reduced. Particle growth after the pulse does not occur via further $Ag^+$ reduction as in ordinary colloid formation, but via consecutive condensations of the atoms produced in the pulse. In the absence of a stabilizer, these condensations first lead to oligomeric silver species, which finally build-up larger and larger particles by agglomeration. The condensation and agglomeration reactions occur in the millisecond till second or even minute time range. By measuring the absorption spectrum at various times after the pulse, one obtains information about the nature of the intermediate species. **Figure 3** shows the absorption spectrum of the solution at different times. The nature of the species, to which the absorption bands are attributed, is also indicated. Note that the species carry a positive charge, which makes the condensation reactions relatively slow because of electrostatic repulsion between the particles. When the experiment is carried out with a solution of

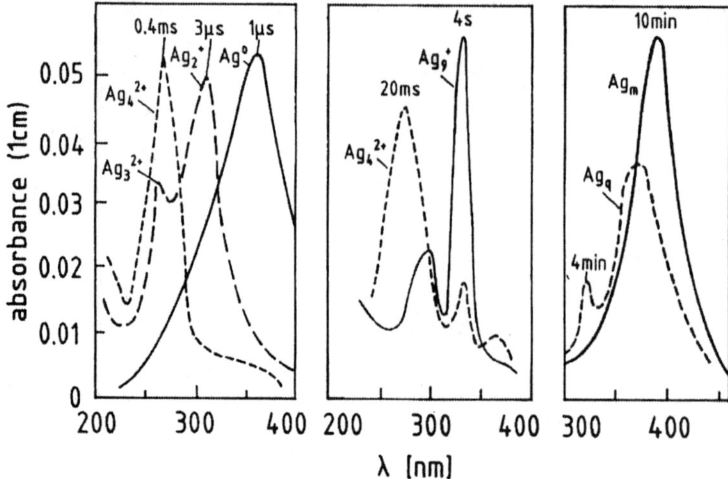


**Figure 3**. The atom-to-metal transition in silver: absorption spectrum at different times of a $10^{-4}$ M AgClO$_4$ solution, in which $10^{-6}$ M silver atoms are produced at t=0.The absorption spectrum of the silver atom is present immediately after the pulse. The growth of the particles occurs through mutual interaction of the intermediates until "quasi metallic" particles Ag$_q$ (q ~ 20) and real metallic particles Ag$_n$ (n > 1000) are present.[6,7]

Ag$_2$SO$_4$, all reactions are drastically accelerated. The sulfate anion complexes the silver intermediates, the result being a reduction of their mutual repulsion.

At first sight it might be surprising that the spectra of Figure 3 essentially consist of only one intense absorption band. If all the intermediate species would equally easily undergo reaction with each other, the solution should always contain clusters with a broad size distribution, and the absorption spectrum should therefore be smeared out. However, certain intermediate species are especially stable. These "magic clusters" accumulate in the solution and dominate the absorption spectrum. At 4 minutes in Figure 3, the absorption band is located at 360 nm, i.e. close to the position of the plasmon absorption of real metallic silver particles. The "quasi-metallic" particles, Ag$_q$, present here consist of approximately 20 atoms. As they aggregate further, the plasmon band shifts toward longer wavelengths and reaches its normal position at about 380 nm after 10 minutes.[6,7]

## 5. Redox Potential and Particle Size

Pulsed particle formation has been used to study redox reactions on the surface of the growing particles as a function of particle size. The standard   redox potential of metallic silver,

$$Ag_n \leftrightarrows Ag^+ + e^- + Ag_{(n-1)} \qquad (5)$$

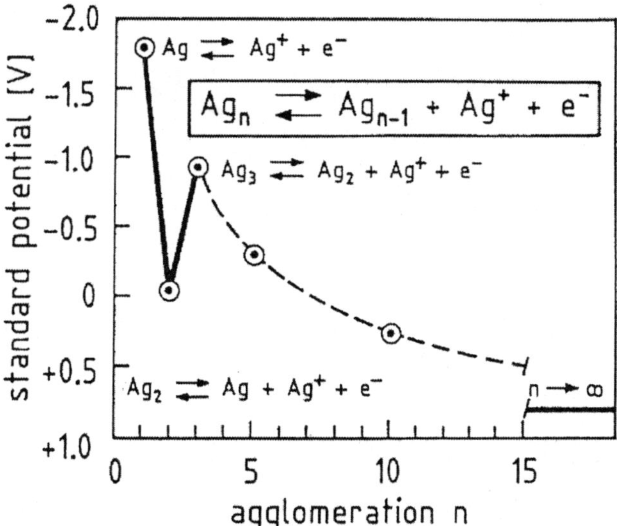


**Figure 4**. Standard redox potential of the silver nano-electrodeas a function of the atom number n.[8] Full line: calculated values. Dashed line: from growing nano-electrode experiments.[9,10]

is equal to +0.799 V for large values of n, i.e. for the compact silver electrode, where $\Delta G_f(Ag_n)$ is practically equal to $\Delta G_f(Ag_{(n-1)})$. However, this is not the case for small values of n. For example, the standard potential of the free silver atom, amounts to -1.8 V; it is therefore a strong reducing agent. **Figure 4** shows the dependence of the redox potential as a function of the atom number n; up to n=3, the potentials have been calculated making use of the free enthalpies of formation of the silver species involved that had been derived from vapor equilibrium measurements.[8]

Apart from the initial oscillations, the redox potential becomes more and more positive with increasing particle size. Regard the "growing-nano-electrode" experiment of **Figure 5** where a solution of $Ag_2SO_4$ was pulsed in the absence and presence of $Cu^{2+}$ ions: the 380 nm absorption was recorded as a function of time.[9] As mentioned above, the growth processes are accelerated in the presence of sulfate; the absorption of larger particles in the 360-380 nm range appears therefore much earlier than in the experiment of Figure 3. In the presence of copper ions, a second increase in the absorption occurs at 9 ms. The silver and copper ions in the solution compete for the hydrated electrons that are formed in the pulse of radiation. Part of the electrons react to produce silver atoms that subsequently condense as in the absence of copper ions. Another part of the electrons produce $Cu^+$ ions. The $Cu^+$ ions are non-reactive toward the growing silver particles as long as the latter have a redox potential more negative than +0.32 V, i. e. the potential of the redox system $Cu^+ \leftrightarrows Cu^{2+}+e^-$. As soon as the particles have reached a critical size, $Ag_c$, where their potential exceeds that of the $Cu^+/Cu^{2+}$ system, the process

$$Cu^+ + Ag^+ + Ag_c \rightarrow Cu^{2+} + Ag_{(c+1)} \qquad (6)$$

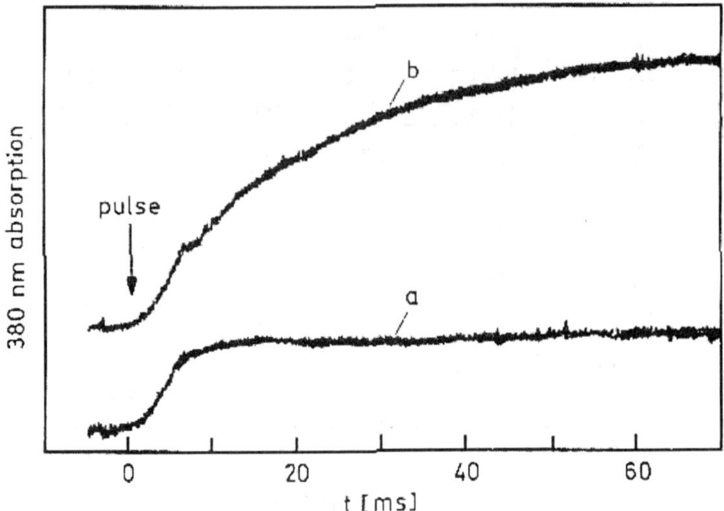


**Figure 5.** Growing-nano-electrode experiment: 380 nm absorption vs. time in (a) copper-free and (b) copper II-containing solutions of $Ag_2SO_4$ (arrow: pulse applied at t=0).The compositions of the two solutions were matched to assure equal initial concentrations of $5\times10^{-7}$ M silver atoms formed during the pulse. The first step in the built-up of absorption is the same for both solutions. In the copper-containing solution, $Cu^+$ ions were produced at a two times higher concentration than $Ag^0$ atoms.[9] They start to react according to eq.6 at t=9 ms.

takes place. The additionally reduced silver causes the second absorption increase in **Figure 5**. The enlarged particle $Ag_{(c+1)}$ undergoes further enlargement via reduction of an $Ag^+$ ion by $Cu^+$ until all $Cu^+$ ions are consumed. Thus, a situation is reached where particle enlargement occurs via particle-particle interaction below the critical size, but mainly via reduction of excess $Ag^+$ ions beyond the critical size.

More detailed investigations have been reported, in which the second solute has a strong optical absorption in its reduced form. A typical example is sulfonatopropylviologen, SVP. The pulse of radiation produces known amounts of the radical anion $SVP^{-\cdot}$ and of Ag atoms. The optical absorptions of both the growing silver particles and the $SVP^{-\cdot}$ radical are recorded as functions of time. By using reagents of different redox potential, the potentials of silver clusters of various sizes could thus be determined.[10,11] It should also be noted that these studies help to understand the mechanism of the photographic development process:[12] silver clusters of different sizes exist in the latent image produced in the illumination of the photographic plate, but only those clusters are developed whose nuclearity is equal to or exceeds the critical one.

### 6. Polymer Stabilized Clusters

Polymers at low concentration are often used to stabilize the final colloidal nano-particles which result from the reduction of metal ions in solution. In some cases, however, small intermediate clusters are stabilized. For example, whereas metallic nano-particles are formed in the reduction of $Ag^+$ ions in the presence of sodium polyphosphate at low concentration, (Figure 2), the non-metallic cluster $Ag_9^+$ is stabilized at higher concentrations of this polymer.[13] The cluster can be enriched up to more than 90% reduction of the $Ag^+$ ions. It lives several days upon aging of the reduced solution; as its absorptions disappear, the 380 nm absorption of nano-sized particles appears. Clusters are also trapped using sodium polyacrylate as protecting material.[14,15]

At higher concentrations of polyacrylae, 0.01 to 0.1 M, even earlier precursors of colloidal silver are stabilized. These clusters absorb in the visible: at the beginning of reduction, the solution acquires a rose color, which then turns into green and finally blue. All these clusters are stable toward oxygen. Upon further reduction, the absorptions of the colored clusters disappear and the absorptions of $Ag_9^+$ and finally of metallic silver appear. The colored precursors consist of silver atoms and silver ions, which are stabilized by the carboxyl groups of the polymer chains, the absorption shifting to longer wavelengths with increasing size.[16,17] It has been proposed that the positive charge is delocalized in these clusters.[18] The clusters would thus be a cationic counterpart to the clusters that cause the blue color in the iodine-starch interaction; in this case, the negative charge is delocalized in chains of I-atoms and $I^-$-anions stabilized in the helices of amylose.[19]

The cluster $Pd_2^{2+}$ is formed in the reduction of $Pd(NH_3)Cl_2$ in the presence of polyethyleneimine.[20] About 60% of the Pd-II can be reduced to this form of monovalent Pd until further reduction to colloidal palladium occcurs. The stabilized cluster lives for days. It can be titrated with $H_2O_2$ and with methyl viologen, and it reacts with hydrogen sulfide to yield hydrogen:

$$Pd_2^{2+} + 2\,H_2S \rightarrow 2\,PdS + H_2 + 2\,H^+ \tag{7}$$

The absorptions of stabilized oligomeric clusters of copper[21] and of platinum[22] have also been detected in the reduction of Cu-II and Pt-II salts, although no definite structures have yet been assigned to them.

### 7. Electron Donation and Positive Hole Injection

The wavelength at which the plasmon absorption band appears is proportional to the reciprocal square root of the free electron density in a nano-particle.[23,24] The most direct method to change the electron density consists of letting the particles react with reducing or oxidizing free radicals which donate an electron or inject a positive hole, respectively. In the experiment of **Figure 6**, free radicals were produced by a pulse of radiation. The radicals reacted with the silver particles within milliseconds, the rate being diffusion controlled. A particle can react with many radicals, i.e. a large number of electrons or holes can be stored on one particle. The figure shows the accompanying optical changes at

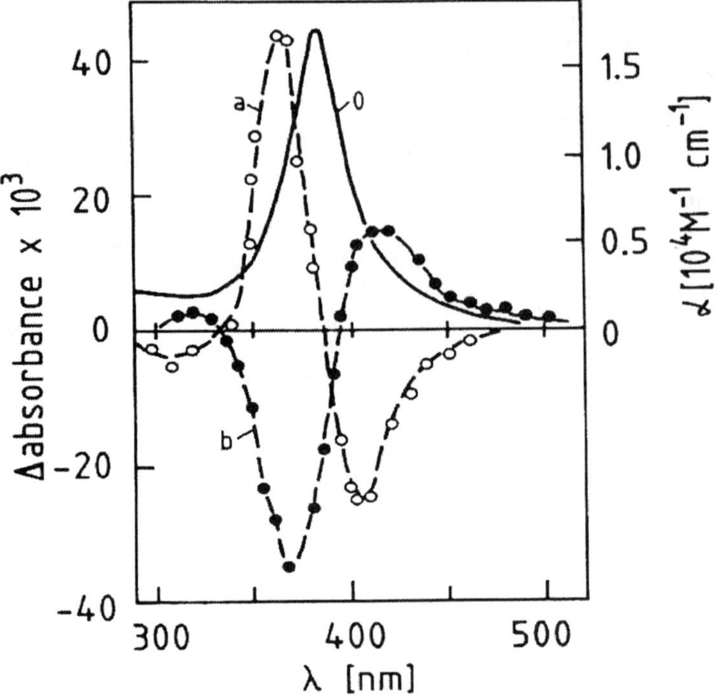

**Figure 6**. (0) Surface plasmon absorption band of a 1.0 x $10^{-4}$ M silver sol (right ordinate scale). Changes in absorption(a) after electron donation and (b) positive hole injection by free radicals[25] (left ordinate scale).

different wavelengths. Curve $\underline{a}$ describes the changes upon electron donation by an organic radical,

$$x\,(CH_3)_2C^{\cdot}(OH) + Ag_n \rightarrow Ag_n^{x-} + x\,(CH_3)_2CO + x\,H^+ \tag{8}$$

and curve $\underline{b}$ upon hole injection by the hydroxyl radical:

$$x\,OH^{\cdot} + Ag_n \rightarrow Ag_n^{x+} + x\,OH^- \tag{9}$$

As expected from theory, negative signals are observed in reaction 8 at wavelengths on the long-wavelength flank of the plasmon band and positive ones on the short wavelength side, corresponding to a blue shift of the plasmon band. In reaction 9, the signals have the opposite signs, which indicates that the band is red shifted upon positive hole injection. The stored electrons can be used to initiate reactions as will be discussed below.

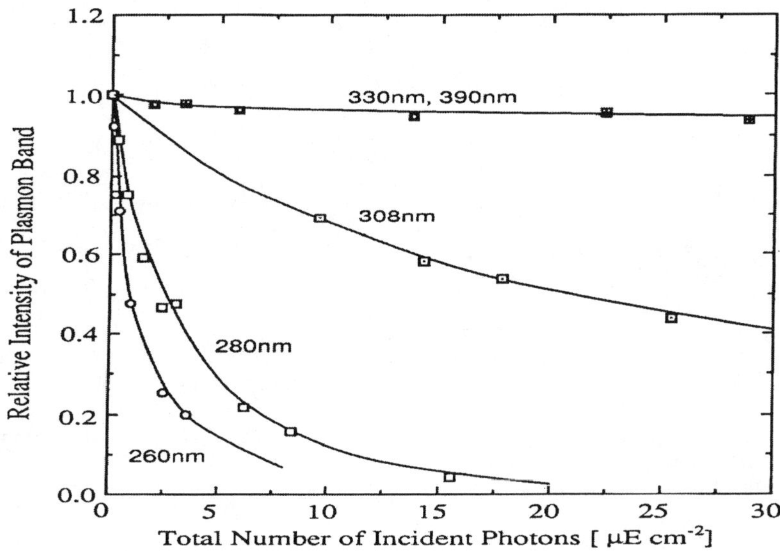


**Figure 7**. UV illumination at different wavelengths of a $1.0 \cdot 10^{-4}$ M silver sol containing $N_2O$: dissolution of silver as the plasmon absorption band decreases in intensity.[27]

## 8. Photoelectron Emission

In the early days of colloid chemistry, it has occasionally been observed that metal solutions degrade when exposed to light. The ongoing photochemistry, however, has rarely been investigated. The question is, whether hot electrons that are produced in a metal by the absorption of photons are able to initiate chemical reactions before they recombine with the holes? Photo-electron emission from compact electrodes has been studied in photo-electrochemistry. The quantum yield of this process is of the order of $10^{-4}$ electrons per absorbed photon. It increases with the negative potential of the electrode.[26] However, electron emission into the solvent from nano-sized particles might be facilitated since 1) the rate of thermalization of hot electrons is decreased due to the lower density of states, and 2) only a short distance has to be traversed by a hot electron in order to reach the surface.

Electrons photo-emitted in an aqueous solution can be detected either optically in a time resolved experiment, as hydrated electrons strongly absorb at 700 nm, or chemically by detecting their reaction products. Typical examples of the latter procedure are shown in **Figures 7 and 8**. When a deaerated silver solution is illuminated with uv light, no changes take place. However, when the illumination is carried out under an atmosphere of nitrous oxide, the particles dissolve, and nitrogen is found as a reaction product. Figure 7 shows the intensity of the plasmon absorption band of the particles as a function of the number of incident photons at various wavelengths. With increasing photon energy, the

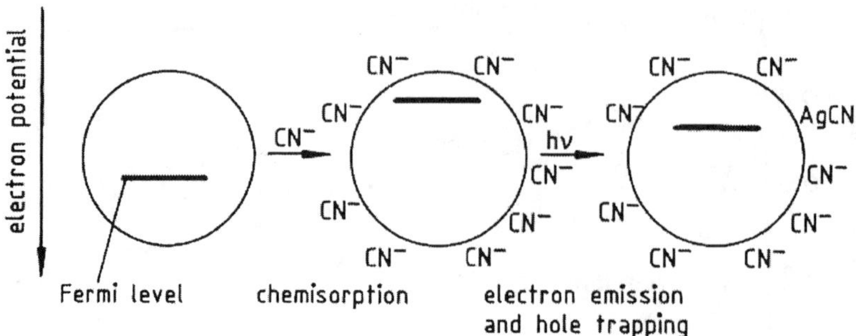



**Figure 8**. Schematic description of photo-electron emission and surface atom oxidation in silver particles carrying adsorbed CN⁻ ions. The shift of the Fermi level in the particles is also indicated. The adsorbed CN⁻ ion acts as a trap of the positive hole.

dissolution of silver becomes faster, the quantum yield at the shortest wavelength used amounting to 0.037 silver atoms oxidized per absorbed photon. The proposed mechanism consists of the following elementary processes:[27]

$$\text{photo-electron emission: } Ag_n + h\nu \rightarrow Ag_n^+ + e_{aq}^- \tag{10}$$

$$\text{back reaction: } Ag_n^+ + e_{aq}^- \rightarrow Ag_n \tag{11}$$

$$\text{electron scavenging: } e_{aq}^- + N_2O + H_2O \rightarrow N_2 + OH^- + OH\cdot \tag{12}$$

$$\text{oxidation of silver: } Ag_n + OH\cdot \rightarrow Ag_n^+ + OH^- \tag{13}$$

$$\text{disssolution: } Ag_n^+ \rightarrow Ag_{(n-1)} + Ag^+ \tag{14}$$

When the solution contains propanol-2 in addition to $N_2O$, nitrogen still is formed, but the silver particles do not dissolve. Under these conditions, the OH radicals from reaction 12 react with the alcohol, forming reducing radicals, see reaction 1, which in turn reduce the $Ag_n^+$ intermediate. The net result is a silver catalyzed photoreaction between nitrous oxide and propanol-2:

$$N_2O + (CH_3)_2CH(OH) + h\nu \rightarrow N_2 + H_2O + (CH_3)_2CO \tag{15}$$

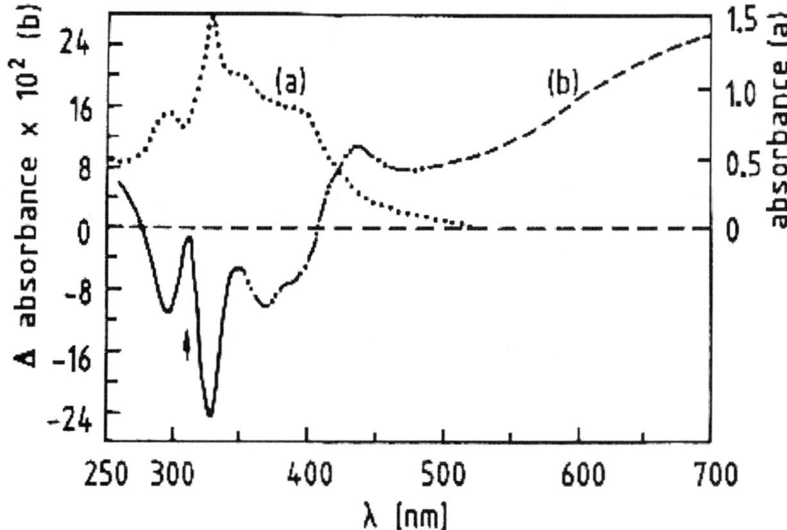


**Figure 9**. Flash photolysis of a solution of oligomeric silver clusters. (a) absorption spectrum of the solution (right ordinate scale). (b) difference spectrum immediately after the laser flash (left ordinate scale). Arrow: wavelength of illumination.[28]

Silver particles also dissolve in deaerated solution when illuminated in the presence of cyanide, the efficiency being even greater than in the presence of $N_2O$. In the presence of $CN^-$, hydrogen is produced, the overall reaction being

$$2h\nu \quad 2Ag + 4\,CN^- + 2H_2O \rightarrow 2Ag(CN)_2^- + H_2 + 2OH^- \tag{16}$$

This effect is explained by a shift of the Fermi level in the silver particles to a higher energy (or a more negative potential) since adsorbed $CN^-$ anions donate electron density into the particles as schematically shown by Figure 8.[27] Electrons are more efficiently emitted, and, in addition, the back reaction 11 is retarded by the negative charge on the particles.

Electron emission from illuminated oligomeric silver clusters has also been observed.[28] Quantum yields larger than 0.1 were found, which is understood in terms of the rather negative redox potentials of these clusters. **Figure 9** describes a typical laser flash experiment: (a) is the spectrum of the solution; it contains several absorption bands of stabilized clusters, and (b) is the difference spectrum immediately after a 308 nm laser flash. The difference spectrum contains the broad absorption at longer wavelengths of the hydrated electron. In addition, it contains negative absorptions where the bands of the clusters were positioned. The latter effect indicates that the clusters (which are positively charged) break down upon electron ejection, possibly because of the accumulation of positive charge upon electron emission.

## 9. Nano-Electrochemistry

When reducing free radicals are continuously produced in the solution of a colloidal metal, the particles become more and more negatively charged. They constitute a "nano-electrode", the stored electrons being able to initiate redox reactions. In the absence of another electron acceptor in the solution, the aqueous solvent itself is reduced to yield hydrogen:[29]

$$2\ (CH_3)_2C^{\cdot}(OH) + Ag_n^{x-} \rightarrow Ag_n^{(x+2)-} + 2\ (CH_3)_2CO + 2\ H^+ \tag{17}$$

$$Ag_n^{(x+2)-} + 2\ H_2O \rightarrow Ag_n^{x-} + H_2 + 2\ OH^- \tag{18}$$

$$2\ H^+ + 2\ OH^- \rightarrow 2\ H_2O \tag{19}$$

The net result of these processes is the reaction:

$$2\ (CH_3)_2C^{\cdot}(OH) \rightarrow 2\ (CH_3)_2CO + H_2 \tag{20}$$

which is highly exoergic. However, it does not occur when the radicals are produced in homogeneous solution, as their sterically more favored reactions, such as combination and disproportionation, prevail. The silver particles merely act as a catalyst of the unusual radical reaction 20. This catalysis is not limited to silver.[30] It has also been observed for many other colloidal metals, such as platinum, iridium,[31] gold,[32,33] cadmium[34], and bismuth.[35] Charging of the silver pool with subsequent $H_2$ formation has also been observed for the $CO_2^-$ radical and many other organic radicals, as well as for inorganic radicals such as $Ni^+$.[36] On the other hand, radicals of less oxidizing properties are reduced on the charged silver pool, such as the $\beta$-hydroxyalkyl radical which is formed in the presence of t-butanol in the irradiated solution:

$$HOC(CH_3)_2(^{\cdot}CH_2) + Ag_n^{x-} + H_2O \rightarrow HOC(CH_3)_3 + Ag_n^{(x-1)-} + OH^- \tag{21}$$

The charging reaction 17 is accompanied by an increase in conductivity of the solution due to the formation of hydrogen ions. In fact, the number of stored electrons can be calculated from this increase. The discharge reaction 18 (followed by reaction 19), on the other hand, is accompanied by a decrease in conductivity. In **Figure 10**, the conductivity of a colloidal silver solution, in which organic radicals are produced by continuous irradiation, is shown as a function of time. The conductivity is expressed here as the number of Coulombs of electrons stored per Liter of solution. At the beginning, there is fast storing. As the discharge reaction becomes faster with increasing negative potential of the nano-electrode, the rate of charging decreases, and finally a plateau is reached where the rates of charging and discharging are equal. At 2.5 minutes, the irradiation was stopped. The stored charges then flow out at a decreasing rate. The shape of these curves is quantitatively understood in terms of conventional electrode kinetics.[37]

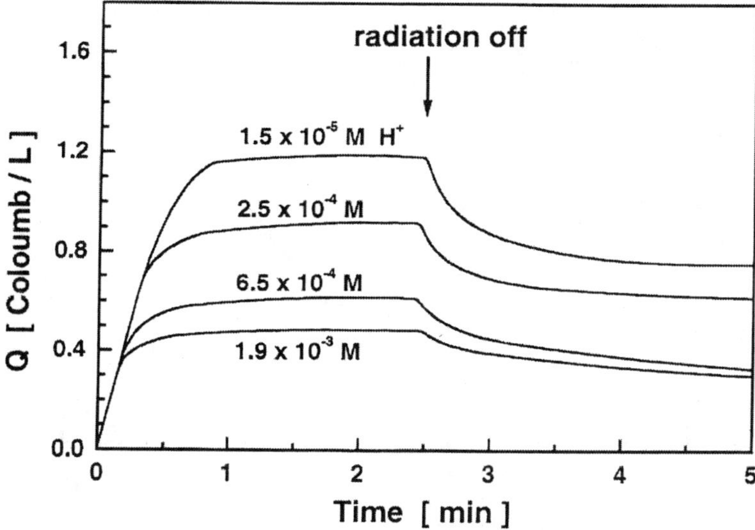



**Figure 10**. Stored charge in an irradiated solution of colloidal silver of different pH as a function of time. At 2.5 minutes, the irradiation was stopped. The plateau reached at this time decreases with increasing $H^+$ concentration as the discharge reaction becomes faster.[37]

**Figure 11** describes an example for competing processes at the nano-electrode. The silver particles are charged by $(CH_3)_2C^\cdot(OH)$ radicals in the presence of different amounts of methylene dichloride. The latter is not directly attacked by the radicals. However, chloride and methyl chloride are found in the presence of silver particles. The hydrogen yield is decreased with increasing $CH_2Cl_2$ concentration, for each $Cl^-$ formed, one molecule less of $H_2$ being formed. As two electrons are required to bring about the formation of one $H_2$ molecule, it is concluded that two electrons are also consumed in the reduction of $CH_2Cl_2$:[38]

$$Ag_n^{x-} + CH_2Cl_2 + H_2O \rightarrow Ag_n^{(x-2)-} + Cl^- + CH_3Cl + OH^- \tag{22}$$

Another example of two-electron transfer is the reduction of $N_2O$ which leads to the formation of $N_2$:[36]

$$Ag_n^{x-} + N_2O + H_2O \rightarrow Ag_n^{(x-2)-} + N_2 + 2\,OH^- \tag{23}$$

## 10. Bimetallic Particles

The ions of a second metal can be reduced on the charged nano-cathode to form a shell around the primary particle. Numerous bimetallic combinations have been

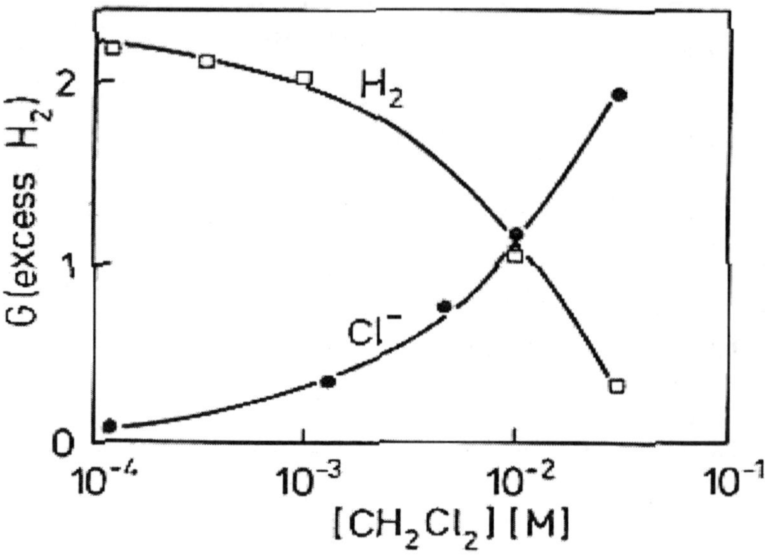

Figure 11. Competitive reduction of water and methylene dichloride on the silver electron pool: yields of hydrogen and of chloride as functions of the concentration of methylene dichloride.[38]

synthesized this way. In this chapter, only a few examples are described to give the reader a general overview of this application of nano-electrochemistry.

**Figure 12** describes the optical changes of a solution of silver particles upon the deposition of an increasing amount of cadmium.[39] Pure cadmium colloid has a strong absorption band at 260 nm.[40] This band does not appear in the first stages of cadmium deposition. Instead, the silver plasmon absorption shifts to the blue. **Figure 13** shows the peak wavelength as a function of the time of cadmium deposition. After 75 minutes, the plasmon band is located at 260 nm, i. e. at the wavelength where pure cadmium absorbs. Figure 13 also shows the amount of cadmium deposited at various times. This amount was determined chemically by adding methylviologen, $MV^{2+}$, to the irradiated solution. The deposited cadmium is oxidized,

$$Cd_{shell} + 2MV^{2+} \rightarrow Cd^{2+} + 2\,MV^+ \tag{24}$$

and its amount could be calculated from the observed 600 nm absorption of the stable radical cation $MV^+$ formed. The m-values in Figure 13 give the number of monolayers of deposited cadmium. Note that the cadmium first deposited in 0.4 monolayers does not react with $MV^{2+}$, i.e. it is more "noble" than the cadmium in the outer layers. This effect is known as underpotential deposition in conventional electrochemistry.[41] The cadmium atoms in the first layer donate electron density into the silver particles, thus creating a

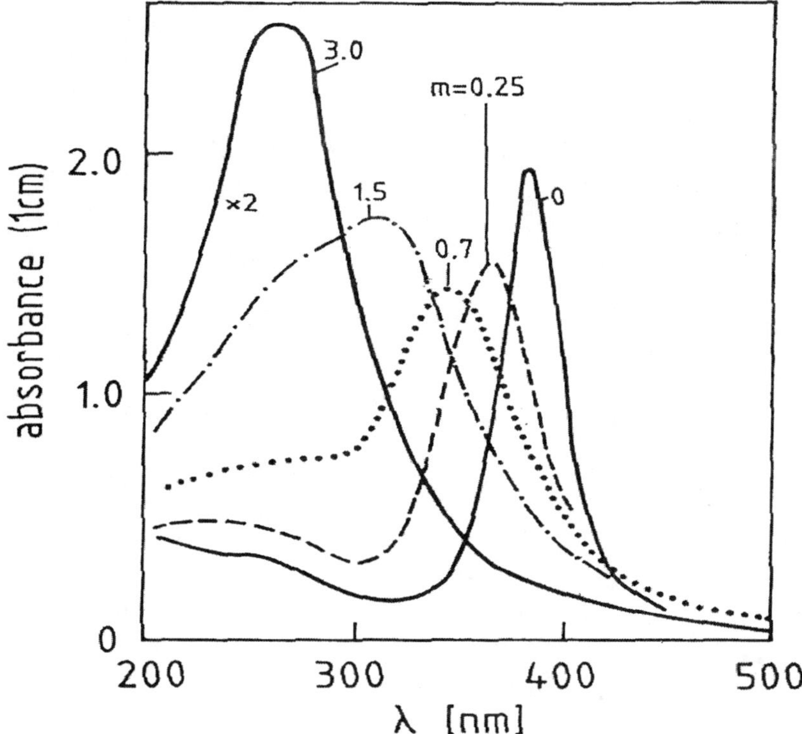


**Figure 12**. Absorption spectrum of a $1.0 \cdot 10^{-4}$ M silver sol containing $1.9 \cdot 10^{-4}$ M cadmium perchlorate after deposition of different amounts of cadmium. The m-values indicate the number of atomic Cd layers deposited.[39]

double layer (or polar Ag-Cd bonds) at the boundary between the two metals. It should also be noted that calculations using extended Mie theory and using the dielectric constants of the bulk materials gree with the experimental spectra in Figure 12.

A thorough investigation of the optical spectra and structure of gold particles carrying a concentric lead shell has also been reported. Trimetallic $Au_{core}Pb_{shell}Cd_{shell}$ particles were synthesized by reducing $Cd^{2+}$ ions on the surface of $Au_{core}Pb_{shell}$ particles.[42]

The deposition of indium on silver occurs in a more complex manner.[43] When the irradiated silver sol contains $In_2(ClO_4)_3$, an underpotential deposition of a thin indium layers occurs at the beginning. Upon further reduction, the deposited indium undergoes partially the surface dismutation:

$$In_{shell} + In^{3+}{}_{aq} \rightarrow 2\,In^{+}{}_{aq} \tag{25}$$

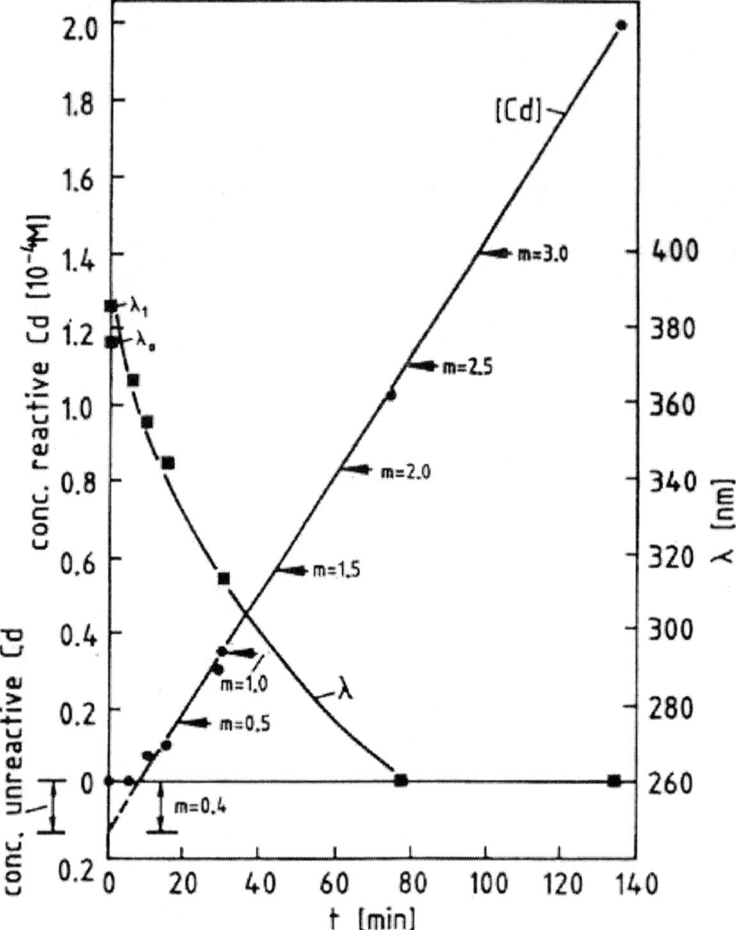



**Figure 13**. Amount of reduced cadmium and wavelength of the plasmon band as functions of time of deposition. m-values: number of atomic layers of deposited cadmium.[39]

This leads for a while to an accumulation of $In^+$ in the solution, which, however, is finally reduced on the pool as the $In^{3+}$ concentration becomes exhausted. Whereas the two metallic phases in the above $Ag_{core}Cd_{shell}$ particles are distinctly separated, it may happen in other cases that alloying takes place at the interface. An extreme case is depicted in **Figure 14**. Mercury ions, $Hb^{2+}$, were reduced on silver particles and the absorption spectrum recorded for different amounts of deposited Hg. Pure mercury colloid has a plasmon absorption at 260 nm. It can be seen that the plasmon band of silver moves to

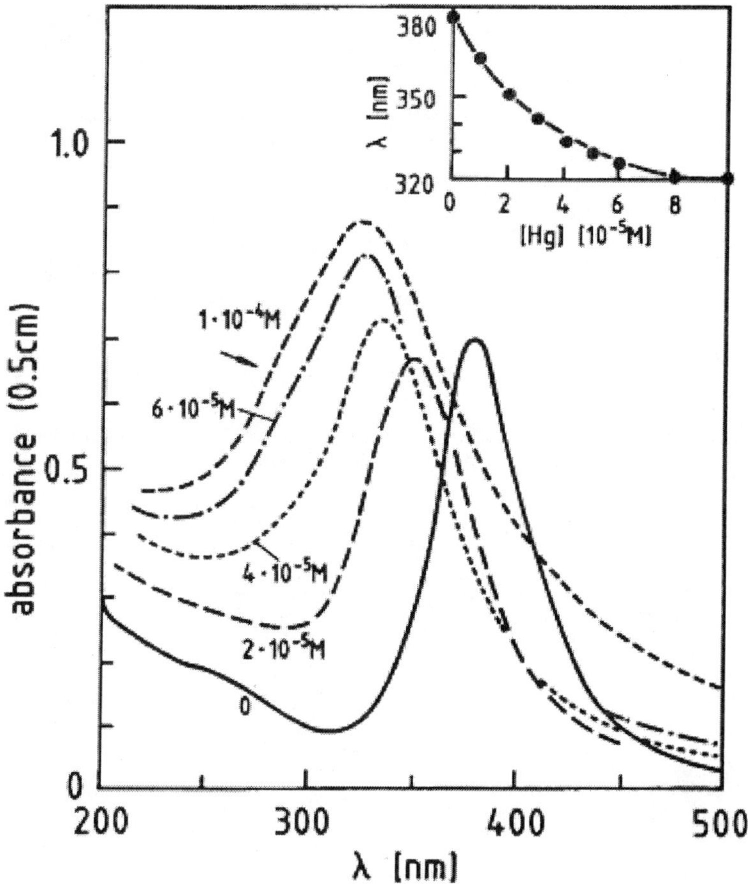


**Figure 14.** Absorption spectra of a $1.0 \cdot 10^{-4}$ M silver sol before and after reduction of $Hg^{2+}$ ions at various concentrations.[44] Inset: wavelength of the plasmon absorption band as a function of the amount of deposited Hg.

shorter wavelengths with increasing Hg:Ag ratio of the particles. The particles were shown to consist throughout of an Ag-Hg amalgam.[44,45] Alloying has also been observed in the deposition of tin on gold particles.[46]

Alloys have also been made by first synthesizing bimetallic core-shell particles and then heating them up by intense laser illuminetion. **Figures 15** and **16** show the absorption spectra of $Au_{core}Ag_{shell}$ and $Ag_{core}Au_{shell}$ particles of different molar composition.[47] The typical features of the pure metals, i.e. the gold plasmon band at 520 nm and the silver band at 380 nm, show up in the bimetallic particles, the two contributions depending on the composition. Thus, it is concluded that two types of

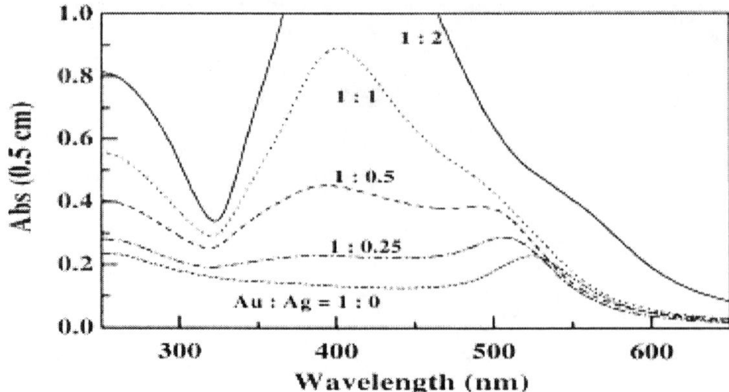

**Figure 15**. Absorption spectra of $Au_{core}Ag_{shell}$ particles of different molar compositions. The gold concentration was $1.7 \cdot 10^{-4}$ M, and silver was deposited in different amounts.[47]

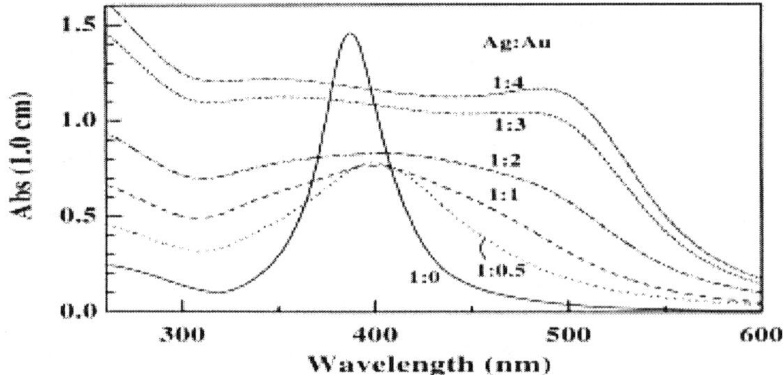

**Figure 16**. Absorption spectra of $Ag_{core}Au_{shell}$ particles of different molar compositions. The silver concentration was $8.5 \cdot 10^{-4}$ M, and gold was deposited in different amounts.[47]

collective electron oscillations occur.[47,48] On the other hand, alloy particles, which are made by simultaneous reduction of silver and gold compounds in solution, possess only one plasmon band.[49,50,51] **Figure 17** shows absorption spectra of $Au_{core}Ag_{shell}$ particles before and after illumination with 30 ps laser pulses. As can be seen, the double plasmon band characteristic of core-shell particles disappears and a single band appears. The latter is positioned where an Au-Ag alloy particle of the same molar composition absorbs. It is interesting to note that strong inter-diffusion of the two metals occurs at temperatures caused by the laser far below the melting point.[47]

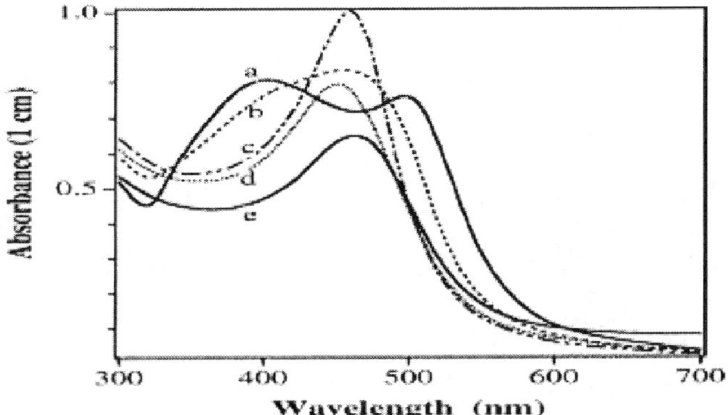



**Figure 17**. Spectra of $Au_{core}Ag_{shell}$ particles (Au:Ag=1:0.5) following photoexcitation with 532 nm, 30 ps laser pulses. (a) non irradiated; (b → e) increasing dose. At the highest doses (d and e), the absorption of the alloy particles decreases as the particles fragment.[47]

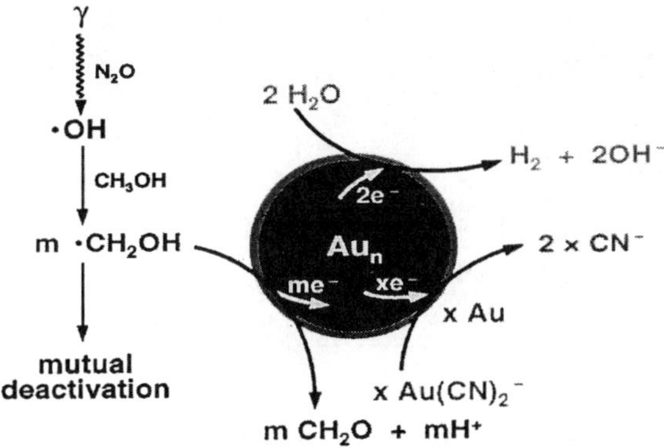



**Figure 18**. Schematic presentation of the radiolytic enlargement of gold particles.[52]

The electron pool method can also be used to grow existing nano-particles into larger ones. **Figure 18** shows the reaction scheme for the enlargement of gold particles.[52] The starting solution contains Au particles made conventionally by reducing $NaAuCl_4$ with citrate, as well as methanol, $N_2O$, and $NaAu(CN)_2$. The hydroxymethyl radicals produced by γ-radiation cannot reduce the Au-I complex in solution, as this reaction would be highly endoergic due to the large free energy of formation of the free Au atom. Thus the radicals are left to transfer electrons to the colloid particles. The stored electrons are able

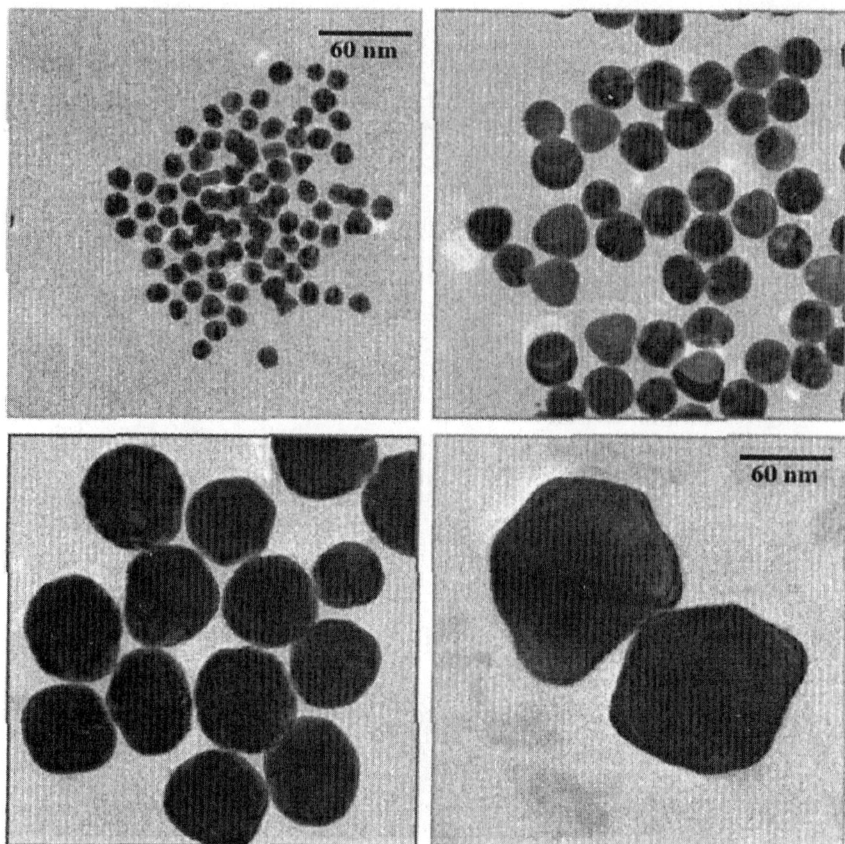

**Figure 19.** Electron micrographs showing the size-doubling of gold particles.[52]

to reduce $Au(CN)_2^-$ drectly onto the surface of the gold particles. Irradiation is carried out until all the Au-I complex is reduced. **Figure 19** shows electron micrographs of Au particles which have been made by successive steps of enlargement by a factor of two, always using the particles of the preceding step as nuclei.

## 11. Fermi Level Equilibration in Mixed Colloids

When two dissimilar metals are connected, there is momentary flow of electrons from the metal with the smaller work function to the other. The result is the build-up of a contact potential at the interface so that the Fermi level is the same in both metals. When

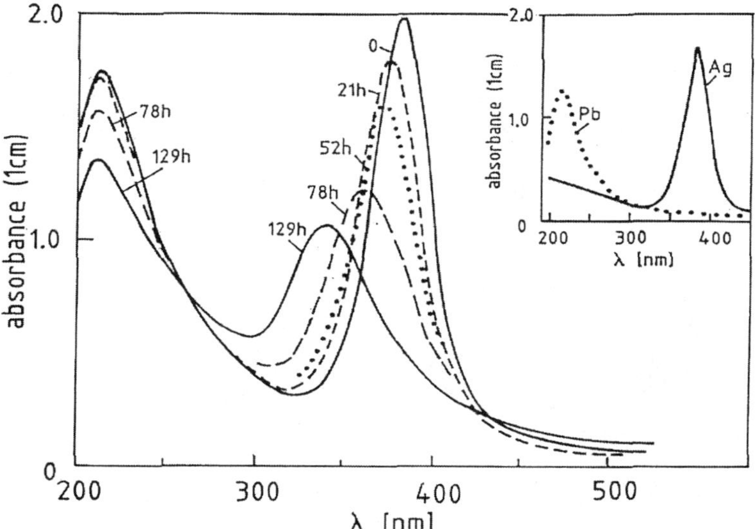


**Figure 20**. Absorption spectrum of a solution at various times after mixing a silver sol with a lead sol. Inset: spectra of the separate sols.[53]

the two metals are present as colloids in a common solution, there exists an electronic instability in the system, and the question arises how the system equilibrates.

The only case that has yet been studied seems to be the silver-lead system.[53] **Figure 20** shows the spectral changes that occur with time in a solution containing both silver and lead particles. Immediately after mixing the individual solutions, the absorption bands of silver at 380 nm and of lead at 215 nm are present. Within hours and days, the lead band slightly decreases in intensity, whereas the silver band shifts gradually to shorter wavelengths. These changes occur until the silver band is located at 337 nm. It turned out that the shift of the silver band is due to the formation of a lead shell, and that this shell consisted of two atomic Pb layers. It is thus concluded that in the mixed colloid experiment Fermi level equilibration is achieved via transfer of Pb atoms from lead to silver particles. Two monolayers of lead are sufficient to practically shift the Fermi level of the silver particles to the position of the Fermi level in the lead particles.

The Fermi level in lead (which is a base metal) lies at an higher energy than in the silver particles. A mechanism has been proposed in which electron tunneling from lead to silver occurs in close encounters of the particles. $Pb^{2+}$ ions are then detached from the lead particles into solution. They are subsequently reduced on the surface of the negatively charged silver particles. The rate of Fermi level equilibration was found to be drastically increased by the presence of small concentrations of methyl viologen, $MV^{2+}$.

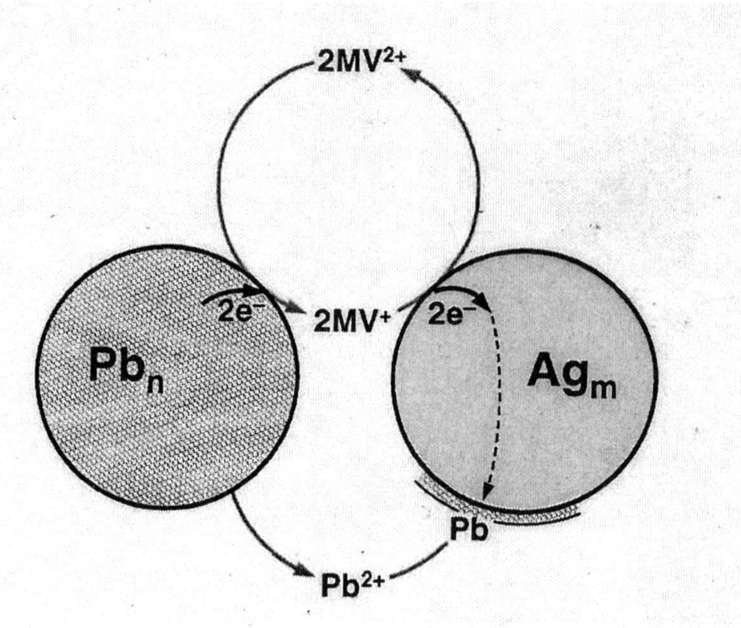


**Figure 21**. Relay mechanism of Fermi level equilibration between Ag and Pb particles in the presence of methylviologen.[53]

The methyl viologen was not consumed, it simply acted as a catalyst. **Figure 21** shows the mechanism of catalysis, in which $MV^+$ acts as an electron relay.

Electron tunneling between particles may in some cases be the reason for Ostwald ripening, i.e. the conversion of smaller particles into larger ones upon aging. A colloidal solution contains particles of different redox potential, the Fermi level in the smaller particles generally lying at an higher energy than in the larger ones. Thus, according to the above principle, the smaller particles dissolve and the larger ones become even larger.

## 12. Adsorption of Electrophiles

Adsorption of ions or neutral molecules on the surface of metal particles can lead to a change in optical absorption and chemical reactivity. The optical changes are most strongly produced in the wavelength range of the plasmon absorption band. Most remarkable effects are observed for silver particles whose plasmon absorption is especially intense, i.e. little damped. The adsorption of substances can lead to a destabilization of particles, i.e. to agglomeration. This effect generally is accompanied by a redshift of the plasmon absorption band due to dipole-dipole interaction between near-by

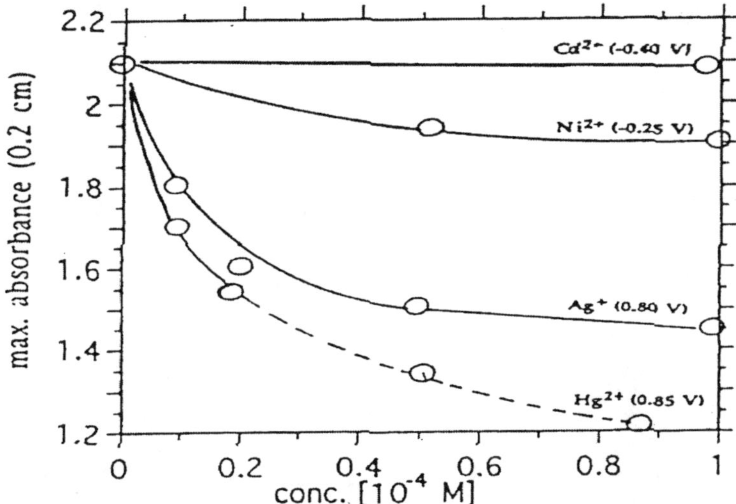


**Figure 22.** Absorbance in the maximum of the plasmon band of $4.0 \cdot 10^{-4}$ M silver sol as a function of the concentration of added metal perchlorates. The electrochemical potential of the metal is also indicated.[55]

particles.[54] It should again be emphasized that adsorption experiments should be carried out under the strict exclusion of oxygen, as the latter often affects the particle surface. Contradicting observations in the literature are often due to the fact that the experiments had not been performed under suffiently anaerobic conditions.

Adsorption of cations on silver particles leads to a decrease in intensity and a redshift of the plasmon absorption band.[55] This is explained by the withdrawal of electron density from the particles by the adsorbed cations. In the experiments of **Figure 22**, various metal salts were added to a $4 \cdot 10^{-4}$ M silver sol, and the absorbance in the maximum of the plasmon band is plotted versus the concentration of added cations. The decrease in absorption is the stronger the more positive is the electrochemical potential of the cation. It should also be emphasized that the charge on the cation is not important as the monovalent silver ion influences the absorption band more strongly than the divalent ions of cadmium and nickel. Thus, ionic strength effects of the added metal salts cannot be responsible for the observed changes; rather the electronic interaction of the cations with the silver particles causes the changes. The plasmon band in the experiments of Figure 22 shifts to the red as long as the curves are fully drawn. The dashed part of the curve for $Hg^{2+}$ at concentrations above $2 \times 10^{-5}$ M indicates a region where the plasmon band moves to the blue. This effect has been explained as the transition from chemisorption at low $Hg^{2+}$ concentrations,

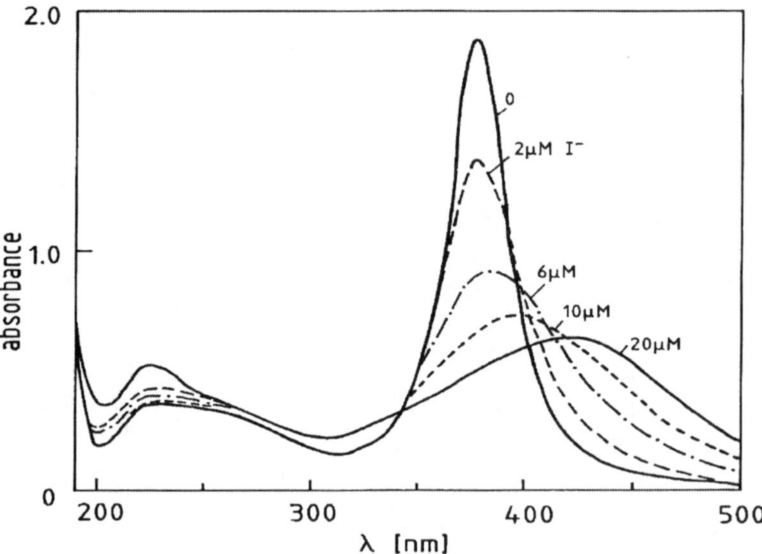



**Figure 23**. Absorption spectrum of a $1.0 \cdot 10^{-4}$ M silver sol before and after addition of various concentrations of NaI.[56]

$$x\,Hg^{2+} + Ag_n \rightarrow Ag_n(xHg^{2+})_{ads}, \tag{26}$$

to chemical reaction, i.e. amalgam formation, at higher concentrations.:

$$Hg^{2+} + Ag_n(xHg^{2+})_{ads} \rightarrow Ag_{(n-2)}Hg(xHg^{2+})_{ads} + 2\,Ag^+ \tag{27}$$

The blueshift of the absorption band upon amalgam formation has already been mentioned above (Figure 14).

### 13. Adsorption of Nucleophiles

The plasmon absorption band of silveris also changed upon the adsorption of anions. **Figure 23** shows the absorption spectrum of a $1 \times 10^{-4}$ M silver sol before and after the addition of various concentrations of NaI. The band is decreased in intensity and broadened. As long as the concentration of added iodide is below 20μM, the charge-to-solvent (CTTS) band at 225 nm of $I^-$ does not show up in the spectrum. It is concluded that all the added iodide is adsorbed. Above 20 μM, the silver band does no longer change, but the CTTS band appears and increases with increasing iodide concentration. Obviously, at 20μM, all available adsorption sites of the colloid are occupied by $I^-$.[56]

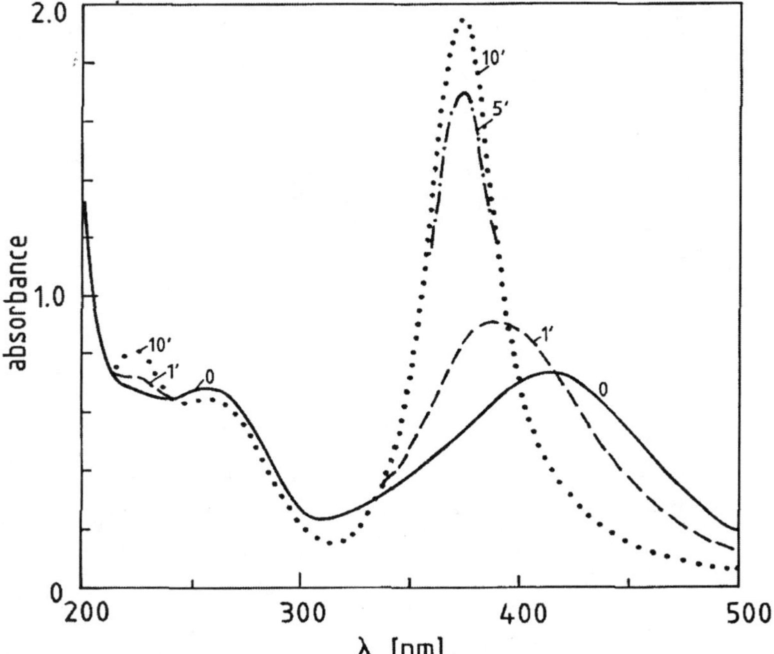



**Figure 24**. Absorption spectrum of a silver sol carrying adsorbed iodide before (0) and after various times of electron deposition via free radicals.[56]

The adsorption is understood in terms of electron donation from the anions to the silver particles, which means that the Fermi level in the particles is shifter to a higher energy, i.e. to a more negative potential. It may thus be stated that the establishment of an adsorption/desorption equilibrium on a metal particle is accompanied by a shift of the Fermi level in the particle. On the other hand, one may expect that a change in the position of the Fermi level in the particle would affect the adsorption/desorption equilibrium. A typical experiment of the latter type is shown in **Figure 24**. The starting material was a 1.0 x $10^{-4}$ M silver colloid carrying 20μM of adsorbed iodide. Free organic radicals were now produced by irradiating the solution. As the radicals transfer electrons to the particles, the Fermi level is shifted to a more and more negative potential. One can see that the narrow absorption band of silver growths in as the iodide anions are detached. At the same time the CTTS band of free iodide appears. At 10 minutes of electron deposition, all the iodide is desorbed. When the solution was now aged for several hours, the plasmon band became broader again. This is due to the loss of the deposited electrons as they slowly react with solvent molecules to produce hydrogen, and, as the Fermi level is anodically shifted, $I^-$ is readsorbed.

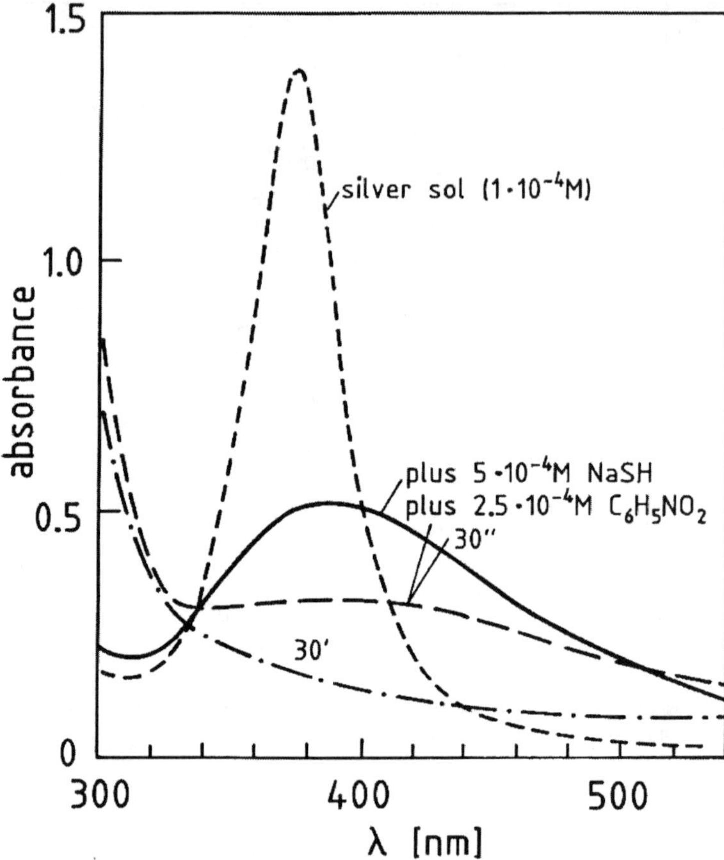



**Figure 25**. Spectrum of a silver sol before and after addition of NaSH, and 30 sec and 30 min after the subsequent addition of nitrobenzene.[57]

Metal particles carrying nucleophilic adsorbates on their surface are very reactive toward oxygen. $O_2$ picks up the negative charge from the particle, thus allowing further complexation of surface atoms by the adsorbed nucleophile until the whole particle is oxidized. For example, a silver sol containing iodide dissolves within seconds when exposed to air. Organic electron acceptors often are also able to oxidize metal particles carrying a nucleophile. This is demonstrated by **Figure 25**, where the absorption spectrum of a silver sol before and after addition of $SH^-$ ions is shown as well after the subsequent addition of nitrobenzene. One can see how the plasmon band is broadened by adsorbed $SH^-$ and finally disappears within seconds and minutes upon the addition of nitrobenzene. The final spectrum is that of colloidal $Ag_2S$.[57]

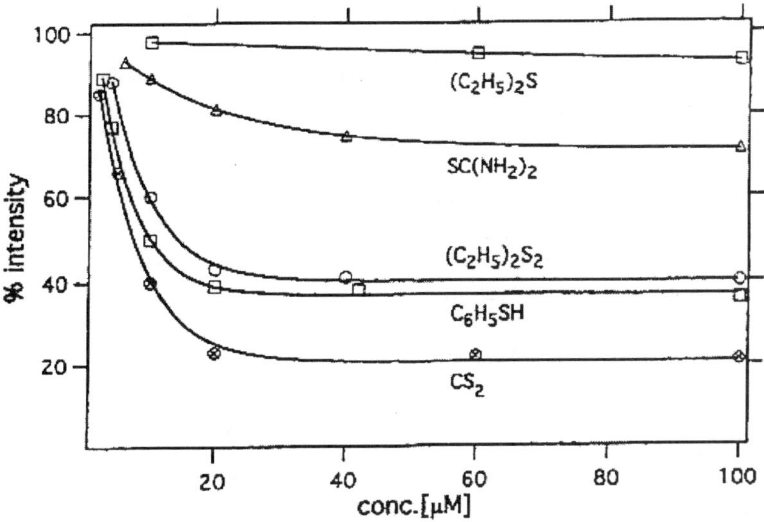


**Figure 26**. Intensity of the plasmon band of a $2.0 \cdot 10^{-4}$ M silver sol as a function of the concentration of various sulfur-containing compounds.[58]

**Figure 26** shows how the plasmon absorption band of silver is affected by various organic sulfur compounds.[58] Carbon disulfide is irreversibly adsorbed, as a S-C bond is broken. A chemical transformation has also been proposed in the adsorption of diethyl disulfide, where the S-S bond is broken. A heterolytic mechanism has been formulated for the adsorption of organic thiols, i.e. an electrolytic dissociation according to

$$R\text{-}SH_{(ads)} \rightarrow RS^-_{(ads)} + H^+, \tag{28}$$

the adsorption free energy of $RS^-$ being substantially greater than the energy of protolytic dissociation of R-SH.[58] Two mechanisms for the adsorption of thiosulfate on silver particles have been discussed, 1) an ion exchange,[59]

$$S_2O_3^{2-}{}_{(ads)} + OH^- \rightarrow S^{2-}{}_{(ads)} + HSO_4^-, \tag{29}$$

and 2) a redox process,[58]

$$Ag_n + S_2O_3^{2-}{}_{(ads)} \rightarrow Ag_n^{\delta+} + S^{\delta-}{}_{(ads)} + SO_3^{2-}. \tag{30}$$

The mechanism of the damping of the plasmon absorption band by adsorbed anions is not yet fully understood. A chemical bond with a silver surface atom is preformed with the adsorbed species, the atom acquiring a small positive charge $\delta+$, and a negative charge $\delta-$ is shifted into the interior of the particle. Thus, the electronic interaction between the adsorbed anion and the particle surface results in a change in the electron density of a thin layer. This change is believed to be the main reason for damping by shortening the life time of the collective oscillation in the particle, as the frequency of the oscillating electron gas in the surface layer is expected to be different from that in the interior of the particle.

The adsorption of thiolates on metal particles has often been used to modify the properties of nanomaterials. For example, colloidal gold particles that carry long aliphatic chain-thiols form a crystal lattice, where the lattice constant is determined by the length of the chain.[60] Highly oriented silver-thiol particles have also been reported.[61,62,63] The binding of gold particles to sulfur-containing sites on modified DNA[64,65] and in sulfur-group-containing silicon-organic glasses are other appllictions.[66]

Adsorbed phosphine suppresses the surface plasmon band of silver particles most strongly.[67] The adsorption is accompanied by an decrease in the pH of the solution. These effects are more pronounced in alkaline than acidic solution. In alkaline solution, the adsorption is irreversible. It has been proposed that $PH_2^-$ is the actively adsorbed species, as the equilibrium

$$PH_3(ads) + OH^- \rightleftharpoons PH_2^-(ads) + H_2O \qquad (31)$$

is shifted to the right side. Upon aging of the solution, the adsorbed phosphine undergoes chemical reaction within hours, as hydrogen gas and phosphate are developed. At the same time, the narrow plasmon band of the silver particles. recovers. The silver particles obviously catalyze the reaction:

$$PH_3 + 4 H_2O \rightarrow H_3PO_4 + 4 H_2 \qquad (32)$$

## 14. Competitive Adsorption and Displacement Processes

When two substances, such as $I^-$ and $SH^-$, are adsorbed, the stronger nucleophile displaces the weaker one. For example, when $I^-$ is first adsorbed and covers one-third of the surface of a silver particle, the addition of $SH^-$ leads to further broadening of the plasmon band and the appearance of the CTTS absorption band of free $I^-$. The displacement of $I^-$ from the surface by $SH^-$ before the surface is saturated indicates that chemisorption on a particle is a cooperative phenomenon. The displacement cannot be explained by missing adsorption sites for $SH^-$, but must be due to an electronic effect. In the case of $SH^-$, the stronger nucleophile, the Fermi level in the particle is pushed to such negative values that the adsorption of $I^-$, the weaker nucleophile, takes place.[56]

When a particle is stabilized by a weak nucleophilic polymer such as polyphosphate, the adsorption of a stronger nucleophile such as $I^-$ may lead to the detachment of the particle from the polymer chain. Only one experiment has yet been made to study this effect, using stopped flow technique.[68] When a silver sol is mixed with a NaI solution under the conditions of **Figure 27**, the plasmon band of silver decreases within 0.2 s. The light scattering of the solution was also recorded. As can be seen, it decays within the same time interval. This effect has been attributed to the detachment of the particles from the polymer chain. However, at very much longer times, i.e. in the 100 s range, the

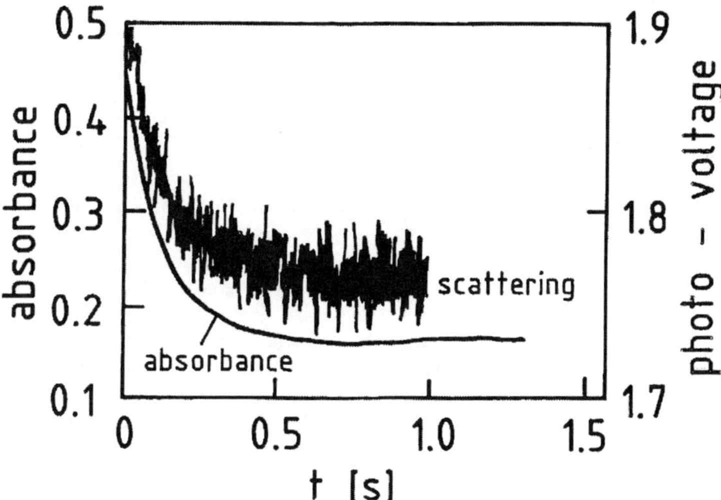


**Figure 27**. Stopped-flow experiment by mixing a $5 \cdot 10^{-5}$ M silver sol with a $1 \cdot 10^{-5}$ M NaI solution. Intensity of the plasmon absorption band at 380 nm and the scattering of the solution at 275 nm as functions of time.[68]

scattering intensity increases again, which is explained by the agglomeration of the detached silver particles.

## 15. Final Remarks

Various disciplines such as colloid chem.istry, electrochemistry, electron microscopy, and solid state physics come together in the investigation of metallic nano-particles in solution. In the studies described above, photo- and radiation chemical methods play an important role in both the preparation of nano-particles and the initiation of surface chemical processes. In fact, the field of free radical chemistry has been enriched by the studies on the interaction of radicals with the surface of finely dispersed metals. The examples mentioned here are to give the reader an impression of the many aspects which one encounters in these investigations. For more details and aspects, the reader's attention is called to recent reviews.[11,69,70]

## 16. References

1.    Buxton, G. V.; Greenstock, C. L.; Helman, W. P.; Ross, A. B. Phys. Chem. Ref. Data **1988**, 17, 513.
2.    Schwarz, H. A.; Dodson, R. W.J. Phys. Chem. **1989**, 93, 409

3.  Henglein, A., Giersig, M.J. Phys. Chem. B **1999**, 103, 9533
4.  Gutierrez. M.; Henglein, A.J. Phys. Chem. **1993**, 97, 11368
5.  Kreibig, U.; Vollmer, M."Optical Properties of Metal Clusters", Springer-Verlag, Berlin, **1995**
6.  Ershov, B. G.; Janata, E.; Henglein, A.J. Phys. Chem. **1993**, 97, 339
7.  Henglein, A.Ber. Bunsenges. Phys. Chem. **1997**, 101, 1562
8.  Henglein, A.Ber. Bunsenges. Phys. Chem. **1990**, 94, 600
9.  Henglein, A.; Tausch-Treml, R.J. Colloid Interface Sci. **1981**, 80, 84
10. Mostafavi, M.; Marignier, J. L.; Amblard, J.; Belloni, J.Radiat. Phys. Chem. **1989**, 34, 605
11. Belloni, J.; Mostafavi, M.in "Radiation Chemistry, Present Status and Future Trends", Jonah, C. D.; Madhava Rao, B. S. eds., Elsevier, Amsterdam, **2001**,411-452
12. Belloni, J.;Mostafavi, M.; Marignier, J.-L-; Amblard, J.J. Imaging Sci. **1991**, 35, 219
13. Henglein, A.Chem. Phys. Lett. **1989**, 154, 473
14. Mostafavi, M.; Keghouche, N.; Delcourt, M.-O.; Belloni, J.Chem. Phys. Lett. **1990**, 167, 193
15. Henglein, A.; Linnert, T.; Mulvaney, P.Ber. Bunsenges. Phys. Chem. **1990**, 94, 1449
16. Mostafavi, M.; Delcourt, M.-O.; Keghoche, N.; Picq, G. Radiat. Phys. Chem. **1992**, 40, 445
17. Remita, S.; Orts, J. M.; Feliu, J. M.; Mostafavi, M.; Delcourt, M.-O. Chem. Phys. Lett. **1994**, 218, 115
18. Ershov, B. G.; Henglein, A.J. Phys. Chem. **1998**, 102, 10663
19. Grätzel, M.; Henglein, A.; Scheffler, M.; Bössler, H. M.; Schulz, R. C.Ber. Bunsenges. Phys. Chem. **1972**, 76, 72
20. Michaelis, M.; Henglein, A.J. Phys. Chem. **1992**, 96, 4719
21. Khatouri, J.; Mostafavi, M.; Amblard, J.; Belloni, J.Chem. Phys.Lett. **1992**, 191, 351
22. Keita, B.; Nadjo, L.; de Cointet, C.; Amblard, J.; Belloni, J.Chem. Phys. Lett. **1996**, 249, 297
23. Doyle, W. T. Phys. Rev. **1958**, 111, 1067
24. Doremus, R. H.J. Chem. Phys. **1965**, 42, 414
25. Henglein, A.; Mulvaney, P.; Linnert, T. Faraday Discuss. Chem. Soc. **1991**, 92, 31.
26. Sass, J. K.; Sen, R. K.; Meyer, E.; Gerischer, H.Surf. Sci. **1974**, 44, 515
27. Linnert, T.; Mulvaney, P.; Henglein, A.Ber. Bunsenges. Phys. Chem.**1991**, 95, 838
28. Linnert, T.; Mulvaney, P.; Henglein, A.; Weller, H.J. Amer. Chem. Soc. **1990**, 112, 4657
29. Henglein, A.J. Phys. Chem. **1979**, 83, 2209
30. Grätzel, C. K.; Grätzel, M.J. Am. chem. Soc. **1979**, 101, 7741
31. Mills, G.; Henglein, A.Radiat. Phys. Chem.**1985**, 26, 385, 391
32. Meisel, D.J. Amer. Chem. Soc. **1979**, 101, 6133
33. Westerhausen, J.; Henglein, A.; Lilie, J.Ber. Bunsenges. Phys. Chem. **1981**, 85, 182
34. Henglein, A.; Lilie, J.J. Phys. Chem. **1981**, 85, 1246
35. Gutierrez, M.; Henglein, A.J. Phys. Chem. **1996**, 100, 7656
36. Henglein, A. Ber. Bunsenges. Phys. Chem. **1980**, 84, 253
37. Henglein, A.; Lilie, J.J. Amer. Chem. Soc. **1981**, 103, 1059
38. Henglein, A.J. Phys. Chem. **1979**, 83, 2858
39. Henglein, A.; Mulvaney, P.; Linnert, T.; Holzwarth, A.J. Phys. Chem. **1992**, 96, 2411
40. Henglein, A.; Gutierrez, M.; Janata, E.; Ershov, B. G.J. Phys. Chem. **1992**, 96, 4598
41. Kolb, D.M.; Przasnyski, M.; Gerischer, H. J. Electroanal. Chem. **1974**, 54, 25
42. Mulvaney, P.; Giersig, M.; Henglein, A.J. Phys. Chem. **1992**, 96, 10419
43. Henglein, A.; Mulvaney, P.; Holzwarth, A.; Sosebee, T. E.; Fojtik, A.Ber. Bunsenges. Ges. Phys. Chem. **1992**, 96, 754
44. Katsikas, L.; Gutierrez, M.; Henglein, A.J.Phys. Chem. **1996**, 100, 11203
45. Henglein, A.; Brancewicz, C.Chem. Mater. **1997**, 9, 2164
46. Henglein, A.; Giersig, M.J. Phys. Chem. **1994**, 98, 6931
47. Hodak, J. H.; Henglein, A.; Giersig, M.; Hartland, G. V.J. Phys. Chem. **2000**, 104, 11708
48. Hodak, J. H.; Henglein, A.; Hartland, G. V.J. Phys. Chem. 2000, 104, 9954
49. Link, S.; Wang, Z. L.; El-Sayed, M. A.J. Phys. Chem. B **1999**, 103, 3529
50. Link, S.; Burda, C.; Wang, Z. L.; El-Sayed, M. A.J. Chem. Phys. **1999**, 111, 1255

51. Papavassiliou, G. C.J. Phys. F **1976**, 6, L103
52. Henglein, A.; Meisel, D. Langmuir **1998**, 14, 7392
53. Henglein, A.; Holzwarth, A.; Mulvaney, P.J. Phys. Chem. **1992**, 96, 8700
54. Weitz, D. A.; Lin, M. Y.; Sandroff, C.Surf. Sci. **1985**, 158, 147
55. Henglein, A.Chem. Mater. **1998**, 10, 440
56. Linnert, T.; Mulvaney, P.; Henglein, A.J. Phys. Chem. **1993**, 97, 679
57. Mulvaney, P.; Linnert, T.; Henglein, A.J. Phys. Chem. **1991**, 95, 7843
58. Henglein, A.; Meisel, D.J. Phys. Chem. B **1998**, 102, 8364
59. Sal'sedo, S. K. A.; Tsvetkov, V. V.; Yagodovskii, V. D. Russ. J. Phys. Chem. **1990**, 64, 993
60. Giersig, M.; Mulvaney, P.Langmuir **1993**, 9, 3408
61. Harfenist, S. A.; Wang, Z. L.; Alvarez, M. M.; Vezman, I.; Whetten, R. L.J. Phys. Chem. **1996**, 100, 13904
62. Wang, Z. L.; Harfenist, S. A.; Whetten, R. L.; Bentley, J.; Evans, N. D.J. Phys. Chem. B **1998**, 102, 3068
63. Markovich, G.; Collier, C. P.; Heath, J. R.Phys. Rev. Lett. **1998**, 80, 3807
64. Mirkin, C. A.; Letsinger, R. L.; Mucic, R. C.; Storhoff, J. J.;Nature **1996**, 382, 607
65. Alivisatos, P. A.; Johnsson, K. P.; Peng, X.; Wilson, T. E.; Loweth, C. J.; Bruchez, M. P.; Schultz, P. G.Nature **1996**, 382, 609
66. Schmidt, H.; Lesniak, C.; Schiestel, T. In "Fine Particle Science and Technology", Pelizzetti, E., Ed.; NATO ASI Series 3, High Technology; Kluwer Academic Publishers: Dordrecht, **1996**, Vol. 12, p. 632
67. Strelow, F.; Fojtik, A.; Henglein, A. J. Phys. Chem. **1994**. 98, 3032
68. Strelow, F.; Henglein, A.J. Phys. Chem. **1995**, 99, 11834
69. Katz, E.; Shipway, A.; Willner, I." Chemically Functionalized Metal Nanoparticles: Synthesis, Properties and Applications", in"Nanoscale Materials", Liz-Marzán, L. M.; Kamat, P. V., Eds., Kluwer Academic Publishers, Boston, **2002**
70. Toshima, N. "Metal Nanoparticles for Catalysis", ibid.

5

# BIOANALYTICAL SENSING USING NOBLE METAL COLLOIDS

C. Mayer[1] and Th. Schalkhammer[1-4]

## 1. BIO-NANOTECHNOLOGY

Bio-Nanotechnology is a world determined by atoms, molecules and clusters and their complex assemblies integrated in a biological matrix. Whereas machining and handling at the hundred-nanometer level is already state-of-the-art in semiconductor industry, a number of new phenomena become important in the bio-nano world.

Bionanoengineering aims to develop techniques for characterizing, processing and organizing molecular structures via biomolecules. Biomolecular recognition principles are the basis for novel functional elements of molecular circuits. Main research activities are biorecognition-induced assembly techniques, the direct detection of single molecules, biomolecule-nano-cluster structures, unique opto- and electronic phenomena, elaborated deposition techniques for nano-particles, nano-waveguides, techniques of nanoscale fabrication, nano-imprinting, handling of individual molecules, and tuning of bioprocesses at molecular level.

In bionanotechnology artificial nanostructured material is either a template for assembling biological materials or the other way round, a biological structure which allows to arrange, couple and modify nano-scale opto/electronic circuits. The bio-chemistry included in this field is broad, ranging from fundamental bioorganic processes, DNA and complex protein networks up to living cells on a chip.

Among all areas of nanoscale science colloid-sized particles gained most attention due to a number of properties making them most suitable as ultimate building blocks. These building blocks enable the construction of materials and devices with novel mechanical, electrical and optical properties.

[1] C. Mayer, Analytical Biotechnology, Technical University of Delft, Julianalaan 67, 2628 BC Delft, The Netherlands.
[2] Nanobioengineering, Vienna Biocenter, Universität Wien Dr. Bohrgasse 9, 1030 Wien, Austria.
[3] Attophotonics Bioscience-Schalkhammer KG, Klausenstrasse 129, 2534 Alland, Austria.
[4] To whom correspondence should be addressed. Schalkhammer KG, Klausenstrasse 129, 2534 Alland, Austria. Phone: 0043-2258-76190, Fax: 0043-2258-76290, Email: Schalkhammer@bionanotec.org.

*Topics in Fluorescence Spectroscopy, Volume 8: Radiative Decay Engineering*
Edited by Geddes and Lakowicz, Springer Science+Business Media, Inc., New York, 2005

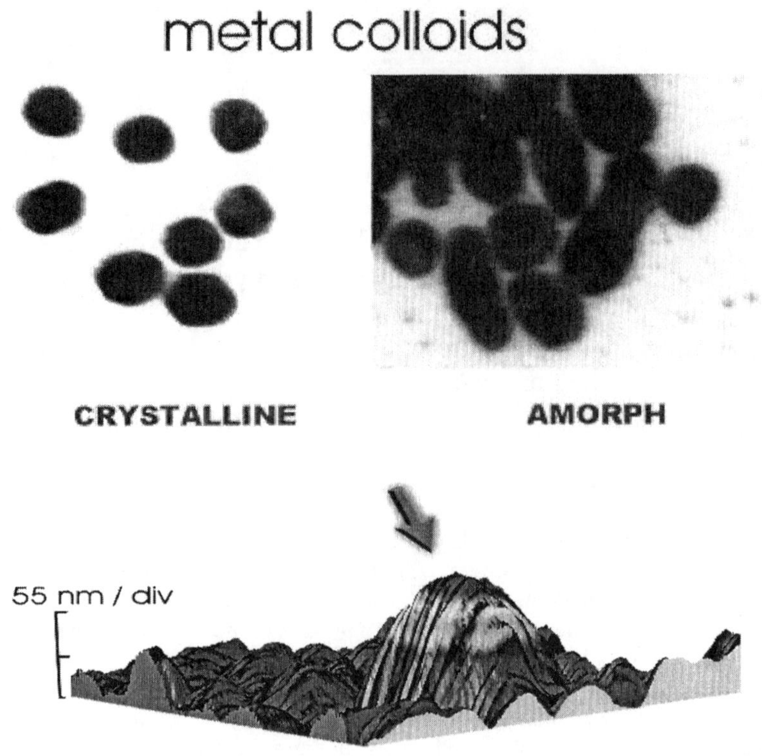



Figure 1. EM and AFM of metal nano particles (~ 10-20 nm) .

## 1.1. Metal Colloids

Metal colloids are assemblies of atoms with unique physical and chemical properties. These colloidal particles are nano-crystals or nano-amorphous lumps composed out of 2 to $10^8$ atoms, which are bound together by crystal energy. Within the state of matter described by cluster-type behavior we define a nano-particle, nano-cluster, nano island, precipitate or- whatever name is used for these assemblies as a colloidal particle not specifying the size, shape or property of this assembly (Figure 1).

Colloids can be formed out of a wide variety of materials just limited by their chemical stability in air, water or biological environment. Even biomolecules such as proteins or DNA are essentially colloidal particles in the nm size range. Contrary to metal colloids bio-nanoparticles are primarily produced via bio-synthetic routes or are made

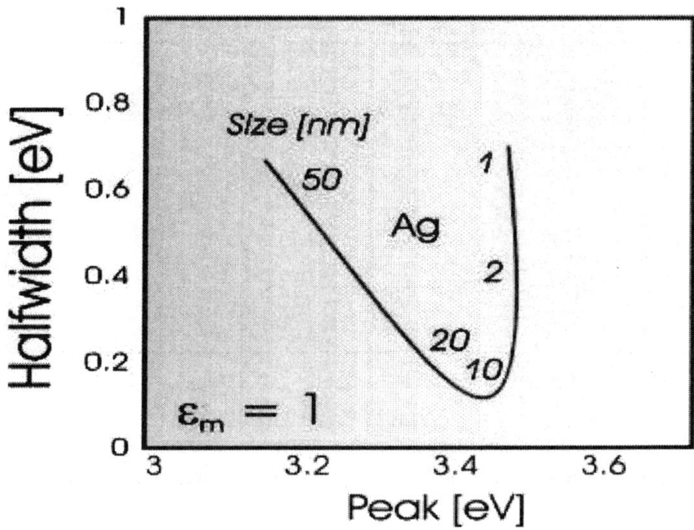


**Figure 2**. Optical spectrum of silver nano-cluster versus size.

using e.g. electro-spraying technique followed by electrostatic classification and manipulation.

For electronics and nano-optical application particles are preferably composed out of a conducting metal such as gold, silver, copper, platinum, or at least a semi conductor such as silicon, cadmium sulfide, cadmium selenide or zinc oxide. To achieve efficient collective behavior noble metal colloids of either a high homogeneity in size and shape are required or a deep understanding of excitation modes (particle plasmons and extended plasmons, Figure 2) within particle assemblies is necessary.

Clusters composed out of a few atoms are better described and modeled as supra-molecular structures exhibiting no cluster type behavior such as e.g. a well-defined particle plasmon resonance (Figure 3). Ultra-large metal colloids of 300 nm or more exhibit macroscopic behavior such as metallic luster and bulk conductivity (Figure 4). Colloidal structures made up of a plurality of individual clusters are often referred to as mesoscopic systems, cluster matter or nano-crystalline materials. Thus, colloidal particles cover an extreme wide range in size [1,2,3,4].

Within the last decade the use of nanoclusters for novel (bio)electronic and (bio)optical devices gained world-wide attention due to a number of key-developments describe in this article. Furthermore, attempts to reduce the size of electronic devices with top down techniques will encounter insurmountable barriers due to limits of UV-lithography. The use of colloidal particles as elements and building blocks for new devices is a novel and cost efficient route. Bio-nano-assemblies will enable us to construct three-dimensional electronic circuits of ultra-high packing density. Manufacturing of bio-nano-devices will require the development of a completely new set of techniques to arrange, manipulate and couple colloidal particles.

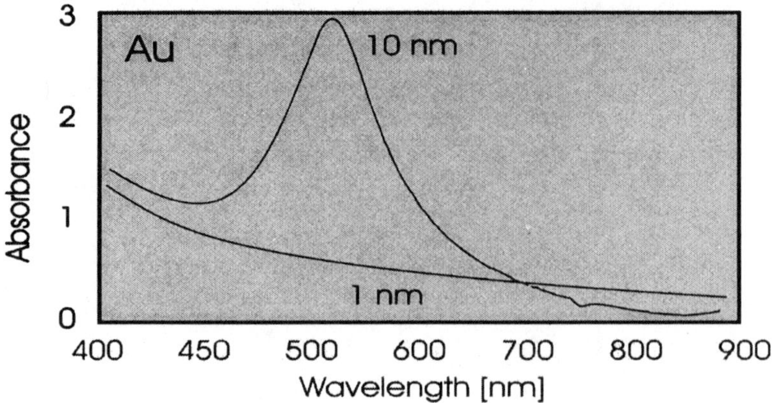


**Figure 3**. Optical spectrum versus size (Au). Note: for Au a distinct plasmon band is only obtained at a size of > 5nm.

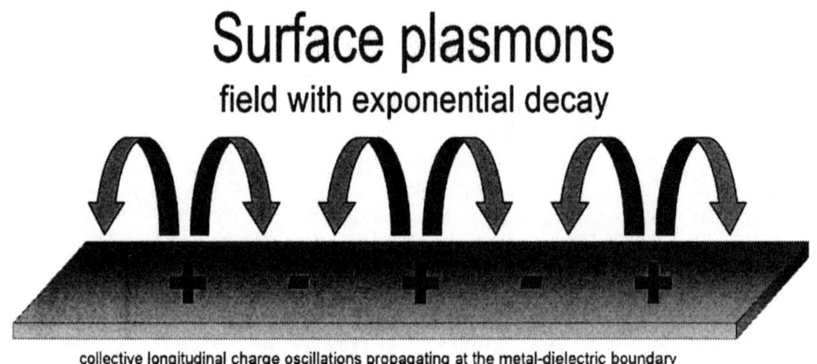


**Figure 4**. Surface plasmon in metal films.

To understand the colloidal state of matter it is first necessary to focus on the behavior of isolated clusters and later to describe more complex cluster-cluster, cluster-molecule and cluster–resonator assemblies.

A wide variety of colloids are found in everyday life forming silver-centers in photographic films, the red color of glass, aerosols, exhausts, precipitates or just dust and dirt. Thus, the discovery of the colloidal state of matter already dates back hundreds of years based on inventions and theories from Faraday, Mie, Ostwald, Seitz, Svedberg and Zsigmondy. Since 1986 clusters as a base for nanotechnology, nano-engineering or nano-

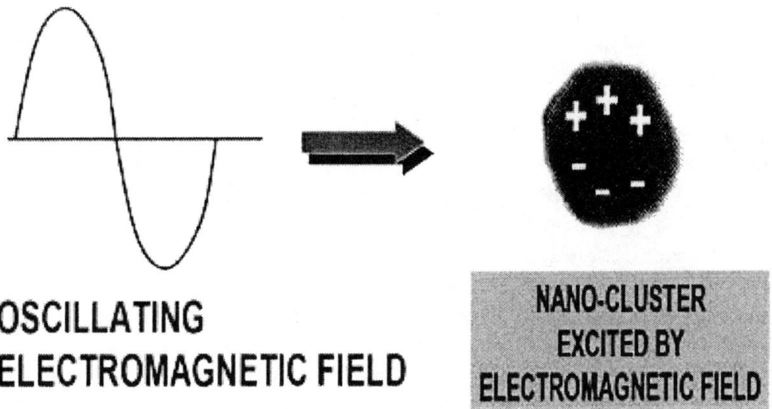


**Figure 5**. Particle plasmon of a cluster.

devices gained a central role in the upswing caused by STM and AFM techniques followed by the nano-boom at the end of the nineties. Cluster are not really a fifth state of matter but combine rather specific properties of metals and semiconductors including unique catalytic properties as well as well-defined surface and particle plasmons.

Colloid properties are intrinsic and extrinsic, the first being e.g. chemical behavior, ionization potential or crystallographic structure, the second collective phenomena such as electron gas and lattice resonance. The scheme of electron-energy-levels highly dependent on cluster size and cluster shape is often cited as based on a quantum-size effect. Still most of the behavior of isolated and assembled metal nanoparticles could be deduced from the classic electromagnetic theory without any use of quantum behavior or statistic (Figure 5). Only ultra-small or semi-conductor clusters are not readily described by collective phenomena due a the well-defined assembly of atoms in nano-crystals or due the low number of electrons respectively.

A fundamental change in behavior is observed when reducing the size of the colloidal particle as a transition from macroscopic metals to semi-conductor-type colloids and further on to atom assemblies and isolated atoms . This transition to the isolating state is observed for most metals in between 2 and 100 atoms. A unique behavior is often observed if a magic number of atoms forms a defect-free nano-crystal. Gold in a perfect nano-crystal is even more stable against oxidation than a macroscopic gold metal sheet.

Phenomena such as quantum dots, fluorescent nano clusters, REF (resonance enhanced fluorescence), SERS (surface enhanced Raman-spectroscopy), SEA (surface enhanced absorption), or unique catalytic effects are the major technological quantum leaps of nano technology.

The high surface to volume ratio of nano-colloidal matter offers novel pathways to use otherwise inefficient surface-confined energy transduction. In nanometer sized clusters the boundary region reaches deep beyond the surface and even penetrates the whole grain. Thus, the state of every atom is influenced in a more or less distinct way.

Even colloids with thousand or more atoms still contain up to 25% of surface near layer. At $10^5$ atoms the behavior can be described by the bulk property due to the fact that less than 1% of atoms are near to the surface.

Nano-particles had been synthesized by various techniques either as free colloids in solution or gas phase, deposited or bound to a surface, embedded within a material or as nano-powder. Electrical and optical properties are primarily determined by material and size but are also modified by shape, by surface bound molecules, and the embedding medium. Although it is possible to gain cluster-matter by a break down and milling process out of macroscopic matter it is more convenient to use a synthetic approach to build cluster *de novo* from solution.

Clusters a few atoms in size can be modeled by molecular techniques using quantum based ab-initio algorithms. At the other end of the scale metal grains are to be described using solid state physics of metals or semi-conductors.

A variety of techniques might be applied to study fundamental properties of typical colloids including optical spectra, photo-ionization, magnetic moment, or polarizability. Due to the high chemical reactivity - based on the high surface to volume ratio - small colloidal particles are often examined using high-vacuum beam techniques. Up to a few hundred atoms nano-crystals with ´magic numbers´ are targets of major interest. Quite often mass spectrometry is used to obtain insight into size and setup of these ultra small colloids.

Clusters deposited on a substrate surface are often either obtained via chemical reduction of metals in a surface catalyzed technique or by deposition from solution or gas phase. These colloids are of a typical size around 1-100 nm. To study these clusters electron microscopy (EM), surface tunneling microscopy (STM), atomic force microscopy (AFM) or optical near field techniques (e.g. SNOM) directly access size, shape and electro-optical properties of individual colloidal particles.

To describe the behavior of clusters the classical electrodynamics may be used for colloidal particles whereas for nano-crystals of less than 100 atoms the optical functions become size dependent. Simplified models can be set up for ideal clusters such as sodium or potassium not being disturbed by interband interference. Nevertheless, it should be clear that these calculation are of reduced value due to the fact that almost all clusters of practical applicability are made of noble metals or semiconductors with a far more complex behavior.

## 1.2. Metal Colloid Devices

Metal colloids as signal transducers of molecular interaction and biorecognition are chosen due to their superior extinction coefficients compared to any organic or inorganic chromophores. Thus, metal colloids are more intensely colored than even the most intensive organic dyes. Based on this fundamental property colloid capture assays enable to visualize the binding of biomolecules at a given surface by a bound layer of bio-modified metal clusters. A direct approach of detection with chromophores turns out to be either very insensitive without using additional amplification by e.g. an enzyme (well known and established as ELISA) or requires large particles resulting in severe limitations due to diffusion speed and unspecific background.

The first introduction of gold and silver colloid staining methods dates back into the mid 20th century. Following decades of simple bio-assays based on deposition and aggregation of nano particles the signal transduction by metal nano clusters has undergone an enormous evolution. The basis was the development of a wide variety of techniques to synthesize clusters by physical and chemical means. Rapid and simple synthesis, a narrow size distribution and a straightforward surface modification with bio-ligands enabled application of colloidal particles as transducers of biorecognitive binding and molecular structure.

The fundamental key-techniques and properties are:

- preparation as mono-disperse particles,
- simple one pot chemical synthesis,
- efficient coating by thiols or phosphines,
- coupling with various biomolecules,
- electron dense core,
- highly resonant particle plasmons,
- direct visualization of single nano clusters by scattering of light,
- catalytic size enhancement by deposition of other metals as e.g. nickel or silver
- no photo-bleaching under illumination, a fundamental advantage compared to fluorophores.

Based on these key properties nano-cluster or colloids became the basis of new devices employing cluster resonance, cluster-cluster interactions and cluster field enhancement:

| Nano property | Application in |
| --- | --- |
| Mono-disperse particles | nano-electronic building block |
| Arrays or colloids | molecular filters |
| Large surface area | heat-exchange |
| Large specific surface area | catalysis |
| Lower sintering temperature | fabrication via sintering |
| Plastic behavior of nano granules | ductile ceramics |
| Single magnetic domain | ultra-dense magnetic recording |
| Small mean free path of electrons | quantum dots, resonators |
| High & selective optical absorption | biochips, colors, filters, solar absorbers, photovoltaic |
| Metal-ceramic mixture | new materials |
| Grain size too small for dislocation | high strength and hardness |
| Cluster coating | metallization |
| Multi-shell particles | optical devices |

Whereas the first generation of colloidal particles was based on spherical clusters within the last decades rod-shape metal particles, tubes, prisms, cubes and a variety of other nano building blocks had been designed.

## 2. NANO-CLUSTER BASED TECHNOLOGY

### 2.1. Properties

The unique interaction of metal nano clusters with an electromagnetic field and with radiation is among the most exciting properties of colloidal particles (Figure 6). In a macroscopic metal electron move unconfined by any material border. This free

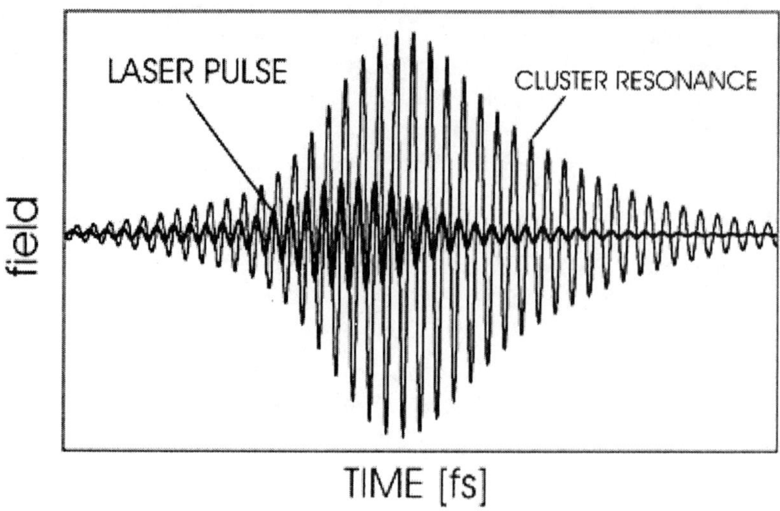


**Figure 6**. Resonance of the cluster driven by a laser pulse at resonant frequency

movement of the electron gas results in strong, unspecific reflectivity, well known as metallic luster. Contrary to that, for colloidal particles of less than 300 nm in size the optical properties are primarily determined by the strongest oscillation process of the electron gas thus the particle plasmon. This electromagnetic resonance of nano particles is dominated by the collective oscillation the electrons within the cluster either forming plasmons or plasmon-polaritons. Most of the novel phenomena are based on this excitation found in the visible and infra red part of the spectrum whereas the behavior in the UV and blue range of the spectrum is heavily deteriorated by interband energy transfer.

Plasmon based optical spectroscopy of metal clusters is best described by electrodynamics. To describe the behavior of the plasmon oscillation it is appropriate to apply a quasi-static regime only for a cluster of around 10-20 nm. The static regime assumes that the phase shift in the colloidal particle is small enough to be neglected reducing the cluster oscillation to a simple dipole.

Optical spectra of thousands to millions of individual clusters show a broad and smeared out resonance due to the cluster-size and shape-distribution. Furthermore, clusters deposited at a substrate are influence by material boundaries, the spectra of colloids larger than 30 nm are determined by finite penetration depth and multi-polar modes of the electromagnetic resonance.

Whereas for small clusters the field attenuation inside the cluster is small, in larger clusters (> 30 nm) only surface regions are excited by the field. Thus, the electron mean free path plays a key role for the optical properties of large metal colloids. The skin depth at 2 eV is smallest in aluminum (13 nm) and larger for most noble metals such as gold or silver (Ag: 24 nm, Au: 31 nm).

The electromagnetic theory is based on the assumption that electrons are allowed to move freely being excited by an external field. Contrary to that, the positively charged ion cores are immobile. Thus, it is clear from this precondition that metals with free electrons exhibit the strongest optical effects and are best suited to study the electronic properties of resonant particle plasmons. These metals have partially filled conduction bands but completely filled valence bands. Ideally free electron metals response is based on the conduction electrons whereas in reality most metals exhibit various inter-band transitions.

Noble or semi-noble metals colloids from copper, silver or gold exhibit both free electron and inter-band transition behavior. As a clear consequence from the band model of molecular and cluster excitation the deterioration is depending on the frequency range. Often the IR and red part of the spectrum exhibits ideal behavior whereas at the inter-band level (higher energy) spectra a dominated by inter-band excitation and plasmons are quenched rapidly. For theoretical studies alkali metals are ideal, but due to their instability in an ambient environment, only noble metal clusters are applicable in air and under aqueous solvents.

A metal cluster is excited by the electromagnetic field of a light wave. The theory describing this phenomenon as the dielectric function $\varepsilon(\omega)$ was developed by Drude-Lorentz-Sommerfeld (DLS). The internal field of a metal cluster is calculated by adding the boundaries of the sphere surface. Based on DLS the static electric polarizability of a simple single spherical cluster of radius r is given by:

$$\alpha_{cl} = 4\pi\varepsilon_0 r^3 (\varepsilon - \varepsilon_m)/(\varepsilon + 2\varepsilon_m).$$

This formula is applicable for metal colloids within the size range of 4-40 nm. Using this formula it is straightforward to determine the resonance conditions for a single cluster, thus the plasmon resonance energy. For a small $\varepsilon_2$ a resonance is obtained at:

$$\varepsilon_1 = -2\varepsilon_m$$

with a resonance frequency at an $\varepsilon_m$ of one at:

$$\omega = \sqrt{(e^2/(m_e 4\pi\varepsilon_0 R^3))}.$$

The plasmon frequency ($\omega$) is optically obtained as the resonance of a colloidal particle. To clarify:

- The conduction electrons in colloid particle act as an oscillator system.
- The conduction electrons in a bulk metal act as a relaxator system with no excitation at the Eigen-frequency $\omega$.

To obtain a more realistic model of the electron gas the interaction with the more or less immobile ion core can be described by introducing an effective optical mass instead of the electron mass. This important development was the Mie-Gans-Happel theory based on a solution of Maxwell's equations with the appropriate boundary conditions.

Mie divided the problem into two parts:

- the electromagnetic one and
- one based on material properties (using a dielectric function $\varepsilon(\omega,R)$).

The function introduced includes all electronic and atomic behavior of the colloidal particle as e.g. dipolar, multipolar or magnetic modes of absorption and scattering. The success of this theory was based and is based on extracting data from real experiments and not predicting them by "ab-initio" calculations. In short:

- $\varepsilon_1$ is correlated to $\Delta\lambda_{max}$,
- $\varepsilon_2$ with the peak height, and
- n with band-width.

As a simple consequence a steep $\varepsilon_1$ versus $\omega$ relation results in a narrow resonance peak whereas a flat $\varepsilon_1(\omega)$ results in a broad band.

## 2.2. Metal Colloids and Quantum Dots

In an ideal metal the visible spectral region is dominated by the dipolar surface plasmon since damping is small. Due to the fact that there is not a single metal stable in air and exhibiting ideal behavior we need to switch to noble metal colloids as the most suitable material for nano-technology. In contrast to ideal metals, noble metals (e.g. Ag and Au) show significant deviation from free electron behavior due to inter-band transitions in the visible region. The plasmon energy shifts from the free electron value of around 10 eV to 1-4 eV. The threshold for silver is 3.9 eV and for gold 2.4 eV. This resonance is no longer a free electron resonance but a cooperative effect based on conduction and mostly d electrons. Whereas silver is near to an ideal metal in copper the plasmon energy limit shifts from more than 9 eV to around 2 eV. In air and aqueous environment only silver has a sharp well-defined resonance peak near 300-400 nm. Gold exhibits a significantly broader peak at around 520 with an inter-band shoulder to higher energy. Copper exhibits no resonance at $\varepsilon_m \sim 1$. A sharp resonance peak is observed in media of high refractive index as a clear consequence of the formulas and resonance conditions given above. An extreme narrow peak occurs at low frequencies with a slight red shift for metals such as V, Y, Zr, Ti or Ta. Most of the other bands found in various metals have nothing to do with plasmons and thus will not fulfill the basic criterion of $\varepsilon_1 = -2\varepsilon_m$.

Indirect band-gap materials exhibit no significant photoluminescence (PL) at room temperature. As a specific nano-effect the quantum confinement in semiconductor clusters induces visible room-temperature PL. Thus, nano-colloidal silicon, molybdenum sulfide, and pyrite shows intense luminescence.

The famous quantum dots are nanometer-size colloidal particles smaller than the exciton radius. These quantum dot materials are around 20 times brighter and more than 100 times more stable against photo-bleaching compared to organic fluorescent dyes. Within recent years these luminescent nano-particles found wide-spread use for multicolor staining, homogenous assays, and tissues imaging.

Colloidal gold is orange, red, or red-violet, depending on the preparation method and thus on size and aggregation state. Orange indicates the lack of a plasmon band and is a direct indicator of ultra-small gold colloids up to a few nm in size. A red gold sol indicates a cluster with a well developed plasmon resonance. A blue color indicates either a non-crystalline core thus an amorphous state of the cluster or an aggregation of the colloids. Both states can be characterized by their typical spectral behavior (see below).

Colloidal gold is prepared by reducing gold salts with numerous reagents in aqueous or organic solution. The most common reagents are sodium borohydride, citrate, phosphorus or ascorbic acid. Depending on the method and the exact reagent composition of the reaction mixture, colloidal gold preparations vary in particle size and size distribution expressed as the coefficient of variation of the gold particle diameter. A sol is often referred to as mono-disperse if the coefficient of variation is smaller than 15 %.

A direct technique to prepare colloidal gold particles bound to a substrate surface is described by Natan et al [5]. Natan developed a surface-catalyzed reduction of $Au^{3+}$ by $NH_2OH$. This solution chemistry allows to grow existing nanoparticles and surface bound clusters into larger particles of a size determined by the amount of $Au^{3+}$ added. The resulting clusters exhibit improved monodispersity. The technique provides an attractive route to prepare colloidal gold monolayers and nano grids. The oldest technique, used in thousands of laboratories all over the world, is the electron-dense staining with colloids in cytochemistry.

## 2.2.1. Techniques to prepare noble metal colloids

**Gold cluster via sodium citrate reduction method [6]**

The most well know method based on the protocol developed by Frens provides monodisperse colloidal gold. The size of the clusters can be adjusted in between 6 and 60 nm by adding varying amounts of reducing agent.

Protocol for 40 nm colloids:
1. 0.01% (w/v) tetra-chloroauric[III]acid trihydrate ($HAuCl_4.3H_2O$ in 100 ml of water) is heated to boiling.
2. 1 ml of 1% (w/v) trisodium citrate dihydrate is added to the boiling solution under constant stirring.
3. In about 25 sec the slightly yellow solution will turn into faintly blue (nucleation).
4. After approximately 70 sec the blue color then suddenly changes into dark red, indicating the formation of monodisperse spherical particles.
5. The solution is boiled for another 5 min to complete reduction of the gold chloride.
6. The optical density, measured at 540 nm (OD540) of such G40 suspensions will be about 0.9.
7. Supplemented with 0.05% (w/v) sodium azide, the obtained G40 suspensions can be stored at +4°C for several months.

**Table 1**. Nano-particle size versus amount of trisodium citrate to be added

| Average diameter in nm | Amount of 1% trisodium citrate added (ml) |
|---|---|
| 15 | 4,8 |
| 25 | 1,9 |
| 30 | 1,6 |
| 40 | 1 |
| 50 | 0.75 |

### Gold colloid preparation accordi ng to Slot-Geuze [7]

Slot and Geuze developed a method that combines tannic acid with tri-sodium citrate as reducing agents. They described the optimal conditions to obtain mono-disperse sols of 3 to 15 nm, varying the amount of tannic acid to be added.

Solution A: Gold chloride solution (250 ml Erlenmeyer flask with a stirring bar), 79ml double distilled water and 1 ml 1% $HAuCl_4$ in double distilled water

Solution B: Reducing mixture (250 ml Erlenmeyer flask) 40 ml of filtered 1% tri-sodium citrate in distilled water, a variable volume (x) of 1% tannic acid and 25 $\mu$molar $K_2CO_3$ (to correct the pH of the reducing mixture) if the tannic acid volume is above 1 ml.

1. Solutions A and B are heated to 60°C.
2. B is quickly added to A under vigorous stirring.
3. A red sol is formed very rapidly, depending on amount of tannic acid added.
4. The colloid is then heated until boiling and gently boiled for another 5 minutes.
5. Sometimes the red color develops only upon boiling, which does not imply a lower quality of this preparation.

### Preparation of gold colloid of around 25 nm in size with aspartic acid

1. Prepare 90 ml of a $10^{-4}$ M solution of $HAuCl_4$ in double distilled water; boil
2. Add 10 ml of a $10^{-3}$ M solution of aspartic acid under boiling conditions
3. A red sol is formed very rapidly, depending on the amount of aspartic acid added.
4. The solution is boiled for another 5 min to complete reduction of the gold chloride.

### Preparation of gold particles in toluene

This technique yields highly concentrated gold colloidal suspension with particle diameter in the range 5-10 nm. Colloid can be suspended in both polar and nonpolar solvents.

Solution A = Hydrogen tetrachloroaurate = 0.3537g of $HAuCl_4$ in 30mL of $H_2O$
Solution B = Tetraoctyl ammonium bromide = 2.187g in 80mL of toluene

Prepare solution B

1. Add solution B to solution A
2. Stir for 10 min.
3. Add $NaBH_4$ (0.38 g in 25 ml of $H_2O$) drop wise over a period of ~30 min under continuous stirring.
4. Ensure that organic and aqueous phases are being mixed together.
5. Stir solution for an additional 20 min.
6. Extract organic phase and wash once with diluted $H_2SO_4$
7. Extract organic phase five times with distilled water.
8. Dry organic phase with $Na_2SO_4$.

Platinum colloid: use $H_2PtCl_6$ C $H_2O$
Iridium colloid: use $H_2IrCl_6$ C $4H_2O$

Based on: Brust, M.; Walker, M.; Bethell, D.; Schffrin, D. J. Whyman, R., *J. Chem. Soc., Chem. Commun.*, 1994, 801-802 and George Thomas, K. Kamat, P. V. *J. Am. Chem. Soc.* 2000, **122**, 2655.

## Silver Colloids by sodium citrate reduction

This method yields relatively large size silver nano-crystallites of 60-80 nm With a plasmon resonance wavelength of ~420 nm. Sodium Borohydride as a reductant will give smaller sized silver nanoparticles with plasmon absorption around 380 nm.

1. Make 125 ml of a solution of 1.0 x $10^{-3}$M in $AgNO_3$ in deionized $H_2O$.
2. Make a solution of 1% sodium citrate.
3. Heat the $AgNO_3$ solution until it begins to boil.
4. Add 5 ml of 1% sodium citrate solution.
5. Continue heating until a color change to pale yellow.
6. Remove the solution from the heater
7. Stir until it has cooled to room temperature.

   *J. Phys. Chem. B*, 1998, **102**, 3123

## Colloidal silver via citric/tannic acid reduction

1. 45 ml of 0,1 M $AgNO_3$ solution are diluted in 220 ml of double distilled water.
2. After adding 1ml of 1% (w/v) tannic-acid solution the mixture is heated to 80°C.
3. 5ml of 1% (w/v) Na-Citrate are added under rapid stirring, the solution is kept at 80°C all the time until the colloidal color becomes visible (yellowish).
4. Colloid formation can be followed spectroscopically, an absorption peak develops with a maximum at 400 nm.
5. The colloids prepared in this way are poly-disperse with an average particle diameter of 30nm and a standard deviation of 20nm. They adsorb readily to most surfaces.

## Preparation of Au coated Ag

1. Take 125 ml of 1.0 x $10^{-3}$M solution of Ag colloids
2. Heat this solution to boiling.
3. Add the appropriate amount of 5.0 x 10-3M $HAuCl_4$.
4. Boil for 10 minutes
5. Remove the solution from the heater
6. Stir until it has cooled to room temperature.

## Thiol coated silver colloids synthesis in organic medium

1. Prepare 30 ml 5.0M $NaNO_3$ in deionized water.
2. Prepare 50 ml 50mM TOAB (tetraoctylammonium bromide) in toluene.
3. Add the TOAB solution to the $NaNO_3$ solution.
4. Stir vigorously for 1 hour.

5. Discard aqueous phase.
6. Prepare 5 ml 30mM $AgNO_3$ in water.
7. Add 5ml of 30mM $AgNO_3$ solution to the organic solution.
8. Stir vigorously for 45 minutes.
9. Discard aqueous phase.
10. Add 0.16mg (-0.189ml) of a thiol e.g. 1-dodecanethiol to organic solution
11. Stir vigorously for 15 minutes.
12. Prepare 0.4M $NaBH_4$ in water.
13. Add 6.25ml of the $NaBH_4$ drop wise over a 35min. period, to the solution containing the silver (organic layer), while stirring vigorously.
14. Stir overnight.
15. Discard aqueous phase.
16. Wash organic phase 3 times with aqueous ethanol.
17. Store organic phase with thiol coated colloid in closed container.
    Based on: Korgel, B.A.; Fullam, S.; Connolly, S.; Fitzmaurice, D. *J. Phys. Chem. B* 1998, **102**, 8379-8388.

## Preparation of gold coated $TiO_2$ Colloids

1. Prepare 10% titanium(IV) isopropoxide in 1-propanol (or iso-propanol):
2. 3 ml of 10% titanium(IV) isopropoxide are slowly added to 200 ml of water, 1 drop at a time, while stirring vigorously, to obtain a clear, colorless solution. The pH of water is adjusted to 1.5 using 1M $HClO_4$ prior to addition, to stabilize $TiO_2$ colloids.
3. Do not store!
4. Dissolve 0.1697g of $HAuCl_4$ in 100ml water.
5. Add desired volume of gold solution drop wise to the $TiO_2$ colloids while stirring vigorously (e.g. [TiO2] at 2-4mM and [Au] of 0.06-0.3mM )
6. Stir for 5-10 min to allow all gold to bind to $TiO_2$ surface.
7. Prepare ~10 mM $NaBH_4$ in water.
8. Add $NaBH_4$ drop wise to solution with vigorous stirring until a color change to blue and finally to red is seen (some gas will be released).
9. Continue stirring until gas is no longer forming.
10. Top solution to desired volume using water with pH adjusted to 1.5 (with $HClO_4$).

## Various techniques

- Graphite coated metal nano-clusters are synthesized using electric arc techniques. The metal of interest and carbon are co-evaporated by an electric arc between a tungsten cathode and a graphite/metal composite anode. An encapsulation occurs in-situ.
- Metal colloids can be synthesized within the pore structure of zeolite. The coordination of metal atoms with the framework of the zeolite leads to formation of stable nanoclusters with a structural geometry imprinted by the host matrix..
- Colloids are prepared by solution phase thermolysis of organo-metal precursor compounds.
- Nanoclusters have been synthesized using organometallic reagents in an inverse micellar solution.
- Gold nanoclusters are fabricated using metal vapor co-deposition techniques with styrene or methyl-methacrylate.

- Metal nanoclusters are produced via depositing a metal on the surface of a microphase separated PS-PMMA block copolymer. After deposition, the film is annealed.

## Stability

In the absence of stabilizing agents, aqueous dispersions are stabilized only by ion repulsion due to an adsorbed layer of a charged compound (e.g. citrate) and are thus inherently unstable. The strong tendency of coagulation and further coalescence is caused by attractive van der Waals forces, which act at short distances. The primary van der Waals minimum for colloidal particles is much deeper than the thermal energy. Therefore, two nano clusters which collide by Brownian motion will adhere, aggregate and precipitate.

The process of aggregation was already modeled by Smoluchowski a century ago based on dimer formation in an initially stable dispersion. A detailed description is based on the hydrodynamic behavior and the finite range of the van der Waals attraction. Without repulsive stabilization dispersed colloidal particles will aggregate within seconds. A dispersion of charged nano-clusters is usually stable at low salt concentrations. At a high concentration of salt ions, the particle charge is shielded, and the repulsive electrostatic force is outperformed by the attractive van der Waals force. Thus, aggregation is induced by an increase in the ionic strength and changes in the solution pH.

## Non-spherical, core-shell and phase-boundary cluster

Based on the theory of Mie modified concepts deal with non-spherical cluster shape and core shell particles. Clusters of a increasing eccentricities exhibit a splitting of the peak to two or three more or less distinct peaks. Moreover, the oblate cluster shape induces a well-defined splitting of the polarization. Thus, by choosing the appropriate polarization of the exciting beam selective excitation of modes can be done. Clusters randomly oriented with a Gaussian size distribution result in a flat and broad peak (Figure 7) [8].

Colloids composed from more than one metal can often be treated by combining the functions of the pure metals resulting in a peak split or peak shift of the plasmon resonance. For a very thick overcoat spectra converge to the spectra of the top-layer metal due to the finite skin depth of the resonance (Figure 8).

Amorphous metal clusters (nano-lumps) are broadened in resonance to the long-wave-absorption of metal-cluster-films.

Mono-molecular surface layers exhibit strong effects in small colloidal particles. In the visible region even sub-monomolecular ad-layer of sulfides, iodides, oxides or borates lead to a significant modification of the plasmon resonance [9] (Figure 9). Ordered arrays of metal-columns deposited on, imprinted in or etched out of a substrate surface allow the excitation of extended plasmons similar to colloid arrays. Sharp tip structures allow high local fields and thus enable efficient energy transfer applied e.g. in surface enhanced Raman scattering (SERS).

Due to a size dependence of light scattering two-color or multicolor scattering is feasible. This technique allows to discriminate in between multiple interaction family e.g. in combinatorial library screening.

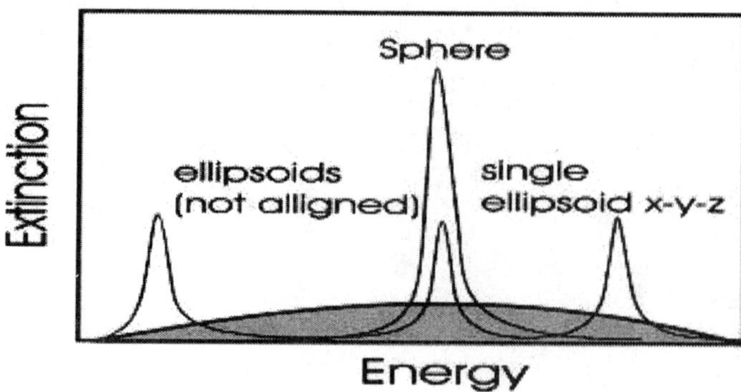


**Figure 7**. Splitting of the plasmon resonance peak into three distinct peaks in an aspherical cluster.

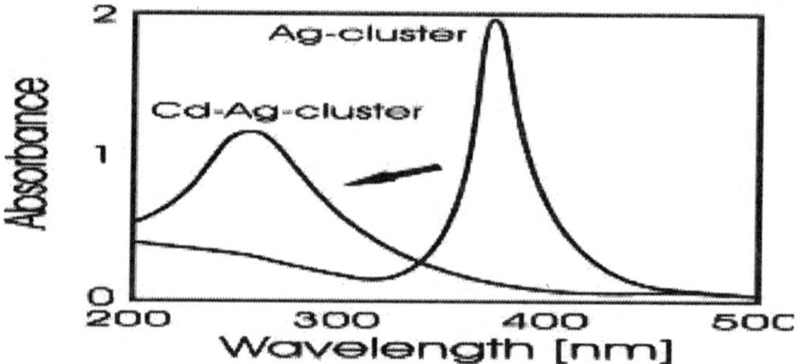


**Figure 8**.  Shift of plasmon frequency of single-metal and dual-metal clusters.

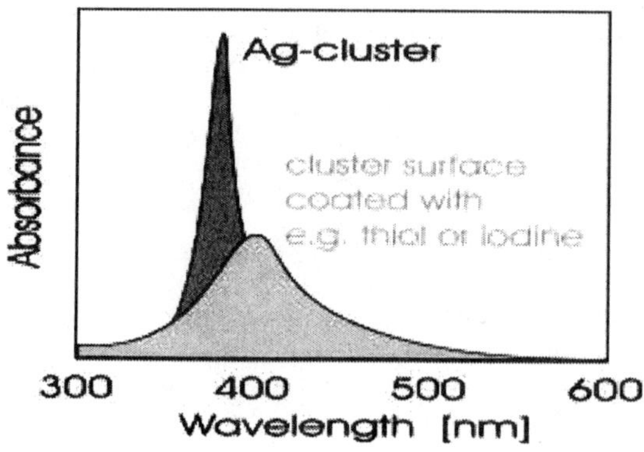


**Figure 9**. Coating of a silver cluster with a monolayer of iodine and a strong damping  of the plasmon peak

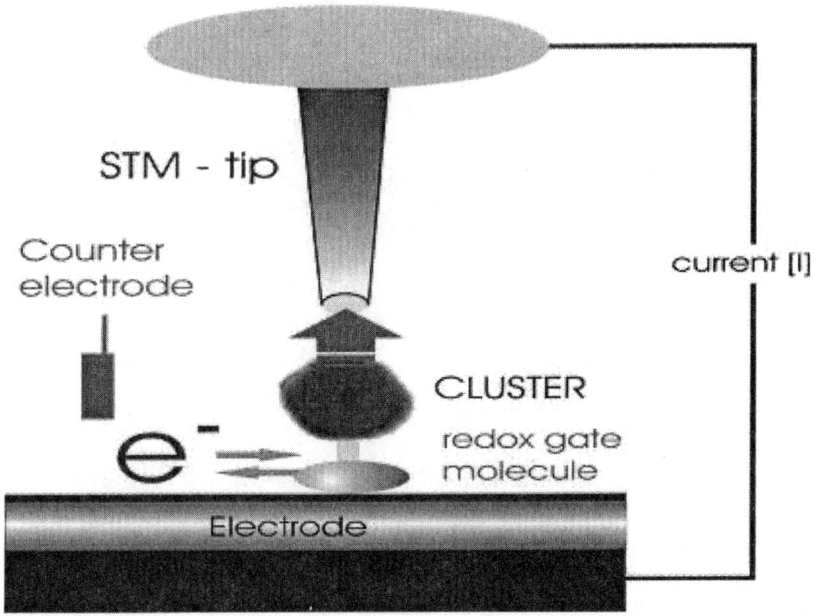


**Figure 10**. Cluster based nano cluster device for electronic readout.

Bimetallic cluster assemblies and nano-structures act as unidirectional light pipes. The effect is attributed to plasmon propagation. The combination of 2 nano-clusters with non identical band gaps are an optical analogue to a diode, and thus may be used to construct e.g. optically addressable memory elements or nano-sensors.

## 2.3. Nano-Switches

Ultimately electronics will consist of networks of molecules. Single molecule conductivity measurement was first done by using a scanning tunneling microscope (STM) tip positioned above a molecule adsorbed to a metallic or semi-conducting surface [10] (Figures 10, 11).

A number of research-groups passed electricity through single molecules such as DNA or organic molecules often less than one nanometer long. Measurements of the DNA conductivity were recently obtained by several groups [11,12].

For testing, most groups use a technique based on breaking a gold bridge around 50 nm wide. First real metal-molecule-metal junctions were based on mechanically controllable break (MCB) junctions, created by Reed, Bourgoin and Magoga [13,14]. They molecules are attached via sulfur atoms to the gold surface - a technique well established in metal cluster devices. The sulfur atoms at the ends of a molecule are precisely bonded to the two gold electrodes or an electrode and a metal cluster to form the electrical contacts. Varying the voltage across the gap either allows to characterize the behavior of the single molecule trapped within or other way round the redox molecular

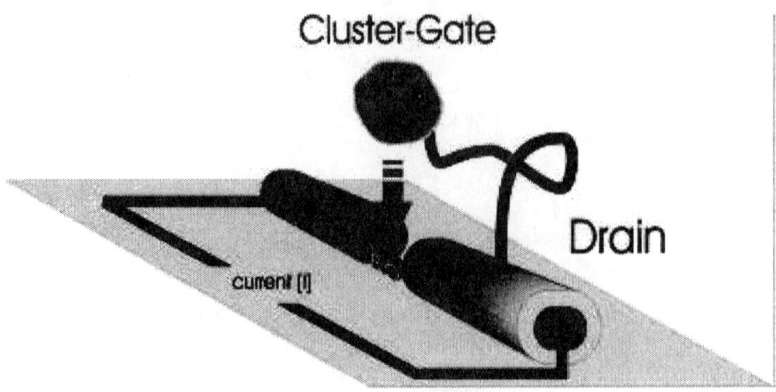


**Figure 11.** Cluster based bio-transistor.

state of the molecule within allowed to switch the current in the device. This result suggests that DNA-based biomolecules can be applied to electric circuits, trapped in between a cluster and a conducting electrode.

A new approach for wiring of nanostructures is linking nanocluster produced via metallized DNA or metallized microtubules to pre-structured microelectrodes. Briefly, the proteins or DNA molecules are activated by adsorption of Pt-catalyst particles using $K_2PtCl_4$ solution, followed by a metallization using Ni acetate [15]. Biomolecular metallization techniques are well established for gold or silver coating. Electron beam-induced deposition (EBD) of highly doped conducting diamond type carbon was used for connecting the bio-nano-structures to nano-electrodes. A metallized molecule yields a resistance below 50 Ohm / micron.

EBD-line writing is an add-on to simplify the electrical connection of single molecules (Figure 12). Repeated line scan induces the polymerization of highly doped conducting diamond type carbon along the path of the electron beam on the substrate, which appears as a dark line in the SEM contrast. A high conductivity of the substrate enhances the growth. It is essential to use a SEM with minimal drift, otherwise a broadening of the lines occur.

First step towards synthesizing reversible molecular switches, and integrating these switches into an electronic circuits, had been done pioneered by Rotaxane-based-technology developed by HP. Any digital microelectronic device is based on memory elements and switches, set up as capacitors, diodes and transistors. Complex elements, such as bi-stable flip-flops, are created from these simple elements combining e.g. two transistor to set up a Schmitt trigger.

In a recent development novel nano-cluster based devices enabled to switch the conductivity of a non-junction by changing the oxidation state of a bridging molecule. Some redox-active molecules contain a molecular center where reduction or oxidation can be achieved more or less reversibly supporting quite large currents. A fundamental prerequisite is the overlapping of the electron energy bands of the molecule with those of

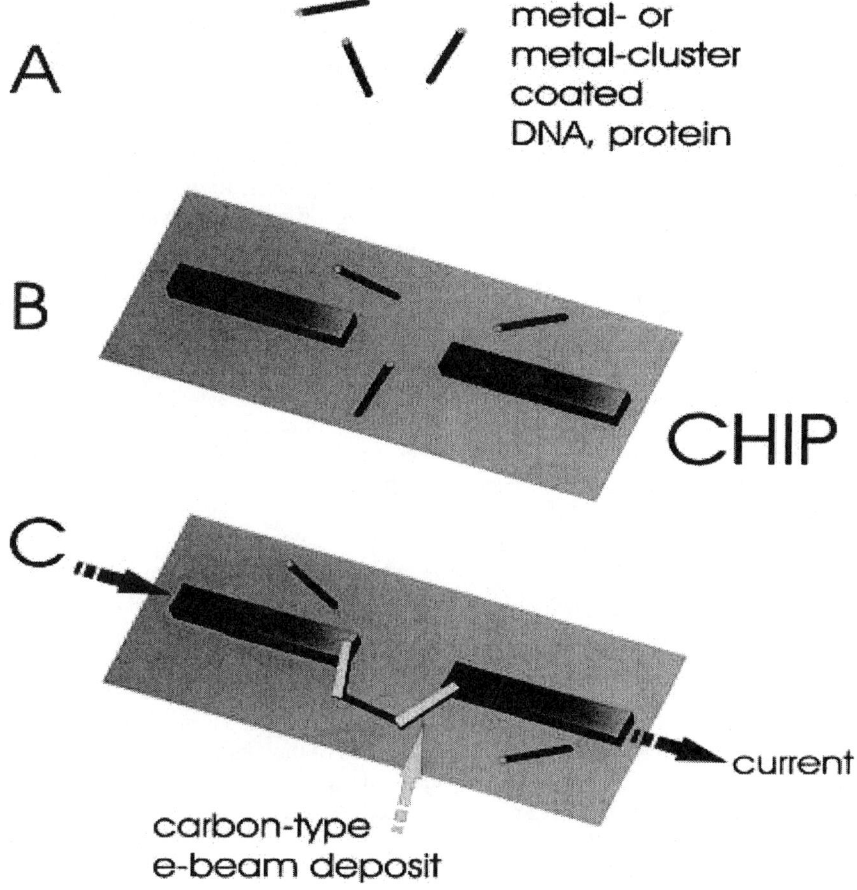



Figure 12. E-beam line writing to contact nano-tubes or metal-coated DNA strands.

the metal. A molecule sandwiched between a metal electrode and gold nano-cluster serves as the molecular switch. A convenient setup can be obtained by attaching phosphines or thiols to the metal electrode as well as to the gold nano-cluster. For lab-test devices the read-out is done via the tip of a scanning tunneling microscope positioned on respectively over the cluster and recording the current through the nano-cluster. Changing the redox state of e.g. a bipy molecule trapped in between the chip and the cluster can control electron transport between the gold electrode and the nano-cluster.

A cluster-switch-array might work with fewer than 30 electrons. In e.g. the reduced state, the cluster-switch exhibits high conductivity at a low voltage drop. At a certain threshold voltage the tunneling current decreases markedly. The threshold voltage is chosen to change the redox state of the molecule used as a switch.

A direct biomolecular trigger would operate too slowly to be applied as an electronic circuit but is a very promising technology to construct novel (bio)analytical devices with ultra-high sensitivity actually at single molecule level. Further on, we could demonstrate that individual gold nanoclusters can be controlled by adding protons to molecules bound at the cluster surface.

A very promising field of colloidal cluster application are nano-optical devices. Whereas nano-electronic devices already take advantage from single cluster behavior the ultimate nano-optical devices are based on coupling of surface optical waves (plasmon waves) in cluster-metal strip structures. A point defect created in a regular cluster array will pull a light mode into the band gap and because such a state is forbidden from propagating in the bulk layer. Thus, this mode decays exponentially into the bulk. Biorecognitive interactions such as DNA-hybridization or protein-ligand interaction are ideally suited for opening or closing these defined spots via biorecognitive interactions with metal nano clusters. An array of resonant dot-cavities can be coupled with a set of waveguides to produce a very sharp filter through resonant tunneling. By modifying the spatial arrangement, its frequency can easily be tuned to any value within the band gap. Point defects are already in use in many non-bio devices, such as channel drop filters or resonant cavities.

Artificial membranes had been constructed that contain a parallel layer-type assembly of gold nano-tubules that span the complete thickness of a natural or artificial membrane. These nano-tube-doped membranes are e.g. prepared via a template method depositing gold within the pores of a template membrane. The inside diameters of the Au nano-tubules can be adjusted by controlling the metal deposition time from fully open to completely closed. Based on the template (e.g. Nucleopore membranes) the nano-tubules have inside diameters down to molecular dimensions. Thus, these nano-filters can be used to cleanly separate small molecules on the basis of molecular size.

Based on hollow-tube like nanoclusters a novel nano-analytical device had been constructed. If nano-tube-membranes are placed in a salt solution and a potential is applied across the membrane a trans-membrane current will be induced caused by the migration of ions through the nanotubes. Reversible or irreversible blocking of the tubes by an analyte molecule such as a DNA or a protein will induce a trans-membrane current drop. To achieve the selectivity in this type of sensor a biorecognition molecule is bound to the nano-tube-cluster. This approach is an artificial cell membrane gate similar to an ligand-gated ion channel and thus can be assayed by patch clamp, AC-impedance or potential-step methods.

## 2.4. Cluster-Cluster Aggregates

A single well-defined plasmon peak is typical for stable and isolated gold or silver clusters. Standard plasmon behavior can only be observed as long as no significant coagulation of the clusters takes place. The origin of a multiple (mostly two) peak spectrum observed on aggregate formation can be deduced from theory. Experiments clearly show that the low frequency peak is reduced with increasing the p-polarization in contrast to the high-energy peak which is nearly insensitive to a change of polarization of light. Thus, the low-frequency peak is caused by l-mode excitation whereas the high-frequency peaks corresponds to the t-mode of the multi particle assembly.

The stability of a cluster system against coagulation has been discussed above. Colloidal sols are stabilized against coalescence due to electrostatic repulsion of the

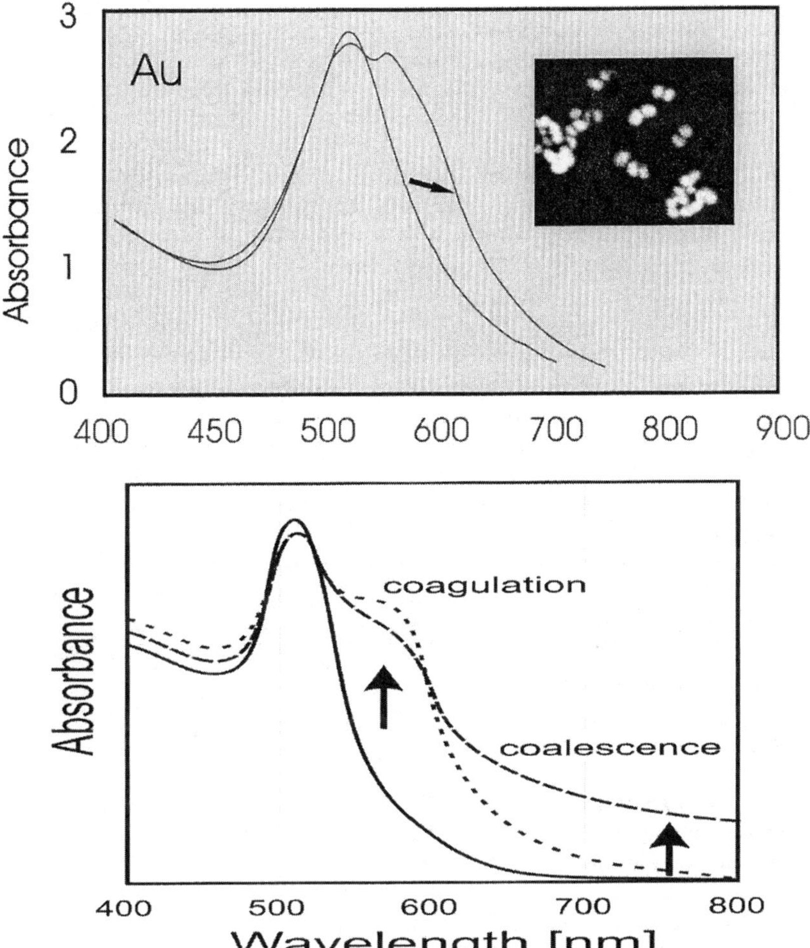



Figure 13. Spectral change induced by cluster aggregation and coalescence.

particles. To obtain ultimate stability repulsion is often combined with steric inhibition by polymers, proteins or DNA. These macromolecules prevent a metal-to-metal contact induced by thermal motion.

Distinct color changes results from the formation of aggregates of the nano clusters out of individual colloidal particles and further precipitation of the aggregated nano clusters. A gold sol will turn from red to violet on aggregation and further on to blue on coalescence and precipitation (Figure 13). This color change can be followed by eye or spectroscopically.

Aggregation of the nano clusters can be imaged by electron microscopy and quantified by light scattering techniques. A wide variety of bio-assays was constructed based on the aggregation of particles - often gold colloids - due to antigen-antibody interaction. All these assays have a unique advantage over ELISA being quick and simple single step assays. A disadvantage is the lower sensitivity compared to the ELISA approach.

The observation of a color change is facilitated by spotting a sample on a solid white surface, e.g. an RP-18 TLC plate surface. Upon drying a blue spot develops if the colloids are linked by biochemical interactions, otherwise the spot appears red. Silver enhancement kits - well established in biochemical research - greatly increase the sensitivity of the reaction.

Surfaces coated with aggregated or deposited cluster layers exhibit unique catalytic and chemical properties. Colloidal arrays made from cluster with a size from a few nm up to 5 μm had been employed to form lithographic masks on surfaces. The masks consist of hexagonally arranged monolayers of these particles. A droplet of the colloidal suspension placed on a substrate starts a self-organization process. The liquid film causes an attractive force between the nano-particles due to the lateral capillary force. Nano-clusters protruding out of the fluid induce an attractive force between the nano-particles, which exceeds the thermal energy kT.

Another simple and straightforward technique makes use of immiscible solvent to form nano structures upon solvent evaporation at the interface. Using a floating technique on a liquid interface the particle arrays are transferred to almost any arbitrary substrate. These nano-arrays have been utilized as masks for vacuum deposition of metal cluster arrays, plasma etching, or as print masters.

## 2.5. Coating Clusters with Biomolecules

A number of experiments point out that gold nano-clusters are composed out of a gold-zero core and a gold-one shell. It is mostly agreed that this is a direct result of the incomplete reduction at the outer surface or the high chemical reactivity of non-crystal embedded gold atoms. Similar behavior is found in most other cluster systems such as silver, copper or platinum. Thus, all reactive reagent used in metal-salt reduction coat the surface via physisorption or chemisorption (e.g. citrate or chloride ions used in gold colloid synthesis). As described above salts added to a gold sol turn the color to blue, inducing clusters precipitation. However, if polymers such as proteins are added under proper conditions, these molecules bind via adsorption to the metal cluster surface. Surfaces are modified and the sols remain stable in the presence of added electrolytes protected by steric stabilization. Binding of proteins to metal clusters is more or less irreversible due to the multiple attachment sites including SH, COOH, $NH_2$ or imidazol moieties. Most of the proteins retain their biological activities, at least in part. Stabilization via protein (or DNA) shell layers is proven in the salt precipitation test. Whereas un-stabilized clusters precipitate at less than 10 mM of sodium chloride-stabilized cluster will tolerate 200 mM without any aggregation.

Due to the ease of preparation and the chemical as well as optical robustness protein and DNA coated clusters are useful tools in various assays and biochips. The degree to which biological activity is lost depends upon the structural integrity and will vary from protein to protein. A conformational change induced by tight interaction with the metal surface may either stabilize the protein or induce protein denaturation. Binding of proteins to metal clusters is pH-dependent. Thus, the pH used in cluster coating - versus the isoelectric point of the protein - has to be considered. Though, there is no general rule a pH of around 1 step above the iso-electric point proved to be best for most of the proteins. No or just a minimal amount of buffer should be added to keep the ionic strength low, since salt induces cluster aggregation (a red gold sol turns to blue).The pH of a colloidal sol should be adjusted by the addition of 0.1 M NaOH, diluted potassium carbonate or by addition of 0,1 N HCl. Note: Glass electrodes are sometimes irreversibly coated by colloids!

On the average one 10- 20 nm colloid is covered with tens to hundreds of proteins. Due to this "over kill" even poor immobilization or biological activity results in quite useful cluster reagent. For Proteomics very tiny amounts of the proteins are available. In this case it is vital to determine the minimal amount of protein needed to stabilize the colloid.

Some fundamental quality assessment criteria are:

| | |
|---|---|
| Homogeneity | > 80 % single particles |
| Aggregates | No clusters greater than triplets |
| Particle Sizes | Un-conjugated colloidal gold sols 2-50 (250) nm |
| Optimal Size for optics | 40 nm ideal compromise in between signal and diffusion time |
| Assay -Buffer | TRIS Buffer 20mM, NaCl 200mM, pH 8.2. |
| Stabilizer | 20% glycerol (diffusion !) |
| Stability | Stable if stored at < -25°C. |
| Storage-Buffer | 20mM TRIS-base, 20mM $NaN_3$, 200mM NaCl, pH 8 + 1% BSA, 20% glycerol. |

Whereas choosing ideal buffer conditions protein-coating can often be done within 20 minutes, coating with protein in a 2-3 fold excess for several hours may help with low binding affinity. Anyway, the excess of the protein has to be removed by washing.

## Coating of gold colloid with proteins

1. 50 ml cluster suspension (OD540 ~ 1) is adjusted to pH 8.5 with a 0.2 M potassium carbonate solution.
2. Coating is performed by incubating the purified protein e.g. antibody for 20 min at ambient temperature while being gently swirled on a shaking platform.
3. While adding the protein antibodies, the sol should be stirred vigorously.
4. After coating, the sol is further stabilised by adding 5 ml 1% (w/v) skimmed milk powder that has been adjusted to pH 8.5 with 0.2 M potassium carbonate.
5. The mixture is gently swirled for another 60 min.

## Isolation and purification of protein coated gold colloid

1. To purify the protein coated colloid the obtained mixture is centrifuged through a 50% (v/v) glycerol layer that is mounted onto a 80% (v/v) glycerol cushion.
2. The centrifugal speed and/or running time is adjusted such that the right sized coated antibody clusters are found almost entirely in the 50% glycerol layer.

3.   After centrifugation, antibodies and/or blocking proteins that are not bound to the nanoclusters will be found in the upper water layer, whereas the large, purple colored colloid aggregates will be present in the 80% glycerol layer and on the bottom of the centrifuge tube.

4.   In order to avoid contamination of the detector reagent with free protein and/or large antibody nanocluster complexes, the coated cluster is removed sideways out of the 50% glycerol layer with a syringe.

5.   The obtained 4-5 ml of coated colloid in glycerol is diluted to 15 ml with water that has been adjusted to pH 8.5 with 0.2 M potassium carbonate.

6.   The solution is then transferred into a centriprep-30 concentrator.

7.   The glycerol is removed in about five centrifugational runs in which the retentate after each run is filled up to 15 ml with water of pH 8.5.

8.   If required, the suspension is concentrated to 2 ml.

9.   For storage 400 Fl of a 5% (w/v) skimmed milk powder solution in water of pH 8.5 (1% (w/v) end concentration) and 12 Fl of a 10% (w/v) sodium azide (0.05% (w/v) end concentration) are added

10.  The suspension is stored at +4°C.


## Conjugates with proteins:

| Size | OD [520nm] | Protein (100 kD) μg/ml | Proteins/ cluster |
|---|---|---|---|
| 5 nm | 3 | 36 | 3 |
| 10 nm | 3 | 30 | -10 |
| 20 nm | 4 | 30 | -50 |
| 30 nm | 5 | 15 | -100 |
| 40 nm | 5.5 | 12 | -150 |

Expensive proteins, proteins at ultra-low concentration or very tiny amounts of proteins are coated in sub-equimolar ratio to the cluster surface. In addition to the functional protein in a second step an nonreactive protein (e.g. BSA) is applied for the stabilization of colloids.

Similar to biochip technology a poly-L-lysine monolayer can be bound via adsorption onto a gold cluster coated with a self-assembled monolayer of a carboxylated thiol e.g. mercapto-succinic acid. Using a biotinylated lysine-polymer proteins are immobilized via avidin conjugation. The poly-L-lysine monolayer including all the proteins can be rinsed from the surface with a low or high pH buffer.


## 2.6. AFM

Ultra-sensitive analysis and high throughput screening require novel techniques with single molecule sensitivity. Microdotted arrays of oligonucleotides hybridize with DNA-functionalized gold or silver nanoparticles. The procedure enables the detection of single base-pair mismatches but suffers from limited sensitivity by a simple absorbance readout. To overcome this limitation either surface enhanced procedures had successfully been developed (see chapter 3) or scanning microscopes had been employed.

The position of individual colloidal gold particles is easily detected by AFM, or other scanning probe microscopy techniques along the straightened DNA molecule (Figure 14). DNA-oligonucleotides labeled with colloidal gold particles hybridizing to

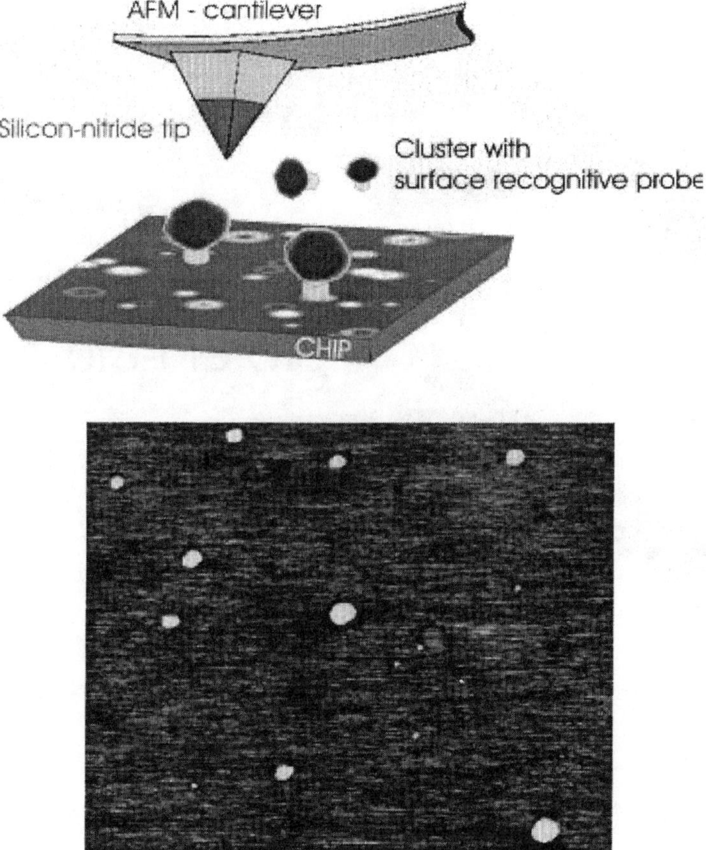



**Figure 14.** Bioassay via cluster counting with AFM. See Fig. 5.14 in the color insert at the end of this volume.

their appropriate targets allow sequence recognition as well as imaging of the DNA contour via AFM and STM devices. DNA molecules are labeled with metal nano clusters in a two step procedure by first hybridizing synthetic nucleotides conjugated with biotin. In a subsequent step coupling with colloidal gold particles coated with avidin allow imaging without any steric inhibition sometimes observed with large metal cluster reagents. A novel way to trace e.g. RNA in extremely large ribonucleoproteins (RNP) is the synthesis of RNA molecules that are covalently derivatized with gold nano-clusters, and thereby can be visualized by EM or by atomic force microscopy.

Single and multi Cantilever sensors (actually multi-AFMs) are developed to measure specific interactions between surfaces at the molecular scale (Figure 15). These forces acting on the sharp tips of a multi-cantilever chip are measured by a micro-AFM.

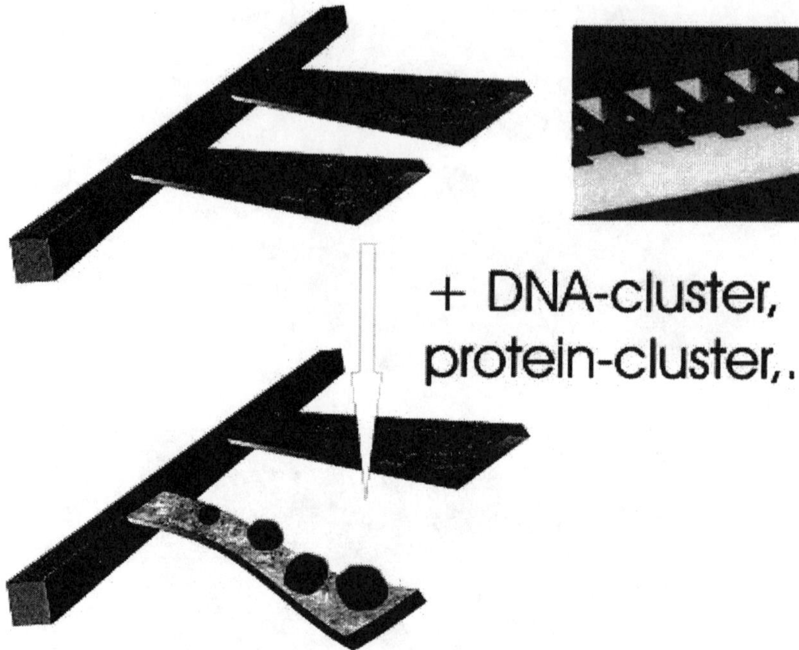


**Figure 15.** Resonance-frequency shift based detection of colloid binding.

Downsizing the cantilever structure to a microscopic size results in fast response times at high sensitivity and a high resonant frequency, additionally such sensors are quite immune to external mechanical noise.

Operating a micro-cantilever in the oscillating mode, enables e.g. to count the number of colloidal particles adsorbed onto the cantilever by antibody binding, due to significant shifts in resonance frequency. Coating the cantilever surface with specific receptors and labeling the analyte with beads results in an extremely sensitive single molecule sensor.

The Scanning Near-Field Optical Microscope (SNOM) operates on a quantum mechanical phenomenon, namely optical tunneling, in brief light crosses propagation barriers due to photon delocalization. SNOM scans a small spot of "light" over the specimen, thereby detecting the reflected, transmitted or scattered light. The resolution of the image obtained by SNOM is defined by the size of the aperture, therefore nano-apertures with diameters in the range of ~50 nm are used. The aperture is scanned at a distance of about 10 nm from the sample, providing a optical image with high resolution and the possibility of single molecule detection. SNOM proved to be particularly suitable for detection and characterization of the optical response of nano-colloid clusters and to investigate details as e.g. shape - resonance relation in nano-clusters extensively studied by Aussenegg and Leitner at the University of Graz, Austria.

## 2.7. Immune Colloidal Techniques

Conjugates, prepared by coupling of antibodies to colloids are used in a solution particle immunoassay. Various proteins are quantified via assays based on cluster coagulation, resulting in a color shift or precipitation. The best visibility to the eye is obtained using big particles of 40-50 nm or more. The upper limit is given by stability of the colloidal preparation and by diffusion limits. Higher concentrations resulted in steeper calibration curves by increasing the unspecific background of the assay. Addition of selected polymers sometimes increased the agglutination rate considerably. Whereas humane serum is a tricky environment the dose-response curves in buffer or urine are almost identical. Cluster based aggregation-assays have a high practicability, and are easy to automate by robots.

As a POC (point of care) device semi-quantitative flow-immunoassay methods for the detection of antibodies or antigens had been developed and are well establish in e.g. pregnancy tests. This format is also known as strip test, one step strip test, immunochromatographic test, rapid flow diagnostic, rapid immunoassay or lateral flow immunoassay (LFI). The label is anyway a colored colloidal particle. Additionally to metal colloids, carbon (black) and silica or latex (various colors) are in use.

Colloidal particles of a diameter around 25-40 nm are the most commonly applied in strip tests. The colloidal particles are covered with an organic reagent which is attached to the antibody by covalent bonds. Thus, antibodies or antigens - for a sandwich type immune-assay - are immobilized on the colloid surface and onto the surface of the porous chromatography matrix. Thus, a cluster-chromatography device consists of a porous strip that is mounted in a plastic cassette. Additionally, the device includes: a sample pad, a conjugate pad, an absorbent pad and a membrane that contains the capture reagents. In micro-fluidic immune-chromatography devices the immune reaction proceeds directly at the wall of the nano-channel.

First devices developed in the early 80s had been built on paper strips developed with an appropriate gold reagent in a plastic tube in the presence of urine or serum [16,17]. The analyte-reagent mixture migrates up the strips towards the top end. A purple dot or band develops which indicates the presence of the corresponding antigen in 5 minutes or less.

Strip tests are simple, rapid and cost efficient. A disadvantage is the limited sensitivity of the assay and lack of multi-dimensional multi-analyte capabilities. Recently, micro-fluidic cluster chromatography devices proved to be promising candidates to overcome these limitations.

## 2.8. Binding and Assembly of Functionalized Colloids

Any biointeraction e.g. a standard hybridization is a useful tool to assemble colloidal particles in solution or at surfaces. By use of this technique arrays of colloids are easily obtained at a substrate surface. These particle arrays deviate in their optical and electrical behavior from single isolated colloidal particles.

In optical devices regular arrays of nano-clusters give a very sharp resonance signal due to the large number of cooperative interactions in between the particles. Whereas this effect allows precise resonance tuning the same cooperative effect limits the dynamic range of the reaction if used for direct analyte detection. E.g. DNA-covered nano-clusters

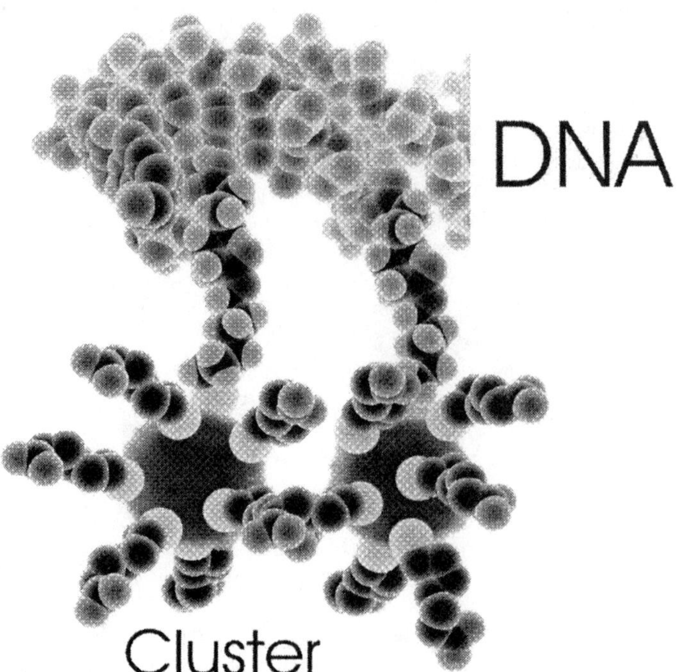


Figure 16. Coupling and assembly of DNA-functionalized clusters.

will only assemble in the presence of a complementary DNA strand. The change in optical, electrical and material properties induced by assembly can be used as an indicator of whether a particular DNA is present in a sample or not (Figure 16).

Based on the detection of a well-defined optical peak induced via colloid aggregation a new analytical device was developed by Mirkin et al. [18-20]. The method is based on a colorimetric detection of DNA and RNA down to femtomolar level using mercapto-oligonucleotide modified gold colloids. Oligonucleotide-clusters assembly in solution via hybridization results in the formation of a cluster network. Two preparations of colloids are coated with DNA sequences specific for the same target molecule. Upon addition of the target the colloids bind to the target. This aggregation is visible to the eye as a red to purple-violet shift of the solution (see chapter on cluster stability). Dotting of the aggregated cluster-solution to a silica matrix (e.g. TLC-plate) results in a shift of the signal to blue.

Immobilization of short synthetic oligonucleotides to gold nanoparticles follows slightly different rules than the attachment of proteins. A gold colloid 15-20 nM in particle concentration is reacted with a 200 fold access of a thio-functionalized oligonucleotide (3.5 µM) in water. In order to allow reorientation at the surface the reaction is allowed to stand for 24 hours at room temperature. The excess of reagent is removed by centrifugation. Up to 100 molecules are immobilized to the surface of a 15 nm colloid.

Whereas thiol bonds are stable at room and slightly elevated temperatures the stability at 70-100°C is low. To increase the chemical stability either multivalent thiols or silanes are applied. Both strategies clearly target at multiple attachment sites in between the oligonucleotides and the cluster. Using silane coated colloids the DNA is bound covalently to functional groups at the surface of the coated particle.

Nature provides a complete toolbox of specific biomolecular reagents, such as ligases, endonucleases, methalyses and another DNA modifying enzymes, which allow for the processing and handling of DNA-nanocluster material with atomic precision and accuracy.

## 2.9. Bio-Templating

Microlithography is in a severe need for novel lithographic techniques applicable to pattern semiconductors at the nanometer scale. Bio-templating is among the most promising strategies for the fabrication of such molecularly engineered nanostructures [21,22,23].

Protein recognition and hybridization of DNA are excellent tools to directly and precisely assemble nanometer-sized structures. These biotool-kits allow the fabrication of materials with defined nano-properties. Assembling techniques are applicable towards the construction of novel nanostructured materials made from any material whatsoever.

Nano-devices are e.g. set up via synthesizing appropriate DNA linkers and colloidal building blocks and assembling them via hybridizing and dehybridizing. Well-defined protein layers deposited at the chip surface are used as templates for the build-up of organic and inorganic nano-assemblies. Biomolecular templating focuses on complex structures employed for the fabrication of nano-chips including well-defined nano-dots connected via metallic nanowires to large macromolecular assemblies built from identical subunits with a regular structure.

The periodicity arises from the self organization of biomolecules [24]. The 2-dimensional nano-grids are a precise spatial modulation of chemical surface (Figure 17). Nano-technology makes use of this grid to accomplish site-specific chemistry.

Contrary to UV- or ion beam lithography biotemplating-techniques provide the advantage of parallel fabrication. Nano-arrays from e.g. gold, silver, platinum or cadmium sulfide clusters deposited onto the 2D protein template had been constructed by several groups. These arrays are nano-masks which in a further processing step are used to e.g. set up array of nano holes by etching techniques. The topology of the nano-pattern is imprinted by the geometry of the template representing the intrinsic spatial structure of the underlying protein film.

Nature demonstrated the use of regular 2-dimensional protein grids in bacterial surface S-layers (5 nm to 15 nm thick). These layers are protein crystals forming the outermost cell envelope of many prokaryotes. A variety of lattice symmetries from p1, p2, p3, p4 to p6 is found in bacterial strains. Sometimes even a single strain is able to switch from one type of S-layer to another induced via a change in the micro-environment. The pores and spacing in between the units varies from 3-30 nm. S-layer proteins are handled dissolved or suspended in a buffer at a concentration of up to 2 mg/ml. Their stability and unique property to coat two-dimensional arrays with perfect uniformity makes them an ideal nano-template.

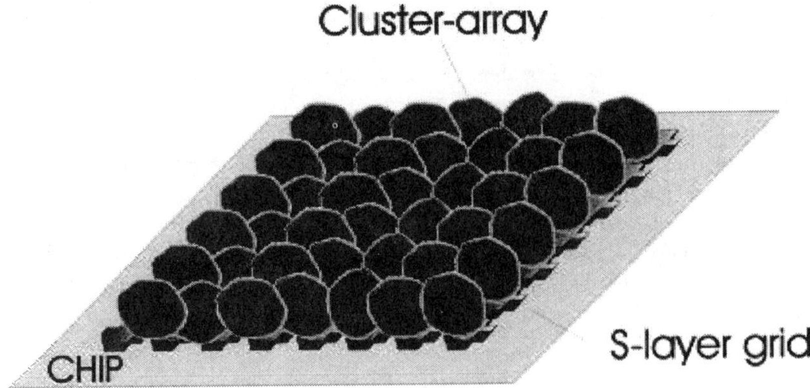


**Figure 17.** Array of nano-clusters arranged via bio-self-assembly.

Well-defined sites either found in the natural protein or inserted via genetic engineering are decorated with metal particles deposited from the aqueous or gas phase. Dieluweit et al. constructed a super-lattice of uniform gold colloids with a precise repetitive distance of 13 nm based on the S-layer grid of *Bacillus sphaericus* using chemical modification with thiol-groups as artificial anchor points.

For *in situ* generation of clusters protein might be activated by molecular deposition of metal catalyst (e.g. Pd or Pt) adsorbed from a metal salt solution. The activated protein layer in immersed in the metal salt. The chip surface turns from colorless to light yellow (or red) getting darker with increasing reaction time, indicating metal deposition. The catalyzed chemical deposition of metal onto the protein crystal leads to highly ordered arrays of metal clusters aligned along the crystalline structure of the protein template. Another related technique makes use of a recrystallization process induced by the electron beam while scanning over a thin gold film deposited on a S-layer lattice.

Biomolecules are an enormously broad class of reagents. The electrostatic and topographic properties of RNA or DNA can be employed in template synthesis of supramolecular assemblies. The negatively charged phosphate backbone of the DNA-double helix enables to accumulate multivalent metal ions, which are subsequently used as a catalyst for gold, silver or nickel deposition. TEM analysis reveals that these particles are homogeneous in nature with a typical size around 5-10 nm. Exchanging alkali metal ions for silver ions and using them as a catalytic center for further silver deposition allowed the construction of nano-wires based on DNA templating based on a chain of silver nano-granules. Similar experiments had been done to deposit palladium colloids on nucleic acid templates. Based on electrostatic attraction the assembly of poly-lysine coated colloidal particles had been reported. Not only nucleic acids and S-layer proteins but also e.g. microtubules and even viruses had been used for bio-templating.

Microtubules are highly ordered protein structures. The tubes are composed out of regularly arranged alpha-beta tubulin dimers (l=8 nm, d=5 nm). Micro-tubes are useful templates for assembling small metal clusters into nano-fibers. As typical for all protein molecules, the surface of the tubulin mo lecules exposes a well-defined pattern of amino acids. These charged micro-domains provide active sites for nucleation, organization, and binding of metal clusters. Structural analysis revealed histidines to be the sites of metal clusters binding in the tubulin molecule. Clusters deposited on microtubules formed organized chains reflecting the regular arrangement of the tubulin subunits within the structure. Taxol, a well established drug for cancer treatment, helps to stabilize two-dimensional tubulin sheets.

## 2.10. Colloidal Particles and Electrodes

Electrochemical and micro-conductivity sensors can be set up by *in situ* assembly of colloids onto microelectrodes. Nanospheres suspended in solution are collected via electrophoresis or dielectrophoresis (based on the field gradient force) in a micrometer-sized electrode-gap. Biorecognitive molecules on the nano-particle surface bind the analyte molecules. Aggregated micro-spheres disassemble when the field is turned off. The critical step to bind the nano particles is to decrease the repulsive interactions and to coagulate the particles. The attracted nano-particles are thus fixed by e.g. changing the colloidal interactions or by washing away the protective layer. Aggregated micro-spheres, however, do not exhibit enough signal. Thus, coating of the nano-particles with colloidal gold particles bridges the electrode gap with a conducting layer. This step is carried out by immunological tagging of colloidal gold clusters to the nano-sphere layer. If necessary, further enhancement could be achieved using silver enhancement. The signal to be quantified is the resistance between the electrodes. The method is ideally suited to set up disposable chips for proteins or DNA [25].

The reconstitution of an apo-flavoenzyme (apo-GOD) on a 1.4 nm gold nano colloidal particle functionalized with the cofactor FAD has been done recently by Xiao et al. The electron transfer rate observed in this system was up to 5000 s-1 and thus significantly high than ever observed with the natural substrate oxygen (700 s-1). The noble metal particle acted as an electronic relay wiring the charge from the active site of the enzyme to the supporting electrode.

One of the outstanding issues is the use of colloids to follow DNA-hybridization on electrodes via conductivity changes. First experiments had been done by the group of Mirkin et al. interconnecting oligonucleotides of 24 to 72 bases length. The assemblies of nano-particles behave like a semiconductor and retain their discrete multi-particle structure.

## 2.11. SPR-Transduction

Surface plasmons are excited if an electromagnetic wave is coupled into a metal/dielectric interface. These plasmons are charge density oscillations propagating at that surface of the metal. The wave is a perturbation of the electron plasma extending several tens of nm into the metal (depending on the skin depth) and decays exponentially up to 500 nm into the adjacent medium.

To couple a light wave into a metal surface plasmon a well-defined resonant angle is necessary. Any over-coating of the metal surface varying n and thus will change the coupling angle of the resonant system. This change induces a shift in resonant angle. A read out of the angle or a color-coded detection is the basis of various devices commercially available for about one decade and well established in the pharmaceutical industry to monitor drug-receptor interaction [26, 27, 28]. Whereas the first instruments had been single channel instruments new devices run up to 12 channels in parallel. Some companies are testing a parallel detection of SPR-on a chip array.

Whereas a direct transduction is very useful to study biorecognitive processes of pharmaceutical importance the overall-sensitivity is low. Thus based on the large difference of $\varepsilon_{/20}$ and $\varepsilon_{metal}$ a significant increase in assay sensitivity is obtained using a metal nano-cluster-label. For biomedical application the significant increase in sensitivity is outweighing the disadvantage of a 2-step labeling procedure. The immune-colloid-reaction leads to a large shift in resonance angle, a broadened plasmon band, an increase in minimum reflectance, and to a boosting of the sensitivity up to 3-orders of magnitude. Using colloidal labels in SPR allows to drive the detection limit near to the traditional fluorescence based techniques.

## 2.12. Electroluminescence

In contrast to standard colloidal particles ultra-small cluster of a few atoms in size exhibit unique luminescent properties. To produce these clusters e.g. a thin film of silver oxide is treated with a high electrical current of about 1 A. This creates a thin line of silver clusters. Under AC these clusters exhibit strong electroluminescence, emitting light at a frequency that depends on their band-gap and thus on their size. Driving the cluster with high-frequency (> 100 MHz) alternating current produces an electroluminescence response that is 10,000 times as great as by DC.

Whereas bulk materials cannot respond quickly enough to an alternating current of such a high frequency, AC creates rapid electron-hole recombination within nano-clusters. Similar fluorescence emission phenomena had been observed with a variety of ultra-small metal nano-clusters.

## 3. NANO-CLUSTER AND FIELD EFFECTS

## 3.1. Surface Enhanced Optical Absorption (SEA)

### 3.1.1. Physical Principles

The basis of surface enhanced optical absorption is the so-called "anomalous absorption". To observe anomalous absorption an absorbing colloid or colloid layer is positioned in a defined distance to a metal mirror and illuminated from the colloid side. At a certain distance of the colloid or absorbing layer to the mirror the incident fields has the same phase as the electromagnetic field that is reflected by the mirror at the position of the absorbing colloid particle (or colloid particle layer). The set-up is described as a reflection interference system, which feedback mechanism strongly enhances the absorption coefficient of the absorbing colloid (layer).

At a given colloid mirror distance and defined path of the light only one wavelength is optimally in phase, resulting in a relatively narrow spectrally reflection minimum. Changes in the path length (colloid – mirror – colloid) either by varying the colloid mirror distance or the angle of the incident light (in case of a colloid layer) result in a spectral shift of the reflection minimum [29,30] (Figure 18) accompanied by a change of the color impression of such a colloid layer.

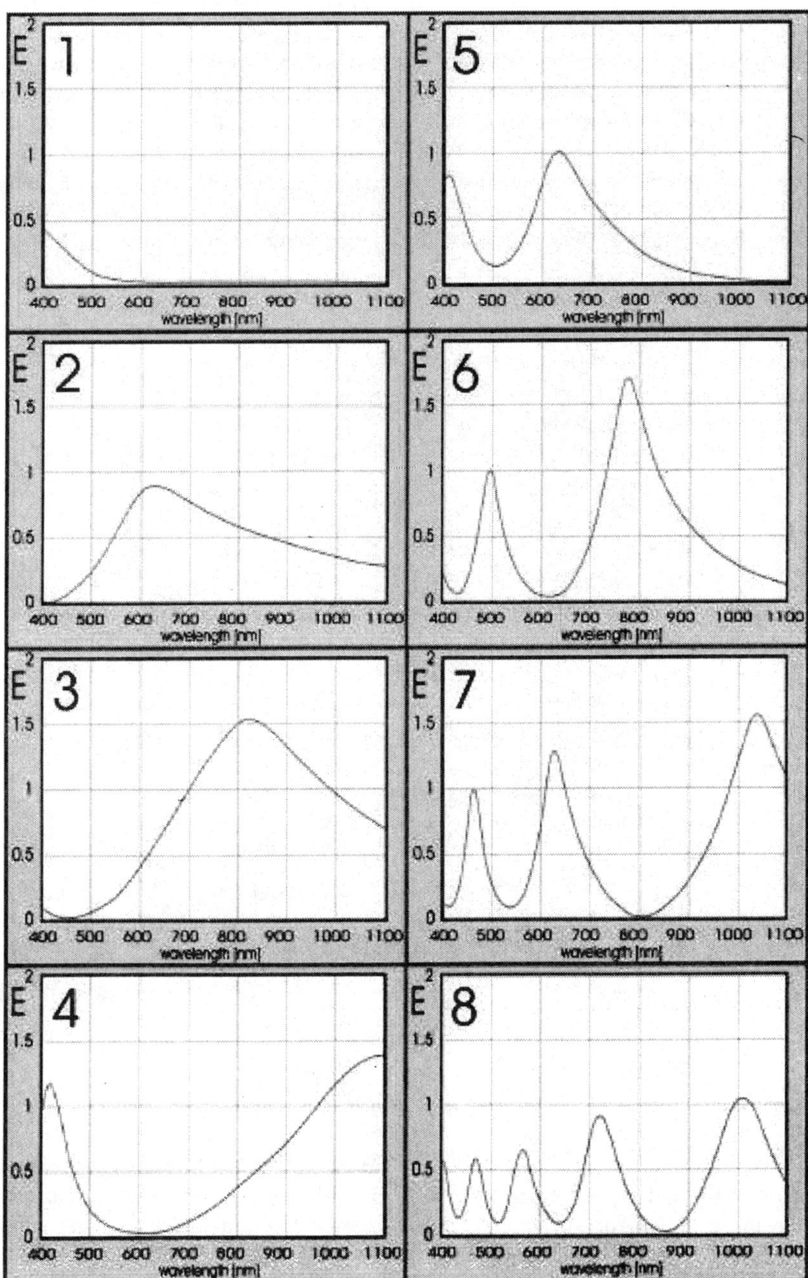

**Figure 18.** Spectral response of a SEA-chip with increasing distance layer.

The above-mentioned dependence of the reflection minimum enables to tune the optimum wavelength, the dynamic range of the system and the thickness of the "distance" layer as well as using changes of the distance layer's thickness to change the other parameters, in most of the cases resulting in an altered color impression of the reflected light (when using white-light illumination). The possibility to apply these changes as detection principles are discussed below.

Competing effects to the colloid-resonance generated color are unspecific background absorption as well as thin film colors. Thin film colors are dependent on the differences in the refraction indices of the used layers, and generally cause problems with high differences. Techniques to overcome that problem are costly or time consuming - they either include changes in the layers (height, comp osition) or the exchange of air (as the top medium) with a substance with a proper refraction index. Contrary to that unspecific background absorption of the sample measurement can be avoided easily by using two angles of observation. Whereas absorption of chromophores is independent of the angle of observation, the specific signal strong shifts are dependent on the angle. Thus by simple subtraction of both signals the background resulting from matrix effects can be eliminated.

A theory to describe the optical characteristics of a colloid - metal surface system is the CPS-theory developed by Chance, Prock and Silbey. Another theory more focusing on colloid films interacting with metal surfaces is the stratified medium theory (SMT).


### 3.1.2. Applications

There are various options to use the SEA set-up for detection and investigation of nucleic acids, enzymes or receptor proteins. The possible parameters are:

- induced changes in the packing density of the colloid layer or
- induced changes in the spatial arrangement of bound colloids
- distance changes of the colloid layer to an electron conducting surface
- local variation of $\varepsilon_m$.

The result of these changes are changes of the sensor surface's optical appearance, thus the intensity and/or color of the surface is altered e.g. due to binding, catalytic activity or a structural rearrangement of the analyte or a biorecognitive active component added to the mixture. Moreover the sensor setup can transduce changes in the extent of surface coverage with bound colloids via:

- a strong change in absorption at a defined wavelength or
- a spectral shift of an absorption maximum.

However, the most obvious technique is to couple a bio-component to the colloid and make the presence or absence of the colloid on the surface dependent on a bio-recognition process. This means that the analytes either induce binding (colloid is bound to analyte in solution) or dissociation (colloid is bound at the surface) of metal colloids at the proper distance to a reflecting substrate surface. These binding or dissociation events are transduced by the system into an optical signal, which in a good setup and proper concentration conditions can be observed with the naked eye.

Dependent on the application, colloids from 5-100 nm can be used. Usually the diameters are well below 100 nm, due to problems of instability of the sol and due to slow diffusion. Generally, diffusion works faster with smaller colloids, on the other hand the signals are stronger with bigger colloids. The choice of the size of colloids will on one hand depend on the averaged diffusion way the colloid-bio-component couple has to find its target (small are preferred) and on the other hand on the concentration of the bio-component, which is coupled to the colloid. A low bio-component concentration will result in fewer colloids at the targets position (thus, big colloids are preferred). Other reasons influencing the choice of size are the suppression of multipole peaks in the spectrum and storage time due to precipitation and coagulation of the colloidal reagents. According to all these considerations colloids mostly in between 12 and 50 nm in size are chosen.

The technique is simple, reproducible and more cost effective compared to standard ELISA methods. The one step procedure reduces handling time. Additionally the test can be stored for years without losing signal intensity due to the photo- and environmental stability of nano particles.

This technology can provide rapid and sensitive one-step-test-kits for clinical use, lab use and even field tests (in cases, where the result is visible to the naked eye). Possible fields of application include e.g. the screening of allergens, especially food allergens, food-component identifications, detection of pathogenic bacteria and diagnosis of urinary tract infections.

Another way to use the technique is to fix the colloid onto the distance layer's surface. If the height of distance layer itself is changed specifically via. an enzymatic assay shrinking or swelling the distance layer, for example, a color shift is observed. By this, the presence of an enzyme substrate or a reactive chemical is transduced by the system into an optical signal. The effect of moderate to high analyte concentrations can be observed with the naked eye. Applications and details of this technique are discussed below.

### 3.1.3. Distance Layer and Colloid Layers

Innumerable variation of colloidal assay formats are available all of them positioning the nano particles at a defined distance to a mirror. Glass, metal oxides, nitrides, metal fluorides or just polymers can act as transparent distance layer, which can be activated with various chemicals (Figure 19). In the case of polymers the active groups for biomolecular attachment can be present "ab initio". The colloids can be bound by biochemical linkers, which couple either to the mirror layer itself or to the distance layer, at a defined distance to a mirror. A detectable signal will result if these linkers are either formed or cut by biochemical recognition or by catalysis, or if their spatial arrangement is altered. An example for that are oligonucleotides acting as linkers which can then be cut by specific restriction enzymes from microorganisms. Due to the fact that many microorganisms express specific restriction endonucleases their presence can be identified due to color loss of the sensor's surface.

Cluster layers can be applied via sputtering or precipitation processes directly onto a surface, but in most cases they will bind via the above described linker application. Neither way do these colloid layers or cluster films form a barrier of diffusion for fluids or gases. The reason for this is the particle structure of these cluster films enabling an "easy" passage for fluids and gases. This allows measurement of analyte concentrations

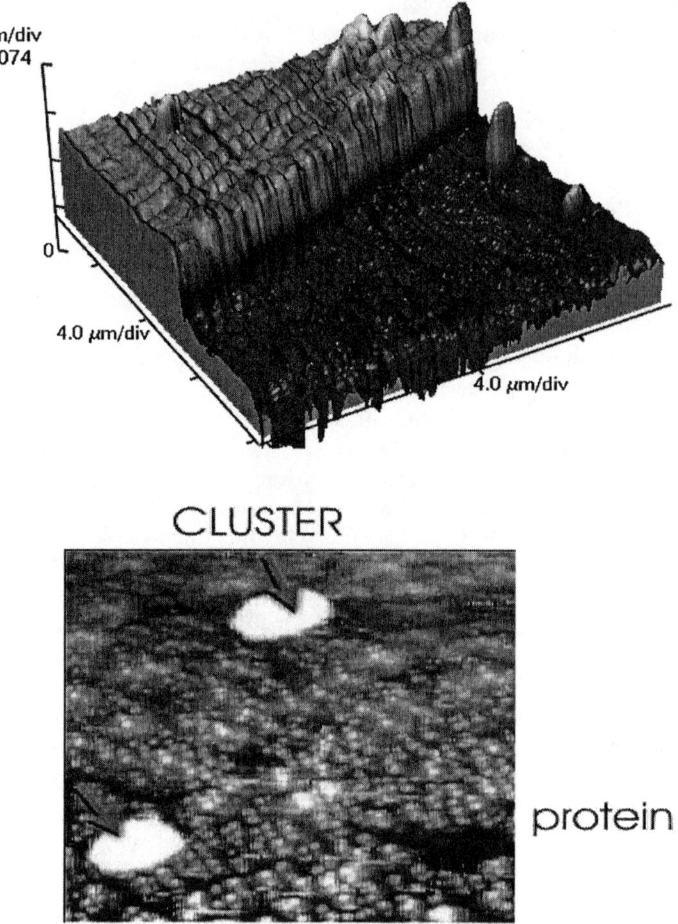


**Figure 19**. Chip surface: (a) the interlayer (glass-type SnNx),  (b) cluster bound to protein  coating on interlayer.

with high sensitivity via visual inspection of the sensor surface as described above. Due to the fact, that the intensity of the absorption band is directly proportional to the number of interacting colloids, a quantitative measurement can be applied. Any reduction or increase of the number of colloids results in a lowering or increase of the absorption of the system and additionally at a high surface coverage in a spectral shift due to a change in colloid-colloid interactions. However, most of the POC systems are focusing on a semi-quantitative measurement.

It should be noted that in a SEA-setup, up to a wavelength of around 600 nm the optical properties are equivalent for chemically synthesized colloids and sputtered metal colloids. At any wavelength above 600 nm, sputtered cluster layers show dramatically different optical properties than crystalline colloids. This results from the difference in shape (high asymmetry in sputter coated clusters) as well as from varying degrees of crystalline nano-structure.

### 3.1.4. SEA-Biochips

*3.1.4.a. The SEA Chip.* The SEA biochip makes use of surface enhanced absorption, thus the increased optical absorption of colloids in nanometric vicinity to a wave reflecting surface is influenced via biochemical recognition processes. The chips are set up with a mirror layer on a substrate flat surface (mostly metal), an inert and transparent distance layer e.g. deposited by spin coating or chemical vapor deposition, on top of which individual biorecognitive molecules are bound. These biocomponents can either cover the whole or big parts of the surface, or are applied via dotting robots. The introduced arrays enable such a chip to investigate a big variety of proteins, oligonucleotides or microorganisms at the same time in a one-step procedure.

Depending on the nature of the biorecognitive interaction the colloids are applied to the surface. One possibility is to incubate the colloids with the analyte forming protein or DNA coated colloidal particles. At incubation on chip this coating interacts with the biorecognitive molecules on the chip surface and leads to a binding of the colloid to the surface in case of biorecognition. Another possibility is to incubate the chip with the analyte, which leads to a binding of the analyte to the biorecognitive molecules on the chip surface. Pre-synthesized particles with a second biorecognitive molecule as their "shell" targeting either the analyte or the complex of analyte- biorecognitive molecule also binding the colloid to the chips surface during incubation. If required (in case of reversible binding events) these binding steps can be monitored in real time. Such direct and sensitive kinetic monitoring enables to transduce a number of molecular binding events into an macroscopic optical signal, under ideal circumstances even visible to the eye.

The enhancement in absorption of spherical metal colloids is approximately 8 fold, relative adsorption enhancement is reported above 100 fold. The enhancement is independent of the way of application of colloids but depends strongly on the attachment of colloids under given conditions [31-36]. The relation between the number of colloids and the obtained optical signal obtained is investigated using scanning techniques based on STM and AFM.

*3.1.4.b. Applications and General Requirements.* On principle any DNA hybridization assay and conventional ELISA can be adapted to the colloid detection protocol, transducing either DNA-DNA or antigen-antibody interactions. An example for that with *E. coli* proteins and the corresponding antibodies is discussed below. To choose the primary support in a standard micro-slide format, in the easiest case a microscopic glass slide itself, gives the advantage that many standard dotting and scanning devices can be used without time consuming modifications.

To set up a very cost and time efficient interlayer, polymers are deposited via spin coating and activated via plasma chemistry. However for large numbers and high quality, interlayers manufactured by sputter techniques or CVD techniques are an excellent choice. Using printing techniques on surfaces similar to a CD-ROM give an advantage

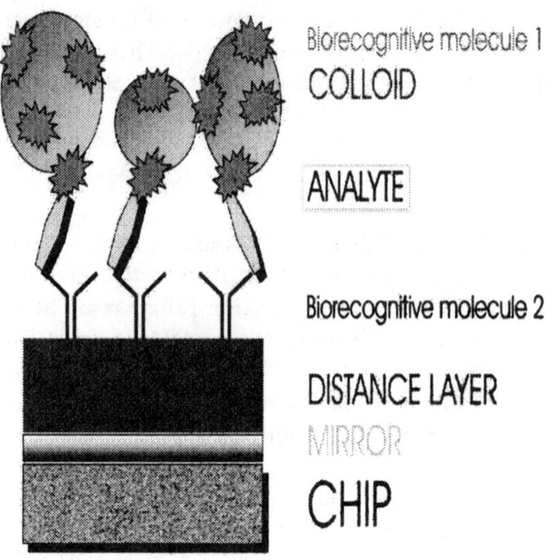


**Figure 20**. SEA-Capture assay.

due to the ultra-high density which can be achieved. Areas of 1 x 1 μm easily can be read out with rather cheap and simple devices based on a commercial CCD-camera chip technology. Using such ultra-flat surface materials array-based sensor chips are easily set up using standard micro-dotting or inkjet-printing devices or photolithography. With help of these techniques most biological important molecules such as. ligands, proteins or oligonucleotides can be immobilized in a well-defined array on a SEA-chip fully compatible with glass-slide based systems (Figure 20).

As described above the procedures according to Frens is a suitable technique to produce colloid solutions. However the aimed red color is obtained after color changes from yellow to black, dark blue, violet, and finally shifts suddenly to red. These earlier stages of colloid synthesis absorb light substantially above 600 nm and can be isolated by a rapid cool-down or addition of metal ions blocking crystallization of the gold colloid. With such glass-type non-crystalline colloids also resonance enhancement at $\lambda > 600$ nm is achieved, which enables to shift the desired absorption peak all over the visible spectrum to IR.

*3.1.4.c. Setup*

**Substrate**

As substrates standard microscope glass slides, aluminum foils, metallized plastic foils like prefabricated aluminated or titanium-coated polyethylene-terephthalate proved to be suited. A wide variety of polymers as e.g. polycarbonate or polyethylene-terephthalate sheets after hydrophilization by oxygen-plasma etching might serve as chip substrates. Moreover, any highly polished metal foils or sheets can be used provided that their surface is sufficiently reflective. The disadvantage of such approaches is the poor compatibility to many standard arraying and scanning devices.

## Mirror layers

On a plastic chip or glass slide the deposition of a mirror-layer is necessary. Silver, aluminum, gold or gold-palladium is deposited by DC- or RF-sputter-coating. For metal coating mainly argon (for some noble metals nitrogen is possible) is used as a sputter gas, bled into the chamber during the process. Sputtering must be optimized for the optical quality of the metal film, means the highest reflectivity, as well as adhesion of the metal film on the support and chemical and mechanical stability of the metal film for further treatment processes. Attachment of silver, platinum and gold on glass is rather poor, thus, adhesion layers of tungsten, chromium or titanium are required. One should keep in mind that many of these adhesion layers are chemically reactive, e.g. the standard adhesion layers used for gold or silver mirrors such as chromium or tungsten are dissolved easily by chloride based buffers. The chemical stability of the mirror is excellent for gold whereas silver darkens quickly and aluminum is often etched in aqueous environment. The problems of the sensitivity and chemical stability can be solved by applying a proper protecting distance layer quickly.

Due to poor adhesion properties most plastic based chips need a treatment to achieve sufficient adhesion of metal to their surface. The surfaces of such materials can be functionalized by oxidation in plasma. The chip is placed in the vacuum chamber of the sputter coater. Aluminum targets are used because of their "bad" sputter rates. The chambers oxygen pressure is adjusted to 0,02 mbar and voltage is applied at a low power setting, which does not "ignite" the aluminum targets. A plasma is formed between the targets and the metal bottom of the sputter coater. The activated oxygen atoms bombard and oxidize the chips surface. Thereby any surfaces can be cleaned from organic substances. In the case of polymers the oxidation reaction results in a wide variety of active groups like hydroxyl, carbonyl and carboxyl groups. The etching process with oxygen plasma is usually 10-30 seconds.

## Distance layers

To obtain the desired resonance an distance layer tuned to the appropriate optical thickness is coated on the mirror. This can be achieved via sputtering glass or metal glasses such as e.g. $SnN_x$ or $ZnO$. Another way to achieve this is by spin-coating of a polymer-layers, whereby the adhesion of the polymer film onto the metal surface is most critical for chip stability. E.g. a distance layer of hexyl polymethacrylate is applied by spin-coating at 4000 rpm using varying dilutions of the polymer in AZ 1500 Photothinner (Hoechst) and n-decane or n-octane. The optical thickness of the interlayer can be obtained pretty accurate by using a color step ladder (with known thickness of the steps) of a SEA system, which enables the calculation of the real thickness, provided that the refraction index is known. Without such a color step ladder the thickness is best studied using atomic force microscopy, and the optical thickness is calculated.

## Activation of distance layers

To attach (mostly covalently) biomolecules onto the distance layer's surface a chemical activation is needed. One way to achieve that activation is the etching of polymer layers by oxygen plasma. The polymer surface is e.g. etched with oxygen plasma for 10-30 sec with the above described procedure and settings, which will result in strong hydrophilicity of the distance layers surface. Another way for chemical activation of the polymer surface is the chemical break-down of the polymer surface, introducing chemical reactive groups. E.g. functional –COOH groups can be created at the surface by hydrolyze the bonds of an ester polymer, usually using strongly alkaline solutions (like 4 M KOH).

The introduced groups can be further modified and activated, e.g. with carbodiimide or divinylsulfone chemistry, which enables the surface to bind to e.g. amino groups of proteins. However, also the activation of the protein (or any other amino or thiol-group carrier) using the same chemistry is a means to couple covalently to the surface. On the other hand for many applications simple non-covalent absorption to a plasma-etched polymer is sufficient to bind the ligand-layer to the polymer surface adapting know-how from the ELISA-technology.

Another powerful technique to functionalize the distance layer so that DNA, proteins or thin films can be covalently attached is silane based technology. A variety of silane derivatives can be bound and crosslinked to form a stable monolayer introducing a variety of chemically reactive groups. To achieve a thin and very homogenous coverage, vapor coating is the correct choice, provided that the silanes applied have a considerable vapor pressure to guarantee a sufficient adsorption rate. The chip surfaces must be cleaned, either by washing with iso-propanol or via reactive oxygen plasma etching. Chips and silane (mostly a silane solution) are placed in a vacuum chamber and vacuum is applied for several minutes. After the pump is switched off, silane fills the evacuated chamber and attaches to all the surfaces therein. The adsorption process is usually done overnight. Subsequently, the chips are heated for 1 hour to 60°C or 10 minutes at 105°C to cross-link silane molecules with each other and the surface. Thick and crosslinked silane layers are obtained via coating in solution. The surfaces are cleaned as described above. The silanes are diluted in ethanol at concentrations from 0,01% to 5% and an equivalent amount of distilled water is added to the solution. The chip is immersed into the solution, whereby concentrations and incubation time determine the layer thickness. The chips are washed with iso-propanol to remove access silane or dried immediately, what strongly increases the tendency to form multilayers. Subsequently, the chips are heated as described above.

**Colloid covering**

As mentioned in 2.7 unlabeled or not fully labeled colloids are unstable, especially sensitive to electrolytes, therefore an efficient labeling strategy is needed: The protein of interest is dissolved in 200 Fl of water at 1 mg/ml. Serial dilutions (1:5 to 1:10) of the protein in distilled water are prepared with 100Fl of volume each. 500 Fl of the pH-adjusted gold sol is added to each tube, and after 10 minutes 100 Fl of 10% NaCl (electrolyte) in distilled water are added. Tubes with stabilized, means sufficiently covered colloids will maintain a red color, unstable gold sol will turn to violet and blue and finally flocculate. The second tube containing more protein than the one whose color changes to blue is sufficiently covered. With the optimal pH and amount for adsorption determined, the protein-colloid solution is prepared in the desired amount. Excess protein is removed by centrifugation, the pelleted colloid particles are re-suspended in a small volume to prepare more concentrated solutions. Often the use of protein concentrators (Centriprep spin columns) is preferred to remove protein and concentrate the colloids,

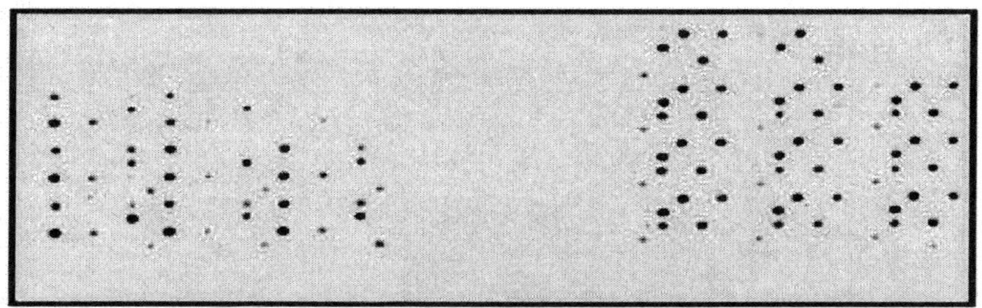

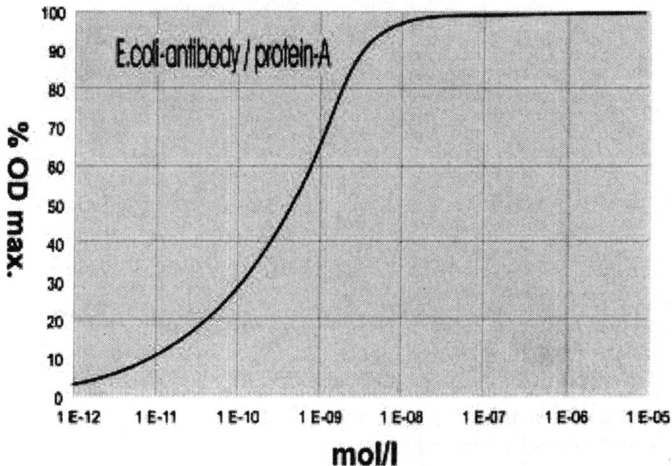


Figure 21. Optical and AFM-scan of a cluster biochip and calibration graph for a protein A based assay.

because pelleting and subsequent loss of precipitated colloids is avoided using that method. Most protein-gold conjugates retain their biological activity for months at 4°C when protected from microorganism growth by 0,02% sodium azide added to the buffer solution. Storage for a longer period is possible at –20°C in the presence of 45% glycerol.

*3.1.4.d. Example and Results.* Example to set up and use a polymer-interlayer-SEA-biochip: Onto metallized plastic foils a 6 % solution of polyhexyl metacrylate is applied by spin coating (4000 rpm). The surface of the thin-film is functionalized under oxygen-plasma as described above. E. coli proteins are micro-dotted onto the chip SEA-surface and bound via EDC coupling in a moist chamber (2h, RT). The chip is washed several times with neutral buffer. Gold colloids with 30 nm diameter, coated with anti-*E. coli* (IgG) are added in a buffer containing 0.1% Tween 20 (or similar reagents) to suppress unspecific binding. Anti-*E. coli* coated colloids bind to the sensor surface and the proteins develop a color visible to the eye. The binding of the colloid to the surface is followed either visually or via a CCD camera (Figure 21). After subsequent washing the

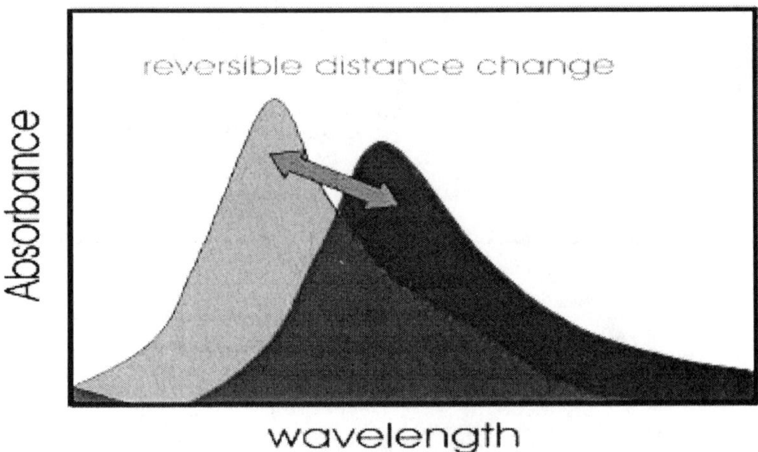


**Figure 22.** Shift in optical response induced by a change in the reactive interlayer.

chip is dried and scanned in direct reflection mode. The detection limit of *E. coli* protein is in the range of fmol/mm$^2$, enabling the setup to be used to test for *E. coli* proteins as impurities in commercial drugs produced via genetic engineering in *E. coli*.

### 3.1.5. Nano-distance transduction via SEA biochips

*3.1.5.a. How It Works.* The 'metal island coated on reactive interlayer system' (MICORIS) is a type of SEA chip, whereby the transparent distance layer is chosen in a way, that it responds to the presence or absence of an analyte by shrinking or swelling, which is equivalent to changing the colloid-mirror distance. The gold colloid film is permanently linked to the distance layers' surface. As described above, the response of the SEA setup is strongly dependent on the thickness of the transparent interlayer (Figure 22), changing its color with a varying thickness (when illuminated with white light). The MICORIS system can be used for distance layers with an optical thickness in the range of 15 nm up to 1 μm, whereby changes of a few nanometers (<10) can be followed, and using an elaborate equipment even sub-nm changes are detectable. Measured is the increase in absorbance at a certain wavelength, which corresponds to a decrease in reflectance at that wavelength, whereby a linear relation between absorbance and thickness change is mostly given for small thickness variations. The response of a MICORIS sensor is fully reversible if the interlayer can perform a reversible conformational change. Glass and similar materials are clearly of no use to this kind of sensor, polymers and biopolymers, especially proteins are the proper choice, whereby the analyte can but not necessarily must change the thickness of the distance layer. It is also possible to integrate an enzyme into the distance layer reacting to the presence of the analyte with an output, which changes the thickness of the layer, these output can be e.g. a change of the pH. The advantages of this approach is the possibility to boost up the signal and the fact that the gel affecting molecule is produced inside the gel without the need of diffusion to reach its point of action.

*3.1.5.b. Polyvinylpyrrolidone as Distance Layer.* Polyvinylpyrrolidone crosslinked with sulfonated bisazidostilbenes is known to exhibit ion dependent shrinking and swelling. This is also observed by using this polymer as distance layer in a MICORIS setup, whereby the response of the sensor depends on the type of the ion as well as its charge and concentration. The thickness changes respond to analyte variations very fast due to the thin layer, and are fully reversible. The setup with polyvinylpyrrolidone as distance layer was used to monitor changes in the concentrations of different ions, pH, organic solvents and polyphenols [37-42].

While for monovalent cations and anions sigmoid logarithmic calibration curves are obtained the results for bivalent cations turned out to be more complex. E.g. $Ca^{++}$ showed a swelling effect at high ionic concentrations overlaying the sigmoid logarithmic shrinking curve.

For medical application ion variations were established using parameters and buffer systems comparable to the human blood. Variations of chloride induced within the pathological concentration could be monitored, whereby no or minor pH-induced changes were observed at blood-pH. Therefore ion effects could be well separated from pH effects. To prove the applicability of the sensor, more than 500 measurement cycles were performed, whereby the reversibility of the polyvinylpyrrolidone distance layer was fully sustained.

Due to its related polyamide-structure polyvinylpyrrolidone is also thought to be a model for protein behavior. With help of the MICORIS system it was shown that the effects of chaotropic agents on polyvinylpyrrolidone are very similar to the effects on proteins, due to the fact the these agents do not only interact with the surface of a polymer but increase the free volume of polymer chains comparable to unfolding a protein. A comparative study of the ion effect demonstrated a fundamental correlation of the polyvinylpyrrolidone swelling and shrinking properties with the 'Hofmeister serie' of chaotropic agents. The observed ion effects are in good accordance with the molecular theory of chaotropic agents.

*3.1.5.c. Proteins as Distance Layer.* Not only a wide variety of polymers crosslinked for gel-formation are suitable sensor distance layers, also many protein films can be used [43,44] (Figure 23). However, to keep the protein functional its 3-dimensional structure must be maintained, meaning the proteins should perform volume changes, but not dissolve. Therefore the protein must be cross-linked carefully to form a thin-film gel pad. The proteins can be arrayed on a chip, or the whole chip surface can be covered with a protein gel layer. Arraying enables the possibility to investigate various protein film behaviors at the same time while a total surface cover, e.g. applied via spin coating is superior in homogeneity. Arrayed protein-gel-pads are applied by standard arraying devices, whereby the concentration of the protein solution is about 5%. After dotting a cross linking procedure within the deposited protein dots leads to the formation of a small (0-250 micron) gel pad.

**Crosslinking:**
Reactive proteins are cross-linked by exposure to the UV-light of a Stratagene DNA cross-linker at 25 mJ for 5 minutes. The UV-light activates aromatic amino-acids (Tyr,

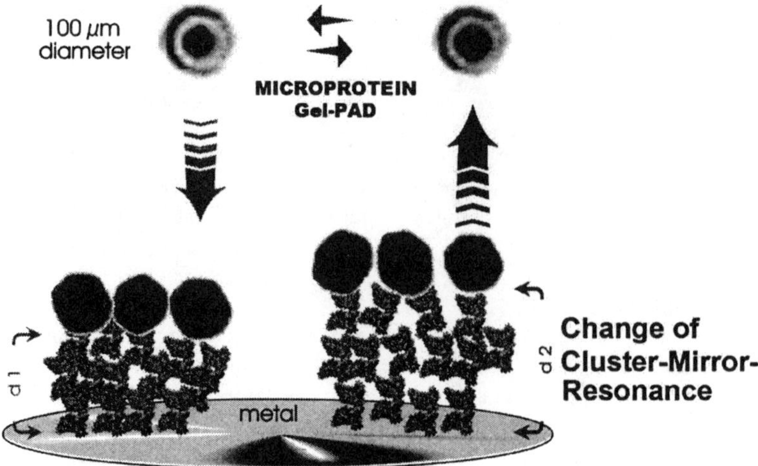


**Figure 23**. SEA response of a thin protein nano-pad (with a cluster top layer). See Fig. 5.23 in the color insert at the end of this volume.

Trp, Phe) and thiols on the outside of the protein, which leads to coupling to the slide surface as well as crosslinking between proteins.

Many standard proteins lacking sufficient aromatic amino acids and thiols on their surface as e.g. urease need reactive photo cross-linking. For that purpose the protein stock solutions 5% (w/v in distilled water) is mixed with a solution of 3% (w/v) 4,4'-diazidostilbene-2,2'disulfonic acid disodiumsalt tetrahydrate (DIAS) in a ratio of 10:1 (v / v). The mixture is spin-coated onto the mirror and activated with a monolayer 1-(3-aminopropyl)-methyl-diethoxysilane. Amino-silane, an efficient adhesion promoter, is applied via vapor silanization as described above. Finally the spin-coated protein film is cross-linking via irradiation with UV-light for 30 seconds (350 nm, 60 W). The application of wavelength less than 350 nm will lead to considerable DIAS decomposition, therefore DNA cross-linker devices as used above are not recommended for this method.

*3.1.5.d. Spin-coating of DNA.* Besides polymers and proteins, DNA can also be used to create a distance layer. Due to the poor cross-linking possibilities of DNA itself, a cross-linking procedure like for standard proteins is applied: Herring sperm DNA is dissolved under vigorous stirring in deionized water (5% (w/v)). 100 Fl of this solution is mixed with 40 Fl of a 3% DIAS solution, the mixture is spin-coated onto the mirror surface at 4000 rpm for 30 seconds and dried. To introduce hydroxyl groups the surface is oxygen plasma treated and subsequently activated with 1-(3-aminopropyl)-methyl-diethoxysilane via vapour silanisation as describe above. The spin-coated DNA film is cross-linking via irradiation with UV-light for 30 seconds (350 nm, 60 W). The application of wavelength less than 350 nm will lead to considerable DIAS decomposition, therefore DNA cross-

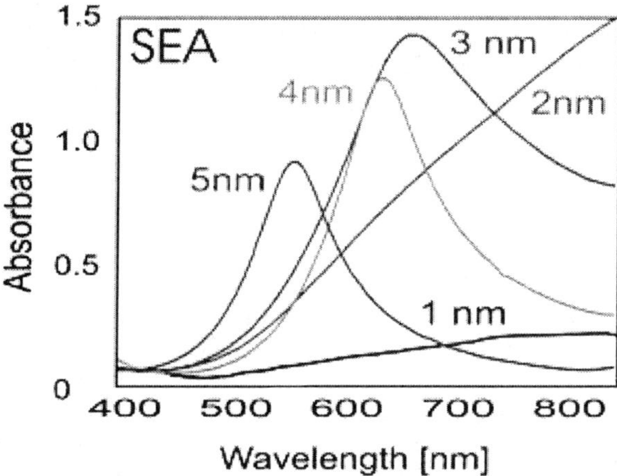


**Figure 24**. SEA-spectra versus mean mass-thickness of cluster layer (in nm) deposited via sputter coating. Whereas thin gold layers are extremely asymmetric - thick layer are more of the spherical cluster type.

linker devices as used above are not recommended for this method. Incubation with saturated NaCl for 30 min after the top layer colloid deposition results in a more stable conformation of the polymer increasing the MICORIS-chips stability.

*3.1.5.e. Setup of a MICORIS Chip.* As discussed in 3.1.4.c many solid surfaces and mirrors can be used, whereby the washing and activation steps of the support are as described in 3.1.4.c. For low resolution cost limited applications highly polished aluminum chips (diameter: 12 mm, thickness: 0.5 mm) served both as support and mirror layer; for high performance applications tungsten or chromium as adhesion layer (approximately 10 nm height) as well as silver or gold as reflection layer (100-200 nm height) are deployed onto glass slides via sputter coating techniques. Important is the deposition of the swelling polymer distance layer with homogeneous thickness, which is best achieved by spin-coating (3000-5000 rpm). Therefore stock solutions containing 2.5% (w/v) PVP (polyvinylpyrrolidone MW = 360.000) or 5% (w/v) MA-PVP (polyvinylpyrrolidone-dimethylaminoethyl methacrylate quaternized) in distilled water and 0.75% (w/v) DIAS (4,4′-diazidostilbene-2,2′-disulfonic acid disodiumsalt tetrahydrate) in distilled water are mixed (3:2 v/v). The mixture is spin-coated onto the mirror and dried. Crosslinking is done by exposing the chip to the light of a UV-50W-Hg-medium-pressure lamp for 5 seconds. The crosslinking procedure has to be done carefully, due to the fact that to long light exposure induces cracks in the layer, to short exposure times will lead to dissolution of the film. The top colloid layer can be applied via adsorptive coupling of colloids to the polymer surface; in many cases the top layer, a gold colloid film, is sputtered onto the polymer surface (Figure 24). The latter technique is faster and easier to handle but not applicable for all materials due to the fact that colloid formation is dependent on the mobility of single gold atoms on the polymer surface.

The same considerations for support, mirror, cleaning and washing of the surfaces are valid for arrayed MICORIS protein chips. 5% (w/v) protein solution is applied onto the slide surface by microdotting using e.g. a pin and ring arrayer. The dotting is done in the dark, otherwise the protein of interest is already crosslinked in solution in the support plate, thus precipitates and subsequently will either not be printed to the slide or not attach there. The proteins are crosslinked as described above and finally the top colloid layer is applied via adsorptive coupling of colloids or sputter techniques following the same considerations as above.

## 3.2. Resonance enhanced fluorescence (REF)

### 3.2.1. Physical Principles

Fluorophores brought into near distance of a metal cluster are exposed to the strong local field surrounding every cluster instead of the general incoming light. This local field modifies all optical properties of a fluorescent molecule. This effect is well known, because metals are known to be very efficient quenchers. However if this very local near field metal effect of quenching (dominating up to about 5 nm with $d^{-3}$ dependence) can be avoided the fluorophore will show a enhancement in its fluorescence compared to free fluorophores. Fluorescence is a two-step process starting by absorption of light at wavelength A and as the following step emission of light at wavelength E. If the fluorophore is positioned in the resonance distance to the cluster then both steps are driven on interaction between the resonant film and the fluorophore: In absorption fluorophores and metal plasmons are oscillators driven by irradiated light. In emission, the fluorophore is the origin of radiation and an oscillator with a resonant structure in its vicinity. These 2 mechanisms result in an effective increase in the measured fluorescence signal occurring when fluorophores are positioned within a distance of 5–45 nm to the cluster.

The same effect, although weaker is working when the fluorophore is brought in the vicinity of a cluster film, enabling to set up slides, which enhance the fluorescence signal (Figure 25). Enhanced fluorescence is a relatively new technique used for chip technology, nevertheless there are several setups of such nano-resonant multi-layer films known to provide an increased local field resulting in a fluorescence amplification from 2 to 200 times, among them nano-particle layers, rough surfaces, surface plasmons in thin metal films or multi-layer waveguides. Compared to standard fluorescence assays, the intensity as well as the signal to noise ratio are increased using this field boosting technology [45-49].

The effect of such metal cluster and cluster related but also plain metal films is explained on one hand by the strong enhancement of the field strength of light due to surface plasmons and Mie plasmons (not for plain metals), which are driven by the irradiation of the cluster layer. The resonant interlayer now is driven by the plasmons, and the fluorophore is exposed to this enhanced light field of the interlayer and has consequently an enhanced absorption probability. On the other hand during emission the fluorophore acts as radiating dipole and oscillator, thereby influenced by the nearby resonant layer. Thus, an additional charge is induced and the emission process is enhanced.

In summary, it can be said, that fluorophores feel more exciting light compared to standard glass slides and emit more photons per unit time. While signal amplification will reach up to 8 on smooth films enhancement factors up to 200 times have been observed for cluster structures (Figure 26). The reason for this is the fact, that more energy can be coupled into cluster structures due to the additional presence of Mie plasmons, into which most of the energy is coupled.

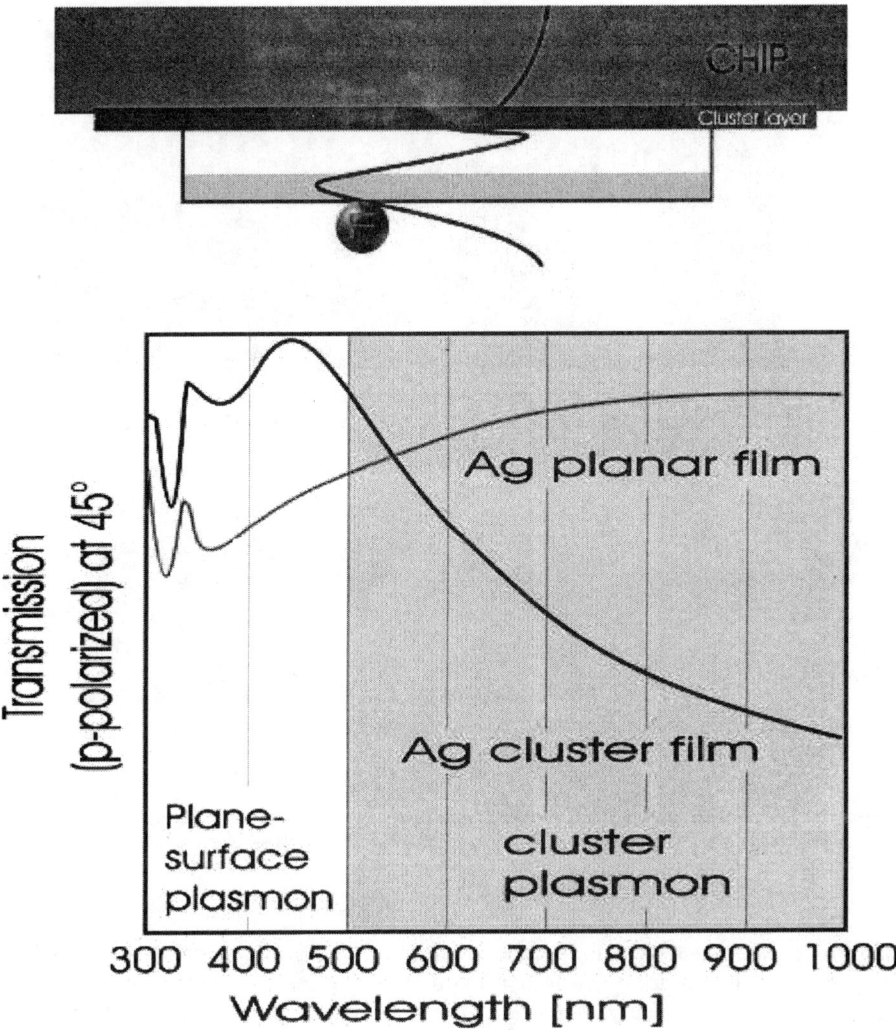


**Figure 25**. Electromagnetic field in a multilayer system and fluorescence amplification in a Ag-cluster-layer – SnNx-interlayer system (0 – 700 nm) and plasmon resonance of a cluster film.

As already pointed out an additional radiative rate results in an increase in the total radiative decay rate. If a dye has a high quantum yield near one, the additional radiative decay rate cannot substantially increase the quantum yield. The more interesting case is for low quantum yield fluorophores. In this case the enhancement can be boosted to near one and thus an overall higher fluorescence amplifications is obtained. This suggests the emission from weakly fluorescent substances can be increased if they are positioned at an

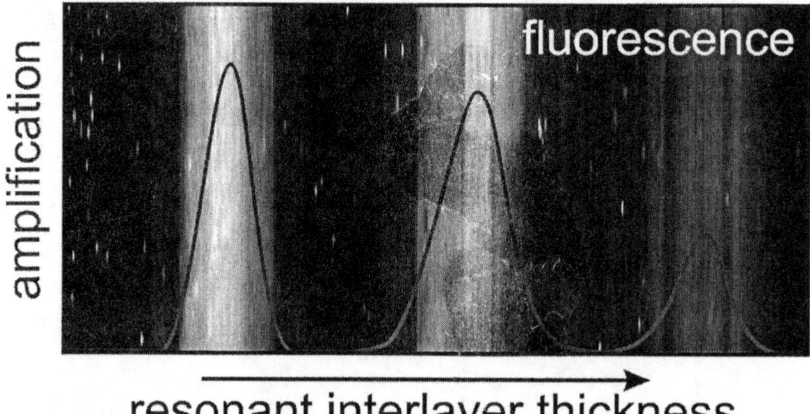

Figure 26. Fluorescence amplification in a Ag-cluster-layer – SnNx-interlayer system (steps from 0 – 700 nm) , photo (fewer steps than scan above). See Fig. 5.26 in the color insert at the end of this volume.

appropriate distance to a colloid coated surface. The ability to increase the radiative decay rate enables to drive nearly non-fluorescent species to become fluorescent. This phenomenon has extensively been studied by Aussenegg, Leitner, Schalkhammer, Mayer and Lakowicz, further investigations and applications are to come. For example DNA could become brightly fluorescent without any labeling [50]. Similarly, weakly fluorescent species such as tryphtophan, bilirubin, fullerenes, metal-ligand complexes, or porphyrins could display usefully high quantum yields when appropriately adjacent to a metal surface.

### 3.2.2. Applications

With the potential to amplify the local field and thus to enhance the fluorescence signal such nano-resonant multi-layer films there are several possibilities to use the principle for detection and investigating several bio-components like nucleic acids, enzymes all kind of proteins or any pharma-ligand. All what is needed is a biochemical recognition process, which results in fluorophores bound within the field of the resonant

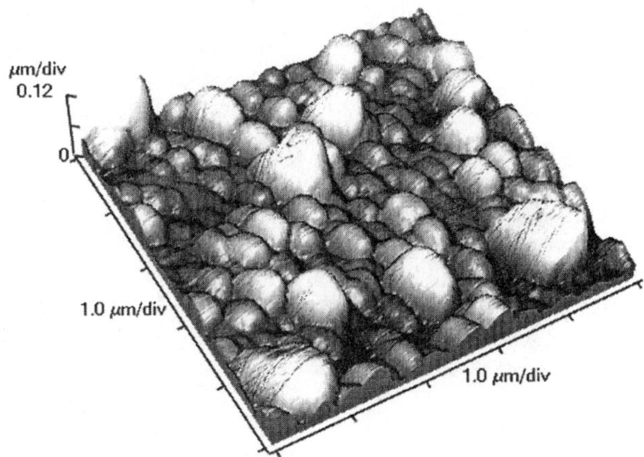


**Figure 27**. AFM scan of a silver cluster layer prepared via thermal curing at 400°C.

structure, resulting in enhanced absorption and emission of the fluorophore driven by the resonant field. In general biochips have a low absolute signal and suffer often from a poor signal to noise ratio, limiting them to relatively high concentration ranges and often introduce the need for long incubation times.

As discussed above biochips based on metal particles layers are among the best in enhancing fluorescence. The REF-chip is a slide with a coating of designed metal nano-particles and an overcoat of a transparent inert non-metal distance layer and topmost an array of biorecognitive components, which do bind the fluorophores in the proper distance during an assay.

The metal nano-particles layer of a plurality of nanometric clusters, islands or colloids of electrically-conductive material (silver or gold for best plasmons !) can be applied via sputter or evaporation coating or by adsorption of metal colloids from an aqueous solution. Annealing of metal films deposited via sputter coating or similar techniques on a glass surface leads to self-organized metal cluster films (Figure 27). It was shown by Aussenegg, Schalkhammer and Strekal that these films work for REF as well as for Surface enhanced Raman scattering (SERS).

The resonant interlayer can be set up by a broad variety of transparent materials, glasses or metal glasses deployed via sputter coating are used as well as spin coating of polymers. The layer couples the energy from the metal film to the fluorophore, whereby 2 to 4 enhancement zones are found at well-defined distances from the cluster-film located at multiples of the wavelength of the desired fluorophore (Fig. 26). Due to the fact that on general the enhancement is strongest in the first, means nearest, zone the setup of assay chips is usually tuned to this optical distance at the desired wavelength .

The biorecognitive components are arrayed and bound onto the chips distance layer surface using standard arrayers, for the binding process practically all immobilization methods can be used, like silane or carbodiimide chemistry.

To make the chip useful first the compound of interest has to be labeled with fluorophores, usually requiring a random labeling of the sample, but also analyte-specific labeling is possible. After incubating of such a chip with the now labeled molecule of interest weak interactions are washed away, an excitation radiation which is suitable for excitation of the analyte-bound fluorescent compound is applied, and the fluorescence radiation emitted by the fluorophore is measured, determining the analytes absence or presence and if necessary its concentration.

Another approach avoiding the random labeling of the sample is to incubate the chip directly with the sample, resulting in the binding of the molecule of interest. Afterwards the chip is incubated with a analyte specific fluorophore or with a fluorophore labeled compound, which is analyte specific, e.g. an antibody. Washing, excitation and measurement steps are executed as described above.

### 3.2.3. REF in Microtiter-Plates

The REF technology is not limited to chip and slide format, the technique was applied to microtiter-plates, performing kinetic single step biorecognition assays with fluorophore labeled compounds. Introducing the REF technique was done by binding of silver clusters to standard microplates via silanization of the plastic surface, making the method compatible with conventional microplate processing and reading devices. Excitation with coherent laser light is not required, the measurement is performed using a standard fluorescence microplate reader.

Due to the plate geometry sputter techniques are unemployable to introduce clusters into the wells of the microtiter plates, the coating is done via adsorption of silver colloids from an aqueous solution to amino-microplates. Using 2-4 mM colloid solutions, saturation of the surface is achieved after 1 h. Until saturation the amount of colloid bound to the microtiter-plate surface strictly increases with time and is monitored by reading the absorption increase at 400 nm. Increasing the concentration of the silver colloid solutions leads to a loss of spherical shape of the clusters as well as an increase in particle size. Consequences are on one hand longer diffusion times and on the other hand a shift in the absorption maximum of the Ag-clusters from 400 nm to 470 nm. The coating with colloidal clusters prepared from 6 mM silver solution is therefore monitored at 470 nm, saturation of the surface is achieved after 2 h. Studying the quality of the structure and the relative silver amount at the surface of the colloid-films is done by atomic force or scanning electron microscopy.

Using such REF-microtiter-plates a detection limit in the pico-molar range for a standard protein bioassay was achieved. The ease of colloid-surface preparation and the increased sensitivity makes REF-microtiter-plates a promising tool for rapid single-step immunoassays or the study of reaction kinetics.

### 3.2.4. Cluster-Layer Enhanced Fluorescence DNA Chip Setup

Standard microscope slides are coated with silver or gold clusters by absorption or sputtering. The latter requires an adhesion layer of chromium, titanium or tungsten (also sputter coated) and produces on top of the adhesion layer a metal film, 'clustering' is achieved by film curing at a temperature of 200 to 300°C (silver) or 400-600°C (gold). The inert and preferably hard distance layer (glass or metal glasses as e.g. SnN or ZnO) is also deposited by sputter coating using reactive sputtering, means Sn is sputtered in nitrogen plasma as Zn is sputtered in oxygen plasma.

To produce binding possibilities for DNA the chip is covered with poly-lysine (0.1% of poly-lysine in phosphate buffered saline (PBS), pH 7.4). Alternatively it can be coated with BSA (1mg/ml in PBS pH 7.4), but this approach introduces the need of further activation. Therefore after removing unbound BSA via washing with MQ water the chip is incubated with amino-silane (10 mg/ml N-(3-Dimethylaminopropyl)-N-Ethyl-carbodiimide (EDC) in PBS pH 7), cross-linking and activating the BSA-surface.

Immediately after EDC activation amino-modified single stranded DNA probes (5 pmol/Fl in PBS, pH 5.4) are arrayed onto the chip. To enable the EDC chemistry to work the chip is stored in a humidity chamber for 30-60 minutes (preferably at 60°C, but also RT is possible), afterwards washed with MQ water and air-dried. Crosslinking with UV-light can be done to further stabilizing the DNA.

Hybridisation is executed in 3-5*SSC or Dig Easy hybridisation buffer (1-16 h), if the analyte DNA is labeled with a fluorophore (e.g. a PCR product) the chip is washed (3-5*SSC and 0.3*SSC), air-dried and scanned. In many cases the analyte DNA will come unlabelled, then it is bound as above, afterwards a complementary "reporter" DNA (2 pmol/Fl), which is fluorescently labeled (e.g. Cy3 or Cy5) is hybridized to a free sequence of the analyte DNA in 3-5*SSC or Dig Easy hybridization buffer for 15 minutes to 3 hours. It should be noted that the hybridization temperature of the second step must be below the one of the first step. Finally the chip is washed (3-5*SSC and 0.3*SSC), air-dried and scanned.

### 3.2.5. Clusters Layer Fabrication Methods

Due to the more important role the preparation of cluster layers plays in the REF technique compared to other techniques, and the need to prepare various clusters and layers by different strategies (like the above discussed REF-microtiter plates) the manufacturing strategies of such layers are discussed here. A straight forward approach to produce cluster layers of many materials like silver, gold, palladium, copper, tin and indium is the sputter coating technique. Especially suited for metals with low evaporation temperatures (as the materials mentioned above have) films of all heights and materials are deposited within seconds or minutes. The coating is usually done in an inert argon plasma at $10^{-2}$ to $10^{-1}$ mbar, whereby the thickness of the metal layer is adjusted mainly by sputter time, but also adjustment via pressure and current (power) is possible. The metal is deposited as atoms, and on general the surface temperature is high enough to enable migration of the atoms to assemble into clusters on the surface. However, if and how that occurs depends on the interaction of the metal atoms with the surface they are deposited on. If that interaction is stronger than the interaction between that adjacent metal atoms a *layer-by-layer growth mode* is observed, resulting in a film instead of clusters. If the interaction between neighboring metal atoms exceeds the interaction with the surface an *island growth*, means the formation of clusters, is the result. An intermediate between these 2 cases is possible, whereby first a layer is formed and island formation occurs growing on top of this layer (*layer-plus-island growth*).

All in such way produced clusters are small, flat and asymmetric. Upon heating to 100 to 350°C (Au up to 600°C) the clusters as well as layers melt, resulting in round shaped well-defined nano-clusters films.

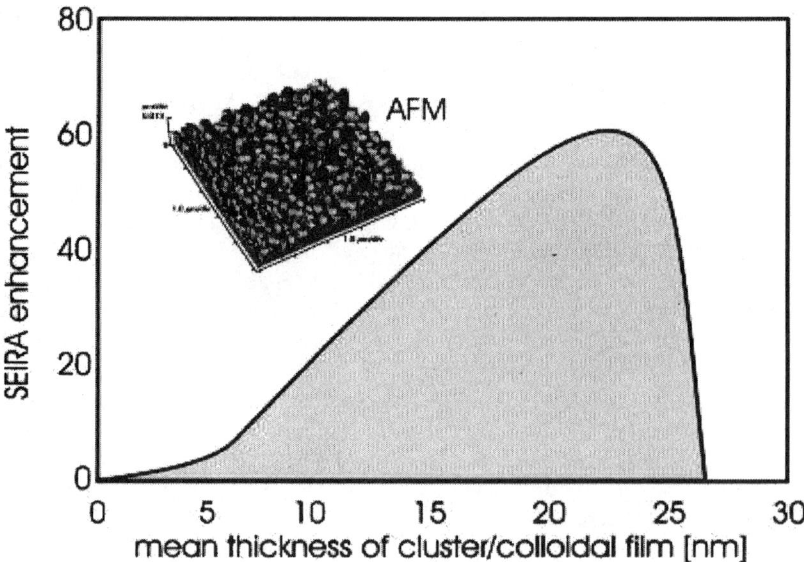


**Figure 28**. SEIRA - active surface - mean metal thickness versus amplification factor (nano-structure effect).

The preparation of gold clusters in solution for colloidal absorption is discussed above. Silver clusters are produced in a similar way, whereby silver nitrate is dissolved in boiling double distilled water, sodium citrate as reducing reagent is added and the solution is kept boiling for 5 min. The size of the produced clusters is dependant on the ratio of silver to sodium citrate. Such silver clusters are useful for coating of aminated-surfaces, if sputter coating is because of size, geometry, plastic material, heat sensitivity or other reasons not an option.

A novel technique to produce silver clusters films dispersions is the reduction of $Ag^+$ ions by N,N-dimethylformamide (DMF) [51]. Either thin films of silver nanoparticles electrostatically attached onto surfaces are produced, or in the presence of a silane (e.g. 3-aminopropyltrimethoxysilane) stable dispersions of silver nanoparticles.

### 3.3. Surface-Enhanced Infrared Absorption (SEIRA)

Thin film of metal clusters enable other types of surface enhancement, one of these is the surface enhanced infrared absorption (SEIRA) spectroscopy. The enhancement is an addition of electromagnetic field effects and chemical effects, therefore highly dependent on the distance to the cluster layer and on the nano-structure of the cluster layer. In fact most studies done point out that the effect is very short ranged and only applicable to the first monolayer adsorbed to the surface. Compared to standard infrared spectroscopy and Surface enhanced Raman scattering SEIRA has a remarkable signal to noise ratio and the advantage of non-destructivity. However, a full understanding of the SEIRA technique is still required [52-54].

The SEIRA phenomenon was discovered due to the enhanced infrared absorption of chemisorbing molecules on Ag and Au surfaces, nowadays silicon or germanium are the preferred substrates on which an appropriate nano-structure of the gold cluster film is produced (Figure 28). After recording a IR spectrum, the analyte (an organic material) is then dispersed on the chip in the ideal case forming a monolayer. The spectrum is recorded again and compared with the former. Results show that only some molecular vibrations are enhanced, but in dramatic exchange, many remain un-amplified. An explanation for that behavior is given by some models assuming that the molecule needs to be bound chemically with an appropriate orientation to the metal surface.

An example of a specific application of the technique is the detection of *Salmonella* sp., whereby antibodies to the organism are bound to the chip surface. After recording the spectrum the chip is incubated in a *Salmonella* containing solution, binding the *Salmonella* bacteria to the cluster surface. The second spectrum is recorded and overlaid with the original spectrum. In such an experiment a band shift from the original bands at 1085 and 990 cm$^{-1}$ to 1045 cm$^{-1}$ was obtained, which was assigned to a phospholipid in the cell wall of the captured cell.

### 3.4. Scattered Evanescent Waves (SEW)

Scattered evanescent waves (SEW) measurements rely upon the detection of back scattered light from an evanescent wave disturbed by the presence of a colloidal gold on the interface. The evanescent wave at the interface is the result of a totally internally reflected incident light wave. A superior signal-to-noise ratio is achieved due to placement of the detector at a back angle above the critical angle. SEW-measurements depend on the identification of the critical angle associated with total internal reflectance. This angle is a function of the refractive indices of the material through which an incident light wave is directed, e.g. glass or plastic, and the bordering material, which will be in most cases a solution with a lower refractive index. The angle is measured from a line perpendicular to the interface between the two materials (90 degree = plane of the interface). Light directed through the chip toward the interface formed by the solution and chip surface layer at the critical angle results in total internal reflectance of the light within the layer. In practice, an angle several degrees greater than the critical angle is chosen, making sure that the incident light is totally internally reflected within the chip layer. As light source lasers e.g. He-Ne laser are used, but other light sources such as light emitting diodes are possible.

As mentioned above the scattered light is influenced by the presence of clusters at the interface, due to a much higher refractive index than that of the solution, and preferably also higher than the first light transmissive material, means this material is chosen to have a lower refraction index than the cluster material. Consequently light is de-coupled from the evanescent field due to the interaction of the colloidal gold particles with the evanescent wave. Due to the clusters higher index of refraction compared to the underlying solid the scatter light is increased.

The phenomenon becomes an analytical tool, when the absence or presence of the cluster is dependent on a biological recognition reaction, means that the cluster label is brought to the interface e.g. by an immunological reaction [55]. The interface is therefore coated with a biorecognitive component, colloidal gold is used as a label for the solution phase immunologically active component. The detector is placed in a location whereby only light scattered backward toward the light source is detected, avoiding the detection of scattered light within the bulk liquid medium. Due to the possibility of adjusting the particle size such assays are much more flexible with respect to light wavelength compared to fluorescence.

### 3.5. Surface-Enhanced Raman Scattering (SERS)

At the single molecule level Raman scattering seems to be of no importance due to the small cross sections involved. However, this situation is different on surfaces, where molecules in close proximity to metal-cluster-coated surfaces exhibit large (typically $10^6$ fold) enhancements in vibration spectral intensities, named surface-enhanced Raman scattering. Enhancement factors up to $10^{14}$ have been reported, which corresponds to effective Raman cross sections of $\sim 10^{-16} cm^2$ / molecule, which are even higher than fluorescence cross sections. Similar to SEIRA the enhancement process of SERS is due to two effects, whereby the short range or molecular or chemical enhancement seems to rely on adsorption induced changes in Raman susceptibility. The second effect is long-range electromagnetic enhancement, due to the enhanced fields present at clusters and rough metal surfaces, whereby this effect dominates the overall effect in most situations. Nevertheless to perform SERS the substance being studied has to be placed in direct or close contact with the surface. Au colloids unfortunately reveal a significantly lower SERS enhancement with visible excitation compared to silver colloids. Similar to SEIRA SERS is highly dependent on the nano-structure of the metal surface, small variations in the nano-structure of the surface due to the manufacturing process may influence the effect by orders of magnitude. Particle size and shape are responsible for the differences, but even more influence comes from grain boundaries within the particles supporting local plasmons, whereby this effects are found mainly at larger grain sizes of 30-200 nm. On the other extreme also smooth metal films were proven to give some small enhancement, but have the disadvantage that many biomolecules are denaturated during absorption. Contrary to that it is well known that most biomolecules retain their biological activity when conjugated to colloidal clusters, e.g. gold clusters.

Many techniques are used to fabricate the required nano-structures including aggregated colloidal metal sols, electrochemically roughened electrodes, evaporated and or cured metal films.

The enhancement can even be pushed further by 3 orders of magnitude by using laser light, which is in resonance with an electronic transition of the substances. This technique is called surface enhanced resonance Raman scattering ("resonant SERS"). Applications of SERS include not only the detection of molecules, but also the investigation of the structure and function of large biomolecules as well as the analysis of chemical processes at interfaces. SERS is particularly well suited for sandwich-type immunoassays and hybridization assays.

The SERS intensities for standard colloids or often quite weak, but can be easily enhanced by binding a strong chromophore near the surface of a metal cluster. Such "labeled" metal-clusters are adsorbed to a metal chip surface, preferably in a sandwich type assay. With a protein having a strong chromophore itself it can act as the label and analyte at the same time, binding the colloidal particle to the SERS active surface. In an ideal setup the protein is then sandwiched between two SERS active clusters. If the

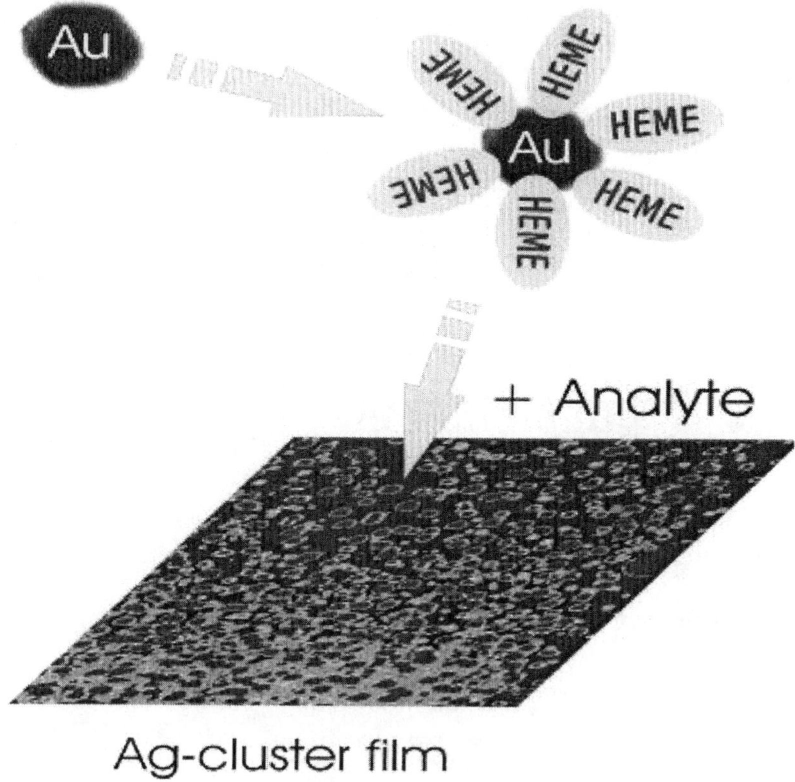



**Figure 29**. Setup of a chip type SERS device combining well defined and highly reproducible gold colloids with the high amplification factor of silver-cluster surfaces using a HEME-protein for energy transfer.

protein lacks a strong chromophore, like e.g. antibodies it is necessary to introduce additionally to the protein a chromophore to the particle, e.g. a Haem-protein being bound to the nanoparticle via a lysine rich micro-domain (Figure 29).

A setup of such an assay could be as follows. The SERS active surface is coated with a capture antibody. The SERS active particle is first coated with the detector antibody and second with a protein containing a Haem-group. The analyte is captured by the antibody on the SERS active surface, and the second antibody docks the cluster in near distance to that surface. This approach makes use of the effect that within two nanoparticles the local field reaches a maximum. This strong inter-colloid electromagnetic coupling is deduced from wavelength-dependent SERS enhancement and the existence of a wavelength-shift of single colloid plasmon bands in this way proving a proof for the existence of the ultra-strong near-field in cluster assemblies. Moreover, SERS can be applied to array-type assays provided that the devices is capable of scanning with two-dimensional resolution.

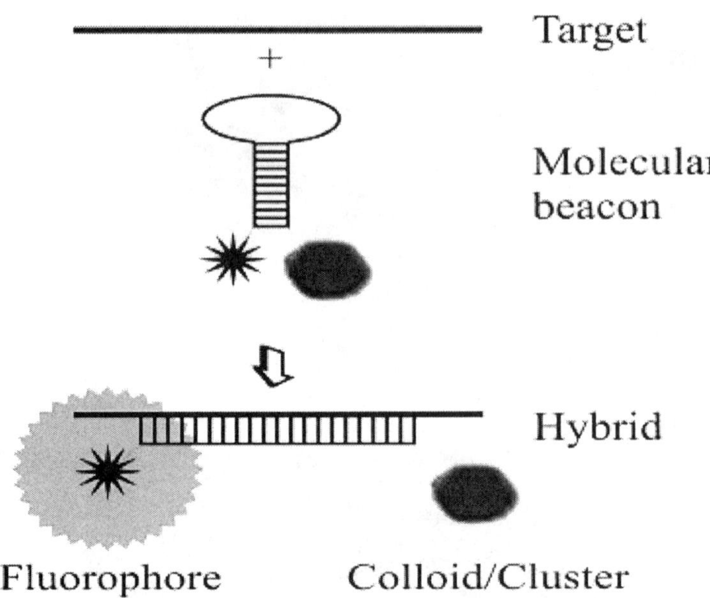


**Figure 30.** Molecular beacons with ultra-efficient metal cluster-quenchers.

## 3.6. Cluster-Quenched Fluorescence

As mentioned in the REF section metals are generally known to quench fluorescence, bringing a metal or a metal cluster to near to a fluorophore destroys the fluorophores ability to emit radiation. On surfaces such designs are avoided, but in solution it is of great use in form of the molecular beacon, which is set up of a small gold nano-cluster and a fluorescent dye, bound to the opposite ends of a ssDNA. This ssDNA forms a hairpin-shaped structure in which the ends are self-complementary, bringing a fluorophore and a quencher into close nanometric proximity. In this conformation the metal particle quenches the fluorescence (Figure 30). When the proper (complementary DNA) probe hybridizes to the ssDNA the hairpin structure opens, the fluorophore and cluster are brought into greater distance, what disables the quenching effect, means that the fluorescence is restored (Figure 31).

Although not the cheapest reagents, molecular beacons became quickly famous due to the possibility of an easy DNA detection without the need of labeling the sample DNA. The early beacons suffered from the limited quenching efficiency of organic quenchers, as e.g. 4-((4'-(dimethylamino) phenyl)azo)benzoic acid, leaving some 1% of non-quenched fluorescence. The ratio of fluorescence to background fluorescence defines the sensitivity as well as the dynamic of the molecular beacon, therefore more efficient quenchers can enhance the sensitivity greatly. Due to near field de-excitation metal nano-clusters can quench fluorescence up to 100 times more efficient, and therefore replace the

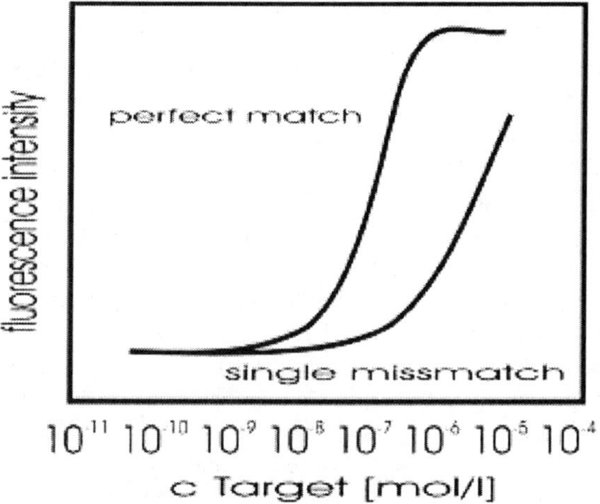


**Figure 31**. Hybridization assay using a metal cluster quenched beacon for the discrimination of perfectly matching versus a mutated DNA sequence.

organic quenchers. This extremely high quenching efficiency opens new perspectives for beacon-probes in fluorescence-based assays [56].

The quenching is also dependent on the fluorophores ability to be quenched, meaning a fluorophore has to be chosen, which has a high fluorescence yield and is highly sensitive to quenching. Nowadays Rhodamine 6G is known as one of the best quenched dyes, with an average quenching efficiency of ~99.5%, but also quenching efficiencies for other fluorophores of >99.9% have been reported. As mentioned above efficient quenching determines the sensitivity and dynamic of the molecular beacon, by use of highly quenching sensitive fluorophores the dynamic range of molecular beacon can be strongly expanded.

The **Quenching mechanism** of metals clusters is a non-radiative energy transfer from the molecule to the cluster, due to the metals ability to provide alternative non-radiative relaxation mechanism by the creation of electron-whole pairs or surface plasmons. These mechanisms favor the non-radiative to the radiative relaxation pathways of the fluorophore. Involved are dipole-dipole interactions which are described by a Foerster-mechanism, explaining why the quenching is so strongly dependent on the distance of cluster to fluorophore. The cluster itself relaxes radiative as well as non-radiative. The above mentioned ability to create surface plasmons is strongly dependent on the clusters size, so that very small gold clusters show significant less quenching effects. The plasmon resonance is also dependent of size, shape and composition of the cluster regarding the absorbing wavelength, enabling to tune the clusters resonance wavelength as required.

The **design of a beacon** is in general a 25-30 base synthetic oligonucleotide (ssDNA) equipped with a primary amine at its 3' end, and a sulfide at its 5´ end, enabling to covalently link the fluorophore as well as the cluster (or an organic quencher). The hairpin-loop structure is achieved by complementary sequences of the 3' and 5´ end, whereby the length of these complementary sequences determines the temperature stability of the hairpin structure (usually room temperature). Poly-TTTTTTT tails in the middle of the sequence permit the proper back-folding of the DNA molecule.

**The preparation of a molecular beacon** starts by coupling an amino-reactive fluorophore (e.g. rhodamine-6G) to the terminal 3′-amino group of the oligo. After purification using a Sephadex column and fractionation on a reverse-phase HPLC, the fluorescent DNA is dried, and re-suspended to an oligonucleotide concentration of 15 µM. The protection group from the 5′-end is removed by cleaving the disulfide bond and the resulting free sulfhydryl is covalently attached to small 1–3 nm gold clusters. Uncoated but also coated clusters, as e.g. N-propylmaleimide coated ("Nanogold") are used, whereby multiple attachment is avoided by an excess of clusters.

**Application:** Beacons are used to show the presence of a target DNA in a sample, means if the target hybridizes to the hairpin- loop, the loop opens and the constantly monitored fluorescence intensity increases. Hybridization temperature and beacon composition are chosen to allow the method to discriminate between a perfectly matched probe and a mismatched one, by varying the hybridization temperature even the number of mismatches can be obtained. For some colloidal preparations the high hybridization temperature of around 50-70°C is critical due to a chemical instability of the Au-thiol-bonds at that temperature. To work with cluster quenched beacons buffers with low salt concentrations must be used decreasing the hybridization temperature. A buffer containing e.g. 90 mM KCl and 10 mM TRIS, pH 8.0 enables highly sensitive hybridizations assays using cluster quenched molecular beacons at room temperature.

### 3.7. Cluster-Emission Devices (CED)

A standard technique to follow bio-recognitive interactions is labelling an involved component with a fluorophore. However, with their limited lifetimes of $< 10^7$ emission processes, their signal is declining during consecutive measurements. A novel alternative lacking this limitation is the cluster emission device (CED) capable of detecting biomolecules at single molecule resolution [57].

Figure 32 shows the principal setup of a CED. Via a biorecognitive interaction (shown is a ligand receptor interaction) a 30 nm gold cluster is bound on the chip surface, which is placed inside a chamber. After evacuation a high voltage is applied between the 20 nm thin bottom chromium and the top aluminum electrode. As a result the protein complex stops acting as an insulator (electric breakdown) and the cluster emits electrons towards the 20 nm thin aluminum film (the "extractor"). Consequently the thin aluminum film heats up and evaporates locally within milliseconds, clearing an area of about 1 :m². To avoid overheating of the whole device pulsed voltage is applied. The process is monitored via a microscope, whereby the system is illuminated through the semitransparent chromium electrode, displaying each cluster as a bright spot.

With help of the Wentzel-Kramers-Brillouin approximation the field dependent transmission coefficient was calculated for the vacuum potential barrier. Integrating over all energies for a free electron gas at room temperature results in an exponentially increasing I-V curve that reaches 1 nA at 0.7 kV. With a band gap well above 2 eV proteins are good insulators, their electric breakdown is believed to be a conduction mechanism, which occurs at a voltage of about 200 V. The necessary potential for tip-emission can be further decreased to about 100 V using nano-fibers. To create transparent

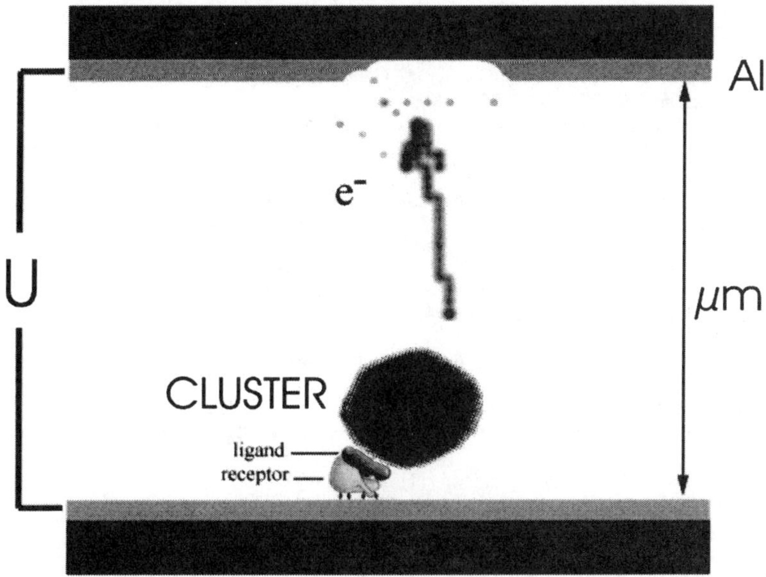


**Figure 32**. Cluster emission device - proteins are insulators: (band gap of ~2 eV), electric breakdown at ~500 V/μm, nano-cluster-enhanced breakdown at about 3-200 V.

areas within seconds, thermal oxidation in air may be used as an alternative; plasma formation could further decrease the required extraction potential.

## 4. ACKNOWLEDGEMENTS

Part of this work was supported by the LifeTech program at TUDELFT / The Netherlands. We want to thank Prof. Fritz Pittner, Vienna Biocenter, Austria, Dr. Bauer, November AG-Erlangen, Germany and Prof. Aussenegg and Prof. Leitner from University of Graz, Austria for helpful discussions and support.

## 5. REFERENCES

1. E. Garbowski in *"A Short Textbook of Colloid Chemistry"* Pergamon, Oxford (1954).
2. H. Dörfler in *"Grenzflächen und Kolloidchemie"* VCH, Weinheim (1994).
3. U. Kreibig, M. Vollmer in *"Optical properties of metal clusters"* Springer, Heidelberg (1995).
4. H. Adair in *"Applied Colloid & Surface Chemistry"* ISBN: 0849386632 (1999).
5. K. Brown, M. Natan, Langmuir **14**, 726 (1998).
6. G. Frens, Nature Phys. Sci. **241**, 20 (1973).
7. J. Slot, H. Geuze, Eur-J-Cell-Biol. **38**, 87 (1985).
8. F. Caruso Nanoengineering of Particle Surfaces Adv. Mater. **13** (1), 11 (2001).
9. M. Valina-Saba, G. Bauer, N. Stich, F. Pittner, Th. Schalkhammer, Supramol. Science / Material Science and Engineering C **8-9**, 205 (1999).

10. L.A. Bumm, J.J. Arnold, M.T. Cygan, T.D. Dunbar, T.P. Burgin, L. Jones II, D.L. Allara, J.M. Tour, P.S. Weiss, Science **271**, 1705-1707 (1996).
11. H.W. Fink, C. Schonenberger, Nature **398**, 407-410 (1999).
12. P.J. de Pablo, F. Moreno-Herrero, J. Colchero, J. Gomez Herrero, P. Herrero, A.M. Baro, P. Ordejon, J.M. Soler, E. Artacho, PRL **85**, 4992-4995 (2000).
13. C. Kergueris, J.-P. Bourgoin, S. Palacin, D. Esteve, C. Urbina, M. Magoga, C. Joachim, Phys. Rev. B **59**, 12505-12513 (1999).
14. M.A. Reed, C. Zhou, C.J. Muller, T.P. Burgin, J.M. Tour, Science **278**, 252-253 (1997).
15. R. Kirsch, M. Mertig, W. Pompe, R. Wahl, G. Sadowski, E. Unger. Thin solid films **305**, 248-253 (1997).
16. J. Leuvering, B. Goverde, P. Thal, A. Schuurs, J Immunol Methods. **60**, 9 (1983).
17. J. Leuvering, P. Thal, M. Van der Waart, A. Schuurs, J Immunol Methods **45**, 183 (1981).
18. R. Elghanian, J. Storhoff, R. Mucic, R. Letsinger, C. Mirkin Science. **277**, 1078 (1997).
19. C. Mirkin, R. Letsinger, R. Mucic, J. Storhoff, Nature **382**, 607 (1996).
20. C. Mirkin, "Towards DNA Based Technology for Preparing Nanocluster Circuits and Arrays," MRS Bulletin **25**, 43 (2000).
21. M. Mertig, R.Kirsch, W.Pompe, and H. Engelhardt, Eur. Phys. J. D **9**, 45 (1999).
22. K. Bromann, M. Giovannini, H. Brune and K. Kern, Eur. Phys. J. D **9**, 25 (1999).
23. D. Pum, A. Neubauer, E. Györvary, M. Sara, U. Sleytr, Nanotechnology **11**, 100 (2000).
24. U.B. Sleytr, P. Messner, D. Pum, M. Sara, Crystalline bacterial cell surface proteins, Academic Press, San Diego (1996).
25. O. D. Velev and E. W. Kaler, Langmuir **15** (11), 3693 (1999).
26. H. Takei „Biological Sensor based on localized surface plasmon associated with surface bound Au/polystyrene composite microparticles". SPIE **5515**, 278 (1998).
27. H. Takei, Technical Digest of the 16th Sensor Symposium, 123 (1998).
28. M. Natan, IBC-Conference on Biosensor Technology, Boston (1998).
29. G. Bauer, N. Stich, Th. Schalkhammer, Chapter 6: Nanoclusters and colloids in bioanalysis, MTBM Volume on Analytical Biotechnology ISBN 3-7643-6589-7 and ISBN 3-7643-6590-0, Birkhäuser Verlag, Switzerland (2002).
30. A. Leitner, Z. Zhao, H. Brunner, F. Aussenegg and A. Wokaun „Optical properties of a metal island film close to a smooth metal surface". Applied Optics **32**, 102 (1993).
31. Th. Schalkhammer, Chemical Monthly **129**, 1067 (1998).
32. Th. Schalkhammer, G. Bauer, F. Pittner A. Leitner, F. Aussenegg, SPIE **3253**, 12 (1998).
33. G. Bauer, F. Pittner, Th. Schalkhammer, Mikrochimica Acta **131**, 107 (1999).
34. C. Mayer, N. Stich, R. Palkovits, G. Bauer, F. Pittner, T. Schalkhammer, Journal of Pharmaceutical and Biomedical Analysis **24**, 773 (2001).
35. C. Mayer, R. Verheijen, Th. Schalkhammer, SPIE **4265**, 134-141 (2001).
36. C. Mayer, N. Stich, T. Schalkhammer, G. Bauer, Fres. Anal. Chem. **371**, 238 (2001).
37. F. Aussenegg, H. Brunner, A. Leitner, F. Pittner, G. Bauer, T. Schalkhammer European patent EP00677738B1 (2000) US-patent US05611998 (1997).
38. Th. Schalkhammer, Ch. Lobmaier, F. Pittner, A. Leitner, H. Brunner and F.R. Aussenegg, Sensors and Actuators B **24** (1-3), 166 (1995).
39. F.R. Aussenegg, H. Brunner, A. Leitner, Ch. Lobmaier, Th. Schalkhammer, and F. Pittner, Sensors and Actuators B **29**, 204 (1995).
40. Th. Schalkhammer, Ch. Lobmaier, F. Pittner, A. Leitner, H. Brunner and F.R. Aussenegg, Mikrochimica Acta **121**, 259 (1995).
41. G. Bauer, S. Voinov, G. Sontag, A. Leitner, F. Aussenegg, F. Pittner, Th. Schalkhammer, SPIE **3606**, 40 (1999).
42. M. Lepek, R. Palkovits, G. Bauer, T. Schalkhammer, F. Pittner, Rec. Res. Devel. Anal. Biochem. **1**, 1 (2001).
43. Th. Schalkhammer, Ch. Lobmaier, F. Pittner, F. Aussenegg, A. Leitner, H. Brunner, SPIE **2508**, 102 (1995).
44. C. Mayer, R. Palkovits, G. Bauer, T. Schalkhammer, Journal of Nanoparticle Research **3**, 361 (2001).
45. F. Aussenegg, H. Brunner, A. Leitner, F. Pittner, G. Bauer, T. Schalkhammer, US patent US05866433 (1999).
46. K. Sokolov, G. Chumanov and T. Cotton, Anal. Chem. **70**, 3898 (1998).
47. Th. Schalkhammer, F. Aussenegg, A. Leitner, H. Brunner, G. Hawa, Ch. Lobmaier, F. Pittner, SPIE **2976**, 129 (1997).
48. C. Mayer, N. Stich, G. Bauer, T. Schalkhammer, SPIE **4252**, 37-46 (2001).
49. N. Stich, C. Mayer, Th. **Schalkhammer**, SPIE 4434, 128 (2001).

50. J. R. Lakowicz, B. Shen, Z. Gryczysnki, S. D'Auria, and I. Gryczynski, Biochem. Biophys. Res. **Commun.**, 286, 875-879 (2001).
51. I. Pastoriza-Santos and L. Liz-Marzán, Pure Appl. Chem. **72**, 83 (2000).
52. Ch. Brown, Y. Li, J. Seelenbinder, A. Rand, St. Letcher, O. Gregory, M. Platek, Anal. Chem. **10**, 2991 (1998).
53. J. Seelenbinder, Ch. Brown, P. Pivarnik, A. Rand, Anal. Chem. **71**, 1963-1966 (1999).
54. Z. Zhang, T. Imae, Coll. and Interf. Sci. **233**, 99-106 (2001).
55. E. Schutt, S. Richard, P. William, B. George, E. Raymond, L. Karin, L. David US05017009 (1991).
56. B. Dubertret, M. Calame and A. Libchaber Single-mismatch detection using gold-quenched fluorescent oligonucleotides. Nature Biotechnology **19**, 365 (2001).
57. M. Dorrestijn, Th. Schalkhammer, Biosensor with single molecule sensitivity using field emission from metal clusters, Lab-on-a-Chip and Microarrys (CHI, Zurich) (2001).

# THEORY OF FLUOROPHORE-METALLIC SURFACE INTERACTIONS

Joel I. Gersten*

## 1. INTRODUCTION

The study of fluorescence of molecules situated near nano-sized metal particles has been the subject of considerable experimental interest in recent years, largely due to important biological applications and the emergence of nano-technology. The spectroscopic properties of molecules interacting with small solid-state particles was investigated theoretically[1] and experimentally[2] over two decades ago. The work pointed to a competition between enhanced radiative processes and additional non-radiative channels introduced by the presence of particles. It was found that the specific optical properties of the solid played an important role, as well as the geometric shape of the particle,[3] and the location of the molecule relative to the particle.

An example of the recent application to biology is in the use of molecular beacons[4] for the detection of polynucleotides. A fabricated oligonucleotide is attached at one end to a fluorophore moiety and at the other end to a quenching moiety, as in Fig. 1. Typical fluorphores are fluorescein, rhodamine 6G, Texas red, Oregon green, and cyanine-5. Originally DABCYL was used as the quencher but it was later found that gold nanoparticles are more effective quenchers, with a 100-fold times better quenching efficiency. A typical size for the gold nonoparticle is 1.4 nm. The structure of the oligonucleotide is given by the sequence is $SNS'$, where $S$ represents a string of nucleotides (e.g., GAGCGA), $N$ is a target nucleotide sequence (e.g., ACGTTGCC.....T), and $S'$ is the set of base-pairs complementary to $S$ arranged in reverse order (e.g., TCGCTC). Here A,C,G and T refer to the usual nucleotides present in DNA. Sequence $S$ typically consists of 5-7 nucleotides. The structure assumes a hairpin-loop shape in which the fluorophore and the nanoparticle are in close proximity. The particle is effective in quenching the fluorescence of the fluorophore and very little fluorescence is observed. If a polynucleotide containing the sequence $T'$, which is complementary to $T$,

*Joel I. Gersten, Department of Physics, City College of the City University of New York, New York, NY 10031

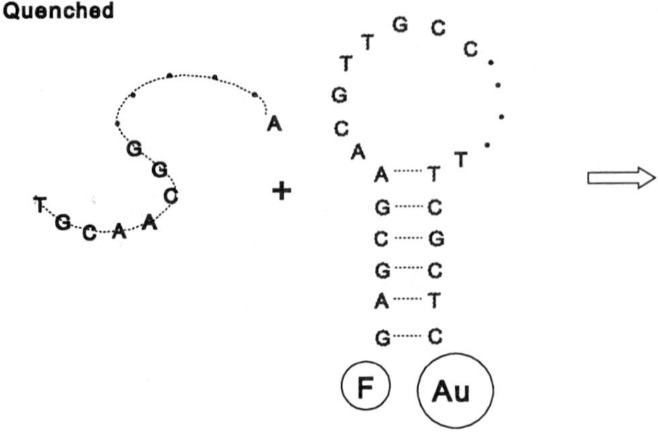

Figure 1. Hypothetical oligonucleotide with an attached fluorophore, F, and gold nanoparticle, Au, exposed to a complementary target polynucleotide. A transition from a quenched fluorescent state to an unquenched state is made upon binding of the target polynucleotide.

is introduced into the solution it binds to $T$ and the hairpin loop is unwound. The fluorophore moves away from the nanoparticle and its normal fluorescence is restored. The distance from the fluorophore to the nanosphere may be controlled by varying the length of the target $N$. The method has proven to be so sensitive that single base-pair mismatches in $T$-$T'$ may be detected[5].

Since most of the modern day applications involve the use of colloidal particles the present article will only refer to spheres. However, it should be noted that the shape-dependence of particles on fluorescent emission has been studied by placing DNA in the proximity of silver island films[6]. In that study the particles were spheroidal in shape.

The article is organized as follows. First a review of the fluorescence process will be given, with a detailed analysis of what happens when a fluorophore is in the vicinity of a particle. It will be shown that both the incident field and radiated fields may be enhanced, thereby tending to produce increased fluorescence. However, we will also see that the particle opens additional non-radiative decay channels and these can serve to quench the fluorescence. Following this a discussion of some recent experiments relating to fluorescence will ensue. A general theorem relating enhanced absorption and emission will be proved in the appendix.

## 2. THEORY

Consider the fluorescence of a fluorophore in the proximity of a solid-state particle. Both the molecule and the particle are assumed to be suspended in solution. The process may be regarded as occurring in a sequence of three incoherent steps: absorption of the light, followed by the rapid relaxation of the excited molecule to some lower electronically excited state, followed by radiative emission. Light of incident intensity $I_0$ and angular frequency $\omega_l$ impinges upon the system. The intensity that the molecule experiences is modified by the presence of the particle to some new value, $I$. The photon may be absorbed, exciting the molecule from its ground state to a higher electronic state. The molecule will then undergo non-radiative relaxation processes. In some fraction of the cases radiation ensues and light is emitted with a spectrum of frequencies. The angular frequency of a particular emitted photon is denoted by $\omega$. In other cases, no radiation occurs and the molecule returns to the ground state by non-radiative processes. The presence of the particle influences both the radiative and non-radiative relaxation of the molecule. In the case where the non-radiative relaxation rate is strongly increased the fluorescence is said to be quenched by the particle.

Let the absorption cross section of the molecule be $\sigma(\omega_l)$ and let the concentration of molecules be $n$. The number of photons absorbed per unit time is $InV\sigma(\omega_l)/(\hbar\omega_l)$, where $V$ is the interaction volume between the light and the system. Let $N_e$ be the number of excited molecules, $\Gamma_r$ be the radiative decay rate, and $\Gamma_{nr}$ be the non-radiative decay rate. The rate equation for $N_e(t)$ is

$$\frac{dN_e}{dt} = \frac{nIV\sigma}{\hbar\omega_l} - (\Gamma_r + \Gamma_{nr})N_e. \tag{1}$$

The number of photons emitted per unit time is

$$\frac{dN_\gamma}{dt} = N_e\Gamma_r. \tag{2}$$

These equations are useful for studying time-resolved fluorescence, in which case an incident pulse of time-dependent intensity $I(t)$ induces a transient response with a decay constant $\Gamma = \Gamma_r + \Gamma_{nr}$. For a short pulse of excitation at $t = 0$ the photon emission rate will decay exponentially as $\exp(-\Gamma t)$ for $t > 0$. However, we will concentrate here on the steady-state case where both $I$ and $N_e$ are constant. Then,

$$\frac{dN_r}{dt} = \frac{nIV\sigma Y}{\hbar\omega_l}, \tag{3}$$

where $Y$ is the fluorescence yield

$$Y = \frac{\Gamma_r}{\Gamma_r + \Gamma_{nr}}.$$

(4)

As we will see, both $\Gamma_r$ and $\Gamma_{nr}$ can be strongly modified by the presence of the particle, and hence so can the value of $Y$.

The solid particle can have an arbitrary shape, but attention for now will focus on the case of a sphere. The radius of the sphere is denoted by $a$ and its complex dielectric function by $\varepsilon(\omega)$. The dielectric function will be assumed to be local in this discussion so there is no dependence on the wave-vector of the photon. The dielectric function for the solvent is denoted by the local function $\varepsilon_s(\omega)$. The particle size is assumed to be sufficiently small compared to the wave length of the relevant photons that the electrostatic approximation to electrodynamics is warranted. Thus retardation effects will be neglected here.

The excited molecule is modeled as a classical dipole of moment $\vec{\mu}(t)$ located a distance $R$ from the center of the particle and defined by a position vector $\vec{R}$. This dipole is taken to oscillate at a frequency $\omega$. The dipole vector makes an angle $\psi$ with respect to the position vector. The dipole sets up an electrostatic potential field in all of space $\Phi(\vec{r})$. This field fluctuates at the emission frequency $\omega$, the factor $\exp(-i\omega t)$ being implicitly understood to be present. From a knowledge of $\Phi(\vec{r})$ it is possible to determine both the radiative and non-radiative decay rates for the molecule, as we will show.

The general geometry is illustrated in Fig. 2. We begin by considering the configuration where the molecular dipole is oriented perpendicular to the surface, i.e. $\psi = 0$. The potential is expanded in an infinite Legendre polynomial series as

$$\Phi(\vec{r}) = \sum_{n=0}^{\infty} A_n r^n P_n(\cos\theta), \quad for\ r < a,$$

(5a)

and

$$\Phi(\vec{r}) = \sum_{n=0}^{\infty} B_n r^{-n-1} P_n(\cos\theta) + \frac{\mu(z-R)}{\varepsilon_s |\vec{r} - R\hat{k}|^3}, \quad for\ r > a.$$

(5b)

Starting with the formula

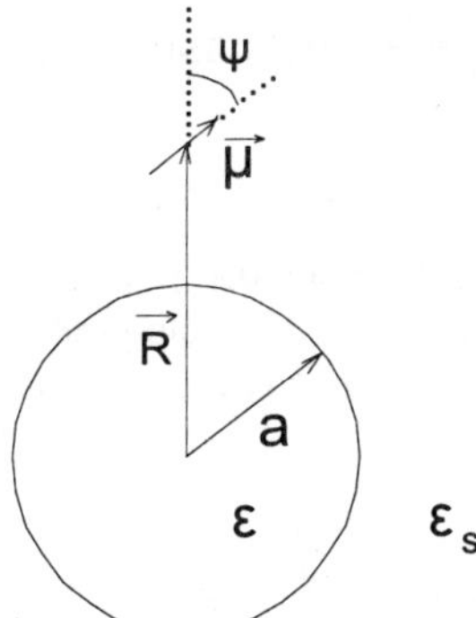



Figure 2. A dipole in the vicinity of a spherical nanoparticle.

$$\frac{1}{|\vec{r} - R\hat{k}|} = \sum_{n=0}^{\infty} \frac{r_<^n}{r_>^{n+1}} P_n(\cos\theta), \tag{6}$$

where $r_< = \min(r, R)$ and $r_> = \max(r, R)$, and differentiating with respect to $z$ (using $\partial/\partial z = \cos\theta \partial/\partial r - (\sin\theta/r)\partial/\partial\theta$) yields

$$-\frac{z - R}{|\vec{r} - R\hat{k}|^3} = \sum_{n=0}^{\infty} (n+1)\frac{r^n}{R^{n+2}} P_n(\cos\theta), \tag{7}$$

so for positions between the particle and the fluorophore

$$\Phi(\vec{r}) = \sum_{n=0}^{\infty} B_n r^{-n-1} P_n(\cos\theta) -$$

$$\frac{\mu}{\varepsilon_s} \sum_{n=0}^{\infty} (n+1) \frac{r^n}{R^{n+2}} P_n(\cos\theta), \quad \text{for } R > r > a. \tag{8}$$

At the surface of the sphere both the potential and the radial component of the electric displacement must be continuous. Using the linear independence of the Legendre polynomials this leads to a set of coupled algebraic equations

$$A_n a^n = \frac{B_n}{a^{n+1}} - \mu(n+1) \frac{a^n}{\varepsilon_s R^{n+2}}, \tag{9a}$$

and

$$n A_n a^{n-1} \varepsilon = -\varepsilon_s (n+1) \frac{B_n}{a^{n+2}} - \mu(n+1) n \frac{a^{n-1}}{R^{n+2}}. \tag{9b}$$

These equations may be solved for $A_n$ and $B_n$:

$$A_n = -\frac{\mu(n+1)(2n+1)}{R^{n+2} [n\varepsilon + (n+1)\varepsilon_s]}; \tag{10a}$$

and

$$B_n = \frac{\mu a^{2n+1} n(n+1)(\varepsilon - \varepsilon_s)}{R^{n+2} \varepsilon_s [n\varepsilon + (n+1)\varepsilon_s]}. \tag{10b}$$

For large $r$ the potential goes as

$$\Phi \to B_1 \frac{P_1(\cos\theta)}{r^2} + \frac{\mu P_1(\cos\theta)}{\varepsilon_s r^2} \equiv \frac{D P_1(\cos\theta)}{\varepsilon_s r^2}. \tag{11}$$

where the parameter $D$ is interpreted as the net dipole moment of the entire system: molecule plus particle. Thus one obtains

$$D = \mu\left[1 + \frac{2(\varepsilon - \varepsilon_s)a^3}{(\varepsilon + 2\varepsilon_s)R^3}\right] \equiv B_\perp \mu \tag{12}$$

The parameter $B_\perp$ will be referred to as the dipole amplification factor for the case where the molecular dipole is oriented perpendicular to the sphere's surface.

In a similar way one may derive the dipole amplification factor $B_{//}$ for the case in which $\psi = 90^o$ in Fig. 2, i.e., where the dipole vector is oriented parallel to the surface. One finds

$$D = \mu\left[1 - \frac{(\varepsilon - \varepsilon_s)a^3}{(\varepsilon + 2\varepsilon_s)R^3}\right] \equiv B_{//}\,\mu. \tag{13}$$

Note that of the three independent orientations of a dipole two are parallel to the surface and one is perpendicular to the surface. Hence, on average, $\overline{B} = 2B_{//}/3 + B_\perp/3 = 1$. Nevertheless, it is possible for $B_\perp$ and $B_{//}$ to be individually large in magnitude, provided they are of opposite sign. This will occur near the dipolar plasmon resonance, defined by the condition $Re[\,\varepsilon(\omega) + 2\varepsilon_s] = 0$. The magnitudes will be particularly large when $Im[\,\varepsilon(\omega) + 2\varepsilon_s]$ is small.

According to classical radiation theory the power radiated by a system which is much smaller in size than the wavelength of the emitted light is determined by the electric dipole moment. If the electric dipole moment is enhanced, so will the radiated power, the enhancement being proportional to $|B(\omega)|^2$. The power per steradian radiated by the system is in a direction making an angle $\theta$ with the dipole moment is

$$\frac{dP_r}{d\Omega} = \frac{\sqrt{\varepsilon_s}\,\omega^4\,|B\mu|^2\sin^2\theta}{8\pi c^3}. \tag{14}$$

Integrating over solid angles and dividing by the photon energy gives an expression for the radiative decay rate

$$\Gamma_r = \frac{\sqrt{\varepsilon_s}\,\omega^3\,|B\mu|^2}{3\hbar c^3}. \tag{15}$$

Since $B$ may be large in magnitude for either the perpendicular or parallel orientation the net value of $\Gamma_r$ may be enhanced over that of a free molecule. One may say that the particle acts as an antenna which assists the radiation process.

Next consider the other part of the fluorescence - the absorption process. Light at frequency $\omega_1$ and electric field vector $\vec{E}_0$ impinges on the system. The local field at the position of the molecule will be denoted by $\vec{E}_1$ and may be different than $\vec{E}_0$. To determine the field we proceed to solve a different electrostatic problem, depicted in Fig. 3. It will be assumed for now that the external field is directed parallel to the vector $\vec{R}$, which may be taken to be parallel to the $z$ -direction. The electrostatic field is given by

$$\Phi(\vec{r}) = -E_0 z + \sum_{n=0}^{\infty} C_n r^{-n-1} P_n(\cos\theta), \quad for\ r > a, \qquad (16a)$$

and

$$\Phi(\vec{r}) = \sum_{n=0}^{\infty} D_n r^n P_n(\cos\theta), \quad for\ r < a. \qquad (16b)$$

When the boundary conditions at the surface of the sphere are imposed one finds a non-zero contribution arises only from the $n = 1$ terms. Thus

$$-E_0 a + C_1 a^{-2} = D_1 a, \qquad (17a)$$

and

$$-\varepsilon_s E_0 - 2 C_1 a^{-3} \varepsilon_s = \varepsilon D_1, \qquad (17b)$$

so

$$C_1 = \frac{\varepsilon - \varepsilon_s}{\varepsilon + 2\varepsilon_s} E_0 a^3, \qquad (18a)$$

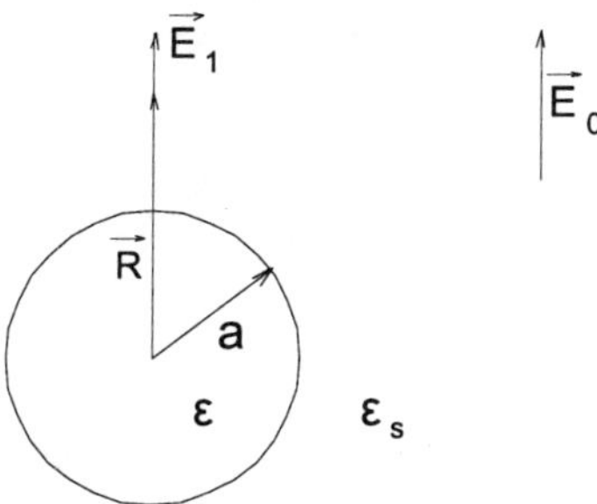


Figure 3. An external electric field is applied parallel to the location vector of the molecule.

and

$$D_1 = -\frac{3\,\varepsilon_s\,E_0}{\varepsilon + 2\,\varepsilon_s}.$$

(18b)

The electric field at the point $(r,\theta) = (R,0)$ is obtained from the formula $E_r = -\partial\Phi/\partial r = E_0\cos\theta + 2\,C_1\cos\theta/\,r^3$ and is

$$E_z = E_0\left[1 + \frac{2(\varepsilon - \varepsilon_s)\,a^3}{(\varepsilon + 2\,\varepsilon_s)\,R^3}\right] \equiv A_\perp E_0,$$

(19a)

where we will call $A_\perp$ the field enhancement factor. Note that $A_\perp = B_\perp$.

In a similar way one may determine the field enhancement factor for the case in which the external electric field is perpendicular to the position vector $\vec{R}$. In that case

$$E_z = E_0 \left[ 1 - \frac{(\varepsilon - \varepsilon_s) a^3}{(\varepsilon + 2\varepsilon_s) R^3} \right] \equiv A_{//} E_0 . \qquad (19b)$$

Again it is noticed that $A_{//} = B_{//}$.

In the fluorescence problem the frequencies at which $A(\omega_1)$ and $B(\omega)$ are evaluated are, of course different, $\omega_1$ being the frequency of the incident light and $\omega$ being the frequency of the radiated light, so they will not be numerically the same. Nevertheless, it is perhaps surprising that the functional forms are exactly the same, i.e., $A(\omega) = B(\omega)$ for the two orientations. In Appendix A a general theorem is derived for an arbitrary geometry. In general the enhancements are described by tensors $\vec{\vec{A}}(\omega)$ and $\vec{\vec{B}}(\omega)$. One tensor is proven to be the transpose of the other. For the case of a sphere the tensors are diagonal and the corresponding coefficients are equal.

Next let us turn our attention to non-radiative decay. The mechanism that will be described is the Joule heating of the particle by the radiating dipole. Crudely speaking, we regard the particle as a resistor and the radiating dipole as providing an AC potential difference across the particle. This induces electrical current to flow throughout the volume and, since there is finite conductivity, there will be energy dissipation. Let the conductivity of the sphere be denoted by $\sigma$ and assume that it is constant throughout the sphere. Local electrodynamics will be assumed so the heating in a given portion of the sphere will be determined by the local electric field. The total dissipated power is

$$P_{nr} = \int d\vec{r} \, \frac{\sigma}{2} |E|^2 . \qquad (20)$$

The conductivity is related to the imaginary part of the dielectric constant by $\sigma = [\omega/(4\pi)] Im(\varepsilon)$. In quantizing the theory one realizes that energy transfer takes place in units of photons, so one may obtain a formula for the non-radiative decay rate by dividing the power by the photon energy. Thus

$$\Gamma_{nr} = \frac{Im(\varepsilon)}{8\pi\hbar} \int d\vec{r} \, |\vec{E}|^2 . \qquad (21)$$

Let us begin by starting with the geometry of Fig. 2 in the case where the dipole is perpendicular to the surface. From Eq. (5a) the electric field inside the sphere is a function of position and is given by

$$\vec{E}(\vec{r}) = -\nabla\Phi = E_r\,\hat{r} + E_\theta\,\hat{\theta} =$$

$$-\hat{r}\sum_{n=0}^{\infty} nA_n\,r^{n-1}\,P_n(\cos\theta) - \hat{\theta}\sum_{n=0}^{\infty} A_n\,r^{n-1}\frac{\partial P_n(\cos\theta)}{\partial\theta}. \tag{22}$$

Thus,

$$\int d\vec{r}\,|\vec{E}|^2 = \sum_{n,n'} A_n\,A_{n'}^*\,2\pi\int_{-1}^{1} d(\cos\theta)\int_0^a dr\,r^2\,r^{n-1}\,r^{n'-1}\,x$$

$$\left[P_n(\cos\theta)\,P_{n'}(\cos\theta) + \frac{1}{r^2}\frac{\partial P_n(\cos\theta)}{\partial\theta}\frac{\partial P_{n'}(\cos\theta)}{\partial\theta}\right]. \tag{23}$$

Using the orthogonality relation for the Legendre polynomials

$$\int_{-1}^{1} dx\,P_n(x)\,P_{n'}(x) = \frac{2}{2n+1}\,\delta_{n,n'}, \tag{24}$$

and the Legendre differential equation

$$\frac{d}{dx}\left[(1-x^2)\frac{d\,P_n(x)}{dx}\right] + n(n+1)\,P_n(x) = 0, \tag{25}$$

and integrating the second term in Eq. (23) by parts leads to the expression

$$\int d\vec{r}\,|\vec{E}|^2 = \frac{4\pi\,|\mu|^2}{a^3}\sum_{n=1}^{\infty}\frac{n(n+1)^2(2n+1)}{|n\varepsilon+(n+1)\varepsilon_s|^2}\left(\frac{a}{R}\right)^{2n+4}. \tag{26}$$

Thus one finally obtains an expression for the nonradiative decay rate

$$\Gamma^{\perp}{}_{nr} = -\frac{|\mu|^2}{2\hbar a^3} \sum_{n=1}^{\infty} (n+1)^2 (2n+1) \left(\frac{a}{R}\right)^{2n+4} Im\left[\frac{1}{n\varepsilon + (n+1)\varepsilon_s}\right]. \tag{27}$$

In a similar way the following expression is obtained for the geometry in which the dipole is parallel to the surface:

$$\Gamma^{//}{}_{nr} = -\frac{|\mu|^2}{4\hbar a^3} \sum_{n=1}^{\infty} n(n+1)(2n+1) \left(\frac{a}{R}\right)^{2n+4} Im\left[\frac{1}{n\varepsilon + (n+1)\varepsilon_s}\right]. \tag{28}$$

We notice that these rates are enhanced near the particle plasmon frequencies defined by $Re[n\varepsilon(\omega) + (n+1)\varepsilon_s(\omega)] = 0$. Values for the plasmon resonance frequencies for Au spheres in water are given in Table 1.

There may be other non-radiative decay channels not described by the optical properties of the solid. For example, if the excited molecule were touching the nanoparticle then there could be energy transfer by the short range contact interaction. There could also be the tunneling of an electron from the excited molecule to the solid, if the energy of the excited electron lies above the Fermi energy of the metal.

The quantum yield for a fluorophore in the absence of the particle, or equivalently, an infinite distance from the particle, is

$$Y^{\infty} = \frac{\Gamma_r^{\infty}}{\Gamma_r^{\infty} + \Gamma_{nr}^{\infty}}. \tag{29}$$

In the presence of the particle the non-radiative rate is augmented by the non-radiative energy transfer to the particle, $\Gamma_{nr}$, so the quantum yield expression becomes

**Table 1.** Plasmon energies, real and imaginary parts of the dielectric constant for Au, and the dielectric constant for water.

| $n$ | $E$ [eV] | $Re[\varepsilon_{Au}]$ | $Im[\varepsilon_{Au}]$ | $\varepsilon_{water}$ |
|-----|----------|------------------------|------------------------|------------------------|
| 1 | 2.43 | -3.58 | 2.79 | 1.783 |
| 2 | 2.49 | -2.67 | 3.28 | 1.785 |
| 3 | 2.51 | -2.43 | 3.46 | 1.785 |
| 4 | 2.53 | -2.22 | 3.67 | 1.786 |
| 5 | 2.54 | -2.13 | 3.76 | 1.786 |
| $\infty$ | 2.58 | -1.77 | 4.26 | 1.787 |

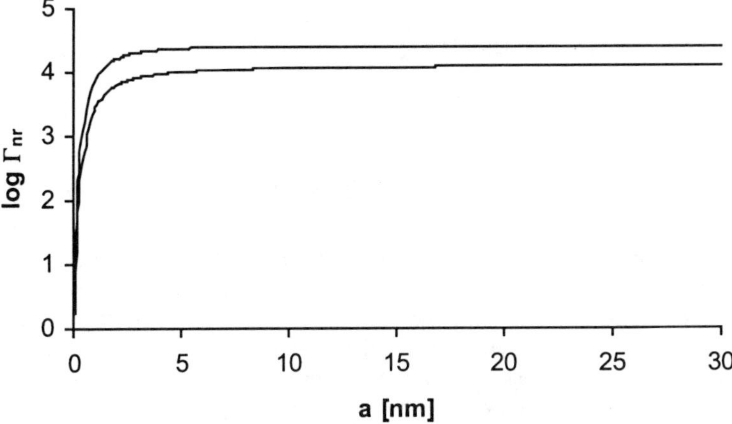

**Figure 4.** Dependence of the non-radiative decay rate on particle radius for fixed distance of the fluorophore to the surface. The data is for a gold particle and photon energy of 3.10 eV. The upper curve is for the perpendicular orientation and the lower curve is for the parallel orientation.

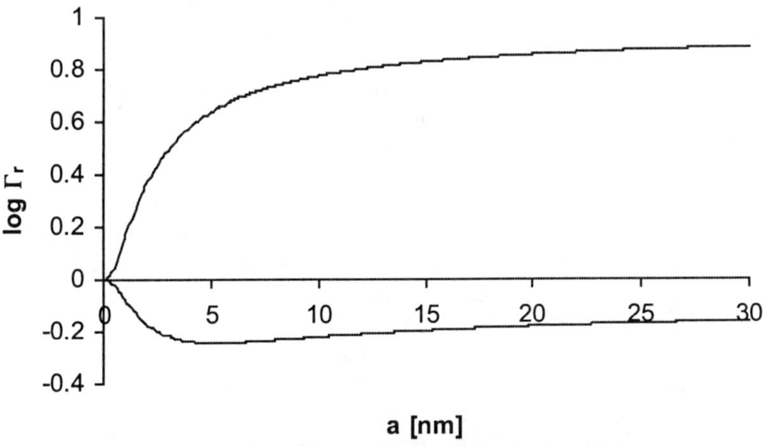

**Figure 5.** Same as in Fig. 4 but for the radiative decay rate. The upper curve is again for the perpendicular orientation and the lower one is for the parallel case.

$$Y = \frac{\Gamma_r}{\Gamma_r + \Gamma_{nr}^\infty + \Gamma_{nr}} = \frac{\tilde{\Gamma}_r}{\tilde{\Gamma}_r + \tilde{\Gamma}_{nr} - 1 + 1/Y^\infty}, \tag{30}$$

where we have introduced the decay rates normalized to the free radiative decay rates $\tilde{\Gamma}_r = \Gamma_r / \Gamma_r^\infty$ and $\tilde{\Gamma}_{nr} = \Gamma_{nr} / \Gamma_r^\infty$.

Typical results are presented in Figs. 4-9 for molecules near gold spheres. In Fig. 4 the normalized non-radiative decay rate is plotted as a function of particle radius, $a$, for a fixed distance from the surface $d = R-a = 1$ nm. The energy of the photon is held fixed at $E = 3.1$ eV, above the plasmon energies tabulated in Table 1. The upper curve is for the case where the transition moment is pointing perpendicular to the sphere, and the lower curve is for the case where the transition moment is parallel to the surface. One notes that for small particle sizes the non-radiative decay is not as strong as for large radii. There is a tendency for the decay rates to saturate to constant values for large $a$. This behavior is expected, since the limiting behavior of a molecule near a planar surface is approached asymptotically.

In Fig. 5 the normalized radiative decay rate is studied as a function of sphere radius. It is expected that the effects of transition dipole orientation will be pronounced. For the perpendicular orientation the image dipole tends to align itself with the molecular dipole thereby enhancing the radiation rate. In the parallel arrangement the image dipole is opposite in direction, leading to a partial cancellation of the dipoles and a weaker radiation. In Fig. 6 the normalized non-radiative decay rate is plotted as a function of the distance of the molecule from the center of the sphere. The sphere radius is 1.5 nm. Again the upper and lower curves are for the perpendicular and parallel arrangements, respectively. As before a value of $E = 3.1$ eV is used. One notices, as expected, a strong falloff of the non-radiative decay rate with increasing $R$. This is in agreement with the general tendency for energy transfer to fall off rapidly with distance.

Figure 7 shows the corresponding behavior for the radiative decay rate. For distances beyond around 10 nm there is little dependence on the orientation of the transition moment and the decay rates assume their values for molecules in solution. For small $R$ the trends are opposite for the two orientations, with the perpendicular orientation again favoring enhanced radiative emission. Figures 8 and 9 are similar to Figs. 6 and 7 except the size of the sphere is increased to 10 nm.

A recent experimental test of the theory was performed[7] using time-resolved fluorescence. Lissamine fluorophores were attached to gold nanoparticles via thio ether groups. Both the radiative and non-radiative decay channels were studied. The sizes of the Au particles were varied in several steps between $a = 1$ and $a = 30$ nm, whereas the distance between the fluorphore and the nanoparticle was kept constant at about $d = r - a = 1$ nm. The optical excitation source was a 120 fs laser pulse at a wavelength of 400 nm (3.10 eV), well away from the dipolar plasmon resonance at 520 nm (2.4 eV) for a gold sphere in water. It was believed that the transition dipole of the fluorophore was parallel to the surface of the particle.

Figure 10 shows the absorption spectrum of the gold particles in solution (dashed curve) and the fluorescence spectrum of a solution of lissamine fluorophores both before and after they are attached to the gold particles. As expected, the gold particles in solution show an absorption peak centered at the dipolar plasmon frequency. It is seen that the fluorescence of the lissamine is strongly quenched by presence of the gold nanoparticles. It was estimated that the quenching efficiency exceeded 99%, when the residual fluorescence of unbound fluorophores was taken into account.

In Fig. 11 they[7] compared the theory with the experiment for the radiative and non-radiative decay rates for spheres of different radii. The qualitative trends of the size dependences of experiment are in agreement with theory. For the radiative decay rate the agreement is somewhat quantitative whereas for the non-radiative rate the experimental rates are too small by two orders of magnitude.

Dulkieth et al[7] suggest some possible causes for the quantitative disagreement between theory and experiment for the non-radiative rates: the need for using a non-local dielectric function rather than bulk optical data[8]; the description of a molecule close to the surface by a point dipole; and additional non-radiative channels such as electron transfer.

Electron transfer has been observed recently for a pyrene thiol bound to a gold nanoparticle[9]. However, such processes are only expected to be active when the excited electronic energy level of the fluorophore is higher in energy than the Fermi level in the metal. If the energy level is not high enough the channel is blocked by the Pauli exclusion principle. Even if the channel is open the fluorophore may lie sufficiently far from the surface so that the overlap of its wave function with the electronic tails of the metal extending out of the solid is very small. Then tunneling would be improbable. In any case, electron transfer would increase the non-radiative decay rate, as pointed out by Dulkieth et al.[7].

The crudity of the point dipole approximation for the transition moment is indeed a concern, especially for small fluorphore-surface distances. However, the effect should not be as large as the two order-of-magnitude effect observed in the experiments.

The non-local dielectric effect can be a more serious concern. Fuchs and Claro[10] investigated the multipolar response of small metallic spheres and derived an approximation for the non-local dielectric function. In Eqs. (27) and (28) the index $n$ has the interpretation of being the angular momentum in units of $\hbar$. Note that the sums extend from one to infinity. They argue that there should be a cutoff in the angular momentum. Electrons in a solid sphere (treated as an isotropic medium) have a maximum angular momentum $n_F$ determined by the Fermi level. For a free-electron solid $n_F \hbar = m a v_F$, where $v_F$ is the Fermi velocity and $m$ is the band effective mass of the electron. For Au the Fermi velocity is $1.40 \times 10^6$ m/s and the band effective mass is very close to the free electron mass. Thus for particle sizes of $a = 1$, 15 and 30 nm, the corresponding values of $n_F$ are 12, 181 and 363, respectively.

The multipolar polarizability of the sphere is defined in the local theory by

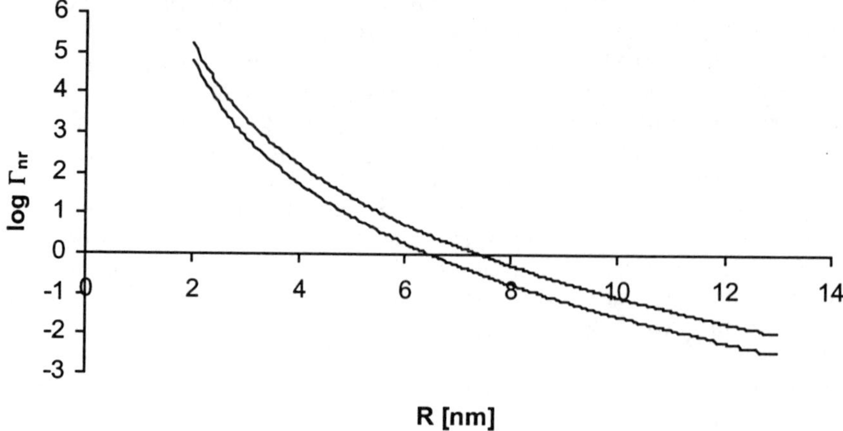

**Figure 6.** Dependence of the non-radiative decay rate on distance from the center of a 1.5 nm radius sphere.

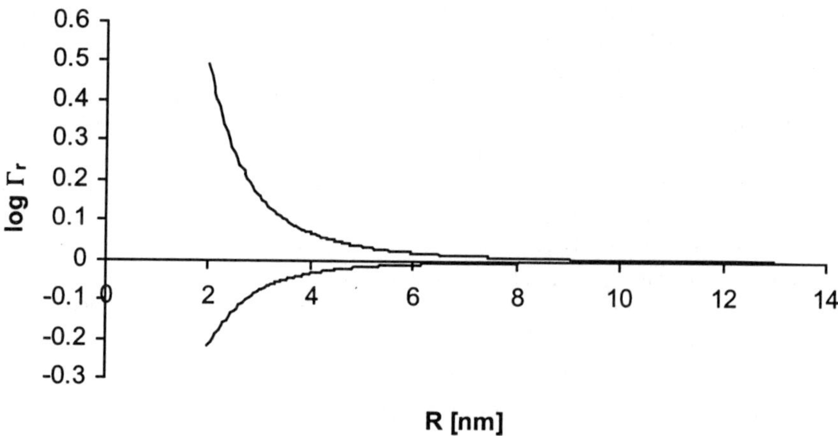

**Figure 7.** Same as Fig. 6 but for the radiative decay rate.

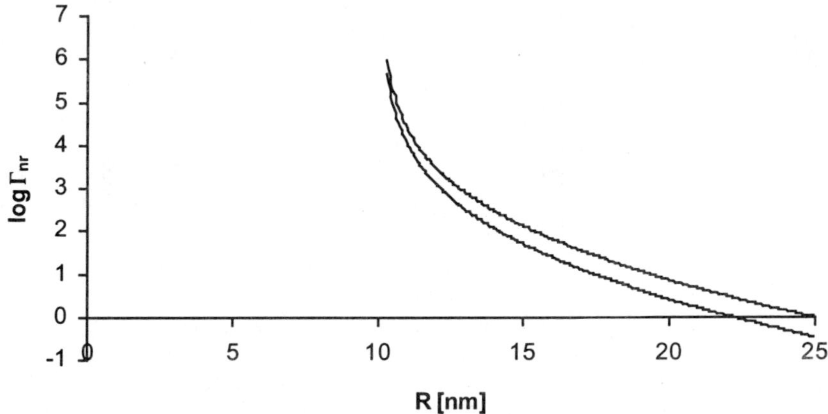

**Figure 8.** Dependence of the non-radiative decay rate on distance from the center of a 10 nm radius sphere. The upper curve is for the perpendicular orientation and the lower curve if for the parallel case.

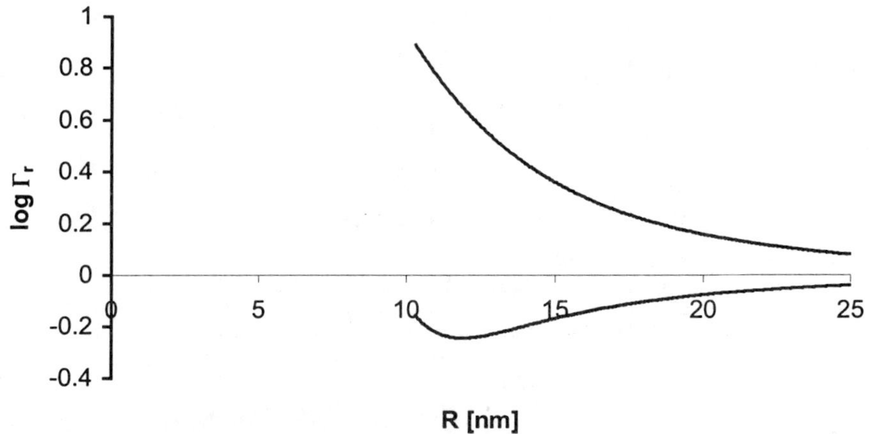

**Figure 9.** Same as in Fig. 8 but for the radiative decay rate. The upper curve is for the perpendicular orientation    and the lower curve is for the parallel case.

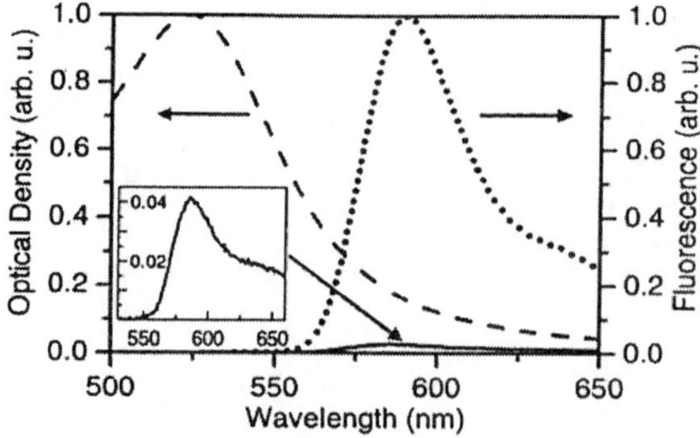

Figure 10.  Optical absorption spectrum for Au nano-spheres of size $a = 15$ nm (dashed curve),  fluorescence spectrum of 0.37 $\mu$ M aqueous lyssamine solution  (dotted curve), and quenched fluorescence spectrum with gold nano-spheres (solid curve and inset).  From E. Dulkeith, A. C. Morteani, T. Niedereichholz, T. A. Klar, J. Feldmann, S. A. Levi, F. C. J. M. van Veggel, D.  N. Reinhoudt,  M. Moller and D. I. Gittins,  Phys. Rev. Letters, 89(20), 203002 (2002).  Copyright (2002) by the American Physical Society.

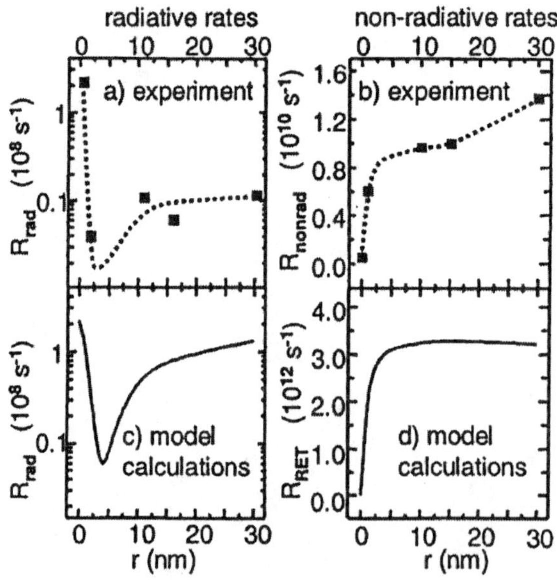

Figure 11.  Comparison of experiment (a and b) and theory (c and d) for the  radiative (a and c) and non-radiative (b and d) decay rates, respectively.  The  particle radius is denoted by $r$ (instead of $a$). From E. Dulkeith, A. C. Morteani, T.  Niedereichholz, T. A. Klar, J. Feldmann, S. A. Levi, F. C. J. M. van Veggel, D.  N. Reinhoudt, M.   Moller and D. I. Gittins,  Phys. Rev. Letters, 89(20), 203002 (2002).  Copyright (2002) by the American Physical Society.

$$\alpha_n = \frac{n(\varepsilon-1)}{n(\varepsilon+1)+1} a^{2n+1} = -(2n+1)a^{2n+1} Im\left[\frac{1}{n(\varepsilon+1)+1}\right]. \qquad (31)$$

This is the same factor that appears in Eqs. (27) and (28) (evaluated for free space). Their expression for this polarizability in the non-local theory is written as

$$\alpha_n = \frac{n(E_n-1)}{n(E_n+1)+1} a^{2n+1}, \qquad (32)$$

with the coefficients $E_n$ given by the an integral over wave vectors involving the non-local dielectric constant and spherical Bessel functions

$$E_n = \left[\frac{2}{\pi}(2n+1)a \int_0^{\infty} \frac{j_n^2(ka)}{\varepsilon(k,\omega)} dk\right]^{-1}. \qquad (33)$$

Several different models for $\varepsilon(k,\omega)$ were explored by them, including the (local) Drude model, a hydrodynamic model, and the Lindhard-Mermin dielectric function. At low frequencies, much below the plasma frequency, they found that the imaginary part of the polarizability was actually enhanced, but that at higher frequencies this enhancement was not as pronounced. The enhancement was attributed to the excitation of particle-hole pairs in the metal.

The calculations were applied to the non-radiative decay of fluorescence by Leung[11] using the hydrodynamic model in which

$$\varepsilon(k,\omega) = 1 - \frac{\omega_p^2}{\omega\left(\omega+\dfrac{i}{\tau}\right) - \dfrac{3}{5}(v_F k)^2}, \qquad (34)$$

where $\omega_p$ is the plasma frequency and $\tau$ is the collision time. The calculations, performed for tin spheres, showed that a suppression of two orders of magnitude for radiative decay rates was possible. However, it was recognized that the hydrodynamic model is an idealization. A modified expression for the hydrodynamic model was used by Corni and Tomasi to examine phosporescence of biacetyl near a Ag surface in an ammonia solvent[12]. Their results showed that a non-local dielectric constant has a substantial effect. In their recipe for the modified dielectric function they used the bulk optical data for Ag, subtracted the AC Drude contribution, and added in the second term in Eq. (34) to approximate the nonlocal effects. They also included the effect of surface roughness and found that it too had a considerable influence on the decay rate.

Other effects can contribute as well to the lowering of the nonradiative decay rate. The Drude collision time for an electron in Au is $\tau = 3.0 \times 10^{-14}$ s. Therefore, the mean free path for de-phasing collisions is $l = v_F \tau = 42$ nm, which is longer than most of the diameters of the nano-particles under study. It is generally believed[13] that collisions with the wall of the sphere constitute de-phasing collisions. Thus, a mean free path given by $l_a = a$ (or $l_a = (4/3)a$) is appropriate. For small values of $a$ this will lead to a considerably shorter mean free path and hence a smaller collision time. The effect will be to increase the size of the imaginary part of the dielectric constant and depress the value of the non-radiative decay rate.

The effects of confinement in small particles and clusters adds another set of issues that need to be considered, but will not be examined here.


## Appendix A


In this appendix a general proof is given of the equivalence of the local field amplification matrix to the transpose of the dipole amplification matrix. Consider a piece of dielectric material of arbitrary shape placed in a uniform external electric field $\vec{E}_0$. Let O be the origin of a coordinate system and $\vec{r}_1$ denote the displacement vector to an arbitrary point in space. The electric field at $\vec{r}_1$ is $\vec{E}_1$. The potential field in all of space is denoted by $\phi(\vec{r})$. The situation is depicted in Fig. 12a.

In Fig. 12b the same shape of dielectric material is placed in the neighborhood of an electric dipole $\vec{\mu}$ located at position $\vec{r}_1$ relative to the origin O. Now there is no external field present. The dielectric develops an induced dipole $\vec{D} - \vec{\mu}$ so the net dipole moment of the system is $\vec{D}$. The potential field in all of space corresponding to this problem is denoted by $\psi(\vec{r})$.

One may introduce two matrices for the two respective problems. The field amplification matrix is defined by $\vec{E}_1 = \overset{\leftrightarrow}{A} \cdot \vec{E}_0$. It relates the local field at position $\vec{r}_1$ to the applied field $\vec{E}_0$. The dipole amplification matrix is defined by $\vec{D} = \overset{\leftrightarrow}{B} \cdot \vec{\mu}$. It relates the system dipole to the dipole that was placed at position $\vec{r}_1$.

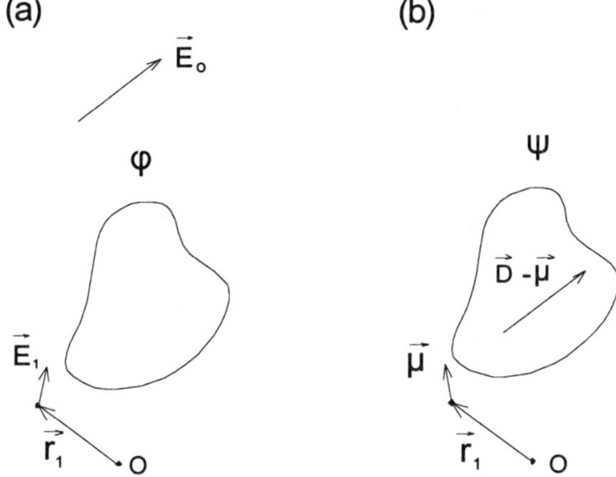


**Figure 12.** a)   A dielectric solid of arbitrary shape is subjected to a uniform external electric field $\vec{E}_0$. The local field at position $\vec{r}_1$ is denoted by $\vec{E}_1$. b)  An electric dipole moment $\vec{\mu}$ is placed at position $\vec{r}_1$ and an electric dipole moment $\vec{D}$ - $\vec{\mu}$ is induced in the dielectric solid.

Theorem: $\vec{\vec{A}} = \vec{\vec{B}}^T$

Proof:   We study the linear response of the system to an electromagnetic wave at frequency $\omega$.   Introduce a space-dependent local dielectric function $\varepsilon(\vec{r},\omega)$ to represent the local dielectric response of both the interior of the solid as well as empty space.  The theorem is general enough to accommodate a graded dielectric function as well.  In the case of a shape of constant dielectric constant $\varepsilon(\vec{r},\omega) = \varepsilon_d(\omega)$ if $\vec{r}$ is inside the solid and $\varepsilon(\vec{r},\omega) = 1$ if $\vec{r}$ is outside the solid.  The potential field $\phi(\vec{r})$ obeys the Laplace equation

$$\nabla \cdot [\varepsilon(\vec{r},\omega)\nabla\phi(\vec{r})] = 0,\tag{A.1}$$

and the potential field $\psi(\vec{r})$ obeys the Poisson equation

$$\nabla \cdot [\,\varepsilon(\vec{r},\omega)\nabla\psi(\vec{r})\,] = 4\pi\vec{\mu}\cdot\nabla\delta(\vec{r}-\vec{r}_1) = 4\pi\nabla\cdot[\,\vec{\mu}\delta(\vec{r}-\vec{r}_1)\,]\,. \qquad (A.2)$$

Multiply Eq. (A.1) by $\psi(\vec{r})$ and Eq. (A.2) by $\phi(\vec{r})$ and form the difference of the two equations. This yields

$$\nabla\varepsilon\cdot[\,\psi\nabla\phi-\phi\nabla\psi\,] + \varepsilon[\,\psi\nabla^2\phi-\phi\nabla^2\psi\,] = -4\pi\phi\nabla\cdot[\,\vec{\mu}\delta(\vec{r}-\vec{r}_1)\,]\,. \qquad (A.3)$$

Introduce a vector field $\vec{S}(\vec{r}) = \psi\nabla\phi-\phi\nabla\psi$ and rewrite this equation as

$$\nabla\varepsilon\cdot\vec{S} + \varepsilon\nabla\cdot\vec{S} = \nabla\cdot(\varepsilon\vec{S}) = -4\pi\nabla\cdot[\,\phi\vec{\mu}\delta(\vec{r}-\vec{r}_1)\,] + 4\pi\delta(\vec{r}-\vec{r}_1)\vec{\mu}\cdot\nabla\phi\,. \qquad (A.4)$$

Next integrate both sides of this equation over the volume of a large sphere centered at O. By the divergence theorem the left hand side of the resulting equation may be converted to a surface integral. If it is assumed that there is vacuum at large distances from O the dielectric constant on the surface of the sphere has the value 1. On the right hand side of Eq. (A.4) the volume integral of the first term also may be transformed to a surface integral. Since the point $\vec{r}_1$ is in the neighborhood of the origin the delta function vanishes when $\vec{r}$ is on the surface of the big sphere and the first term gives zero. The second term, when integrated over the volume of the sphere, gives a contribution determined by the electric field at point $\vec{r}_1$, so

$$\int \hat{r}\cdot\vec{S}\,dA = -4\pi\vec{\mu}\cdot\vec{E}_1\,. \qquad (A.5)$$

The asymptotic dependence of the potential $\psi(\vec{r})$ is dominated by the electric dipole contributions, i.e.,

$$\psi(\vec{r}) \rightarrow \frac{\vec{D}\cdot\vec{r}}{r^3}, \qquad (A.6a)$$

where $\vec{D}$ is the dipole vector of the system, molecule plus particle. The corresponding asymptotic dependence of the potential $\phi(\vec{r})$ has contributions from both the imposed external field $\vec{E}_0$ and the dipole that develops on the dielectric, i.e.,

$$\phi(\vec{r}) \rightarrow -\vec{E}_0 \cdot \vec{r} + \frac{\vec{p} \cdot \vec{r}}{r^3}. \tag{A.6b}$$

Thus, for large $r$ the vector $\vec{S}(\vec{r})$ assumes the form

$$\vec{S} = \frac{\vec{D} \cdot \vec{r}}{r^3} \nabla \left( -\vec{E}_0 \cdot \vec{r} + \frac{\vec{p} \cdot \vec{r}}{r^3} \right) - \left( -\vec{E}_0 \cdot \vec{r} + \frac{\vec{p} \cdot \vec{r}}{r^3} \right) \nabla \left( \frac{\vec{D} \cdot \vec{r}}{r^3} \right) \rightarrow$$

$$-\frac{\vec{D} \cdot \vec{r}}{r^3} \vec{E}_0 + \vec{E}_0 \cdot \vec{r} \nabla \left( \frac{\vec{D} \cdot \vec{r}}{r^3} \right). \tag{A.7}$$

This expression is inserted into the left hand side of Eq. (A.5) and the divergence theorem is used in reverse to evaluate the integral. Thus

$$\int \hat{r} \cdot \left( \vec{E}_0 \vec{D} \cdot \nabla \frac{1}{r} - \vec{E}_0 \cdot \vec{r} \nabla \left( \vec{D} \cdot \nabla \frac{1}{r} \right) \right) dA = \int \nabla \cdot \left( \vec{E}_0 \vec{D} \cdot \nabla \frac{1}{r} - \vec{E}_0 \cdot \vec{r} \nabla \left( \vec{D} \cdot \nabla \frac{1}{r} \right) \right) d\vec{r} =$$

$$\int \left( -\vec{E}_0 \cdot \vec{r} \vec{D} \cdot \nabla \nabla^2 \frac{1}{r} \right) d\vec{r} = \int \vec{E}_0 \cdot \vec{r} \vec{D} \cdot \nabla 4\pi \delta(\vec{r}) d\vec{r} =$$

$$\int \left( \vec{D} \cdot \nabla \vec{E}_0 \cdot \vec{r} \delta(\vec{r}) - 4\pi \delta(\vec{r}) \vec{D} \cdot \nabla \vec{E}_0 \cdot \vec{r} \right) d\vec{r} = \int \nabla \cdot \left( \vec{D} \vec{E}_0 \cdot \vec{r} \delta(\vec{r}) \right) d\vec{r} -$$

$$4\pi \int \delta(\vec{r}) \vec{D} \cdot \vec{E}_0 d\vec{r} =$$

$$\int \hat{r} \cdot \vec{D} \vec{E}_0 \cdot \vec{r} \delta(\vec{r}) dA - 4\pi \vec{D} \cdot \vec{E}_0 = -4\pi \vec{D} \cdot \vec{E}_0 =$$

$$-4\pi \vec{\mu} \cdot \vec{E}_1. \tag{A.8}$$

Inserting the expressions $\vec{E}_l = \overset{\leftrightarrow}{A} \cdot \vec{E}_0$ and $\vec{D} = \overset{\leftrightarrow}{B} \cdot \vec{\mu}$ in Eq. (A.8) one obtains

$$\vec{\mu} \cdot \overset{\leftrightarrow}{A} \cdot \vec{E}_0 = \vec{\mu} \cdot \overset{\leftrightarrow}{B}^T \cdot \vec{E}_0 .  \tag{A.9}$$

Since this must hold for all values of $\vec{\mu}$ and $\vec{E}_0$ it follows that the two matrices are the transpose of each other

$$\overset{\leftrightarrow}{A} = \overset{\leftrightarrow}{B}^T .  \tag{A.10}$$

Q.E.D.

Although the above argument was presented for the case of only a single particle there is nothing in the formalism that restricts it to that case, i.e. the solid doesn't have to be a connected object but can even consist of separated pieces. In the more general case the integrals would get replaced by sums over the particles and integrals over the surface of each particle.

Supported in part by an ECAT research grant from NYSTAR.

## REFERENCES

1. J. I. Gersten and N. Nitzan, Spectroscopic properties of molecules interacting with small dielectric particles, *J. Chem. Phys.* **75**(3), 1139-1152 (1981); Mathematical Appendices, *Physics Auxiliary Publication Service of the American Institute of Physics*, Document No. JCPSA-75-1139-32; Electromagnetic theory of enhanced Raman scattering by molecules on rough surfaces, *J. Chem. Phys.* **73**(7), 3023-3037 (1980).
2. D. A. Weitz, S. Garoff, C. D. Hanson, T. J. Gramila and J. I. Gersten, Fluorescent lifetimes of molecules on silver island films, D. A. Weitz, S. Garoff, J. I. Gersten and A. Nitzan, *Optics Letters* **7**, 89-91 (1982); Raman scattering and fluorescence from molecules adsorbed on a rough silver surface, *J. Chem. Phys.* **78**(9), 5324-5338 (1983); D. A. Weitz, S. Garoff, J. I. Gersten, and A. Nitzan, The enhancement of *Raman scattering, resonance Raman scattering and florescence from molecules adsorbed on a rough silver surface, J. Chem. Phys.* **78**(9), 5324-5338 (1983).
3. J. I. Gersten, The effect of surface roughness on surface-enhanced Raman scattering, *J. Chem. Phys.* **72**(10), 5779-5780 (1980); Rayleigh, Mie and Raman scattering by molecules adsorbed on rough surfaces, *J. Chem. Phys.* **72**(10), 5780-5781 (1980).
4. S. Tyagi and F. R. Kramer, Molecular beacons: probes that fluoresce upon hybridization, *Nature Biotechnology* **14**, 303-308 (1996).
5. B. Dubertret, M. Calame and A. J. Libchaber, Single-mismatch detection using gold-quenched fluorescent oligonucleotides, *Nature Biotechnology*, **19**, 365-370 (2001).
6. J. R. Lakowicz, B. Shen, Z. Gryczynski, S. D'Auria and I. Gryczynski, Intrinsic fluorescence from DNA can be enhanced by metallic particles, *Biochemical and Biophysical Research Communications* **286**, 875-879 (2001).
7. E. Dulkeith, A. C. Morteani, T. Niedereichholz, T. A. Klar, J. Feldmann, S. A. Levi, F. C. J. M. van Veggel, D. N. Reinhoudt, M. Moller and D. I. Gittins, *Fluorescence quenching of dye molecules near gold nanoparticles: radiative and nonradiative effects*, Phys. Rev. Letters, **89**(20),203002 to 203005 (2002).
8. P. B. Johnson and R. W. Christy, *Optical constants of the noble metals*, Phys. Rev. **B6**(12), 4370-4379 (1972).

9.  B. I. Ipe, G. Thomas, K. Barazzouk, S. Hotchandani, and P. V. Kamet, *Photoinduced ˋcharge separation in a fluorophore-gold nanoassembly*, J. Phys. Chem. **B106**(1), 18-21 (2002).
   10. R. Fuchs and F. Claro, *Multipolar response of small metallic spheres: nonlocal theory*, Phys. Rev. **B36**(8), 3722-3727 (1987).
11. P.T. Leung, *Decay of molecules at spherical surfaces: nonlocal effects*, Phys. Rev. **B42**(12)7622-7625 (1990).
12. S. Corni and J. Tomasi, Lifetimes of electronic excited states of a molecule close to a metal surface, J. Chem. Phys. **118**(14), 6481-6494 (2003).
13.  U. Kreibig and M. Vollmer, *Optical Properties of Metal Clusters* (Springer, Berlin,1995).

# 7

# SURFACE-ENHANCEMENT OF FLUORESCENCE NEAR NOBLE METAL NANOSTRUCTURES

Paul J. G. Goulet, and Ricardo F. Aroca[*]

## 1. INTRODUCTION

In 1974, Fleischmann *et al.*[1] observed an unusual experimental result while recording the Raman scattering of pyridine molecules on roughened silver electrodes. They noticed scattering intensities that far exceeded what they expected based on the number of molecules they believed they were probing. This result was interpreted as being due to an increase in the surface area sampled, because of electrochemical roughening, and thus, to an increase in the number of adsorbate molecules available for scattering. In 1977, however, it was discovered that this large increase in Raman intensity could not be explained by surface area increases alone. Jeanmaire and Van Duyne[2], and Albrecht and Creighton[3], independently demonstrated that the observed enhancement was in fact a result of a surface-enhancement process, and the term surface-enhanced Raman scattering, SERS, was coined. This significant scientific discovery led to a flurry of activity in the new field of surface spectroscopy, and soon other surface-enhanced photoprocesses were recognized. Among these was the effect of surface-enhanced fluorescence, SEF. In SEF, as other enhanced processes, energy transfer plays a decisive role, and the classical work on energy transfer[4] to smooth metal surfaces, and coupling of excited molecules to surface plasmons[5], constitutes a relevant pre-SES (Surface-Enhanced Spectroscopy) background for the discussion and understanding of SEF.

Surface-enhanced fluorescence arises from the electromagnetic interaction that occurs between fluorescent molecules and metal nanoparticles that have appropriate enhancing optical properties, and has been described as the "weak cousin of the SERS effect"[6], owing to its rather tiny enhancement factors of generally less than ~100. (There is, however, a report of larger EFs of fluorescence from gold nanorods.[7]) The study of surface-enhanced fluorescence is just more than 22 years old now, and the effect has been well studied, and is well understood.[8-10] However, despite its many possible applications, and its very high cross sections, it is a field that has attracted surprisingly little attention from the scientific mainstream. This is perhaps due to some overshadowing by its "stronger cousin" SERS.[6, 11, 12] Thus, it is the primary focus of this work to discuss SEF within the context of surface-enhanced spectroscopies, and provide a springboard for further work in this field.

---

[*] Materials & Surface Science Group, Department of Chemistry and Biochemistry, University of Windsor, N9B 3P4, Windsor, ON, Canada, R. F. A. email: g57@uwindsor.ca

*Topics in Fluorescence Spectroscopy, Volume 8: Radiative Decay Engineering*
Edited by Geddes and Lakowicz, Springer Science+Business Media, Inc., New York, 2005

The surface-enhanced fluorescence, SEF, and surface-enhanced Raman scattering, SERS, phenomena are inextricably tied to one another due to their common electromagnetic enhancement origin, and thus there is a very strong overlap in their literature. In fact, there is often a direct competition between fluorescence and Raman scattering that, as we shall see, with strategic experimental design, can be exploited. There are, however, a few very important points on which these two effects differ, the most important of which is their distance dependence.

SERS intensities have been shown to decrease as the distance between adsorbates and enhancing metal nanostructures is increased.[13] In other words, SERS is an effect that is strengthened when the adsorbate is placed directly on the metal nanostructure. This is explained by the fact that Raman is a process in which energy losses to the metal can be ignored, because the molecule can be considered a driven oscillator.[6] Nevertheless, photobleaching and photodissociation are common competitors that interfere and hinder the observation of clean SERS spectra. Oppositely, in SEF, molecules are understood as free-running oscillators[6], and intensity is expected to exhibit a maximum enhancement at a certain distance of separation between the molecule and the surface.[10, 14, 15] It is thus common to observe a reduction in luminescence when a fluorophore is placed directly on the surface of a metal (i.e. EF <1). The latter is explained by a direct energy transfer from the dipole to the planar surface, as elegantly shown in the classical work of Chance, Prock, and Silbey[4]. The Förster type dipolar energy transfer is proportional to $d^{-3}$ (where $d$ is the dipole- metal surface separation), and the energy is dissipated into the phonon bath. The distance dependence of EM enhancement indicates that it decreases appreciably when $d$ approaches the dimension of the metal particle. The difference in the distance dependence of these two effects, which will be discussed further later, is integral to the understanding of SEF as a balance between the two opposing processes of surface-enhancement, and non-radiative energy loss to the metal surface.[8]

The competition between these processes demonstrates very clearly that the enhancement of fluorescence is electromagnetic in origin, and therefore, that any chemical enhancement mechanisms can be disregarded when discussing SEF. This conclusion is also strongly supported by work that shows that maximum enhancements of fluorescence will be obtained when the electronic absorption of a dye is in resonance with both the electronic plasmon resonance of the metal nanostructure, and the exciting light.[16, 17] This can partly be attributed to an enhancement in the absorption of the dye. Consequently, surface-enhanced fluorescence can be viewed as a product of three different processes: electromagnetic enhancement of emission, enhanced absorption, and radiationless transfer of energy to metallic surfaces.

In this chapter, a brief theoretical overview is provided that discusses, among other things, EM enhancement of emission, enhanced absorption, quenching to metal surfaces, the distance, coverage, and temperature dependence of SEF, and the effects of quantum efficiencies on enhancement. Also discussed, is the preparation and characteristics of several different nanoparticle metal substrates that have been employed in the collection of SEF, and the surface-enhanced fluorescence of Langmuir-Blodgett (LB) monolayers. Finally, a summary of these concepts is presented, and the future of SEF is discussed.

## 2. ELECTROMAGNETIC ENHANCEMENT

For years the origins of the huge enhancements of SERS have been debated in the literature. The individual contributions of the two accepted enhancement mechanisms, namely the electromagnetic and the chemical, have been discussed heavily with the former being generally accepted as the dominant factor for enhancement of Raman signals. With SEF, however, it was clear from nearly the very beginning that the same EM field enhancements that were at play in SERS were also causing the enhancement of fluorescence.[8, 9, 17, 18]

The excitation of the dipole particle plasmon resonance provides the theoretical basis for the development of the electromagnetic enhancement mechanism (EM).[6, 12, 19] For a recent review on the EM mechanism of enhancement see Schatz and Van Duyne.[20] The theoretical approach allows the calculation of enhancement factors for metal particles of different sizes and shapes.[19, 21, 22] It is, in fact, electromagnetic enhancements that define surface-enhanced spectroscopies.

To begin to understand EM enhancement it is first important to understand the nature of the systems we are considering. Figure 1 depicts a typical SEF experimental setup with a monolayer of fluorescent molecules sitting atop appropriate enhancing metal nanoparticles (not to scale). A wide variety of other possible configurations are possible, and will be discussed later, however this model will suffice for the present discussion. In this model, surface roughness is shown, as it is known to be an essential condition for surface EM enhancement. These features that lead to optimal enhancements of fluorescence, as well as Raman scattering, are on the nanoscale, and have diameters smaller than the wavelength of exciting visible light (typically between 5 and 100 nm), safekeeping the validity of the electrostatic approximation. This size scale encompasses particles produced in metal island films[17, 23, 24], colloidal metal solutions[25, 26], electrode surfaces[2, 3, 27], cold-deposited films[28], and lithographically produced assemblies[29, 30]. As an example, an atomic force microscopy, AFM, image of a mixed Ag/Au island film (5nm Ag + 5nm Au) is shown in Figure 2.

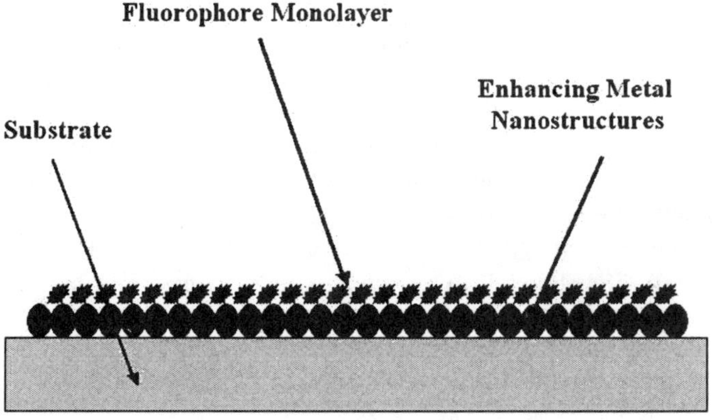


**Figure 1.** Typical SEF experimental setup: monolayer on metal nanostructures.

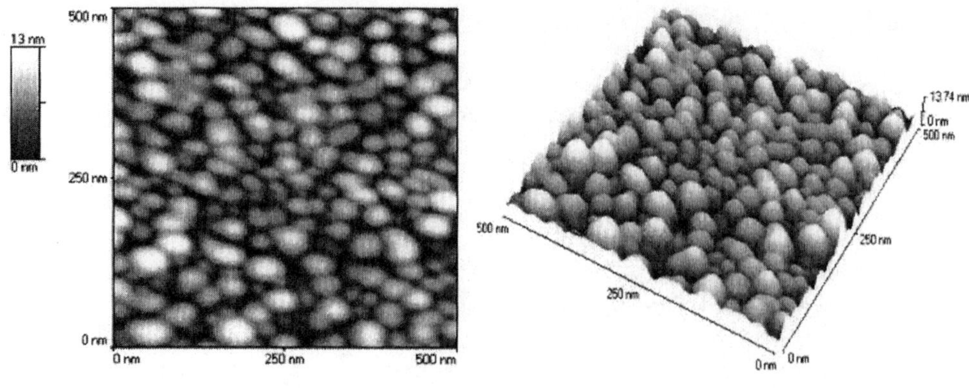


**Figure 2.** AFM image of a mixed Ag/Au island film (5nm Ag +5nm Au).

It is the role of rough surface features to provide strong amplification of local enhanced EM fields at frequencies exciting plasmon resonances.  In a metal particle, a plasmon is the collective oscillation of conduction electrons.  This should not be confused with other plasmons or collective excitations that occur in bulk metal or at smooth metal surfaces.[19]  Notably, without the inclusion of sharp or rough surface features, the plasmon is confined to the surface and is unable to radiate.  This is the case for smooth metal surfaces, where strong EM enhancements of fluorescence and Raman scattering are not observed.  When metal nanoparticles or roughened surfaces are illuminated with light, however, a localized surface plasmon that can radiate is induced.  At frequencies resonant with the surface plasmon absorption of the metal, the electromagnetic field of the laser will cause a polarization of the nanoparticle, leading to the creation of an additional EM field, and resulting in the molecular dipole feeling a much greater EM field than that of the incident radiation.  This enhanced field is not confined to the first layer of the surface and can be considered a field concentration extending from around the outside edge of the metal particle.[6]  Thus, this field can enhance not only the optical signals (including fluorescence) of molecules on the surface, but also of those located within the extended range of the enhanced field.

For a sphere of charge $q=0$, with dielectric function $\varepsilon(\omega)$ (assumed equal to that of the bulk metal), imbedded in a medium with dielectric constant $\varepsilon_m$, and with an incident electric field $E_0$, the most significant part of the local field outside the sphere, is:[31]

$$E_l = \frac{\varepsilon(\omega) - \varepsilon_m}{\varepsilon(\omega) + 2\varepsilon_m} \bullet \frac{a^3}{r^3} E_0 + E_0 \qquad (1)$$

The first part of the total external field is identical to that of a field created by a dipole $p$ at the center of a sphere:

$$p = \frac{\varepsilon(\omega) - \varepsilon_m}{\varepsilon(\omega) + 2\varepsilon_m} a^3 E_0 \qquad (2)$$

Within the dipolar model, the enhancement of the local excitation rate varies as $(a/a+d)^6$, where $d$ is the distance from a metal sphere of radius $a$ to the molecule. When both the excitation and the fluorescence are enhanced by the nanostructure, the total enhancement varies approximately as $(a/a+d)^{12}$.[10, 32] It can also be seen that larger metal particles should provide fields that extend further out. For simplicity, let's consider an incident field $E_i(r,\omega_0)$ in vacuum, where $\varepsilon_m = 1$, then

$$p = \frac{\varepsilon(\omega_0) - 1}{\varepsilon(\omega_0) + 2} a^3 E_i(r, \omega_o) = g_0 a^3 E_i(r, \omega_o) \qquad (3)$$

where the convenient factor $g = [\varepsilon(\omega) - 1]/[\varepsilon(\omega) + 2]$ has been introduced.[33] The local field is maximized when the real part of $\varepsilon(\omega)$ is $-2$ and the resonance condition in $g$ is fulfilled. This expression will give us two very helpful limiting conditions for determining suitable metals for enhancing substrates. The first of these conditions is that surface plasmon resonance will only occur when the real part of the denominator is equal to zero. The second condition is that the effective electric field of the particle will only be large if the imaginary part of $\varepsilon(\omega)$ is very small. It turns out that these conditions are best satisfied in the visible by three coinage metals: Ag, Au, and Cu. These metals are, not surprisingly, also by far the most widely used substrates for surface-enhanced spectroscopies.

The electronic plasmon absorption of a 6 nm Ag island film is shown in Figure 3. In the same figure, the extinction cross section calculated for a silver sphere and a silver prolate with a 3:1 aspect ratio, within the long wavelength limit of Mie theory, are included for comparison. The computation clearly illustrates the considerable shift to the red of the main plasmon absorption of the prolate spheroids in reference to the silver sphere. The broad plasmon absorption indicates a large distribution of sizes and shapes of Ag nanoparticles, and has a maximum at 494 nm. Notably, the SERS excitation profile follows closely the measured plasmon absorption, confirming the EM nature of the observed enhanced intensities.[34]

It has been observed that molecules located close to surfaces of very high curvature (i.e. sharp points) demonstrate increased enhancements. This has been termed a "lightning-rod effect", and is a result of extremely large EM fields that are radiated from these surface features.[21] Another situation in which extremely high fields are realized, is between surface protrusions, or in aggregates of colloidal particles. This is likely responsible for the extremely high enhancement factors that are seen in single molecule colloidal SERS studies.[35-38] However, it is unlikely that surface-enhanced fluorescence would benefit from this kind of experimental setup as quenching to the metal would almost certainly be a large factor.

To summarize, the strength of local electromagnetic fields around the surface of metal nanoparticles depend upon the following properties: the dielectric constant of the

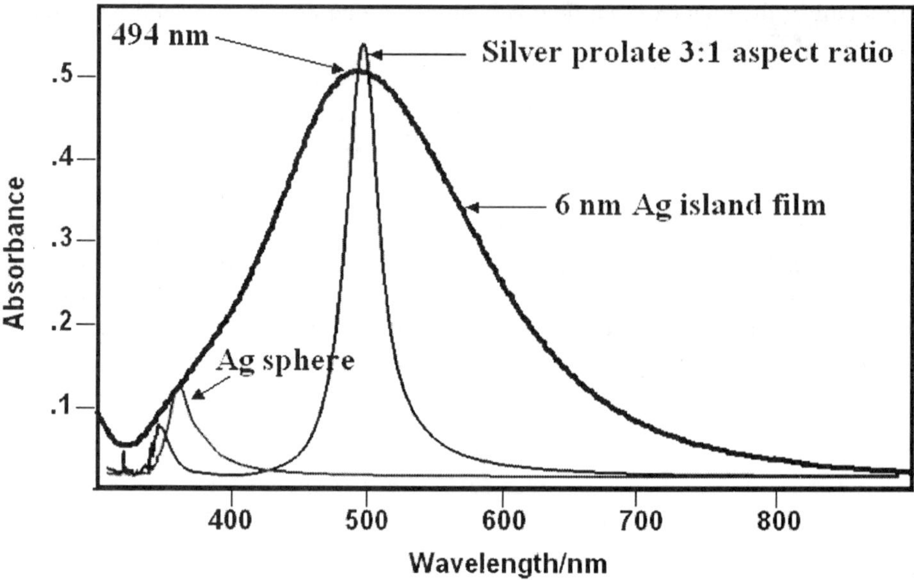


**Figure 3.** Surface plasmon absorption of a 6 nm Ag island film shown with the calculated extinction spectra of a Ag sphere, and a silver prolate with an aspect ratio of 3:1.

metal at the frequency of excitation, the shape of the surface roughness, the size and packing of the surface features or metal nanoparticles, and the dielectric constant of the medium. These factors will control surface enhancement and should be used as a guide for experimental design. In Figure 4, the application of these considerations is demonstrated. The enhancement of both resonance Raman scattering, and excimer fluorescence is shown for a PTCD derivative (bis-benzylimido perylene) in a Langmuir-Blodgett monolayer fabricated on a 6 nm Ag island film. Since excimer fluorescence is observed, the PTCD molecules must be forming stacks within the monolayer, where their most likely molecular orientation is edge-on or head-on to the enhancing surface, i.e. is such that the chromophore does not lie face-on to the metal where maximum quenching would take place.

In addition, it should be understood that the generation of amplified electric fields will increase the likelihood of photochemical reactions on the surface. The photocatalytic decomposition of an adsorbed organic molecule on the surface of a metal nanostructure is a common occurrence in experimental SERS. In fact, the familiar "cathedral peaks" due to graphitic carbon are a reminder of the limitations of the analytical applications of SERS and SERRS. SEF will be maximized at a distance from the nanoparticle where the direct contact of the molecule with the surface can be avoided through the utilization of a spacer layer. This will also lead to the corresponding photodecomposition being minimized.

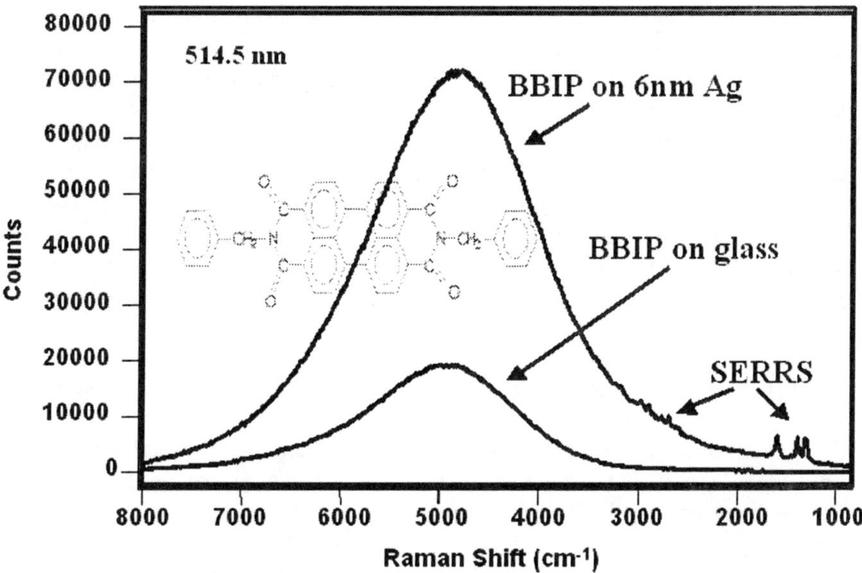


**Figure 4.** The enhancement of both resonance Raman scattering, and excimer fluorescence for a PTCD derivative (bis-benzylimido perylene) in a Langmuir-Blodgett monolayer fabricated on a 6 nm Ag island film.

## 3. ENHANCED ABSORPTION

When an adsorbate is placed near an appropriate enhancing metal nanostructure there are several very strong effects revealed as a result of the coupling of the molecular dipole with the localized electromagnetic field of the metal's surface plasmon resonance. Of these effects, that of enhanced molecular absorption is certainly one of the most important to the phenomenon of surface-enhanced fluorescence. The enhanced absorption of dye molecules near metal surfaces was first reported in 1980 by Glass *et al.*[39], and has been confirmed many times since.[40, 41] Observed, was that the optical absorption of silver island films coated with dye molecules showed very unusual behaviour. When the absorption of the dye was sufficiently different from that of the silver film, the resulting spectra basically showed a superposition of the two separate spectra, except that there was an enhancement of the molecular absorption while the plasmon absorption of the metal was damped. However, in cases of strong overlap between the two resonances, very complex spectra were seen that no longer showed simple superposition, but rather broadening, and complex splitting patterns. The splitting of the plasmon resonance has been attributed to the dispersion in the real part of the dye's refractive index, while the damping of this resonance has been recognized as being due to the absorption of the dye.[41]

As with enhanced fluorescence, the enhancement of the absorption of dyes is able to function at a small distance of separation between the metal nanoparticle and the molecule.  This was demonstrated by Glass *et al.*[39] by the inclusion of an intervening transparent plastic spacer layer between the metal and dye films.  In these experiments it was observed that the spacer layer did not prevent the interaction of the Ag with the dye molecules.  The importance of this result should not be underestimated.  As we know, the enhancement of absorption may lead to enhancement of fluorescence (along with other important processes), but it also will increase the rate of non-radiative energy transfer.  Quenching can in fact be so substantial that it will overpower EM enhancement and lead to a decrease in fluorescence quantum yields.  This can be used in normal Raman experiments to obtain fluorescence-free Raman spectra.  The next section will be devoted to the discussion of this concept in more detail.

## 4. RADIATIONLESS ENERGY TRANSFER AND DISTANCE DEPENDENCE

As mentioned earlier, one of the ideas fundamental to the understanding of the surface-enhanced fluorescence phenomenon is that of it as a competition between the effects of enhancement (both absorption and emission) and non-radiative energy transfer to the enhancing nanostructure.   This is well illustrated in Figure 5, which shows the simplest theoretical distance dependence relationships for the electromagnetic enhancement and quenching of a small spherical metal particle.[14]  It is predicted by the EM mechanism that for a spherical metal particle of radius $a$, separated from a molecule by a distance $d$, enhancement will decay according to $(a/a+d)^{12}$.[32, 42]   It should be noted that this model assumes that both excitation and fluorescence radiation are augmented by the metal surface.  For a monolayer at a distance $d$, enhancement is expected to decrease

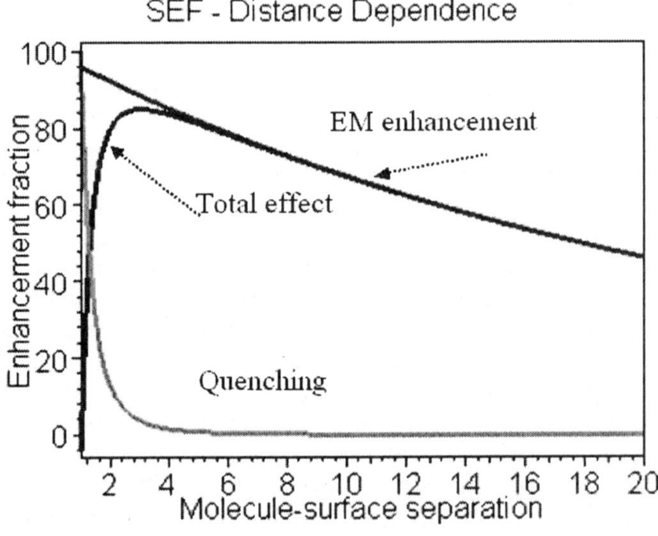


**Figure 5.** Theoretical distance dependence relationships for the electromagnetic enhancement and quenching of a small spherical metal particle.

as $(r/r+d)^{10}$.[43] Competing with enhancement is the short-ranged non-radiative energy loss to the metal that has been shown to decrease as $1/d^3$ as molecule-surface separation is increased.[32] As can be seen in Figure 5, electromagnetic enhancement is a much slower decaying function of molecule-surface separation than that of surface quenching, and as a result of this, maximum fluorescence will occur at distances that are intermediate between the maximum quenching ($d=0$) and "no effect" conditions.

In 1979, Weber and Eagen[44] examined the decay channels responsible for the quenching of molecular fluorescence near metal surfaces. They showed that for a molecule located between 20 and 160 Å from the surface, 80% of its lost energy will lead to surface plasmon excitation in the metal, while at much shorter distances, surface plasmon excitation will go to zero and electron-hole excitation will increase as a result.

Given the strong relationship between radiationless energy transfer and total enhancements, how then can we manipulate our experimental design to obtain the results we would like to? This is a question that has been examined by several groups and has led to results that are beneficial for both those that are practicing fluorescence and Raman spectroscopies. For example, it is a common experience for Raman spectroscopists to have strong fluorescence interfering with the observation of Raman modes. This problem can often be partially or completely alleviated by placing the adsorbate close to a metallic surface where its fluorescence can be quenched. In fact, the literature of SERS shows countless examples of this technique being employed successfully. An example is the use of SERS to study the vibrational spectra of several biological molecules that in Raman measurements show very troublesome luminescence.[45] Also, in single molecule surface-enhanced resonance Raman scattering (SERRS) studies, island films fortuitously remove the interference of dye monomer fluorescence and allow for the observation of overtones and combinations.[46] Fortunately, it is also possible to reduce the effect that quenching has, and thereby increase the enhancement of molecular fluorescence near metal surfaces by using the idea of spacer layers.

Spacer layers can intervene between emitting molecules and enhancing metals in surface-enhanced spectroscopic experiments, and by doing so can prevent fluorescence quenching and allow for the study of distance dependence. These layers should be chemically inert and optically transparent in the region of interest, and should afford a great deal of control over their total thickness. Notably, in the earliest of the reports on SEF, spacer layers were used to study the distance dependence of the effect and to reduce non-radiative decay to the metal.[32, 39] Glass *et al.*[39] used thin films of a transparent plastic as a spacer to determine that enhancement is functional at a distance in their 1980 paper. Since then, several studies have appeared in the literature in which the concept of spacer layers has been employed. Some reports have described the use of thin evaporated $SiO_x$ layers[32, 47], while others have taken advantage of the well controlled architecture of Langmuir-Blodgett (LB) films of fatty acids.[24, 48-51] The general application of LB films as spacer layers for SEF measurements is shown in Figure 6. The LB technique[52, 53] has also been used extensively in the study of SERS and SERRS.[14, 46, 54-57] Figure 7 shows the results of an experiment that demonstrates the distance dependence of SEF[48] with intervening fatty acid LB spacer layers used to increase the enhancement of fluorescence from an emitting phthalocyanine monolayer. The spectra are shown on the same scale, highlighting the fact that the lowest emission intensity corresponds to that of the phthalocyanine (Pc) monolayer on glass, and the highest is that of the Pc monolayer with a spacer layer of a fatty acid between it and the metal nanostructures. The Pc monolayer

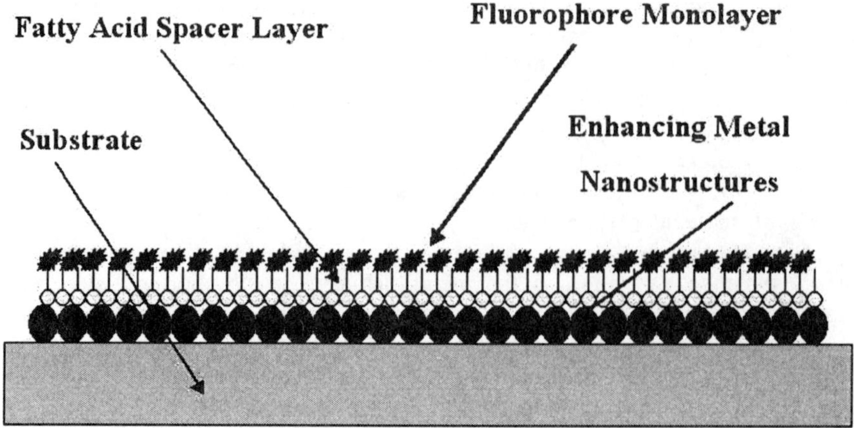


**Figure 6.** The general application of LB films as spacer layers for SEF measurements.

lying directly on the metal islands clearly shows the enhanced Raman scattering on the background of the fluorescence (7a). This fluorescence increases with the addition of the spacer layer, while the Raman signal decreases as the molecules move further away from the metal surface (7c).

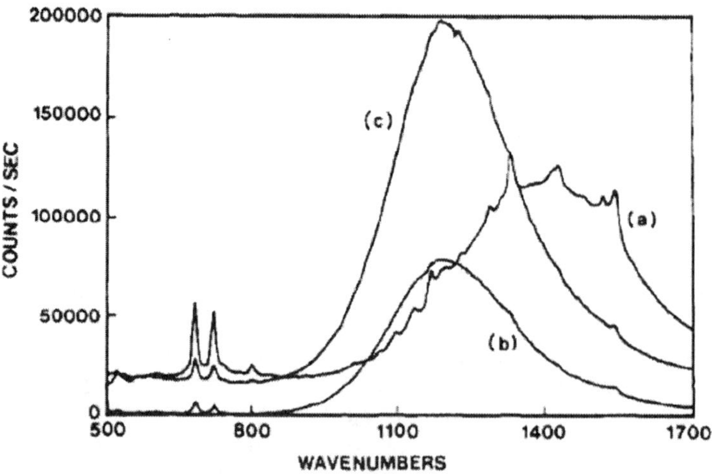


**Figure 7.** Fluorescence spectra of a single LB monolayer of ($t$-Bu)$_4$H$_2$Pc on a 15-nm Ag island film from 647.1-nm krypton laser excitation with (a) no spacer layers and p-polarizd excitation, (b) five spacer layers of arachidic acid and S-polarized excitation, and (c) five spacer layers of arachidic acid and P-polarized excitation. Reprinted with permission from *Langmuir* **1988**, *4*, 518-521. Copyright 1988 American Chemical Society.

## 5. COVERAGE DEPENDENCE

The ability to easily control the architecture of monolayers using the LB technique has also made it very valuable in the study of the coverage dependence of surface-enhanced fluorescence. This technique for thin film fabrication is used throughout our work for the preparation of mixed monolayers containing fluorophores embedded in the films of fatty acids. The concentration of these dyes in the films can thus be varied in order to study the effect of higher surface coverages. Studies on the coverage dependence of SERS have employed this experimental setup, and are very instructive as to some of the general issues of surface coverage.[58-60] An example of a typical experimental setup with a fluorophore dispersed in the matrix of a mixed fatty acid monolayer is sketched in Figure 8. This design has been employed successfully in our group for single molecule SERRS studies.[37, 46, 57, 61] It is also possible to control surface coverage through the manipulation of the thickness of evaporated films, and the concentration of dipping solutions.

Studies examining the effect of the surface coverage dependence of SEF have been much more rare than those for SERS, and show somewhat inconsistent results.[50, 51, 62] However, it does seem clear that in all the systems studied, the relationship between enhancement and surface coverage was a complex one. At low surface coverages molecules behave as isolated molecules, but as concentrations are increased several new factors become apparent. Aggregates of dye molecules (dimers and higher) have been shown to form, leading to excimer fluorescence.[60, 62] These aggregates also show radiationless energy transfer to one another, whether inter- or intralayer. In addition to

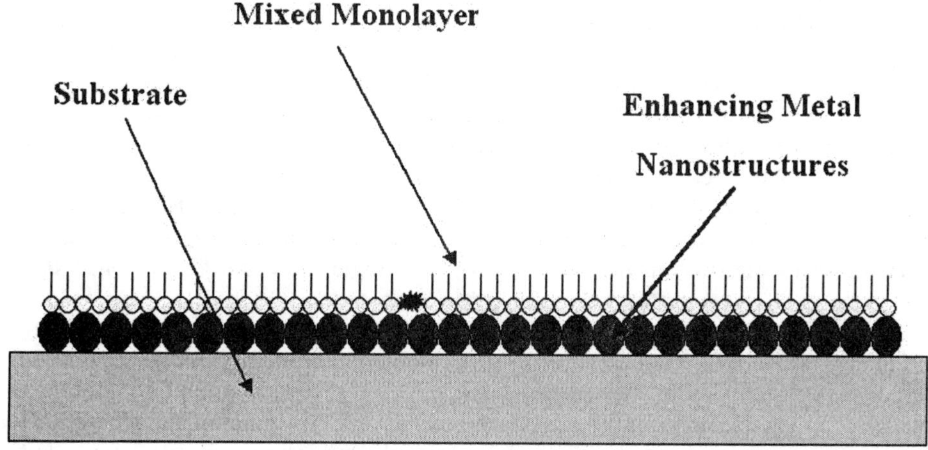


**Figure 8.** Typical experimental setup with a fluorophore dispersed in the matrix of a mixed fatty acid monolayer.

this, it is clear that dye coatings will alter the plasmon resonance of metal nanoparticles, and thus effect the enhancement process. Specifically, it is expected that high surface coverages of resonant dyes would cause the damping of electronic surface plasmons, and result in the reduction of EM enhancement. Similarly, if we consider enhancing substrates as having localized areas of very high EM fields (i.e. "hot spots"), then we would expect these spots would contribute most significantly to the observed SEF intensities. Thereby, we would also expect that at high surface coverage these spots would be saturated, and increasing coverage would not result in corresponding increases in intensity. However, it can be concluded from the experimental evidence that maximum enhancement is achieved before the enhancing metal particle is fully coated with probed molecules. In the case of SERS, signal saturation is completed after a 20% coverage of the total surface.[60] The corresponding experiments for SEF have not been done.


## 6. TEMPERATURE DEPENDENCE

The radiative lifetime of fluorescence is directly connected to absorption coefficients and excited state lifetimes. Lifetimes are determined by several non-radiative deactivation processes including vibrational relaxation, internal conversion, external conversion, and intersystem crossing. Experimentally, quantum yield is defined as the ratio of emitted photons to photons absorbed. Quantum yields can be used in the estimation of non-radiative rate constants. In SEF, the lifetimes and quantum efficiencies of molecules are strongly affected by the presence of metal surfaces. Since the examples selected here are large aromatic molecules, the non-radiative processes in aggregates of molecules should also be included. For a detailed discussion of the fate of excited states in aggregates of molecules we refer the reader to the elegant description of Pope and Swenberg.[63] With increasing temperature, the kinetics of these processes are generally increased, resulting in a reduction in fluorescence. It can often be extremely difficult to determine the extent to which each of these processes is acting, but it is well established that at elevated temperatures molecular collisional rates will be increased leading to deactivation by external conversion. It can be expected that if all other factors remain constant, the quantum efficiency of fluorescence for most molecules will decrease with increasing temperatures. In fact, the majority of single molecule fluorescence studies, until just recently, were performed at cryogenic temperatures.[64, 65]

Of course, when we consider the situation of a fluorophore adsorbed on the surface of a metal nanoparticle, our problem becomes slightly more complex. One of the foremost contributors to this complexity is the effect of temperature on film packing, and stability. Our group has recently published three reports that examine the problem of the temperature dependence of the SEF of Langmuir-Blodgett monolayers deposited on metal island films.[14, 61, 66] In these studies, surface-enhanced fluorescence of three different PTCD derivatives was recorded at several different temperatures ranging between –190 and +250 °C. Two of these papers[14, 67] show clear patterns of decreasing fluorescence intensity with increasing temperatures. Figure 9 shows one set of these results.[14] Also, it was shown that the changes in packing at high temperatures were not reversible. For a third system[66] no significant changes were observed in SEF intensities upon temperature variation, and this would seem to suggest that molecular packing

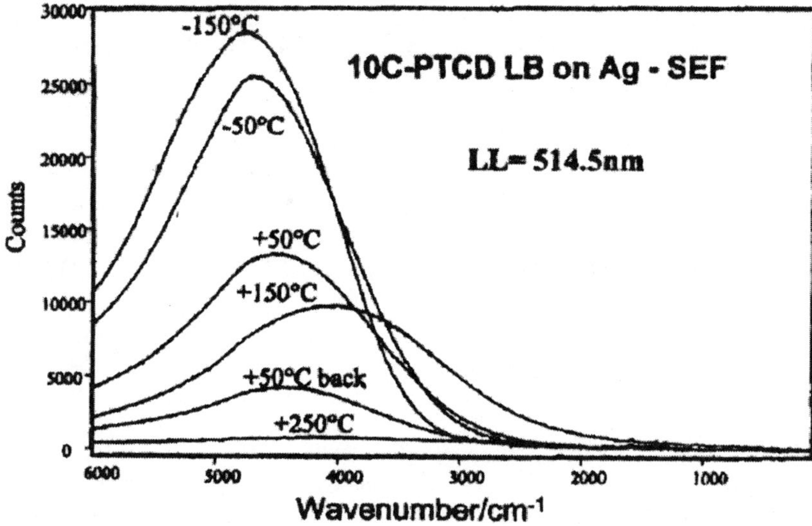


**Figure 9.** SEF at different temperatures of a LB film of 10C-PTCD on silver, excited with 514.5 nm laser line. Reprinted with permission from *Langmuir* **2001**, *17*, 2958-2964. Copyright 2001 American Chemical Society.

remains unchanged within the tested temperature range. Also, one further consideration for the temperature dependence of SEF, is that lower quantum efficiencies of free molecules lead to lower SEF enhancement factors.[8] The relationship between quantum yields and enhancement will be discussed in the next section.

## 7. QUANTUM EFFICIENCY AND ENHANCEMENT

In a series of very elegant experiments[8] Weitz *et al.* demonstrated the relationship between the quantum efficiency (QE) of a fluorophore located in an electromagnetically inert environment and its surface-enhanced fluorescence enhancement. They monitored the fluorescence of two dyes of very different quantum efficiencies on both silica and rough silver surfaces, and found that the dye of high QE, rhodamine 6G, showed a large reduction in fluorescence with an enhancement factor of only ~0.39, while the dye of low QE, basic fuschin, showed an enhancement factor of ~3.1. It was assumed that much of the fluorescence intensity that was seen for the R6G on the silver island film on silica was actually coming from those molecules that were located between islands and sitting on silica. To demonstrate this, a thick layer of Al was placed between the silica and the silver islands. This test revealed that their hypothesis was correct by showing a further reduction in the fluorescence of the R6G. Some of the results of these experiments are shown in Figure 10. To further confirm their results, Weitz *et al.* performed further experiments where they lowered the QE of dyes by increasing oxygen pressures, and temperatures. Through these experiments, it was once again shown that fluorescence enhancements decrease as the QE of a free molecule is increased.

In two more recent papers on single molecule fluorescence near thin metallic layers, Enderlein reminds us that the quenching by metal nanoparticles shortens the lifetime of the excited state and by doing this, increases the number of excitation cycles that the molecule can survive before it is photobleached.[68, 69] This is an extremely important point, especially when dealing with single molecules, as fluorescence quantum yields are not nearly as important as the number of photons that are emitted before photobleaching occurs.

## 8. ENHANCING SUBSTRATES

As has been mentioned previously, the enhancement of fluorescence near noble metal nanoparticles is a result of strong local electromagnetic fields caused by surface plasmon absorption. The intensities of these enhanced EM fields is strongly dependent on local surface morphologies, and this has led to very significant amounts of work being done in search of new and improved substrates. A great deal of very useful information has been obtained from these studies, and considerable advances have been made. However, it is our view, that the pursuit of optimum enhancing substrates that can be made reproducibly, is still the most important problem facing the field of surface-enhanced spectroscopy today. This chapter will not attempt to discuss all the possibilities for enhancing surfaces, but will rather summarize some important considerations for substrate preparation through the very common example of Ag island films.

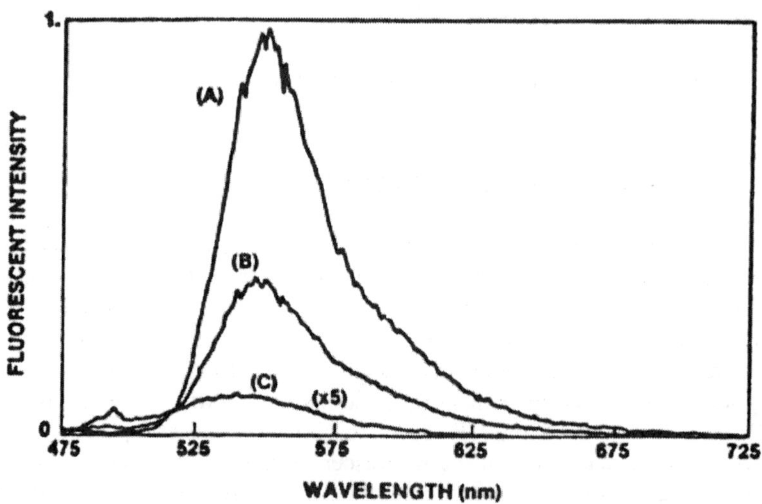


**Figure 10.** Fluorescence spectra of R6G excited at 458 nm on (A) silica, (B) silver-island film on silica, and (C) silver-island film on aluminum. Reprinted with permission from *J. Chem. Phys.* **1983**, *78*, 5324-5338. Copyright 1983 American Institute of Physics.

Silver island vacuum evaporated films are likely the most widely used substrates for the enhancement of fluorescence.[8, 14, 24, 32, 55, 56, 70-73] With mass thicknesses generally ranging between 50 and 150 Å, these films show particle diameters between 10 and 100 Å. This particle size is smaller than the wavelength of visible light used for the excitation of surface-enhanced fluorescence, and thus dipole particle plasmons are generated. The islands provide a surface roughness that makes possible the radiation of localized surface plasmon resonances that generate the huge electromagnetic fields necessary for surface-enhanced spectroscopies. Another valuable feature of vacuum evaporated island films is that they can be made with a certain degree of reproducibility. Achieving reproducibility is an essential feature for analytical applications, and is made possible in the preparation of these substrates by highly controlled deposition procedures, and several accurate methods for film characterization.

The vapour evaporation of thin silver films onto suitable surfaces will produce structures resembling islands for mass thicknesses between 4 and 20 nm. These nanostructures show exceptional stability under normal conditions, but will always have thin oxide layers formed on their surfaces. There are several important experimental factors that will affect strongly the morphology and optical properties of these films and they include; the substrate used, substrate temperature, mass thickness, deposition geometry, and evaporation rate. It is very easy to control all of these factors to attain reproducible results, and so film preparation is reduced to a matter of using optimum conditions, and maintaining consistency in experimental methods.

Glass and quartz are often used as substrates for the deposition of Ag island films for SEF measurements, and have been shown to be quite suitable for this purpose. However, surface chemistry has been shown to play a very important role in the determination of Ag particle growth. Roark, and Rowlen have studied the impact of several different substrates, and of postdeposition treatment on the morphology and optical properties on thin Ag island films.[74] This study examined glass, derivatized glass, mica, and poly(vinyl formal) coated glass and it was found that these different substrates had a large impact on the island's properties. Another important conclusion of this work was that solvent exposure of Ag films also produced changes in these properties. In both cases, these changes were revealed by ultraviolet-visible extinction and atomic force microscopy measurements. Thus, it is possible to manipulate the morphology and optical properties of Ag island films by the use of different substrates and postdeposition treatments. This manipulation is also made possible through the variation of temperature.

The temperature of substrates used for Ag island film deposition is an extremely important factor for determining the properties of the particles produced. Increased temperatures will lead to a corresponding increase in the mobility of Ag atoms on the surface, and eventually to islands of larger size and larger interparticle spacing. These differences cause a blue-shifted band of higher absorption and narrower width in the electronic surface plasmon absorption spectrum.[75] It is also known that the high temperature annealing of Ag island films after deposition strongly affects their morphology and optical properties.[76, 77] A common procedure that is used in our laboratory, is the annealing of 6nm Ag island films at 200 °C for 1 hour after deposition (deposition is also at 200 °C) , and then allowing the film to cool under vacuum for an additional hour.[46] Annealed films generally show blue-shifted maxima, narrower bandwidths, and lower absorbance in their surface plasmon resonances.[75] A temperature controller equipped with feedback capability allows for the maintenance of nearly

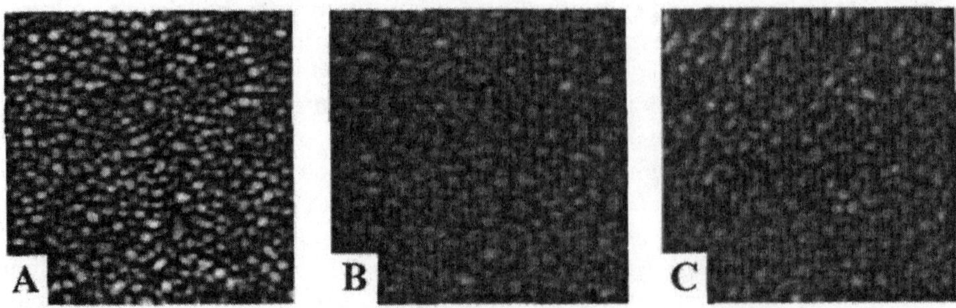

**Figure 11**. Representative AFM images of 5 nm Ag films deposited with varying rates: (A) 0.02, (B) 0.20, and (C) 2.5Å/s. Image sizes are 500 nm *x* 500 nm. All of the AFM images in this work are unfiltered. Reprinted with permission from *Anal. Chem.* **1994**, *66*, 4324-4331. Copyright 2001 American Chemical Society.

constant temperatures, and thus prevents the substrate heating that normally results from evaporation.

Mass thickness, deposition geometry, and evaporation rate are three other factors that have been shown to strongly influence the morphology and optical properties of Ag island films. Increasing mass thickness will lead to a corresponding increase in the size of particles produced. This will cause the surface plasmon absorption of the film to be broadened, shifted to the red, and more absorbing. Deposition geometry has also been studied as an influencing factor in the fabrication of Ag island films.[75] It was found that it can be a very important contributor to the differences that are observed from one laboratory to the next. Thus, great care must be taken to ensure that the angles, and distances between the source, the quartz crystal oscillator, and the sample are maintained. Finally, the influence of evaporation rates on resultant Ag island films will be mentioned. The effect of varying evaporation rates has been studied by several groups[75, 76, 78], and has shown that increasing rates results in films with decreased aspect ratios and interparticle spacings. These changes correspond to red shifts, and decreased absorptions in the UV-visible spectra of the films produced. Figure 11 shows AFM images of films evaporated at three different rates, and reveals these changes.[75] Similarly, at least three recent reports have discussed the effect that morphology has on the optical properties of Ag nanoparticles.[19, 79, 80]

Thus, it is possible to tune surface plasmon resonances according to the needs of any given experiment through the variation of several key factors. This can be particularly helpful for the study of different fluorescing molecules with various laser excitation sources. Changes to Ag island films can be monitored using several very commonly available techniques including UV-visible absorption, atomic force microscopy, scanning tunneling microscopy, and transmission electron microscopy. These methods provide convenient tests of the reproducibility of film deposition procedures, and it is this that is ultimately important, as there is great variation between films produced in different laboratories. It is important to note, however, that a portion of the error that is introduced into surface-enhanced experiments arises from the processes of analyte coating, and obtaining spectra, rather than the deposition of the enhancing substrate.[75] One method of dealing with problems related to coating particles with dyes is the use of the Langmuir-Blodgett technique, which will be discussed in the next section.


## 9. SEF OF LANGMUIR-BLODGETT FILMS

Nearly all of the work on surface-enhanced fluorescence in our group has been done using the Langmuir-Blodgett technique for monolayer fabrication. The LB technique allows for the control of surface uniformity, thickness, and film architecture, and is one of the most successful techniques for the fabrication and study of organized molecular structures (organized dielectric media) such as thin solid films. These qualities make it particularly well suited for studies into the SEF effect. Surface-enhanced fluorescence is an effect that is both distance and coverage dependent, and the literature on LB films provides strategies to manipulate both of these variables in a very controlled way. As was mentioned earlier, fatty acid LB spacer layers can be used to vary the distance between fluorescent molecules and enhancing nanostructures.[48] By employing this idea, it is possible to study the distance dependence of SEF, and to maximize fluorescence intensities by decreasing non-radiative decay channels to the metal surface. To control surface coverage of dye fluorophores in SEF studies, the strategy of using mixed LB monolayers can be used.[54] Dye molecules are dispersed in monolayers of fatty acid molecules that are known to form very stable and easily transferable monolayers. In this way, the concentration of the dye can be controlled by adjusting the ratio of the fatty acid to the dye in the spreading solution used. The use of this technique allows for systematic studies into the coverage dependence of SEF, and makes it possible to maximize SEF enhancements.

In 1988, the distance dependence of surface-enhanced fluorescence was studied for Langmuir-Blodgett monolayers deposited on silver island films.[48] This study was inspired in part by two earlier reports that examined the distance dependence of SERS of LB films on metal surfaces.[13, 43] Varying numbers of spacer layers of arachidic acid were employed in order to probe the competition between EM enhancement and radiationless energy transfer for a phthalocyanine monolayer. In direct contact with the metal surface, a broadened, enhanced, and red-shifted fluorescence spectrum was observed. These spectral changes can be attributed to a drastic decrease in the fluorescence lifetime of the molecule when it is placed in contact with the metal surface. However, an enhanced version of the unperturbed spectrum was observed when intervening spacer layers were introduced. It was found that enhancements on the order of about 400 could be realized when 5 monolayers were placed between the Ag island film and the phthalocyanine monolayer.

For some time now, we have been working with many different perylene derivatives in the study of surface-enhanced spectroscopies such as SERS, SERRS, SEF, and SEIRA (surface-enhanced infrared absorption). These molecules show a very strong tendency towards aggregation at high concentrations, and often display excimer fluorescence in concentrated Langmuir-Blodgett monolayers. The surface-enhanced

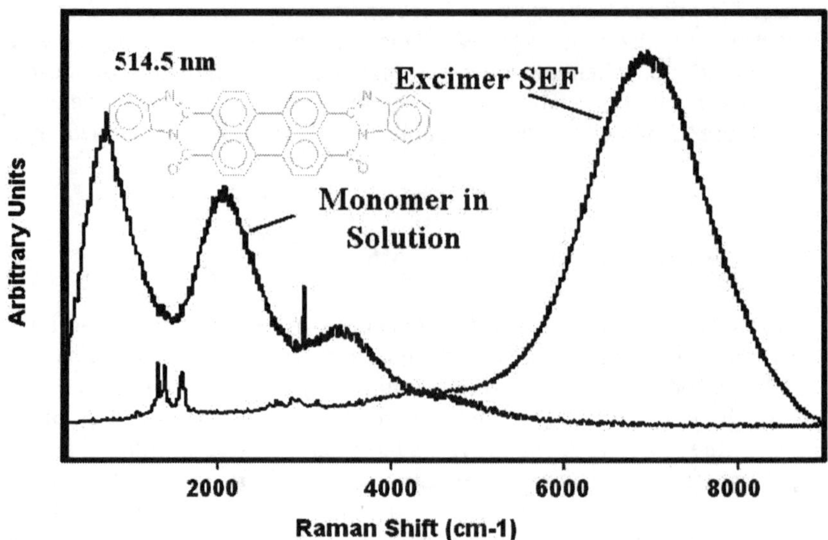


**Figure 12.** Monomer fluorescence from a dilute dichloromethane solution of a PTCD derivative, along with the corresponding surface-enhanced excimer emission of a concentrated LB film of the same molecule on Ag islands. Raman shift values correspond to 514.5 nm excitation.

monomer and excimer fluorescence of several different PTCD derivatives have been studied extensively in our group. In two early reports, the monomer and excimer SEF from LB monolayers of PTCD derivatives were observed and distinguished from one another.[54, 70] This ability to distinguish between monomer and excimer emission has very important implications regarding the physical and optical properties of LB films of these dyes. It also has proven helpful in demonstrating a lack of dye aggregation in our recent work on single molecule SERRS.[37, 46, 57, 61] It has been shown that PTCD excimer fluorescence is strong in concentrated LB films on Ag, and can be seen in the same spectral region as SERRS. As films are diluted to the point where signals are obtained from single molecules, however, surface-enhanced excimer emission vanishes. This is strong evidence that SM SERRS signals are in fact being obtained from monomers. The monomer fluorescence from a dilute dichloromethane solution of a PTCD derivative, along with the corresponding surface-enhanced excimer emission of a concentrated LB film of the same molecule on Ag islands, is shown in Figure 12. The Raman shift values in the figure correspond to 514.5 nm excitation.

A large portion of the work that has been done in our group on the surface-enhancement of fluorescence from LB monolayers has been in conjunction with surface-enhanced Raman studies.[55, 56, 66] In theses studies, SEF has been helpful in providing information about the aggregation, and orientation of dye molecules, as well as information about the thermal stability of the monolayers used. A separate study reported the effects of different variables pertaining to the fabrication of LB films.[71] In particular,

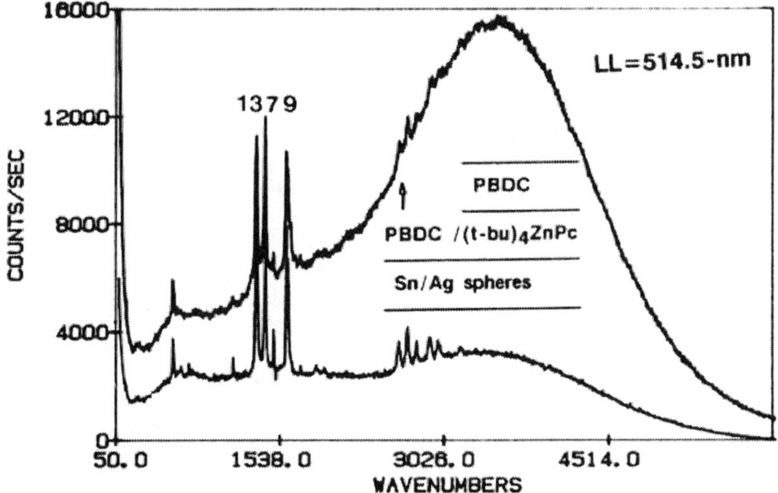


**Figure 13.** Surface-enhanced Raman and fluorescence of PBDC directly on the Sn/Ag substrate (upper trace), and of two-layer structure PBDC-(t-bu)₄ ZnPc on Sn/Ag (lower trace). Reprinted from *Journal of Molecular Structure, 218,* Doriano Battisti, and Ricardo Aroca, Selective Spectroscopic Characterization of Langmuir-Blodgett Monolayers Using SERRS, 351-356, **1990**, with permission from Elsevier.

SEF was used to probe monolayers prepared at different surface pressures (varying the degree of condensation of the monolayer), and on subphases of different pH values.

One particularly interesting application of SEF is its potential application in thin film devices. In 1990 a report was published aimed directly towards this end.[73] In this paper, the selective spectroscopic characterization of Langmuir-Blodgett monolayers of two different materials was accomplished using the techniques of SEF and SERRS. These materials were placed in adjacent monolayers on an enhancing substrate where energy transfer between the two dyes was monitored at different excitation frequencies. It was shown, that at 514.5 nm excitation, the strong excimer fluorescence of the PTCD used was quenched by being adjacent to a monolayer of a phthalocyanine that absorbs in the same region. This energy transfer was facilitated by an enhanced absorption of the phthalocyanine, and is shown in Figure 13. This study also examined the question of energy transfer in horizontal geometries of mixed monolayers of the two dyes, and it was seen that energy transfer is limited in such systems.

It has been mentioned before that the use of spacer layers to prevent quenching has shown success in several surface-enhanced fluorescence studies. However, there are alternative strategies for accomplishing this same result. In two recent studies, the use of long chain alkyl groups substituted on PTCD derivatives to sufficiently separate the chromophore from the enhancing surface, and thus prevent substantial quenching of molecular fluorescence, has been demonstrated.[14, 24] The molecules used in these studies have intrinsically low quantum efficiencies, and so were well suited to SEF studies, as discussed earlier. One of these reports[24] examined the spreading properties of a long alkyl-chained PTCD, as well as the effect of different mass thicknesses of the Ag island films that were used as enhancing substrates. It was found, that of all the substates tested,

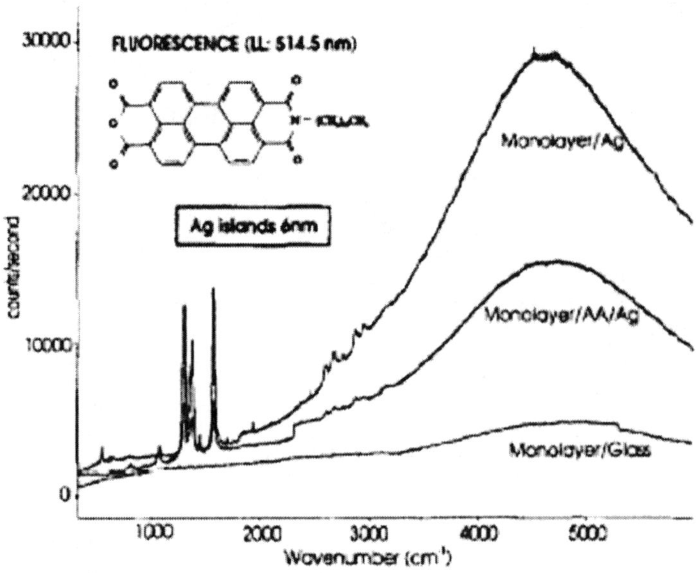


**Figure 14.** Surface-enhanced fluorescence spectra of OD-PTCO: LB film on glass, LB on 6 nm silver island film, and LB film on 6 nm Ag film coated with one spacer layer of arachidic acid. Reprinted from *Spectrochimica Acta Part A: Molecular & Biomolecular Spectroscopy*, *53*, J. DeSaja-Gonzalez, R. Aroca, Y. Nagao, J. A. DeSaja, Surface-enhanced fluorescence and SERRS spectra of *N*-octadecyl-3,4:9,10-perylenetetracarboxylic monoanhydride on silver island films, 173-181, **1997**, with permission from Elsevier.

6 nm Ag islands showed the highest enhancement factors for SEF. Also, in this study it was found that the largest enhancement factors were obtained when the molecule was directly adsorbed on the Ag substrate, as opposed to with an intervening spacer layer. This is shown in Figure 14, and demonstrates the use of the substituted alkyl chain as a spacer to prevent quenching. In the second of these reports [14], the LB film packing and surface-enhanced fluorescence of three different PTCD derivatives of varying alkyl chain length was studied, along with the temperature dependence of SEF. The molecules were found to orient in a head-on fashion (standing on alkyl chains), and showed increased enhancement with increasing chain length, as predicted by the distance dependence of SEF. In the temperature dependence studies, a general trend towards increased enhancement with decreased temperature was observed. The problem of the temperature dependence of SEF will likely prove to be a very important one in the near future, as thin film devices often experience temperature variation that strongly affects their efficiency.

Before concluding this section on the surface-enhanced fluorescence of Langmuir-Blodgett films, two papers that show the expansion of this technique beyond classic dye systems will be discussed. The first of these papers, by Sokolov, Chumanov, and Cotton, was published in 1998 and employed colloidal metal silver films (CMFs) coated on functionalized glass substrates towards the improvement of bioanalytical techniques.[51]  In addition to utilizing the Langmuir-Blodgett technique, this study also used the technique of coating metal surfaces with thick, chemically inert, silica spacer layers.  Importantly, this study represents the first report of biospecific interactions being probed with the increased detection sensitivity afforded by surface-enhanced fluorescence.  The second of these papers, by Constantino *et al.*, demonstrated the SEF of LB azopolymer films that results from the restrictions imposed by the film's structure.[72]  These polymers show photoisomerization when they absorb EM radiation, and thus do not normally display fluorescence.  However, through the spatial hindrance of the LB film structure, photoisomerization can be prevented and SEF can be observed.  This study demonstrates the possibility for detecting fluorescence in systems where it is not normally seen through the combination of SEF and the LB technique.

Thus, the importance of the Langmuir-Blodgett technique in the study and application of surface-enhanced fluorescence has been demonstrated through a variety of examples.  The LB technique provides a unique method for generating organized and well-defined molecular structures at precise distances from metal surfaces with known adsorbate concentrations, and so provides an excellent tool for manipulating important factors such as metal-adsorbate separation distance, and surface coverage for SEF measurements.


## 10. SUMMARY AND OUTLOOK

In this chapter the phenomenon of surface-enhanced fluorescence has been discussed.  Beginning with the discovery of surface-enhanced Raman scattering in the late 1970's, this review has shown SEF in the context of the broader surface-enhanced spectroscopies.  The close relationship of SEF to SERS, due to their common EM enhancement mechanism, in particular, has been stressed.  In order to provide a general overview of this topic, several of the most important considerations for the understanding of surface-enhanced fluorescence spectroscopy have been discussed.  Included in this discussion were the concepts of EM enhancement, enhanced absorption, radiationless energy transfer to metal surfaces, the distance, coverage, and temperature dependence of SEF, and the effects of quantum efficiencies on enhancement.  Also discussed, was the preparation and characteristics of a common enhancing nanoparticle metal substrate for SEF, and the surface-enhanced fluorescence of Langmuir-Blodgett monolayers.

More than twenty years have now passed since the discovery of surface-enhanced fluorescence.  Research into this effect has progressed quickly, and this is largely due to interest in the closely related fields of SERS and SERRS.  SEF is a technique that has not been in the spotlight, but has continued to make steady progress.  It appears now that the effect is largely understood, and is becoming an invaluable analytical technique that will continue to increase in its application.  As it would have been difficult to predict the path of SEF research in 1980, it is now very difficult to foresee future advances that will be made.  Having said this, however, we can gain some clues from the cutting edge work that is being done today.  It is likely that SEF will continue to be applied increasingly in far-flung fields of research.  Biological systems will continue to be further explored, while the study of single molecules[65, 81] with the technique of SEF will likely take off as a viable analytical technique.

## 11. ACKNOWLEDGMENTS

Financial assistance from the Natural Science and Engineering Research Council of Canada is gratefully acknowledged. P. J. G. G. thanks N. Becker for illuminating and stimulating discussions during the writing of this manuscript.

## 12. REFERENCES

1. M. Fleischmann, P. J. Hendra and A. J. McQuillan, Raman spectra of pyridine adsorbed at a silver electrode, *Chem. Phys. Lett.* **26**, 163-166 (1974).
2. D. L. Jeanmaire and R. P. Van Duyne, Surface Raman spectroelectrochemistry. Part I. Heterocyclic, aromatic, and aliphatic amines adsorbed on the anodized silver electrode., *J. Electroanal. Chem.* **84**, 1-20 (1977).
3. M. G. Albrecht and J. A. Creighton, Anomalously intense Raman spectra of pyridine at a silver electrode, *J. Am. Chem. Soc.* **99**, 5215-5217 (1977).
4. R. R. Chance, A. Prock and R. Silbey, Molecular fluorescence and energy transfer near interfaces., *Adv. Chem. Phys.* **37**, 1-65 (1978).
5. M. R. Philpott, Effect of surface plasmons on transitions in molecules, *Chem. Phy.s* **62**, 1812-1817 (1975).
6. M. Moskovits, Surface-enhanced spectroscopy, *Rev. Mod. Phys.* **57**, 783-826 (1985).
7. M. B. Mohamed, V. Volkov, S. Link and M. A. El-Sayed, The 'lightning' gold nanorods: fluorescence enhancement of over a million compared to the gold metal, *Chem. Phys. Lett.* **317**, 517-523 (2000).
8. D. A. Weitz, S. Garoff, J. I. Gersten and A. Nitzan, The enhancement of Raman scattering, resonance Raman scattering, and fluorescence from molecules adsorbed on a rough silver surface, *J. Chem. Phys.* **78**, 5324-5338 (1983).
9. A. Nitzan and L. E. Brus, Theoretical model for enhanced photochemistry on rough surfaces, *J. Chem. Phys.* **75**, 2205-2214 (1981).
10. A. Wokaun, Surface enhancement of optical fields. Mechanism and applications, *Mol. Phys.* **56**, 1-33 (1985).
11. R. K. Chang and T. E. Furtak, *Surface Enhanced Raman Scattering* (Plenum Press, New York, 1982).
12. M. Kerker, in: *Surface-enhanced Raman scattering*, edited by B. J. Thompson (SPIE, Bellingham, 1990)
13. G. J. Kovacs, R. O. Loutfy, P. S. Vincett, C. A. Jennings and R. Aroca, Distance dependence of SERS enhancement factor from Langmuir-Blodgett monolayers on metal island films: evidence for the electromagnetic mechanism, *Langmuir* **2**, 689-694 (1986).
14. P. A. Antunes, C. J. L. Constantino, R. F. Aroca and J. Duff, Langmuir and Langmuir-Blodgett Films of Perylene Tetracarboxylic Derivatives with Varying Alkyl Chain Length: Film Packing and Surface-Enhanced Fluorescence Studies, *Langmuir* **17**, 2958-2964 (2001).
15. J. Kummerlen, A. Leitner, H. Brunner, F. R. Aussenegg and A. Wokaun, Enhanced dye fluorescence over silver island films: analysis of the distance dependence, *Mol. Phys.* **80**, 1031-1046 (1993).
16. A. M. Glass, A. Wokaun, J. P. Heritage, J. G. Bergman, P. F. Liao and D. H. Olson, Enhanced two-photon fluorescence of molecules adsorbed on silver particle films, *Phys. Rev. B* **24**, 4906-4909 (1981).
17. G. Ritchie and E. Burstein, Luminescence of dye molecules adsorbed at a Ag surface, *Phys. Rev. B* **24**, 4843-4846 (1981).
18. C. Y. Chen and E. Burstein, Giant Raman Scattering by Molecules at Metal-Island Films, *Phys. Rev. Lett.* **45**, 1287-1291 (1980).
19. K. L. Kelly, E. Coronado, L. L. Zhao and G. C. Schatz, The Optical Properties of Metal Nanoparticles: The Influence of Size, Shape, and Dielectric Environment, *J. Phys. Chem. B* **107**, 668-677 (2003).
20. G. C. Schatz and R. P. Van Duyne, in: *Electromagnetic mechanism of surface-enhanced spectroscopy*, edited by J. M. C. a. P. R. Griffiths (John Wiley & Sons, Ltd, 2002), pp. 759-774.
21. P. W. Barber, R. K. Chang and H. Massoudi, Electrodynamic calculations of the surface-enhanced electric intensities on large Ag spheroids, *Phys. Rev. B* **27**, 7251-7261 (1983).
22. R. Aroca and F. Martin, Tuning Metal Island Films for Maximum Surface Enhanced Raman Scattering., *J. Ram. Spec.* **12**, 156-162 (1985).
23. S. Garoff, D. A. Weitz, M. S. Alvarez and J. I. Gersten, Electrodynamics at rough metal surfaces: Photochemistry and luminescence of adsorbates near metal-island films, *J. Chem. Phys.* **81**, 5189-5200 (1984).

24. J. DeSaja-Gonzalez, R. Aroca, Y. Nagao and J. A. DeSaja, Surface-enhanced fluorescence and SERRS spectra of N-octadecyl-3,4:9,10-perylenetetracarboxylic monoanhydride on silver island films, *Spectrochimica Acta, Part A: Molecular and Biomolecular Spectroscopy* **53A**, 173-181 (1997).
25. P. C. Lee and D. Meisel, Adsorption and surface-enhanced Raman of dyes on silver and gold sols, *J. Phys. Chem.* **86**, 3391-3395 (1982).
26. S. Sanchez-Cortes, J. V. Garcia-Ramos and G. Morcillo, Morphological Study of Metal Colloids Employed as Substrate in the SERS Spectroscopy, *J. Coll. Int. Sci.* **167**, 428-436 (1994).
27. Z.-Q. Tian, B. Ren and D.-Y. Wu, Surface-Enhanced Raman Scattering: From Noble to Transition Metals and from Rough Surfaces to Ordered Nanostructures, *J. Phys. Chem. B* **106**, 9463-9483 (2002).
28. M. Moskovits, The dependence of the metal-molecule vibrational frequency on the mass of the adsorbate and its relevance to the role of adatoms in surface-enhanced Raman scattering, *Chem. Phys. Lett.* **98**, 498-502 (1983).
29. P. N. Sanda, J. M. Warlanmont, J. E. Demuth, J. C. Tsang, K. Cristmann and J. A. Bradley, Surface-Enchanced Raman Scattering from Pyridine on Ag(111), *Phys. Rev. Lett.* **45**, 1519-1523 (1980).
30. P. F. Liao, J. G. Bergman, D. S. Chemla, A. Wokaun, J. Melngailis, A. M. Hawryluk and N. P. Economou, Surface-enhanced Raman scattering from microlithographic silver particle surfaces, *Chem. Phys. Lett.* **82**, 355-359 (1981).
31. D. J. Jackson, *Classical Electrodynamics* (John Wiley & Sons, Inc., 1999).
32. A. Wokaun, H.-P. Lutz, A. P. King, U. P. Wild and R. R. Ernst, Energy transfer in surface enhanced luminescence, *J. Chem. Phys.* **79**, 509-514 (1983).
33. D. S. Wang, H. Chew and M. Kerker, Enhanced Raman Scattering at the surface (SERS) of a spherical particle, *App. Opt.* 2256-2257 (1980).
34. D. A. Weitz, S. Garoff and T. J. Gramila, Excitation spectra of surface-enhanced Raman scattering on silver islands films, *Opt. Lett.* **7**, 168-170 (1982).
35. S. Nie and S. R. Emory, Probing Single Molecules and Single Nanoparticles by Surface-Enhanced Raman Scattering, *Science* **275**, 1102-1106 (1997).
36. A. M. Michaels, J. Jiang and L. Brus, Ag Nanocrystal Junctions as the Site for Surface-Enhanced Raman Scattering of Single Rhodamine 6G Molecules, *J. Phys. Chem. B* **104**, 11965-11971 (2000).
37. T. Lemma and R. F. Aroca, Single molecule surface-enhanced resonance Raman scattering on colloidal silver and Langmuir–Blodgett monolayers coated with silver overlayers, *J. Ram. Spec.* **33**, 197-201 (2002).
38. K. Kneipp, H. Kneipp, I. Itzkan, R. R. Dasari and M. S. Feld, Ultrasensitive Chemical Analysis by Raman Spectroscopy, *Chem. Rev.* **99**, 2957-2976 (1999).
39. A. M. Glass, P. F. Liao, J. G. Bergman and D. H. Olson, Interaction of metal particles with adsorbed dye molecules: absorption and luminescence, *Opt. Lett.* **5**, 368-370 (1980).
40. S. Garoff, D. A. Weitz, T. J. Gramila and C. D. Hanson, Optical absorption resonances of dye coated silver-island films, *Opt . Lett.* **6**, 245-247 (1981).
41. H. G. Craighead and A. M. Glass, Optical absorption of small metal particles with adsorbed dye coats, *Opt. Lett.* **6**, 248-250 (1981).
42. S. L. McCall, P. M. Platzman and P. A. Wolff, Surface Enhanced Raman Scattering, *Phys. Lett. A* **77**, 381-383 (1980).
43. T. M. Cotton, R. A. Uphaus and D. J. Moebius, Distance dependence of surface-enhanced resonance Raman enhancement in Langmuir-Blodgett dye multilayers, *J. Phys. Chem.* **90**, 6071-6073 (1986).
44. W. H. Weber and C. F. Eagen, Energy transfer from an excited dye molecule to the surface plasmons of an adjacent metal, *Opt. Lett.* **4**, 236-238 (1979).
45. M. E. Lippitsch, Surface-enhanced Raman spectra of biliverdin and pyrromethenone adsorbed to silver colloids, *Chem. Phys. Lett.* **79**, 224-226 (1981).
46. P. J. G. Goulet, N. P. W. Pieczonka and R. F. Aroca, Overtones and Combinations in Single-Molecule Surface-Enhanced Resonance Raman Scattering Spectra, *Anal. Chem.* **75**, 1918-1923 (2003).
47. P. J. Tarcha, J. Desaja-Gonzalez, S. Rodriguez-Llorente and R. Aroca, Surface-enhanced fluorescence on SiO2-coated silver island films, *App. Spec.* **53**, 43-48 (1999).
48. R. Aroca, G. J. Kovacs, C. A. Jennings, R. O. Loutfy and P. S. Vincett, Fluorescence Enhancement from Langmuir-Blodgett Monolayers on Silver Island Films, *Langmuir* **4**, 518-521 (1988).
49. R. Aroca, C. Jennings, G. J. Kovacs, R. O. Loutfy and P. S. Vincett, Surface-enhanced Raman scattering of Langmuir-Blodgett monolayers of phthalocyanine by indium and silver island films, *J. Phys. Chem.* **89**, 4051-4054 (1985).
50. Z. Zhang, A. L. Verma, K. Nakshima, M. Yoneyama, K. Iriyama and Y. Ozaki, Substrate-Dependent Aggregation and Energy Transfer in Langmuir-Blodgett Films of 5-(4-N-Octadecylpyridyl)-10,15,20-tri-p-tolylporphyrin Studied by Ultraviolet-Visible and Fluorescence Spectroscopies, *Langmuir* **13**, 5726-5731 (1997).

51. K. Sokolov, G. Chumanov and T. M. Cotton, Enhancement of Molecular Fluorescence near the Surface of Colloidal Metal Films, *Anal. Chem.* **70**, 3898-3905 (1998).
52. G. Roberts, *Langmuir-Blodgett Films* (Plenum Press, New York, 1990).
53. M. C. Petty, *Langmuir-Blodgett Films: An Introduction* (Cambridge University Press, Cambridge, 1996).
54. R. Aroca, U. Guhathakurta-Ghosh, R. O. Loutfy and Y. Nagao, Surface-enhanced spectroscopy of Langmuir-Blodgett films of N-octyl-, N'-isobutyl-3,4:9,10-perylene-bis(dicarboximide), *Spectrochimica Acta, Part A: Molecular and Biomolecular Spectroscopy* **46A**, 717-722 (1990).
55. R. F. Aroca, C. J. L. Constantino and J. Duff, Surface-enhanced Raman scattering and imaging of Langmuir-Blodgett monolayers of bis(phenethylimido)perylene on silver island films, *App. Spec.* **54**, 1120-1125 (2000).
56. C. J. L. Constantino and R. F. Aroca, Surface-enhanced resonance Raman scattering imaging of Langmuir-Blodgett monolayers of bis(benzimidazo)perylene on silver island films, *J. Ram. Spec.* **31**, 887-890 (2000).
57. C. J. L. Constantino, T. Lemma, P. A. Antunes and R. F. Aroca, Single-molecule Detection using Surface-enhanced Resonance Raman Scattering and Langmuir-Blodgett Monolayers, *Anal. Chem.* **73**, 3674-3678 (2001).
58. U. Guhathakurta-Ghosh, R. Aroca and R. G. Gunther, Vibrational Characterization of Langmuir-Blodgett Monolayers of Tetra(p-methoxyphenyl)porphine, *Makromol. Chem., Macromol. Symp.* **52**, 15-22 (1991).
59. J.-H. Kim, T. M. Cotton, R. A. Uphaus and D. Moebius, Surface-enhanced resonance Raman scattering from Langmuir-Blodgett monolayers: surface coverage-intensity relationships, *J. Phys. Chem.* **93**, 3713-3720 (1989).
60. R. Aroca and D. Battisti, SERS of Langmuir-Blodgett monolayers: coverage dependence, *Langmuir* **6**, 250-254 (1990).
61. C. J. L. Constantino, T. Lemma, P. A. Antunes and R. Aroca, Single molecular detection of a perylene dye dispersed in a Langmuir-Blodgett fatty acid monolayer using surface-enhanced resonance Raman scattering, *Spectrochimica Acta, Part A: Molecular and Biomolecular Spectroscopy* **58A**, 403-409 (2002).
62. S. Garoff, R. B. Stephans, C. D. Hanson and G. K. Sorenson, Energy Transfer and Electronic Interactions Between Dye Molecules at an Interface, *J. Lumin.* **24/25**, 773-776 (1981).
63. M. Pope and C. E. Swenberg, *Electronic Processes in Organic Crystals and Polymers* (Oxford University Press, Oxford, 1999).
64. T. Basche, W. E. Moerner, M. Orrit and U. P. Wild, in: *Single Molecule Optical Detection, Imaging, and Spectroscopy*, edited by (VCH Verlagsgesellschaft mbH, Weinheim, 1997), pp.
65. P. Tamarat, A. Maali, B. Lounis and M. Orrit, Ten Years of Single Molecule Spectroscopy, *J. Phys. Chem. A* **104**, 1-16 (2000).
66. C. Constantino, J. Duff and R. Aroca, Surface enhanced resonance Raman scattering imaging of Langmuir-Blodgett monolayers of bis (benzimidazo) thioperylene, *Spectrochimica Acta, Part A: Molecular and Biomolecular Spectroscopy* **57A**, 1249-1259 (2001).
67. N. Strekal, A. Maskevich, S. Maskevich, J.-C. Jardillier and I. Nabiev, Selective enhancement of Raman or fluorescence spectra of biomolecules using specifically annealed thick gold films, *Biopolymers* **57**, 325-328 (2000).
68. J. Enderlein, Single-molecule fluorescence near a metal layer, *Chem. Phys.* **247**, 1-9 (1999).
69. J. Enderlein, A Theoretical Investigation of Single-Molecule Fluorescence Detection on Thin Metallic Layers, *Biophysical Journal* **78**, 2151-2158 (2000).
70. U. Guhathakurta-Ghosh, R. Aroca, R. O. Loutfy and Y. Nagao, SERS of Langmuir-Blodgett monolayers: N-octyl-3,4:9,10-perylenetetracarboxylic monoanhydride and monoimide, *J. Ram. Spec.* **20**, 795-800 (1989).
71. R. Aroca and E. Johnson, Surface enhanced fluorescence of aggregates in Langmuir-Blodgett monolayers, *Proceedings of SPIE-The International Society for Optical Engineering* **1336**, 291-298 (1990).
72. C. J. L. Constantino, R. F. Aroca, C. R. Mendonca, S. V. Mello, D. T. Balogh and O. N. Oliveira, Surface enhanced fluorescence and Raman imaging of Langmuir-Blodgett azopolymer films, *Spectrochimica Acta, Part A: Molecular and Biomolecular Spectroscopy* **57A**, 281-289 (2001).
73. D. Battisti and R. Aroca, Selective spectroscopic characterization of Langmuir-Blodgett monolayers using SERRS, *J. Mol. Struct.* **218**, 351-356 (1990).
74. S. E. Roark and K. L. Rowlen, Thin silver films: influence of substrate and postdeposition treatment on morphology and optical properties, *Anal. Chem.* **66**, 261-270 (1994).
75. D. J. Semin and K. L. Rowlen, Influence of Vapor Deposition Parameters on SERS Active Ag Film Morphology and Optical Properties, *Anal. Chem.* **66**, 4324-4331 (1994).

76. R. P. Van Duyne, J. C. Hulteen and D. A. Treichel, Atomic force microscopy and surface-enhanced Raman spectroscopy. I. Ag island films and Ag film over polymer nanosphere surfaces supported on glass, *J. Chem. Phys.* **99**, 2101-2115 (1993).

77. F. R. Aussenegg, A. Leitner, M. E. Lippitsch, H. Reinisch and M. Riegler, Novel aspects of fluorescence lifetime for molecules positioned close to metal surfaces, *Surf. Sci.* **189/190**, 935-945 (1987).

78. V. L. Schlegel and T. M. Cotton, Silver-Island Films as Substrates for Enhanced Raman Scattering: Effect of Deposition Rate on Intensity, *Anal. Chem.* **63**, 241-247 (1991).

79. K. C. Lee, S. T. Pai, Y. C. Chang, M. C. Chen and W.-H. Li, Optimum massthickness of Ag-nanoparticle film for surface enhanced Raman scattering, *Mater. Sci. Engin.* **B52**, 189-194 (1998).

80. D. J. Maxwell, S. R. Emory and S. Nie, Nanostructured Thin-Film Materials with Surface-Enhanced Optical Properties, *Chem. Mater.* **13**, 1082-1088 (2001).

81. W. P. Ambrose, P. M. Goodwin, J. H. Jett, A. V. Orden, J. H. Werner and R. A. Keller, Single Molecule Fluorescence Spectroscopy at Ambient Temperatures, *Chem. Rev.* **99**, 2929-2956 (1999).

# 8

# TIME RESOLVED FLUORESCENCE MEASUREMENTS OF FLUOROPHORES CLOSE TO METAL NANOPARTICLES


Thomas A. Klar, Eric Dulkeith, and Jochen Feldmann[*]


## 1. INTRODUCTION

Time resolved fluorescence measurements have become an important tool in applied fluorescence spectroscopy. Recently, it has been pointed out that the controlled manipulation of fluorescence decay rates opens a new dimension in applied fluorescence spectroscopy.[1] The fluorescence decay rate depends on two independent contributions, the pure radiative rate and the nonradiative rate. The latter one can be influenced by the well known Förster-type resonant energy transfer processes, while the radiative rate can be changed if the molecules are embedded or close to media comprising a dielectric constant markedly different from vacuum. Especially metal nanostructures have been used to alter both decay paths of fluorescent molecules. Apart from a change of those two rates, the absorption cross-section might also be altered.

To get a full understanding how the optical properties are changed for molecules close to metal nanostructures, time resolved measurements are essential. It is naturally more difficult to investigate the relevant physical processes for fluorophores placed close to an extended nanostructure rather than for the much simpler situation of fluorophores close to a special metal nanoparticle of well known geometry. Therefore it is advantageous to understand these "simple" systems first, before going on to more complex structures.

In the second section of this chapter we want to introduce briefly the physics of metal-nanoparticle plasmons, their damping mechanisms, and give an overview of typical time constants of scattering processes involved in the dissipation of the particle plasmon energy. In the third section a brief listing of theoretical models describing the fluorescence of dipoles in front of metallic nanostructures will be given, as well as basic considerations how to interpret experimental data. In the fourth section we want to exemplarily discuss recently performed experiments, where Lissamine molecules have


[*] Photonics and Optoelectronics Group, Physics Department and Center for NanoScience, Ludwig-Maximilians-Universität München, Amalienstraße 54, 80799 München, Germany



*Topics in Fluorescence Spectroscopy, Volume 8: Radiative Decay Engineering*
Edited by Geddes and Lakowicz, Springer Science+Business Media, Inc., New York, 2005

been attached to gold nanoparticles.[2] The fifth section finally gives some examples of biophysical applications of fluorophore-nanoparticle systems.

## 2. NANOPARTICLE PLASMONS

The scattering and absorption spectrum of noble metal nanoparticles is dominated in the visible range by the particle plasmon, which is a collective resonance of the conduction band electrons (Figure 1). It is a prerequisite for the excitation of these particle plasmons that the dimensions of the particles have to be in the range of or less than the penetration depth of electromagnetic waves in metal. This is some tens of nanometers for noble metals. An incident electromagnetic wave penetrates the whole nanoparticle and pulls the conduction band electrons to one side which in turn causes a restoring dipole force due to the immobile positively charged ions of the crystal lattice. In that simple picture, we can consider the particle plasmon as a harmonic oscillator. A comprehensive treatment of the optical properties of metal nanoparticles can be found in the textbook of Kreibig and Vollmer.[3]

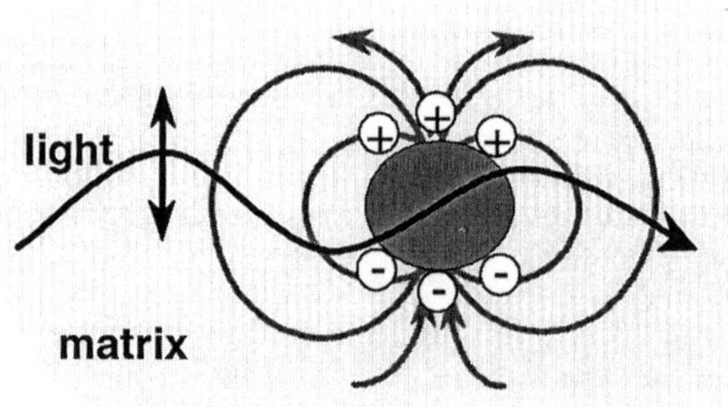


**Figure 1:** An electromagnetic wave penetrates a metallic nanoparticle, driving a collective resonance of the conduction band electrons, which is called a nanoparticle plasmon.

The nanoparticle plasmon oscillation suffers from rapid damping on a timescale of 8 femtoseconds.[4-7] Due to Fourier theory this ultrafast damping corresponds to a broad resonance with a full width half maximum (FWHM) of 160 meV. On the wavelength scale this corresponds to scattering and absorption spectra with a FWHM in the range of 50 nanometers. Two effects are responsible for this ultrafast damping of particle plasmons. The first one is radiation damping, which means that the dipole energy is lost by re-radiation. Note that this is not an incoherent fluorescence process but coherent scattering. The second reason for particle plasmon damping is a loss of energy caused by a decay into electron-hole pairs. This process is called Landau damping and can lead to excitation of intraband electron-hole pairs as well as interband electron-hole pairs. The latter process is energetically allowed only if the particle plasmon is higher in energy than the energetic distance of the d-band to the Fermi edge in the sp-band. This is satisfied, for example, for small and spherical gold nanoparticles. After the nanoparticle plasmon has lost its energy to an electron-hole pair, a series of further scattering processes occur. Due to electron-electron scattering the excited electrons and holes, which have large excess energies as compared to the thermal distribution of the Fermi-sea, thermalize with other conduction band electrons thus heating the Fermi-sea. This process is called thermalization. Using white light pump-probe experiments it is possible to measure the time constant for this thermalization process which is found to be in the range of 1 picosecond.[8] Subsequent electron-phonon scattering then leads to transfer of the thermal energy to the lattice of the nanoparticle. This process occurs on a 4 picosecond timescale. As the electron-electron scattering is faster than the electron-phonon scattering, there is some time where the electron gas is substantially hotter than the lattice. Indeed, the conduction band electrons can easily reach some 1000 K using amplified femto-second laser pulses. These high energetic pulses can even induce vibrations in the nanoparticles.[9]

Coming back to the nanoparticle plasmon resonance we can ask for the parameters that determine the exact resonance position. On the one hand there are intrinsic parameters like the material of the particles, their size and shape. For gold and silver nanoparticles the resonances are within the visible range of the electromagnetic spectrum. For small silver and gold nanoparticles supported on a glass slide, the resonances occur at ~450 nm and 530 nm, respectively (Figure 2). As the nanoparticles become larger, the plasmon resonance shifts to lower energies. Concomitantly the scattering spectrum becomes substantially broader due to an increased radiation damping that scales with the nanoparticles' volume.[10] A similar red shift is observed for gold nanorods with increasing aspect ratio. However, in this case the FWHM of the scattering spectrum is decreased because the red shift of the particle plasmon energetically excludes damping by the interband excitation of d-band electrons.[11] Apart from these intrinsic parameters, the energetic position of the plasmon resonance depends on extrinsic parameters like the refractive index of the surrounding matrix.

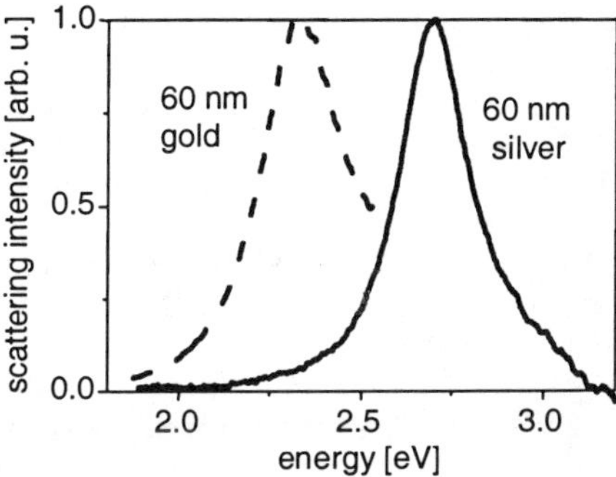



**Figure 2:** Measured scattering spectra of a single gold nanoparticle (left, dashed line) and a single silver nanoparticle (right, solid line). Both nanoparticles are supported by a glass slide. Measurements are taken on air in a evanescent field microscope.

The nanoparticle plasmon resonance is the reason for both, a pronounced resonance in the absorption spectrum as well as a resonance in the scattering spectrum of metal nanoparticles. Both features may be applied for novel devices in optoelectronics or biophysics. The resonance in the scattering spectrum can, for instance, be used in optoelectronics for electrically addressable light scattering devices.[12] The scattering property also leads to promising applications of metal nanoparticles as sensors in biomolecular recognition. Metal nanoparticles, appropriately functionalized with molecular recognition agents, may be used as markers reporting the presence of some molecular species.[13-16] As mentioned above, the energetic position of the plasmon resonance is very sensitive to the refractive index of the immediate surrounding of the nanoparticles. If, for example, biomolecules bind to nanoparticles functionalized with appropriate binding sites one can deduce this binding event by a shift in the absorption spectrum.[17-24] Alternatively, the change in refractive index may be monitored by a shift of the scattering spectrum of metal nanoparticles.[25] Measuring the scattering spectrum instead of the absorption spectrum has the advantage of being more sensitive because the signal is measured against a zero background. Indeed it is sensitive enough to perform single nanoparticle spectroscopy[10, 13, 14] and it has recently been possible to use a single biotinylated metal nanoparticle to monitor streptavidin binding.[26] This opens the way to ultra-sensitive assays reporting the presence of only some tens of target molecules.

In the following sections we now turn to the issue of fluorescence quenching in the vicinity of gold nanoparticles.

## 3. FLUORESCENCE DECAY RATES OF FLUOROPHORES IN THE VICINITY OF METAL STRUCTURES

### 3.1 Theory

It is not our aim to give a comprehensive review of the theoretical work carried out by many groups during the last decades. We rather want to give a selective overview of important effects that have to be considered when a molecule interacts with a nearby metal nanostructure. For a more detailed representation please refer to review articles like those written by Leung and George[27] or Metiu.[28]

Theoretical modeling of the interplay between a dipole and a conducting surface basically dates back to Sommerfeld[29, 30] who investigated the radio problem of a dipole emitter above the soil. Drexhage and Kuhn experimentally observed and theoretically modeled the change of fluorescence lifetime when a molecule is brought from infinite distance towards a flat metal surface.[31-33] However, their experimental data deviate from the given theoretical description in the case of small distances (< 150 nm) because nonradiative energy transfer had not been modeled correctly. In their landmark paper Chance, Prock, and Silbey (CPS)[34] incorporated energy transfer correctly and gave formulas for both, the change of the radiative rate due to mirror effects and the increase of the nonradiative rate due to absorptive losses. With this theory it has been possible to fit the experimental data of Drexhage and coworkers much better.

However, soon after the CPS theory had been established, experimentalists ran into new problems. As handling of nanoscale distances became feasible and molecules were placed only nanometers away from the surface, it became questionable to which extent the assumptions of a perfectly smooth and abrupt surface holds and whether a local theory based on phenomenological Maxwell equations still holds. In the abrupt surface assumption the metal dielectric function is kept constant up to an infinitely thin plane and then jumps suddenly to the bulk medium of the dielectric. Of course, in reality the dielectric constant varies smoothly between the two bulk media.[35, 36] This introduces discrepancies from CPS theory if the molecules are only a few Ångstroms away from the surface.

Random surface roughness on a nanometer scale has been included[37] into the model as well as periodically roughened surfaces.[38] However, on rough surfaces a large variety of different radii of curvatures are present which makes measurements to prove the theory difficult. Hence, it is advantageous to turn to a somewhat radical approach, namely to study the interaction of molecules adsorbed on metal nanoparticles of defined surface curvature. This will lead to a better understanding of the processes involved in molecular interactions with nanostructured surfaces.

Apart from that "top down" approach from smooth surfaces to rough surfaces and finally to nanoparticles, there is a "bottom up" approach motivated by molecular physics and biophysics. In a somewhat crude way, not only the molecule close to a metal nanoparticle can be considered as a point dipole, but also the metal nanoparticle itself. In this limit, the interaction can be considered as a Förster type resonant energy transfer[39, 40] that turns into a Dexter type transfer for small distances where higher order multipole interactions have to be considered[41]. Förster theory yields two well known main results: First, the resonant energy transfer, which is a nonradiative transfer channel, depends on the distance $d$ between the two molecules as $d^{-6}$. Secondly, it is proportional to the overlap integral between the emission spectrum of the donor and the absorption spectrum of the acceptor of energy. This picture gives a vague idea that the interaction between a fluorescing molecule and a metal nanoparticle is expected to depend strongly on the distance and the overlap integral of the fluorophores' emission and the nanoparticles' absorption spectra. However, it fails because of two major shortcomings: First, a nanoparticle, compared to a fluorophore, can no longer be considered as a point-like object and second, the influence of the nanoparticle on the pure radiative rate of the fluorescing molecule is not taken into account by the Förster model.

A model that includes both concerns was set up by Gersten and Nitzan.[42] They treated the molecule-nanoparticle system as an image dipole model relying on static Maxwell equations. Thereby they calculated the dipole moment of the nanoparticle which is induced by the primarily excited molecule. This dipole on the nanoparticle subsequently alters the electric field in its nanoenvironment, which in turn influences the molecular dipole. Taking these mutual interactions into account, the radiative rate is given by inserting the resulting total dipole moment of the composite system into the Hertzian radiation formula. This leads to a radiative rate despite the fact that the model is based on electrostatics. The nonradiative rate is derived by the Ohmic losses caused by the electric field inside the particle, according to the nanoparticle's finite conductivity.

Ruppin[43] derived an alternative theory to tackle the problem, based on a fully electrodynamic theory. Similar to the scattering theory of Mie[44] he derived formulas for the radiative and nonradiative rates based on a linear decomposition of the electromagnetic field into plane wave components. Therefore, if the Ruppin model can be considered as the fluorescence quenching analogy to the electrodynamic scattering theory by Mie, the Gersten-Nitzan model may be seen as the analogy to electrostatic Raleigh scattering. Of course, the mathematical effort to calculate the radiative and transfer rates according to the electrodynamic model outweighs that of the electrostatic model.

Therefore, theoretical investigations were carried out which compare both models and give regimes where the static Gersten-Nitzan model can be safely applied.[45-49] It was found that the Gersten-Nitzan model gives reliable results as long as both, the dimensions of the nanoparticle and the distance between the dipole and the nanoparticle, are in the range of or smaller than the skin depth of electromagnetic radiation in the nanoparticle, i.e. within tens of nanometers.

The above mentioned theories of CPS, Gersten-Nitzan and Ruppin are so called local theories. For molecule-nanoparticle or molecule-surface distances smaller than a few nanometers, however, a local theory is not applicable. This is easily seen by the following considerations.[28] The displacement vector $\vec{D}$ is related to the electric field $\vec{E}$ by the following relation:

$$\vec{D}(\vec{r},t) = \int dt' \int d\vec{r}' \; \vec{\bar{\varepsilon}}(\vec{r}-\vec{r}';t-t') \cdot \vec{E}(\vec{r}',t')$$

or, in its Fourier transformed equivalent:

$$D(\vec{k},\omega) = \vec{\bar{\varepsilon}}(\vec{k},\omega) \cdot \vec{E}(\vec{k},\omega)$$

where $\vec{\bar{\varepsilon}}$ is the dielectric tensor or the dielectric constant in case of homogeneous media. In phenomenological theory a dielectric constant is used which is measured e.g. by ellipsometry, using propagating far field modes. The $k$-vector of these far field modes is small, i.e. $|k| \approx 0$. If we neglect the $\vec{k}$ vector in the equations above, we end up with the well known local result:

$$\vec{D}(t) = \int dt' \; \vec{\bar{\varepsilon}}(t-t') \cdot \vec{E}(t')$$

or, in its Fourier transformed equivalent:

$$D(\omega) = \vec{\bar{\varepsilon}}(\omega) \cdot \vec{E}(\omega)$$

Obviously, dropping the $\vec{k}$-dependence is the Fourier equivalent to removing the integral over space. Hence the non-local theory becomes local.

The validity of phenomenological Maxwell theory for all problems involving propagating photon modes shows, that usually the error introduced by locality is negligible. However, if photons which possess a large $k$-component can not be neglected, local theories should fail. Such photons are present, for example, in near fields. Another way to see why we have to consider large $k$-components is, that very close to a dipole or a metal nanostructure (d < 5nm) the field varies very strongly in space and is therefore composed of planar waves of large $k$:

$$\vec{E}_{dip}(\vec{r},\omega) = \int \frac{d\vec{k}}{(2\pi)^3} \; \vec{E}_{dip}(\vec{k},\omega) \cdot e^{i\vec{k}r}$$

The basic problem of nonlocal theories is to find an appropriate $\varepsilon(\vec{k},\omega)$. Several authors have dealt with this problem for both geometries, a dipole above a surface[50-56] and a dipole close to a metal nanoparticle.[57-60] It is certainly beyond the scope of this chapter to go into detail of those theories. However, let us briefly note that the results are miscellaneous. For example, in the case of a dipole close to a metal nanosphere, Leung[59] predicts one to two orders of magnitude less energy transfer to the nanoparticle in the case where the dipolar transition is energetically lower than the particle plasmon resonance. Ekardt and Penzar[57] predict exactly the opposite.

Experimental results either tend to back the prediction that nonlocal effects weaken the interaction of molecules and metals as compared to local theories[61, 62] or no deviations from local theories are observed at all.[63-67] It is worth noting that all those experiments have been performed with molecules in front of a surface. In this geometry there is inevitably a mixture of all dipole - metal distances from the shortest one possible, i.e. the orthogonal projection, up to infinity. Clearly, it would be advantageous to test nonlocal theories on a system where a molecular dipole is fixed at a well defined distance from a metal nanosphere. In the following we show that experimental results differ from predictions of local theories[2] and suggest the existence of nonlocal effects. In that investigation the dipole was oriented tangential to the nanoparticle and its absorption and emission resonances were lower in energy compared to the particle plasmon resonance.

In the light of the numerous theoretical works available it is certainly necessary to clarify the importance of nonlocal effects in further detail. One reason why this has not been carried out before is probably a lack of nano-engineering expertise in previous decades. As today more elaborate techniques become available, more experiments will become feasible. A powerful tool to gain meaningful data is to perform time resolved fluorescence measurements as it will be discussed in the following.

### 3.2 Time resolved spectroscopy

Up to now, we assumed that the dipole in close vicinity of a nanostructure is already in its excited state and discussed only its radiative and nonradiative decay rates. However, the process of excitation must also be considered. Strictly speaking, if one assumes a two level system, the absorption cross-section of a dipole must change proportionally to the radiative rate. This follows from the proportionality of the Einstein coefficients. Molecules, however, can not be considered as two level systems due to vibronic sidebands. Vibronic degradation in the excited state as well as other sources of Stokes shift like molecular reorganization or solvent effects finally lead to a much more complicated relation between absorption cross-sections and emission rates.[68] Therefore it becomes possible that for example the absorption cross-section may be enhanced but the radiative rate stays constant or is decreased. The opposite effect is also possible. It is clear that time integrated fluorescence spectroscopy is incapable to give details on the change of each rate: The absorption rate, the radiative emission rate and the non-radiative rates such as the energy transfer rate.

To distinguish between the three effects, time resolved measurements have proven to be an important tool. To explain how the three different effects change the fluorescence decay curves, it is useful to consider the consequences of each effect in absence of the other two.[69] Figure 3 gives examples where the solid line always represents the fluorescence decay of the free molecule, while the dashed curve shows the decay

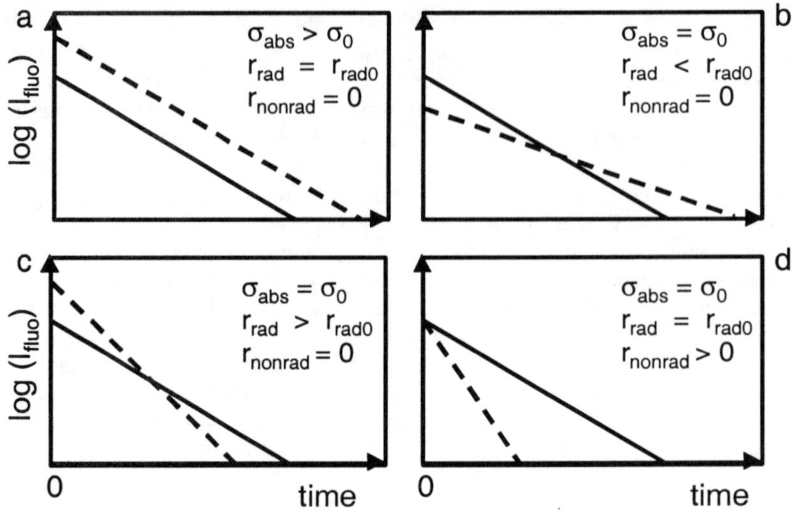



**Figure 3:** Scheme how the fluorescence transients change if (a) the absorption cross-section, (b,c) the radiative rate, and (d) the nonradiative rate is altered by a nearby metal nanostructure. Solid line: free fluorophore, dashed line: fluorophore in front of a metal nanostructure. Only one rate is changed a time while the others are kept constant.

characteristics of a dye-nanostructure composite. The examples are given on a semi-logarithmic scale. Assuming single exponential decays, straight lines result.

Figure 3 (a) shows the change of the fluorescence decay when only the absorption cross-section is increased, but the radiative and the non-radiative rates are unchanged. In that case, the slope of the decay curve on the semi-logarithmic plot remains constant, but the curve is shifted to higher count rates. The total detected intensity which is derived by integrating the curves from time 0 to infinity increases. Please note that a decrease of the absorption cross-section is also possible, depending on the concrete type of molecule-nanostructure composite. In that case the dashed curve would lie below the original decay curve and a decreased total fluorescence intensity would be detected.

In Figure 3 (b) it is assumed that the absorption cross-section stays constant and the radiative rate of the molecule is decreased by the nearby metal nanostructure ($r_{rad} < r_{rad0}$, $r_{rad0}$ being the rate of the free dye). Further, there should be no nonradiative losses ($r_{nonrad} = 0$), which includes a vanishing energy transfer. As the radiative rate is decreased, the fluorescence decays slower leading to a less steep decay curve. As no nonradiative processes are allowed in this case, the integrated fluorescence intensity must be unchanged, which means that the integral must stay constant. Therefore the fluorescence at t = 0 must be less intensive for the composite system (dashed curve) as compared to the free dye molecule (solid line). Figure 3 (c) shows the opposite effect of an increased radiative rate. Again, it is assumed that there are no nonradiative channels and the absorption cross-section is the same for free fluorophores and fluorophore-metal nanostructure composites. The area underneath the curve must stay the same as in figure

3 (b), but the curve now becomes steeper. This can only be achieved, if the fluorescence intensity at t = 0 is increased. Hence, the intercept of the decay curve with the ordinate (t = 0) is a direct measure for the change of the radiative rate.

In Figure 3 (d) it is assumed that neither the absorption cross-section changes nor the radiative rate is influenced when the fluorophore is attached to a metal nanostructure. However, a nonradiative channel is introduced due to energy transfer. Now the total fluorescence does not stay constant and the fluorescence decays faster in the case of the composite system (dashed line) compared to the free fluorophore (solid line). Provided the time resolution of the measurement system is much better than the inverse of the energy transfer rate, the fluorescence at t = 0 should be the same in both cases, the free fluorophores and the composite system.

In summary, figure 3 illustrates how changes of the absorption cross-section, radiative rate and energy transfer each affect the fluorescence decay curve. In practice, more than one effect will be present simultaneously and the interpretation of the data becomes more difficult. If, however, the absorption cross-section is not affected by the metal nanostructure, e.g. because the excitation of the fluorophore takes place far from any resonances of the nanostructure, it is possible to extract both, the radiative and the energy-transfer rate from time resolved decay curves. This is because the change of fluorescence intensity at t = 0 is solely caused by a change of the radiative rate. Once this radiative rate is deduced from the data, one can calculate the energy transfer rate from the total fluorescence rate $r_{fluo}$ due to the simple relation

$$r_{ET} = r_{fluo} - r_{rad} - r_{nonrad\,0}$$

Hereby $r_{nonrad0}$ denotes the other nonradiative processes already present for the free fluorophore.

Time resolved fluorescence measurements have been used for decades[31, 33] because they are such a powerful tool to investigate fluorophore-metal composites. Due to insufficient time resolution, mostly long lived luminescence like that from triplet states has been investigated.[61, 64-66] When fluorophores are attached to metal nanostructures, fluorescence decay times are in the sub nanosecond time range. To measure those decay times accurately, techniques such as time correlated single photon counting,[70-72] frequency domain fluorescence measurements,[73, 74] streak camera measuremets,[2, 69, 75, 76] and femtosecond pump SHG-probe[67] have been used.


## 4. TIME RESOLVED SPECTROSCOPY OF FLUOROPHORES BOUND TO METAL NANOPARTICLES

The preceding chapter showed that many different processes have to be considered if one would like to fully understand the interactions between a fluorophore and a nanostructured metallic template. Depending on the distance regime, classical image theory, electrodynamic theory, nonlocal effects or even wave functions of conduction band electrons leaking out of the metal surface have to be considered. Furthermore, each of the theories gives different results for fluorophores oriented perpendicular or tangential to the metallic surface. Different situations are also expected when either the absorption spectrum or the emission spectrum of the fluorophore overlaps with the plasmon

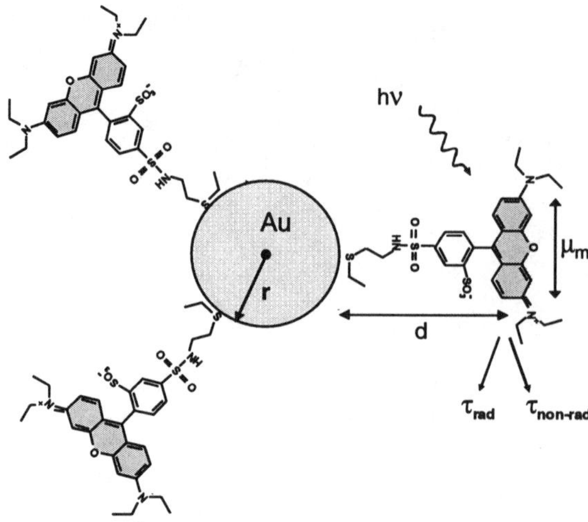

**Figure 4:** Lissamine - gold nanoparticle composite system. Lissamine molecules with dipole moment $\mu_m$ are attached onto gold nanoparticles of radius r via a thioether linker. The distance between the chromophoric part and the nanoparticle surface is d = 1 nm.

resonance. The only solution to get clear answers from experiments is to turn to simple geometries. Therefore, we want to discuss the situation where fluorophores are tangentially attached onto gold nanoparticles of distinct radii *r* at a distance *d* of 1 nanometer. These experiments have been reported recently by our group.[2]

The composite sample under investigation is shown in figure 4. The Rhodamine-type dye "Lissamine" with molecular dipole moment $\mu_m$, has been attached to the gold nanoparticles via a thioether group.[77] The radii of the particles have been 1, 10, 15, and 30 nanometers with a size distribution of less than 15% each.

The surface coverage of the nanoparticles can be calculated from the molar concentrations, the surface area of a sphere with the radius *r* of a particle plus the spacer length of 1 nm, and the assumption that one fluorophore covers ~1 nm². Figure 5 shows the total fluorescence from cuvettes with different Rhodamine concentrations. The fluorescence is heavily quenched for calculated surface coverages of less than 100%. Above 100% surface coverage it increases linearly with fluorophore concentration. Having a close look to the data one recognizes that below 100% surface coverage there is some weak fluorescence. As it will become clear later, this fluorescence does not stem from bound molecules but rather from unbound molecules which are always present due to a thermodynamic equilibrium between bound and unbound dye molecules. Even dialysis cannot remove those unbound fluorophores. The following spectrally and temporally resolved experiments have been carried out with composite systems showing ~50% surface coverage. Experiments with other surface coverages have also been performed, leading to essentially the same results. Hence it can be concluded that dye aggregation effects like J-aggregates play a minor role in time resolved spectroscopy of fluorophores attached to gold nanoparticles. This has also been reported by other investigators.[71, 75]

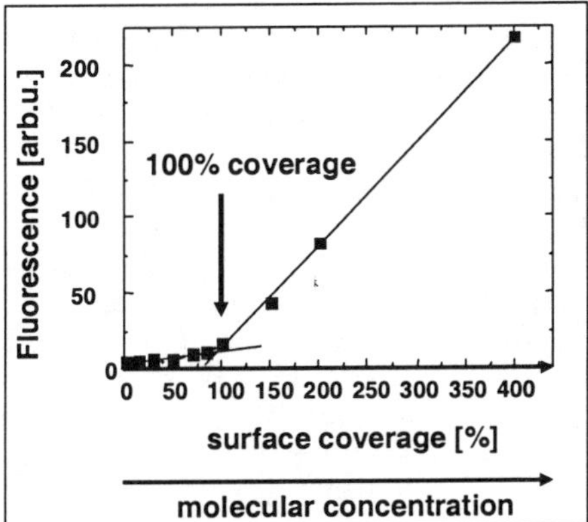


**Figure 5:** Fluorescence signal from cuvettes with constant concentration of gold nanoparticles but altering concentration of tetramethyl-Rhodamine. Assuming that one Rhodamine molecule occupies an area of ~1 nm$^2$ one can calculate the surface coverage for each concentration. Below 100 % surface coverage the fluorescence is quenched, while above the fluorescence intensity increases linearly with Rhodamine concentration.

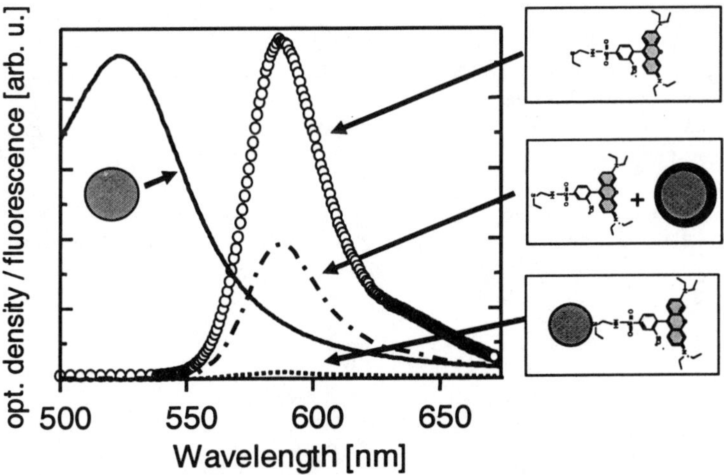


**Figure 6:** Optical density of a 0.029 nM gold nanoparticle solution (solid line) and fluorescence spectra of a 0.18 $\mu$M Lissamine solution (open circles), a solution of 0.18 $\mu$M Lissamine and 0.029 nM passivated gold nanoparticles (dash-dotted curve) and a solution of 0.18 $\mu$M Lissamine bound to gold nanoparticles with a concentration of 0.029 nM (dotted curve).

Figure 6 shows the fluorescence spectra of a pure aqueous Lissamine solution (open circles), of a solution containing unbound Lissamine molecules and passivated gold nanoparticles (dash-dotted curve) and of the composite system (dotted curve). The concentration of the pure aqueous Lissamine solution amounts to 0.18 $\mu$M. The solid line shows the optical density spectrum of gold nanoparticles with a radius of 30 nm. In each case an excitation wavelength of 400 nm has been chosen such that the excitation is not resonant with the nanoparticle plasmon at 520 nm. Therefore, fluorescence enhancement effects by induced absorption are minimized, allowing the influence of the nanoparticle on the radiative and the energy transfer rates to be investigated.

When the Lissamine molecules are added to the nanoparticle solution, two effects have to be considered: On the one hand there are nanoparticle induced changes in the molecules' radiative and nonradiative rates. These are the effects we are interested in. On the other hand, there are trivial effects like a change in excitation power due to extinction of the exciting laser beam by the nanoparticles. Similarly the detection efficiency is also reduced due to extinction of fluorescence light by other particles in the solution. To account for the latter effects, a solution of 0.18 $\mu$M Lissamine molecules and of 0.029 nM passivated gold nanoparticles is prepared. Passivation has been achieved by attaching thio-alkanes onto the nanoparticles such that the Lissamine molecules were not able to bind to the gold nanoparticles. The fluorescence spectrum of this mixed solution of passivated nanoparticles and Lissamine molecules is shown by the dash-dotted curve in figure 6.

The fluorescence spectrum of the composite system, i.e. a solution of 0.029 nM unpassivated gold nanoparticles (r = 30 nm) and 0.18 $\mu$M Lissamine dye molecules, is shown by the dotted curve in figure 6. In fact, it is hardly distinguishable from the abscissa. Compared to the fluorescence of the solution with passivated particles, only ~5 % of fluorescence intensity is left over. From the following discussion it will become clear that most of this residual fluorescence is due to fluorescence from unbound Lissamine molecules inevitably present due to the thermodynamic equilibrium between bound and unbound molecules. Only time resolved measurements are able to distinguish between fluorescence of bound and unbound molecules. Below it will become clear that the nanoparticles quench the fluorescence of the bound molecules by more than 99 %.

Let us note that there is still some near field induced change of the absorption cross-section despite the fact that 400 nm excitation light is used. The Gersten-Nitzan model can be applied to account for that change in absorption cross-section in the following way: The change in the radiative rate of a hypothetical dye molecule emitting at 400 nm can be calculated. We can deduce the change in the absorption cross section of the

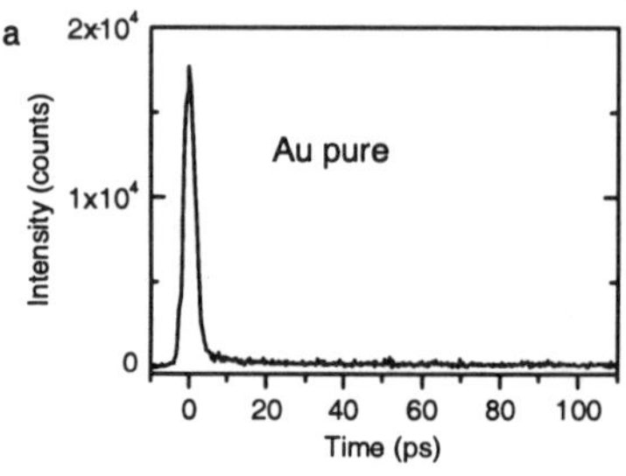


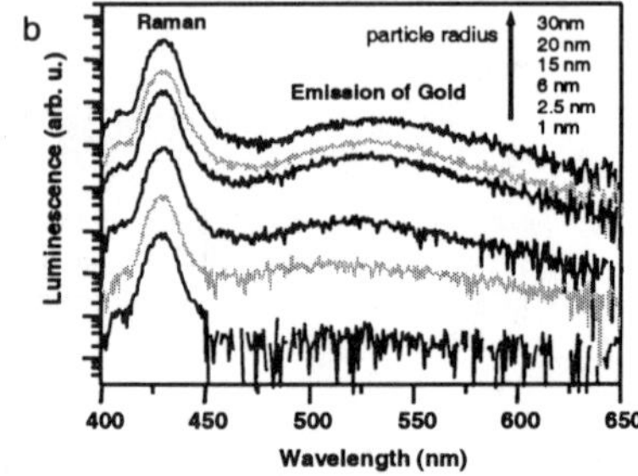


**Figure 7:** (a) Time resolved luminescence signal from a pure gold nanoparticle solution. (b) spectrally resolved signal within the first 5 ps. A pronounced Raman signal at 430 nm and a weak emission following the particle plasmon resonance are observed.

Lissamine molecules in the vicinity of gold nanoparticles because the proportionality of Einstein coefficients of absorption and emission must hold as long as absorption and emission occur without a Stokes shift. We find that the absorption cross section is changed by 30% for nanoparticles of 30 nm radius. This effect is already included in the fluorescence quenching efficiency above and will be included in a similar way in the analysis of the following time resolved measurements.

Time resolved measurements have been carried out using a streak camera (Hamamatsu, C5680) with a temporal resolution of 2 ps. The frequency-doubled output (400 nm) of a Kerr-lens mode-locked titanium-sapphire laser (Mira, Coherent, 120 fs, 76 MHz) has been used as the optical excitation source. Appropriate spectral filters have been used to block any stray light from the excitation beam.

Before reporting on the time resolved fluorescence emission spectra of the hybrid system, we first want to discuss the time resolved optical spectrum of a pure gold nanoparticle solution. As shown in figure 7 (a), an ultrafast spike occurs at $t = 0$, the temporal evolution of which cannot be resolved by the streak camera. Figure 7 (b) shows the spectrally resolved luminescence (Chromex IS250 spectrometer) within the first 5 ps with an excitation wavelength of 375 nm. The spectra are corrected for background, re-absorption, and instrumental response and have been normalized to a common number of particles. As a parameter we changed the particle radius from 1, 2.5, 6, 15, 20, to 30 nm (lines from bottom to top in fig. 7 (b)). Apart from a pronounced Raman line we observe a weak luminescence at ~530 nm (note the logarithmic scale). This luminescence coincides spectrally with the nanoparticle plasmon. Hence we assume that this emissive signal stems from the gold nanoparticles. This signal is also ultrafast, i.e. faster than the time resolution of the streak camera. Similar luminescence from noble metals has been

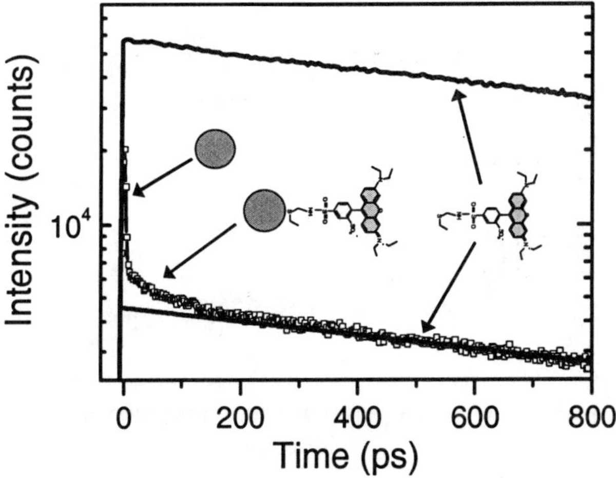



**Figure 8:** Time resolved fluorescence signal from a 0.37 $\mu$M Lissamine solution (upper curve) and a mixed solution of 0.37 $\mu$M Lissamine and of 0.25 nM gold nanoparticles of r = 15 nm radius. The ultrafast spike at t = 0 stems from the gold particles, while the long lived component is from unbound Lissamine molecules. The intermediate decay signal is from Lissamine - nanoparticle composites.

reported recently.[78-86] A more detailed discussion how this gold nanoparticle luminescence depends on nanoparticle size will be published elsewhere.[87]

The time-resolved fluorescence spectrum of a pure Lissamine solution (0.37 $\mu$M) is shown by the upper curve in figure 8. A straight line is observed on the semi-logarithmic plot revealing a fluorescence lifetime of 1.54 ns. As Lissamine molecules have a quantum efficiency of 33 % in aqueous solution,[88] the pure radiative decay rate is $\tau_{rad}^{-1}(r=0) = 0.21 \cdot 10^9 \, s^{-1}$. A vanishing nanoparticle raidus ($r=0$) denotes that no gold nanoparticles are attached to the fluorophores.

Figure 8 also displays the time resolved fluorescence signal from the combined nanoparticle / Lissamine system. The concentrations of the reactants have been chosen exactly the same as for the pure nanoparticle and pure Lissamine solutions, namely 0.25 nM and 0.37 $\mu$M, respectively. A triple decay has been found. The presence of the gold nanoparticles again leads to an ultrafast spike at t = 0 of exactly the same amplitude as for the pure gold nanoparticle solution. For larger time values the fluorescence emission kinetics follows exactly the kinetics of the free dye solution. This can be seen by drawing a straight line representing a fluorescence decay time of 1.54 ns and fitting it to the composite emission at times t > 400 ps. The count rate underneath this straight line therefore represents fluorescence photons originating from unbound dye molecules.

Only the fluorescence of the composite system remains if one subtracts the ultrafast signal originating from the nanoparticles and the long lived component originating from free dye molecules. This remaining signal resembles a monoexponential decay as it is shown in figure 9. In this figure the decay curves of the composite systems with gold nanoparticles of $r = 30$ nm and $r = 1$ nm radius are shown by the steepest and the intermediate curve, respectively. The long lived decay is that of free dye molecules for comparison. The monoexponential decay behavior indicates that the composite systems are structurally homogeneous. This means explicitly that heterogeneities like variations in

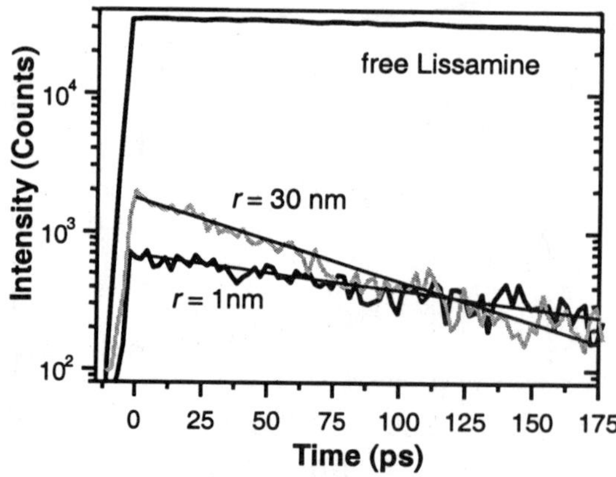


**Figure 9:** Fluorescence decay signal from Lissamine - gold nanoparticle composites after removal of the ultrafast gold nanoparticle spike and the long lived signal from unbound Lissamine molecules. The fastest and the intermediate decay curves are from composites comprising nanoparticles with r = 30 nm and r = 1 nm radius. The slowest decay is that of free Lissamine molecules. Straight lines are single exponential fits.

the fluorophore-nanoparticle distance or orientations of the molecular dipole moment with respect to the nanoparticles' surface can not play a major role. The fluorescence lifetime for the $r = 1$, 10, 15, and 30 nm nanoparticles are 169, 102, 99, and 72 ps, respectively. Compared to the lifetime of 1.54 ns of the free dye solution, a substantial shortening has occurred. As expected, the larger nanoparticles quench the fluorescence more efficiently than the smaller ones.

Apart from the fluorescecne decay rate we can also deduce the radiative rate from the time resolved data as discussed in the previous section. In the discussion of figure 3 (b,c) we assumed a vanishing nonradiative rate, however the argumentation that the fluorescence at $t = 0$ directly gives the radiative rate still holds if the fluorescence rate is much smaller than the inverse of the time resolution of the measuring instrument. This can be seen by considering the quantum efficiency $\eta(r)$ where $r$ is again the radius of the nanoparticle:

$$\eta(r) = \tau_{rad}^{-1}(r) \Big/ \tau_{fluo}^{-1}(r) = g \int_0^\infty I_{fl}(r, t = 0)\ e^{-t/\tau_{fluo}(r)}\ dt = g\ I_{fl}(r, t = 0) \Big/ \tau_{fluo}^{-1}(r).$$

$I_{fl}(r,t)$ is the measured fluorescence transient and g is a collection efficiency factor taking care of the detection efficiency of the instrument. g can be determined from the measurements of free Lissamine molecules with a known quantum efficiency of $\eta = 33\%$, rendering the above equation into a simple proportionality between the radiative rate and the detected fluorescence at t = 0. Once the radiative rate and the fluorescence rate have been obtained from experimental data the nonradiative rate follows via the formula

$$\tau_{nonrad}^{-1}(r) = \tau_{fluo}^{-1}(r) - \tau_{rad}^{-1}(r).$$

In figure 10 (a,b) the experimentally obtained radiative and nonradiative rates are plotted against nanoparticle radius, where r = 0 denotes the rates for the free Lissamine molecules. The radiative rate $\tau_{rad}^{-1}(r)$ of Lissamine molecules drops by more than an order of magnitude for all particle radii. Simultaneously, the nonradiative rate of the fluorophores increases by an order of magnitude when attached to gold nanoparticles. Both effects are responsible for the observed drastic fluorescence quenching.

For Lissamine attached to the smallest nanoparticles of 1 nm as fluorescence quenching of 99.8% is observed. This is made up from a 51-times decreased radiative rate and a 14-times increased nonradiative rate. Hence, for these small nanoparticles the effect of the reduced radiative rate prevails over the effect of an increased nonradiative rate due to energy transfer. The fact that the tiniest nanoparticles of just 1 nm radius quench the fluorescence by more than 99.8% makes them an interesting tool as fluorescence quenchers in biophysical applications,[89] where it is essential that the nanoparticles do not mechanically disturb the sample.

The Gersten-Nitzan model is chosen to compare our experimental results with theory. An electrostatic model like that of Gersten and Nitzan should be accurate enough for a fluorophore-nanoparticle distance of 1 nm and particles small compared to the penetration depth of electromagnetic radiation in gold. A fully electrodynamic calculation is therefore not needed. However, as discussed before, the Gersten Nitzan model is of local nature and hence may fail for distances as small as 1 nm.

Figure 10 (c,d) shows the results of the calculated radiative and nonradiative rates (the latter one being the sum of the calculated energy transfer rate and the known nonradiative rate already present for the free Lissamine molecules). The radiative rate drops by more than one order of magnitude and shows a pronounced minimum at $r = 4$ nm. This drastically reduced emission rate is in reasonable agreement with the experimentally obtained values for the radiative rate and can be explained by two adjacent antennas broadcasting inefficiently when out of phase[42]. Such a phase shift is caused by the complex nature of the dielectric constant of the metal. The less satisfactory agreement of the theoretical and experimentally found rates for larger particles may be due to the electrostatic nature of the model lacking of accuracy for larger particle diameters.[45-49] The nonradiative rates (figure 10 (d)) as calculated by the Gersten-Nitzan model deviate in absolute numbers by two orders of magnitude from the experimental findings, however the shape of the distance dependence of $\tau^{-1}_{nonrad}(r)$ is similar. The increased energy transfer for the $r = 30$ nm particles can again be explained by the electrostatic nature of the Gersten Nitzan model not taking into account retardation effects. These cause a red shift of the particle plasmon resonance and hence lead to an increased overlap of the nanoparticles' absorption spectrum and the emission spectrum of the Lissamine. The two orders of magnitude discrepancy between experiment and theory indicate that nonlocal effects reduce the effects predicted by classical theories. Indeed, Leung has theoretically predicted that a two orders of magnitude reduced decay rate may be expected in the case where dipoles are only 1 nm away from the nanoparticles and the emission frequency is lower in energy than the particle plasmon resonance.[59] Some other experimental work points to the same direction.[61, 62]

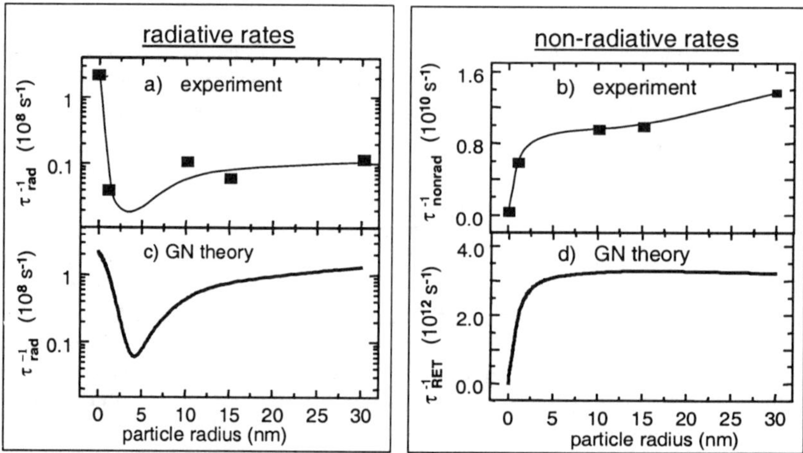


**Figure 10:** (a,c) Radiative and (b,d) nonradiative rates of Lissamine - gold nanoparticle composite systems in dependence of the nanoparticle radius. r = 0 denotes free Lissamine molecules. (a,b) experimental results, lines are guide to the eye; (c,d) rates calculated according to the Gersten and Nitzan model.

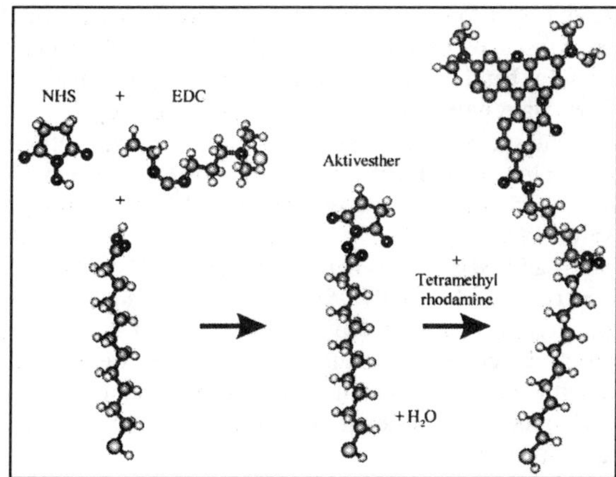


**Figure 11:** Synthesis of tetramethyl-Rhodamine attached to a C$_n$ spacer.

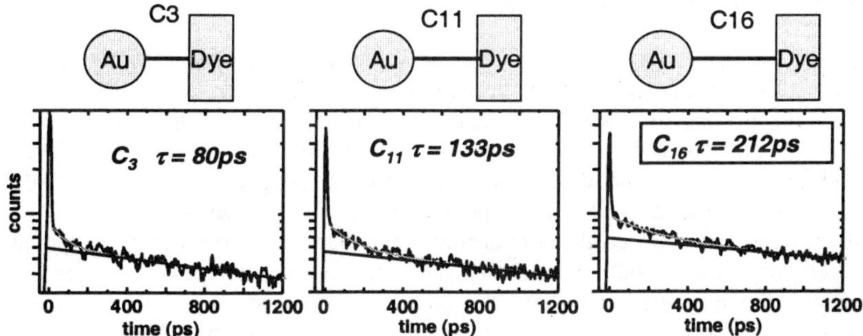


**Figure 12:** Time resolved fluorescence decay of Rhodamine - gold nanoparticle composite systems of varying spacer length.

However, there are experimental results showing no nonlocal effects.[63-67] These investigations have been carried out at samples consisting of molecules in front of a surface. Therefore, a continuum of distances between the molecules and the nanostructures has been measured simultaneously. The more stringent control of parameters like molecule-nanoparticle distance in the experiments presented above may therefore give more meaningful data to answer the question if and how nonlocal effects alter classical theories. A natural next step is altering the distance between molecules and nanoparticles. In the following we want to present some preliminary results.

To study energy transfer as a function of distance between gold nanoparticles and fluorescent molecules a stepwise synthesis has been performed.[90] Alkyl chains have been used as spacers to vary the distance between 1 and 3 nm. The two ends of the alkyl chains have been terminated by a thiol group and by a carboxyl-group. The length of the alkyl chains have been chosen to be $C_3$, $C_{11}$ and $C_{16}$. Longer alkyl chains cannot be used because they tend to bend. Tetramethyl-Rhodamine functionalized with an amino-terminated $C_5$-alkyl chain have been used as fluorophore. Together with the $C_3$, $C_{11}$ and $C_{16}$ -alkyl chains of the spacer this corresponds to a final length between the $\pi$-conjugated system of the Rhodamine and the particles' surfaces of 17.5, 28 and 36 Å, respectively.

The sequence of synthesis is as follows (figure 11): The carboxy-group of the alkyl-spacer is activated and esterified by adding EDC (*1-Ethyl-3-(3-dimethylamino-propyl) carbodiimid*) and NHS (*N-Hydroxy-succinimid*). The Rhodamine is added to the activated alkyl-spacers and finally this fluorophore-solution is added to gold nanoparticles with a radius of 15 nm. Note that the order of subsequent steps of synthesis is essential to prevent direct binding of the amino groups of the Rhodamine to gold and to exclude reaction with the citrate stabilizer of the gold solution. For the final hybrid-samples the Rhodamine concentration has been adjusted to match a 50% surface coverage of the gold nanoparticles.

The results of the time resolved measurements are displayed in figure 12. As it is expected, the fluorescence decay times increase with increasing spacer length. A more detailed study on the distance dependence of fluorescence quenching is in preparation.[91]


## 5. BIOPHYSICAL APPLICATIONS

Systems of fluorescent dye molecules and noble metal nanoparticles are just on the dawn of possible biological and medical applications. In a sense we can distinguish between three groups of desired interaction regimes: Either no interactions between nanoparticles and fluorophores are intended, or the fluorescence intensity should be decreased or increased. The discussed time resolved experiments can give valuable insigths and possible ideas for further improvement for all three cases.

Little interaction between fluorophores and nanoparticles is desirable in applications where cell organelles shall be labeled simultaneously with dye molecules and nanoparticles. Such labeled cells have the double feature that they can be imaged in a fluorescence microscope as well as in an electron microscope. The first imaging technique gives deeper insight due to e.g. multicolor labeling while the second technique provides resolution unattainable with focused light. It is clear that in those applications one tries to attach both, fluorophores and nanoparticles at the same site of interest but striving for only little change of the fluorescence quantum yield. Indeed it has been found by Powell and co-workers, that fluorophores suffer only weakly from the nearby nanoparticles if they are separated by an antibody[92,93] (figure 13). The latter simultaneously serves as a specific binding site to organelles of interest inside a cell. If, however, the tag is designed the other way around, i.e. if the dye molecule is directly attached to the nanoparticle which in turn is fixed on the antibody, the system can not be used as a dual label due to fluorescence quenching.

In the second class of applications, the fluorescence of a donor type molecule should be quenched or un-quenched as the measuring signal. A pioneering experiment has been performed by Dubertret, Calame and Libchaber.[89,94] In a molecular beacon type sensor[95] the donor fluorescence is quenched as long as a loop like single stranded DNA (ssDNA) is closed (Figure 14). Hereby the donor and acceptor are attached at opposite ends of the ssDNA. The loop of the ssDNA is designed to specifically recognize a complementary single stranded target DNA. Dubertret and co-workers now replaced the organic acceptor by a gold nanoparticle. They found, that the quenching efficiency reaches 99.97% in the case of Rhodamine 6G as donor and a gold nanoparticle with a radius of 1.4 nm as acceptor. Similar molecular beacons comprising Rhodamine 6G as donors and the commonly used organic quencher DABCYL show only 97.67% fluorescence quenching efficiency. This shows that the parasitic fluorescence left over by the closed beacon is two orders of magnitude larger in the case of organic quenchers compared to gold nanoparticles, despite the fact that the nanoparticles in those experiments are only 1.4 nm in diameter. As it has been clarified by the time resolved fluorescence measurements discussed before, the extremely high quenching efficiency is due to both, resonant energy transfer and a change in radiative rate, whereby the latter process is dominating for nanoparticles of 2 nm or less in diameter.[2] Dubertret and co-workers argue that the extremely high quenching efficiency is responsible for their finding that the molecular beacon is sensitive to single mismatches in 20-mer ssDNA targets. Apart from biophysics, the quenching efficiency of gold nanoparticles may also be of interest in materials sciences. Recent experiments show that gold nanoparticles also serve as hyper-efficient energy acceptors for conjugated polymers.[96, 97]

In another biophysical application Peleg and co-workers studied nonlinear optical effects originating from dye molecules embedded into membranes onto which gold particles of 1 nm diameter have been attached using antigen- antibody linkers. The presence of the gold particles increases the second harmonic signal generated by the molecular dipoles by a factor of 2.5, presumably due to the enhanced field strength in the vicinity of the nanoparticle, but at the same time decreases the two photon induced fluorescence signal by a factor of 5-10.

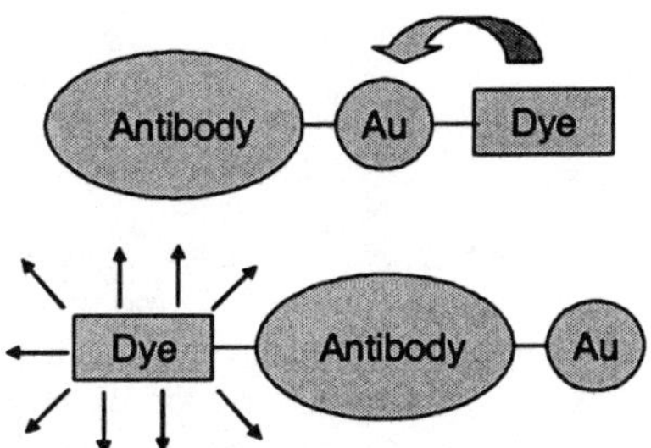


**Figure 13:** Dual label for fluorescence and electron microscopy. The label only works if the distance between fluorophore and gold nanoparticle is sufficiently large such that the fluorophore is not totally quenched. It is advantageous to use an antibody as spacer.

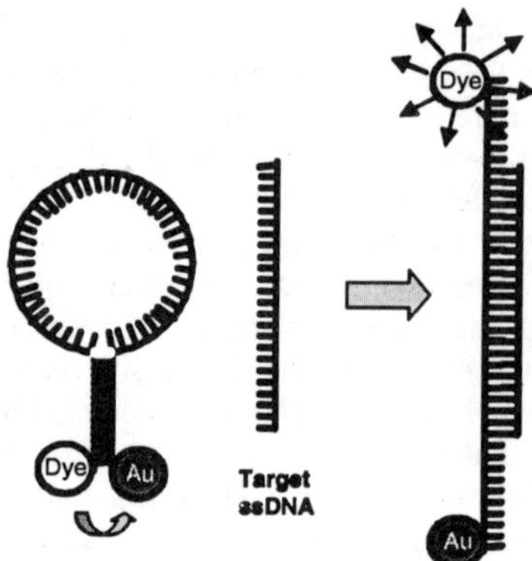

**Figure14:** Molecular beacon utilizing a gold nanoparticle as fluorescence quencher.

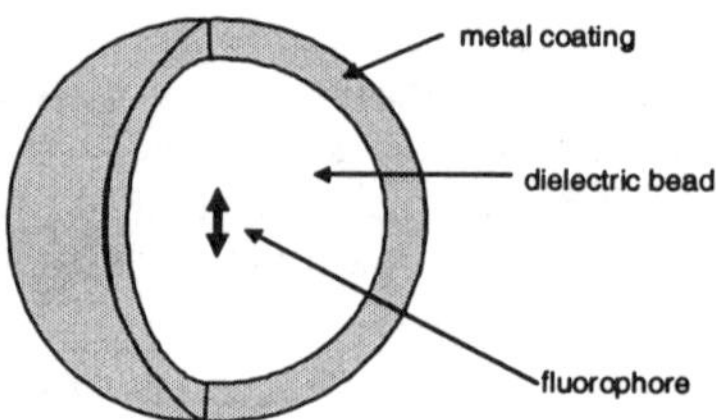



**Figure 15:** Dipole emitter placed inside a metallic sphere

Further intriguing applications of dye molecules interacting with metal nanoparticles rely on the anticipated increase of fluorescence in some very specific geometry. Provided that the emission spectrum of the fluorophore and the nanoparticle's absorption resonance do not overlap, but the excitation spectrum of the fluorophore and the particle plasmon energetically coincide, for certain distances an increase in fluorescence is expected.[98] Experimental evidence of increased fluorescence intensity is given by Wokaun[99] and coworkers, Aussenegg et al.[69] Sokolov et al.[100] or by Kulakovich et al.[101] and has been studied extensively by Gryczynski, Lakowicz and coworkers.[72-74, 98] In those studies, molecules have been placed above metallic particle films. This sample geometry has inevitably the disadvantage that a continuum of molecule-nanoparticle distances are present and uncontrolled resonant effects between adjacent clusters may play a role. Therefore, time resolved experiments at defined molecule-nanoparticle systems with varying distance are extremely desirable.

An interesting concept to achieve the latter regime i.e. to obtain increased radiative rates, has been suggested by Enderlein.[102] As depicted in figure 15, a dye molecule is fixed in the center of a metallic sphere. As it is away from the metal surface, energy transfer is less important, while the increase in field strength increases the molecular absorption, provided the absorption spectrum and the resonance of the sphere coincide in energy. Further, an increase of the radiative rate causes the molecule to statistically emit more photons before being transferred into a triplet state (if the intersystem crossing rate is not enhanced). This increase of the fluorescence rate with respect to the intersystem crossing rate leads to an increased total photon yield within the total lifetime of a molecule before bleaching, because the most prominent bleaching mechanism for organic molecules occurs from the triplet states.

Two photon fluorescence can also be enhanced by some orders of magnitude.[85, 103, 104] In those cases, the increased two photon absorption due to the field enhancement effect clearly outweighs the effects of fluorescence quenching. However, also these experiments had been carried out on submonolayer metallic films. Peleg et al. showed a decreased two photon fluorescence signal from fluorophores close to single gold nanoparticles.[105]

## 6. ACKNOWLEDGMENTS

Our results on, and our current understanding of the fluorescent properties of molecules in the vicinity of metal nanoparticles also rely on the help of and fruitful discussions with many colleagues and friends. We would like to thank all of them, in particular Marco Anni, Franz Aussenegg, David Gittins, Nancy Hecker, Anna Helfrich, Olaf Holderer, Alfred Leitner, Ulrich Lemmer, Stefano Levi, John Lupton, Martin Möller, Arne Morteani, Klaus Müllen, Thomas Niedereichholz, Abraham Nitzan, Wolfgang Parak, Gero von Plessen, David Reinhoudt, Ulrich Schubert, Markus Seitz, Vladimir Shalaev, Carsten Sönnichsen, Melanie de Souza, Joachim Spatz, Werner Stadler, and Frank van Veggel. We also acknowledge financial support by the Deutsche Forschungsgemeinschaft (DFG) through the Sonderforschungsbereich 486 and the Gottfried-Wilhelm-Leibniz Award.

## 7. REFERENCES

1. J. R. Lakowicz, Radiative decay engineering: Biophysical and biomedical applications, *Anal. Biochem.* **298**, 1-23 (2001).
2. E. Dulkeith, A. C. Morteani, T. Niedereichholz, T. A. Klar, J. Feldmann, S. A. Levi, F. C. J. M. Van Veggel, D. N. Reinhoudt, M. Möller, and D. I. Gittins, Fluorescence quenching of dye molecules near gold nanoparticles: Radiative and nonradiative effects, *Phys. Rev. Lett.* **89**, 203002 (2002).
3. U. Kreibig and M. Vollmer, *Optical properties of metal clusters* (Springer-Verlag, Berlin, 1995).
4. T. Klar, M. Perner, S. Grosse, W. Spirkl, G. von Plessen, and J. Feldmann, Surface-plasmon resonances in single metallic nanoparticles, *Phys. Rev. Lett.* **80**, 4249-4252 (1998).
5. B. Lamprecht, A. Leitner, and F. R. Aussenegg, SHG studies of plasmon dephasing in nanoparticles, *Appl. Phys. B* **69**, 419-423 (1999).
6. B. Lamprecht, J. R. Krenn, A. Leitner, and F. R. Aussenegg, Particle-plasmon decay-time determination by measuring the optical near-field's autocorrelation: Influence of inhomogeneous line broadening, *Appl. Phys. B* **69**, 223-227 (1999).
7. F. Stietz, J. Bosbach, T. Wenzel, T. Vartanyan, A. Goldmann, and F. Träger, Decay times of surface plasmon excitation in metal nanoparticles by persistent spectral hole burning, *Phys. Rev. Lett.* **84**, 5644-5647 (2000).
8. M. Perner, P. Bost, U. Lemmer, G. von Plessen, J. Feldmann, U. Becker, M. Mennig, M. Schmitt, and H. Schmidt, Optically induced damping of the surface plasmon resonance in gold colloids, *Phys. Rev. Lett.* **78**, 2192-2195 (1997).
9. M. Perner, S. Gresillon, J. Marz, G. von Plessen, J. Feldmann, J. Porstendorfer, K. J. Berg, and G. Berg, Observation of hot-electron pressure in the vibration dynamics of metal nanoparticles, *Phys. Rev. Lett.* **85**, 792-795 (2000).

10. C. Sönnichsen, S. Geier, N. E. Hecker, G. von Plessen, J. Feldmann, H. Ditlbacher, B. Lamprecht, J. R. Krenn, F. R. Aussenegg, V. Z.-H. Chan, J. P. Spatz, and M. Möller, Spectroscopy of single metallic nanoparticles using total internal reflection microscopy, *Appl. Phys. Lett.* **77**, 2949-2951 (2000).

11. C. Sönnichsen, T. Franzl, T. Wilk, G. von Plessen, J. Feldmann, O. Wilson, and P. Mulvaney, Drastic reduction of plasmon damping in gold nanorods, *Phys. Rev. Lett.* **88**, 077402 (2002).

12. J. Müller, C. Sönnichsen, H. von Poschinger, G. von Plessen, T. A. Klar, and J. Feldmann, Electrically controlled light scattering with single metal nanoparticles, *Appl. Phys. Lett.* **81**, 171-173 (2002).

13. S. Schultz, J. J. Mock, D. R. Smith, and D. A. Schultz, Nanoparticles based biological assays, *Journal of Clinical Ligand Assay* **22**, 214-216 (1999).

14. S. Schultz, D. R. Smith, J. J. Mock, and D. A. Schultz, Single-target molecule detection with nonbleaching multicolor optical immunolabels, *Proc. Natl. Acad. Sci. USA* **97**, 996-1001 (2000).

15. J. Yguerabide and E. E. Yguerabide, Light-scattering submicroscopic particles as highly fluorescent analogs and their use as tracer labels in clinical and biological applications, *Anal. Biochem.* **262**, 157-176 (1998).

16. J. Yguerabide and E. E. Yguerabide, Resonance light scattering particles as ultrasensitive labels for detection of analytes in a wide range of applications, *Journal of Cellular Biochemistry Supplement* **37**, 71-81 (2001).

17. P. Englebienne, Use of colloidal gold surface plasmon resonance peak shift to infer affinity constants from the interactions between protein antigens and antibodies specific for single or multiple epitopes, *The Analyst* **123**, 1599-1603 (1998).

18. P. Englebienne, A. Van Hoonacker, and J. Valsamis, Rapid homogeneous immunoassay for human ferritin in the cobas mira using colloidal gold as the reporter reagent, *Clinical Chemistry* **46**, 2000-2003 (2000).

19. M. D. Malinsky, K. L. Kelly, G. C. Schatz, and R. P. Van Duyne, Chain length dependence and sensing capabilities of the localized surface plasmon resonance of silver nanoparticles chemically modified with alkanethiol self-assembled monolayers, *J. Am. Chem. Soc.* **123**, 1471-1482 (2001).

20. D. Eck, C. A. Helm, N. J. Wagner, and K. A. Vaynberg, Plasmon resonance measurements of the adsorption and adsorption kinetics of a biopolymer onto gold nanocolloids, *Langmuir* **17**, 957-960 (2001).

21. P. Englebienne, A. V. Hoonacker, and M. Verhas, High-throughput screening using the plasmon resonance effect of colloidal gold nanoparticles, *The Analyst* **126**, 1645-1651 (2001).

22. N. Nath and A. Chilkoti, A colorimetric gold nanoparticle sensor to interrogate biomolecular interactions in real time on a surface, *Analytical Chemistry* **74**, 504-509 (2002).

23. A. J. Haes and V. D. R. P., A nanoscale optical biosensor: Sensitivity and selectivity of an approach based on the localized surface plasmon resonance spectroscopy of triangular silver nanoparticles, *J. Am. Chem. Soc.* **124**, 10596-10604 (2002).

24. J. C. Riboh, A. J. Haes, A. D. McFarland, C. R. Y. Yonzon, and R. P. Van Duyne, A nanoscale optical biosensor: Real-time immunoassay in physiological buffer enabled by improved nanoparticle adhesion, *J. Phys. Chem. B* **107**, 1772-1780 (2003).

25. J. J. Mock, D. R. Smith, and S. Schultz, Local refractive index dependence of plasmon resonance spectra from individual nanoparticles, *Nano Letters* **3**, 485-491 (2003).

26. G. Raschke, S. Kowarik, T. Franzl, C. Sönnichsen, T. A. Klar, J. Feldmann, A. Nichtl, and K. Kürzinger, Biomolecular recognition based on single nanoparticle light scattering, *Nano Letters*, in press (2003).

27. P. T. Leung and T. F. George, Molecular fluorescence spectroscopy in the vicinity of a microstructure, *Journal de Chimie Physique et de Physico-Chimie Biologique* **92**, 226-247 (1995).

28. H. Metiu, Surface enhanced spectroscopy, *Prog. Surf. Sci.* **17**, 153-320 (1984).

29. A. Sommerfeld, Über die Ausbreitung der Wellen in der drahtlosen Telegraphie, *Annalen der Physik* **28**, 665-736 (1909).

30. A. Sommerfeld, Über die Ausbreitung der Wellen in der drahtlosen Telegraphie, *Annalen der Physik* **81**, 1135-1153 (1926).

31. K. H. Drexhage, M. Fleck, H. Kuhn, F. P. Schäfer, and W. Sperling, Beeinflussung der Fluoreszenz eines Europium-Chelates durch einen Spiegel, *Ber. Bunsen. Phys. Chem.* **20**, 1179 (1966).

32. K. H. Drexhage, H. Kuhn, and F. P. Schäfer, Variation of the fluorescence decay time of a molecule in front of a mirror, *Ber. Bunsen-Ges. Phys. Chem.* **72**, 329 (1968).

33. H. Kuhn, Classical aspects of energy transfer in molecular systems, *J. Chem. Phys.* **53**, 101-108 (1970).

34. R. R. Chance, A. Prock, and R. Silbey, in *Adv. Chem. Phys.*, edited by I. Prigogine and S. R. Rice (Wiley, New York, 1978), p. 1-65.

35. B. N. J. Persson and N. D. Lang, Electron-hole-pair quenching of excited states near a metal, *Phys. Rev. B* **26**, 5409-5415 (1982).

36. B. N. J. Persson and S. Andersson, Dynamical process at surfaces: Excitation of electron-hole pairs, *Phys. Rev. B* **29**, 4382-4394 (1984).

37. J. Arias, P. K. Aravind, and H. Metiu, The fluorescence lifetime of a molecule emitting near a surface with small, random roughness, *Chem. Phys. Lett.* **85**, 404-408 (1982).

38. P. T. Leung, Z. C. Wu, D. A. Jelski, and T. F. George, Molecular lifetimes in the periodically roughened metallic surfaces, *Phys. Rev. B* **36**, 1457-1479 (1987).

39. T. Förster, Zwischenmolekulare Energiewanderung und Fluoreszenz, *Annalen der Physik* **2**, 55-75 (1948).

40. T. Förster, Experimentelle und theoretische Untersuchung des zwischenmolekularen Übergangs von Elektronenanregungsenergie, *Zeitschrift für Naturforschung* **4a**, 321-327 (1949).

41. D. L. Dexter, A theory of sensitized luminescence in solids, *J. Chem. Phys.* **21**, 836-850 (1953).

42. J. Gersten and A. Nitzan, Spectroscopic properties of molecules interacting with small dielectric particles, *J. Chem. Phys.* **75**, 1139-1152 (1981).

43. R. Ruppin, Decay of an excited molecule near a small metal sphere, *J. Chem. Phys.* **76**, 1681-1684 (1982).

44. G. Mie, Beiträge zur Optik trüber Medien, speziell kolloidaler Metallösungen, *Annalen der Physik* **3**, 25 (1908).

45. P. T. Leung, T. F. George, and Y. C. Lee, Limit of the image theory for the classical decy rates of molecules at surfaces, *J. Chem. Phys.* **86**, 7227-7229 (1987).

46. P. T. Leung and T. F. George, Dynamical analysis of molecular decay at spherical surfaces, *J. Chem. Phys.* **87**, 6722-6724 (1987).

47. P. T. Leung, S. K. Young, and T. F. George, Radiative decay rates for molecules near a dielectric sphere, *J. Phys. Chem.* **92**, 6206-6208 (1988).

48. Y. S. Kim, P. T. Leung, and T. F. George, Remark on the morphology-dependent resonance in the decay rate spectrum for molecules near a spherical surface, *Chem. Phys. Lett.* **152**, 453-456 (1988).

49. Y. S. Kim, P. T. Leung, and T. F. George, Classical decay rates for molecules in the presence of a spherical surface: A complete treatment, *Surface Science* **195**, 1-14 (1988).

50. T. Maniv and H. Metiu, Electron-gas effects in the spectroscopy of molecules chemisorbed at a metal-surface .I. Theory, *J. Chem. Phys.* **72**, 1996-2006 (1980).

51. T. Maniv and H. Metiu, Electrodynamics at a metal surface. II. Fresnel formulas for the electromagnetic field at the interface for a jellium model within the random phase approximation, *J. Chem. Phys.* **76**, 2697-2713 (1982).

52. T. Maniv and H. Metiu, Electrodynamics at a metal surface. III. Reflectance and the photoelectron yield of a thin slab, *J. Chem. Phys.* **76**, 696-708 (1982).

53. G. E. Korzeniewski, T. Maniv, and H. Metiu, Electrodynamics at metal surfaces. IV. The electric fields caused by the polarization of a metal surface by an oscillating dipole, *J. Chem. Phys.* **76**, 1564-1573 (1982).

54. G. Korzeniewski, T. Maniv, and H. Metiu, The interaction between an oscillating dipole and a metal surface described by a jellium model and the random phase approximation, *Chem. Phys. Lett.* **73**, 212-215 (1980).

55. P. Gies and G. R. R., Density.Functional theory for the dynamical response of molecules adsorbed on metal surfaces, *Phys. Rev. B* **37**, 10020-10028 (1988).

56. S. Corni and J. Tomasi, Lifetimes of electronic excited states of a molecule close to a metal surface, *J. Chem. Phys.* **118**, 6481-6494 (2003).

57. W. Ekardt and Z. Penzar, Nonradiative lifetime of excited states near a small metal particle, *Phys. Rev. B* **34**, 8444-8448 (1986).

58. R. Fuchs and F. Claro, Multipolar response of small metallic spheres: Nonlocal theory, *Phys. Rev. B* **35**, 3722-3727 (1987).

59. P. T. Leung, Decay of molecules at spherical surfaces: Nonlocal effects, *Phys. Rev. B* **42**, 7622-7625 (1990).

60. C. Girard and F. Hache, Effective polarizablility of a molecule physisorbed on a spherical metal particle: Nonlocal effects, *Chem. Phys.* **118**, 249-264 (1987).

61. R. Rossetti and L. E. Brus, Time resolved molecular electronic energy transfer into a silver surface, *J. Chem. Phys.* **73**, 572-577 (1980).

62. P. Avouris and J. E. Demuth, Electronic excitations of benzene, pyridine, and pyrazine adsorbed on ag(111), *J. Chem. Phys.* **75**, 4783-4794 (1981).

63. J. E. Demuth and P. Avouris, Lifetime broadening of excited pyrazine adsorbed on ag(111), *Phys. Rev. Lett.* **47**, 61-63 (1981).

64. P. M. Whitmore, H. J. Robota, and C. B. Harris, Mechanisms for electronic energy transfer between molecules and metal surfaces: A comparison of silver and nickel, *J. Chem. Phys.* **77**, 1560-1568 (1982).

65. R. Rossetti and L. E. Brus, Time resolved energy transfer from electronically excited 3b3u pyrazine molecules to planar ag and au surfaces, *J. Chem. Phys.* **76**, 1146-1149 (1982).

66. P. M. Whitmore, H. J. Robota, and C. B. Harris, Electronic energy transfer from pyrazine to a silver(111) surface between 10 and 400 a, *J. Chem. Phys.* **76**, 740-741 (1982).

67. K. Kuhnke, R. Becker, M. Epple, and K. Kern, C60 exciton quenching near metal surfaces, *Phys. Rev. Lett.* **79**, 3246-3249 (1997).

68. S. J. Strickler and R. A. Berg, Relationship between absorption intensity and fluorescence lifetime of molecules, *J. Chem. Phys.* **37**, 814-822 (1962).

69. F. R. Aussenegg, A. Leitner, M. E. Lippitsch, H. Reinisch, and M. Riegler, Novel aspects of fluorescence lifetime for molecules positioned close to metal surfaces, *Surface Science* **189/190**, 935-945 (1987).
70. H. Imahori, H. Norieda, Y. Nishimura, I. Yamazaki, K. Higuchi, N. Kato, T. Motohiro, H. Yamada, K. Tamaki, M. Arimura, and Y. Sakata, Chain length effect on the structure and photochemical properties of self-assembled monolyayers of porphyrins on gold electrodes, *J. Phys. Chem. B* **104**, 1253-1260 (2000).
71. H. Imahori, M. Arimura, T. Hanada, Y. Nishimura, I. Yamazaki, Y. Sakata, and S. Fukuzumi, Photoactive three-dimensional monolayers: Porphyrin-alkanethiolate-stabilized gold clusters, *J. Am. Chem. Soc.* **123**, 335-336 (2001).
72. C. D. Geddes, H. Cao, I. Gryczynski, Z. Gryczynski, J. Fang, and L. J. R., Metal-enhanced fluorescence (mef) due to silver colloids on a planar surface: Potential applications of indocyanine green to in vivo imaging, *J. Phys. Chem.* **107**, 3443-3449 (2003).
73. J. R. Lakowicz, B. Shen, Z. Gryczynski, S. D'Auria, and I. Gryczynski, Intrinsic fluorescence from DNA can be enhanced by metallic particles, *Biophys. Biochem. Res. Comm.* **286**, 875-879 (2001).
74. I. Gryczynski, J. Malicka, E. Holder, N. DiCesare, and J. R. Lakowicz, Effects of metallic silver particles on the emission properties of ru(bpy)3)2+, *Chem. Phys. Lett.* **372**, 409-414 (2003).
75. F. R. Aussenegg, S. Draxler, A. Leitner, M. E. Lippitsch, and M. Riegler, Picosecnond investigations on the fluorescence properties of adsorbed dye molecules, *J. Opt. Soc. Am. B.* **1**, 456-457 (1984).
76. A. Leitner, M. E. Lippitsch, S. Draxler, M. Riegler, and F. R. Aussenegg, Fluorescence properties of dyes adsorbed to silver islands, investigated by picosecond techniques, *Appl. Phys. B* **36**, 105-109 (1985).
77. S. A. Levi, *Supramolecular chemistry at the nanometer level*, PhD thesis (University of Twente, Enschede, 2001).
78. A. Mooradian, Photoluminescence of metals, *Phys. Rev. Lett.* **22**, 185-187 (1969).
79. G. T. Boyd, Z. Yu, and Y. R. Shen, Photoinduced luminescence from the noble metals and its enhancement on roughened surfaces, *Phys. Rev. B* **33**, 7923-7936 (1986).
80. J. P. Wilcoxon, J. E. Martin, F. Parsapour, B. Wiedenmann, and D. F. Kelley, Photoluminescence from nanosized gold clusters, *J. Chem. Phys.* **1998**, 9137-9143 (1998).
81. M. B. Mohammed, V. Volkov, S. Link, and M. A. El-Sayed, The 'lightning' gold nanorods: Fluorescence enhancement of over a million compared to the gold metal, *Chem. Phys. Lett.* **317**, 517-523 (2000).
82. O. Varnavski, R. G. Ispasoiu, L. Balogh, D. Tomalia, and T. Goodson III, Ultrafast time-resolved photoluminescence from novel metal-dendrimer nanocomposites, *J. Chem. Phys.* **114**, 1962-1965 (2001).
83. N. Nilius, N. Ernst, and H. J. Freund, Tip influence on plasmon excitations in single particles in an stm, *Phys. Rev. B* **65**, 115421 (2002).
84. Y. N. Hwang, D. H. Jeong, H. J. Shin, D. Kim, S. C. Jeoung, S. H. Han, J. S. Lee, and G. Cho, Femtosecond emission studies on gold nanoparticles, *J. Phys. Chem. B* **106**, 7581-7584 (2002).
85. I. Gryczynski, J. Malicka, Y. Shen, Z. Gryczynski, and J. R. Lakowicz, Multiphoton excitation of fluorescence near metallic particles: Enhanced and lokalized excitation, *J. Phys. Chem. B* **106**, 2191-2195 (2002).
86. M. R. Beverluis, A. Bouhelier, and L. Novotny, Photoluminescence from sharp gold tips, *CLEO / QELS Proceedings*, QTUD4 (2003 (6)).
87. E. Dulkeith, T. A. Klar, G. von Plessen, and J. Feldmann, to be published.
88. S. N. Smith and R. P. Steer, The photophysics of lissamine rhodamine-b sulphonyl chloride in aqueous solution: Implications for fluorescent proteine-dye conjugates, *J. Photochem. Photobiol. A* **139**, 151-156 (2001).
89. B. Dubertret, M. Calame, and A. J. Libchaber, Single-mismatch detection using gold-quenched fluorescent oligonucleotides, *Nature Biotech.* **19**, 365-370 (2001).
90. T. Niedereichholz, *Fluoreszenzauslöschung von Farbstoffen auf der Oberfläche von Goldnanopartikeln - strahlende und nichtstrahlende Effekte*, diploma thesis (Ludwig-Maximilians-University, Munich, 2002).
91. E. Dulkeith, D. S. Koktysh, W. Parak, T. A. Klar, and J. Feldmann, to be published.
92. R. D. Powell, C. M. R. Halsey, D. L. Spector, S. L. Kaurin, J. McCann, and J. F. Hainfeld, A covalent fluorescent-gold immunoprobe: Simultaneous detection of a pre-mrna splicing factor by light and electron microscopy, *J. Histochem. Cytochem.* **45**, 947-956 (1997).
93. R. D. Powell, C. M. R. Halsey, and J. F. Hainfeld, Combined fluorescent and gold immunoprobes: Reagents and methods for correlative light and electron microscopy, *Microscopy Research and Technique* **42**, 2-12 (1998).
94. B. Dubertret, M. Calame, and A. J. Libchaber, Single-mismatch detection using gold-quenched fluorescent oligonucleotides, erratum, *Nature Biotech.* **19**, 680-681 (2001).
95. S. Tyagi and F. R. Kramer, Molecular beacons: Probes that fluoresce upon hybridization, *Nature Biotech.* **14**, 303-308 (1996).
96. O. Holderer, *Manipulation der optischen Eigenschaften von organischen Halbleitern mit Hilfe von Metall-Nanopartikeln*, diploma thesis (Ludwig-Maximilians-University, Munich, 1999).

97. C. Fan, S. Wang, J. W. Hong, G. C. Bazan, K. W. Plaxco, and A. J. Heeger, Beyond superquenching: Hyper-efficient energy transfer from conjugated polymers to gold nanoparticles, *Proc. Natl. Acad. Sci. U. S. A.* **100**, 6297-6301 (2003).
98. J. R. Lakowicz, Y. Shen, S. D'Auria, J. Malicka, J. Fang, Z. Gryczynski, and I. Gryczynski, Radiative decay engineering 2: Effects of silver island films on fluorescence intensity, lifetimes, and resonance energy transfer, *Anal. Biochem.* **301**, 261-277 (2002).
99. A. Wokaun, H. P. Lutz, A. P. King, U. P. Wild, and R. R. Ernst, Energy transfer in surface enhanced luminescence, *J. Chem. Phys.* **79**, 509-514 (1983).
100. K. Skokolov, G. Chumanov, and T. M. Cotton, Enhancement of molecular fluorescence near the surface of colloidal metal films, *Analytical Chemistry* **70**, 3898-3904 (1998).
101. O. Kulakovich, N. Strekal, A. Yaroshevich, S. Maskevich, S. Gaponenko, I. Nabiev, U. Woggon, and M. Artemyev, Enhanced luminescence of cdse quantum dots on gold colloids, *Nano Letters* **2**, 1449-1452 (2002).
102. J. Enderlein, Theoretical study of single molecule fluorescence in a metallic nanocavity, *Appl. Phys. Lett.* **80**, 315-317 (2002).
103. A. M. Glass, A. Wokaun, J. P. Heritage, J. G. Bergman, P. F. Liao, and D. H. Olson, Enhanced two-photon fluorescence of molecules adsorbed on silver particle films, *Phys. Rev. B* **24**, 4906-4909 (1981).
104. W. Wenseleers, F. Stellacci, T. Meyer-Friedrichsen, T. Mangel, C. A. Bauer, S. J. K. Pond, S. R. Marder, and J. W. Perry, Five orders-of-magnitude enhancement of two-photon absorption for dyes on silver nanoparticle fractal clusters, *J. Phys. Chem. B* **106**, 6853-6863 (2002).
105. G. Peleg, A. Lewis, M. Linial, and L. M. Loew, Nonlinear optical measurement of membrane potential around single molecules at selected cellular sites, *Proc. Natl. Acad. Sci. USA* **96**, 6700-6704 (1999).

# 9

# COPPER-COATED SELF-ASSEMBLED MONOLAYERS: ALKANETHIOLS AND PROSPECTIVE MOLECULAR WIRES

Paula E. Colavita, Paul Miney, Lindsay Taylor, Michael Doescher, Annabelle Molliet, John Reddic, Jing Zhou, Darren Pearson, Donna Chen, and Michael L. Myrick

## 1. INTRODUCTION

The study of the effects of metal overlayers on organic self-assembled monolayers (SAMs) is of interest in many scientific areas. The self-assembly process leads to a well-defined substrate whereas the numerous combinations of end-groups and metals allows for the fine-tuning of the chemistry at the monolayer/metal interface. These unique features make metal/SAM interfaces a valuable tool for probing fundamental processes. SAM/metal interfaces, have been proposed as templates to elucidate details on, for example, metal/"organic host" interactions which are known to occur in biologically active metallic reaction sites [1]; reaction mechanisms at organic/metal/gas interfaces, such as those involved in heterogeneous catalysis and environmental chemistry [2]; the solvation and electron-transfer reactions of metals in molecular solvents or within molecular clusters [3, 4]; and the insertion of zero-valent metals into chemical bonds frequently found in homogeneous catalysis reactions [3, 4, 5, 6].

Metal/organic-molecule/metal "sandwich" structures have also been used as ideal systems to study the fundamental mechanisms behind the enhancement of infrared and Raman signals of organic molecules that has been observed whenever a metallic overlayer is deposited as clusters on the organic surface. The first observation of this phenomenon dates to the early 1980s [7], however, it is only recently that a more detailed investigation into the factors influencing these enhancements has been undertaken [8, 9, 10, 11]. The improved signal to noise ratio can be used to facilitate the study of surface and interface dynamic processes by means of infrared spectroscopy [12]. Such enhancements are relevant also because of the possible analytical applications of these nanoscale sandwich heterostructures.

Of direct technological significance are studies that use SAM/metal interfaces to model polymer/metal contacts [3, 13, 14]. It has been pointed out that it is not possible to obtain simple metal-polymer contacts whenever metal leads are patterned by vapor-deposition in the fabrication of organic electronic devices, as there always is an intermediate layer of varying composition [15]. Functional groups on the surface of

*Topics in Fluorescence Spectroscopy, Volume 8: Radiative Decay Engineering*
Edited by Geddes and Lakowicz, Springer Science+Business Media, Inc., New York, 2005

polymers used for organic electronic devices, for example light-emitting diodes (LEDs), can affect the nature of the adhesion of the evaporated metal to the organic phase [13, 14, 16], thus influencing the resolution of lead patterns in polymer-based devices, as well as their performance. The direct study of polymer/metal interfaces is challenging because of difficulties in preparing well-characterized polymer surfaces that have a known and controllable distribution of functional groups. SAMs of various terminal groups can therefore be used to provide surfaces with the aforementioned qualities, in order to optimize the process of polymer metallization. Furthermore, studies on the stability of SAM/metal overlayer and the permeability of the SAM to metal penetration can be translated into a better understanding of the expected practical lifetime of organic devices [2, 17] and, more importantly, the physical and chemical nature of the interface involved in the charge injection at the polymer-lead contact [15].

In our laboratory, studies in the area of metal/SAM interactions were prompted by our previous work on the charge transport properties of metal/conjugated-oligomer/metal junctions [18]. It has been suggested that conjugated oligomers are good candidates for the fabrication of molecular electronic devices, given their $\pi$-orbital system that extends over almost the entire length of the molecule [19, 20]. This extended conjugation is thought to promote charge mobility in a way analogous to graphite conductive properties. Though there are numerous reports in the literature regarding the fabrication of electronic devices using conjugated oligomers [21, 22, 23, 24], the characterization of the charge transport through these molecules remains challenging. One of the methods of investigating their conductivity is to construct metal-oligomer-metal junctions (MOMs), by self assembling the molecules onto noble metals, such as gold, silver and platinum, via a terminal thiol group, and then forming a second contact by metal evaporation [19, 21, 22, 23, 24]. In our laboratory, the electronic properties of thus assembled Au/oligomer/Cu junctions were investigated by way of electrochemical oxidation of the copper overlayer and the barrier heights to charge injection at the SAM/Cu interface were measured. The results obtained, as well as those of other MOM experiments reported in the literature, strongly depend on the integrity and structure of the SAM after the evaporation process. Perturbations induced by either the metal deposition process or the physico-chemical interactions between the organic molecules and the metal can play an important role in the conductive properties measured through a MOM junction. The conduction properties of a molecular aggregate within such a junction can potentially be different from those in the monolayer as deposited, given that the organization, packing and resulting inter-molecular coupling can be altered by the formation of the second metallic connection. Hence, for both the purpose of characterizing the conductivity of so-called *molecular wires*, and the use of these molecules as part of electronic components, the understanding of SAM/metal interactions is vital.

Several experimental techniques have been applied to the study of metal overlayers on SAMs, providing insights into different aspects of the SAM/metal interactions. X-ray Photoelectron Spectroscopy (XPS) and UV Photoelectron Spectroscopy (UPS) have been widely used as a means to provide information on chemical bonds and elemental composition of the interface at low metal coverage [2 and refs. therein, 3, 4, 13, 17, 25 and refs. therein, 26, 27, 28, 29, 30]. Ion Scattering Spectroscopy (ISS) has been used to supply information on the compositional depth profile, thus showing the solvation of metallic clusters within the organic matrix, or the degree of diffusion through the SAM towards the metallic substrate [2 and refs. therein, 17, 25 and refs. therein, 26, 28, 29]. Ellipsometry has been widely used to characterize freshly deposited self-assembled monolayers [31, 32, 33, 34], however its application to the study of metal overlayers has been somewhat more limited. This technique can also be used to monitor variations in the optical constants and thicknesses of both the organic film [35] and the overlayer upon deposition, thus providing evidence of changes in the SAM coverage and in the nature of the deposited overlayer (dielectric or metallic) [13]. Time-of-flight secondary ion mass spectrometry (ToF-SIMS) has been successfully applied to the study of SAM/metal interactions in the past few years [3, 4, 6, 13, 36]. This technique allows for the detection and identification of the cluster ions that result from the bombardment of the M1/SAM/M2 sample with heavy ions (M1 and M2 are the substrate and overlayer metals respectively). The diffusion of the overlayer metal through the SAM can be revealed by the appearance of cluster ions, of the form $M1_xM2_yS$ or $M2_xS$, in the mass spectrum. The interaction chemistry between the overlayer and the molecule terminal group can be evidenced by the disappearance of cluster ions associated with the pristine terminal group and/or the appearance of cluster ions formed by M2 and the terminal group [3, 4, 6, 13, 36]. Fourier-Transform Reflection Absorption Infrared Spectroscopy (FT-RAIRS) is an extremely versatile technique and has been applied also to the study of SAM/metal interactions. The results that can be obtained using this technique will form the bulk of this chapter and will be discussed in more depth in subsequent sections.

Other experimental techniques have been used to study the SAM/metal interface, but have not been used to the same extent. For example, there are some reports of Scanning Tunneling Microscopy (STM) [2, 37] and Atomic Force Microscopy (AFM) [38, 39] imaging of metal cluster formation on a monolayer upon deposition. STM and AFM provide excellent lateral resolution for the topographical characterization of the interface. The topography can then be correlated to the tendency of the metal to either react with the molecules resulting in the "wetting" of the organic surface, or to form clusters by favoring metal-metal as opposed to metal-organic interactions. Near-Edge X-ray Absorption Fine Structure Spectroscopy (NEXAFS) has also been proposed as a suitable tool, given its ability to probe selectively the local structure in the vicinity of the substrate metal atoms, the atoms of the organic molecule and the overlayer metal atoms. Although to our knowledge, NEXAFS has not been applied to the study of metal overlayers on self-assembled monolayers, it has been successfully applied to the study of closely related system, i.e. the reactivity and penetration of magnesium through ordered organic monolayers formed by vacuum sublimation [40]. Two-photon laser spectroscopy and second-harmonic generation spectroscopy (SHG) have also been used to observe the clustering of metal atoms and their migration through the monolayer. A great advantage of these techniques is that, being non destructive, they can monitor the diffusion of the deposited metal [41, 42, 43] and its cluster growth [38, 44] in real time. Cyclic voltammetry has also been used to study the effect of metal coating on the passivation of gold electrodes by SAMs, though the only report of its use results from the work in our laboratory on alkanethiols [30] and conjugated oligomers [35].

In the following sections the general trends observed for the interactions between metal overlayers and SAMs will be illustrated with reference to the work of several research groups. The results obtained in this laboratory on the perturbations that vapor deposited copper films induce on self-assembled monolayers of dodecanethiol and octadecanethiol on gold, and how these results correlate with previous reports in the literature will also be described. The focus will be mainly on the use of RAIRS to investigate the effect of the copper coating on well-characterized systems, such as dodecanethiols (DT) and octadecanethiols (ODT) [30]. A description of our ongoing work on a conjugated-oligomer/copper system will also be included [35].

## 2. COPPER OVERLAYERS ON ALKANETHIOL SELF-ASSEMBLED MONOLAYERS

### 2.1. General Factors Affecting the Behavior of Metals Deposited onto Self-Assembled Monolayers

There is an abundance of literature on the effects that metal overlayers have on alkanethiol SAMs. Different combinations of metal, terminal group and chain lengths have been investigated using one or more of the aforementioned techniques. Most of the literature focuses on the effect that thin metal overlayers (1-20 nm thicknesses) have on the SAM in terms of the tendency of the metal to "wet" the SAM surface, coalesce on the SAM surface forming clusters, and/or diffuse through the thickness of the organic layer. Which of these occurs depends on the competition among several factors: (1) reactivity of the terminal group; (2) reactivity of deposited metal; (3) tendency of the metal to form clusters; (4) deposition rate, and finally (5) the temperature of the metal-organic-metal structure. The importance of these variables will be discussed in the following paragraphs.

The first issue to consider in rationalizing the interactions between the deposited metal and the SAM, encompasses the choice of both the terminal group (TG) of the assembled thiol, and the metal to be deposited. The reactivity of both the terminal group and the metal will determine the extent of the diffusion of the metal through, and over, the monolayer surface. The observed trend is that of increased penetration through the monolayer whenever the TG-metal interactions are weak, and reduced penetration accompanied by improved wetting in the case of strong TG-metal interactions (see Table 1). Summaries of these interactions can be obtained in [2] and 25. Typical metals considered to be very reactive are titanium and chromium. It has been shown that for the most common alkanethiol TGs, such as $-COOCH_3$, $-CH_2OH$, $CH_3$ etc., these metals display high reactivity. For instance, XPS experiments show that deposition of Ti on carboxyester and hydroxyl groups results in the formation of oxide bonding, whereas its deposition on nitriles leads to the formation of nitrides [14]. These reactions occur at the expense of the terminal groups, whose characteristic XPS signals disappear at even low metal coverages. Correspondingly, penetration of Ti through these layers is minimal, as it tends to react rapidly at the SAM/vacuum interface [14]. The reactivity of the TG is also important; in the case of Ti, for example, whenever the TG is a methyl group the associated XPS signal is only attenuated by the deposition of the metal [14]. However, for any of the TGs investigated it was found that for coverages higher than 0.5 nm, Ti atoms tend to form chemical bonds even with the hydrocarbon chain, as demonstrated by the appearance of carbide XPS peaks [14].

Among the most typical TGs, those that contain heteroatoms are considered of medium or high reactivity, with the carbonyl of acids and esters the most reactive moiety, and with the methyl group being the least reactive one [14, 25]. To our knowledge there are no comparative studies on more complex TGs, such as electroactive ferrocene moieties for example. Among the typically deposited metals, aluminum, copper, and silver are considered of intermediate to low reactivity [25]. In particular, the reactivity of Al is known to vary drastically with differing TGs. It can (a) react at the vacuum/SAM interface yielding the reduction of carbonyl groups [13, 27]; (b) yield a bond insertion in the case of alcohols [4] and (c) can show no reactivity whatsoever with methyl terminated alkanethiols, thus resulting in the penetration of Al atoms to the thiolate/metal interface [13]. An even more remarkable example of the complex interplay between competing

Table 1: Summary of results reported in the literature on reactivity and penetration of metals deposited on alkanethiol SAMs with various terminal groups.

| Metal | Terminal group | Type of bonding | Penetration |
|---|---|---|---|
| Ti[14] | COOCH$_3$, CH$_2$OH | oxide & carbide | low |
|  | CH$_3$ | carbide | no |
|  | CN | nitride & carbide | no |
| Cr[25,45] | COOCH$_3$ | oxide & carbide | no |
|  | CH$_3$ | carbide | no |
| Al[4,13,27] | COOCH$_3$ | Al-O-COCH$_3$ species | no |
|  | COOH | Al-carboxylate | no |
|  | OH | bond insertion O-Al-H | no |
|  | OCH$_3$ | complexation with OCH$_3$ | yes |
|  | CH$_3$ | no reaction | yes |
| Cu[17,26] | COOH | oxide | yes |
|  | COOCH$_3$ | oxide | yes |
|  | CH$_3$ | no reaction | yes |
| Ag[17,25] | COOH | no reaction | yes |
|  | CH$_3$ | no reaction | high |

processes in Al overlayer deposition, is its behavior in the case of methoxy terminated SAMs, where both the formation of metallic clusters at the SAM outer surface, and the penetration through the monolayer take place [4].

The reactivity of copper has also been found to vary depending on the specific TG. FT-RAIRS, XPS and ISS measurements have shown that it can react with the carbonyl of carboxylic acids [2, 17, 26, 37, 45]. This interaction is not strong, however, and, after copper deposition at slow rates (0.1 nm/min), XPS signals of both partially oxidized and metallic Cu atoms appear [26]. The formation of a copper-carboxylate species is also confirmed by the appearance of its characteristic peaks in the SAM FT-RAIRS spectrum [17, 37, 45]. Over time, the C vs. Au and Cu vs. Au XPS signal ratios increase, indicating the copper which does not react with a TG tends to migrate through and underneath the organic layer [26]. This evolution is also in agreement with ISS data [26]. A weaker interaction occurs in the case of methyl terminated SAMs, where the rate of penetration is high enough that 1.0 nm of Cu completely diffuses to the Au surface within 2 h. Carboxylic methylesters interactions display an intermediate behavior, showing formation of small quantities of Cu(I) and penetrating through the SAM in approximately 7 h [17, 26].

Silver displays even faster rates of penetration and lower reactivities than copper [17, 25]; for example, 1.0 nm Ag completely diffuses to the Au surface within 5 min [17]. This contrast has been attributed to the differring tendencies of these two metals to form clusters, which in turn has been ascribed to the different strength of their metal-metal interactions [17]. Smaller clusters or single atoms diffuse more readily through the organic layer, whereas larger clusters tend to remain on the SAM surface. This explanation appears to be confirmed by the importance of deposition rate and temperature on the rate of penetration. In fact, relatively higher deposition rates lead to lower rates of

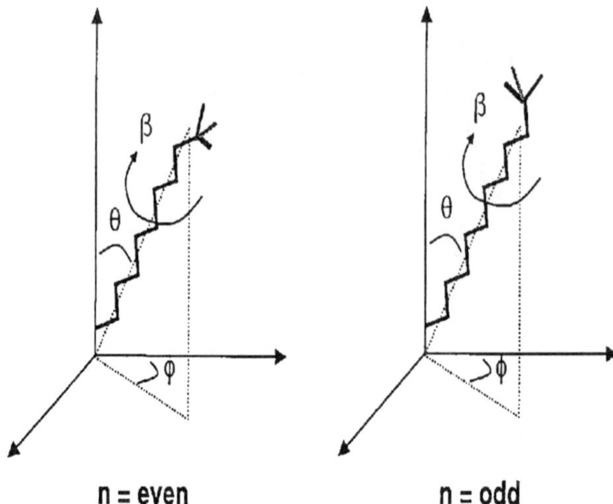



**Figure 1**. Structural parameters that describe the geometry of a chemisorbed alkanethiol $CH_3(CH_2)_nSH$. The tilt angle $\theta$ describe the orientation relative to the surface normal; the azimuth angle $\phi$ describes the orientation relative to the $x$-axis on the surface plane; the twist angle $\beta$ describes the torsion of the acylic chain along its axis. The two figures show the different orientation the methyl end-group adopts according to the parity of the number of atoms in the chain when the thiol self-assembles on Au surfaces [57].

penetration, since cluster growth on the SAM surface is favored [26]. It has also been shown that cooling the sample during metal deposition slows down the penetration rate, whereas the exposure to X-rays, such as those used to collect XPS spectra, accelerates it [26].

Since it will be useful for the discussion of the results obtained in this laboratory with copper overlayers, a brief introduction to the interpretation of RAIRS spectra of alkanethiol SAMs will follow. For a more thorough review of this subject the book by Tolstoy et al. [46] and the work of Parikh and Allara [47] are recommended.

## 2.2. RAIRS Spectra of Alkanethiol SAMs in the C-H Stretching Region

RAIRS can be used as a tool to determine structural parameters in the organization of alkanethiol chains absorbed on a metallic substrate averaged over the probing beam spot-size. Figure 1 shows the three angles that describe the orientation of a chain absorbed on a substrate in an *all-trans* conformation relative the the laboratory reference system: $\theta$ is the tilt angle between the chain axis and the normal surface; $\phi$ is the azimuth angle between the chain projection on the surface plain and the $x$-axis; $\beta$ is the twist angle indicating the torsion of the chain around its axis. Figure 1 shows the RAIRS spectrum of an octadecanethiol chain ($HS(CH_2)_{17}CH_3$) deposited on gold, and Table 2 reports the assignments of these IR peaks, the orientation of the transition dipole relative to the molecular geometry and the position of these bands in the solid and liquide phase for the corresponding $n$-alkane octadecane.

Table 2: Frequencies of the C-H stretching modes of bulk alkanethiols and $C_{12}SH$ and $C_{18}SH$ SAMs on Au. The frequencies for alkanethiols in condensed phases are those reported by Porter et al. [50] for $CH_3(CH_2)_{21}SH$ and $CH_3(CH_2)_7SH$, for solid and liquid respectively.

| | | $C_nSH$ in bulk phases $(cm^{-1})$[56] | | peak positions for $C_{12}SH$ SAM[55] $(cm^{-1})$ | | peak positions for $C_{18}SH$ SAM[80] $(cm^{-1})$ | |
|---|---|---|---|---|---|---|---|
| assignm[49] | TM direction[47] | solid | liquid | as deposited | after Cu evaporation | as deposited | after Cu evaporation |
| $\nu$,i.p. ($CH_3$) | $\perp$ C-CH$_2$ bond, i.p. CCC plane | - | - | 2964 | 2966 | 2964 | 2965 |
| $\nu$,o.p. ($CH_3$)[a] | $\perp$ CCC plane | 2956 | 2957 | . | | . | |
| $\nu_s$ ($CH_3$, FR) | ǀ C-CH$_3$ bond, i.p. CCC plane | - | - | 2937 | . | 2937 | - |
| $\nu_a$ ($CH_2$) | $\perp$ CCC plane | 2918 | 2924 | 2918 | 2924 | 2918 | 2924 |
| $\nu_s$ ($CH_3$, FR) | ǀ C-CH$_3$ bond, i.p. CCC plane | - | - | 2878 | . | 2877 | - |
| $\nu_s$ ($CH_2$) | $\perp$ chain axis, i.p. CCC plane | 2851 | 2855 | 2850 | 2852 | 2849 | 2852 |

[a] This peak appears in alkanethiol SAMs as a shoulder at room temperature; it becomes well-defined only at low temperatures (80 K [56].

$\nu$ = stretching; a = asymmetric; s = symmetric; i.p. = in plane; o.p.= out of plane; FR = Fermi Resonance enhanced; TM = transition moment; $\perp$ = perpendicular; ǁ = parallel.

Because of the reflection of the electromagnetic wave at the metallic surface, only those modes that have a component perpendicular to the metallic substrate will be active in the RAIRS spectrum [46, 48]. The intensity of the corresponding peaks will be proportional to the square modulus of the perpendicular component [46, 49]. Modes that lie along the chain axis will therefore be enhanced by a decrease in $\theta$, the tilt angle of the molecule. Modes that are perpendicular to the planes defined by three consecutive C atoms (CCC plane) in the chain, will be active only if the twist angle $\beta$ allows for a component along the $z$-axis. Modes that are parallel to the CCC plane and perpendicular to the chain axis will follow the opposite trend. The dependence of absorption intensities on the orientation of a chain can be used to determine the tilt [47, 49] and twist [46, 49] angles of the hydrocarbon chains in highly crystalline SAMs.

The C-H stretching region RAIRS spectra usually display five well-defined peaks (see Figure 1) that can be ascribed to the molecular modes on Table 2. The bands corresponding to the methylene vibrations are proportional to the number of $CH_2$ groups in the chain [50] and are of high diagnostic power. It has been observed that for polymethylene chains in the crystalline state, the peak wavenumbers of both the symmetric and asymmetric $CH_2$ stretchings appear at lower values than in the liquid phase, shifting approximately from 2856 cm$^{-1}$ and 1926 cm$^{-1}$, to 2850 cm$^{-1}$ and 1918 cm$^{-1}$ respectively [46, 49, 50, 51, 52, 53, 54]. This difference has been explained by the increased population of *gauche* conformations in the liquid, compared to the solid phase [46, 49, 50, 51, 52, 53, 54]. Rotation around the molecular axis of chain segments draws methylene hydrogen atoms to the CCC plane of neighboring $CH_2$ groups, thus allowing for a more favorable coupling and a consequent increased force constant for the C-H stretching [46]. The same difference has been observed between these bands in the case of long alkanethiols (more than 15 C atoms) and short alkanethiols (14 or less C atoms)

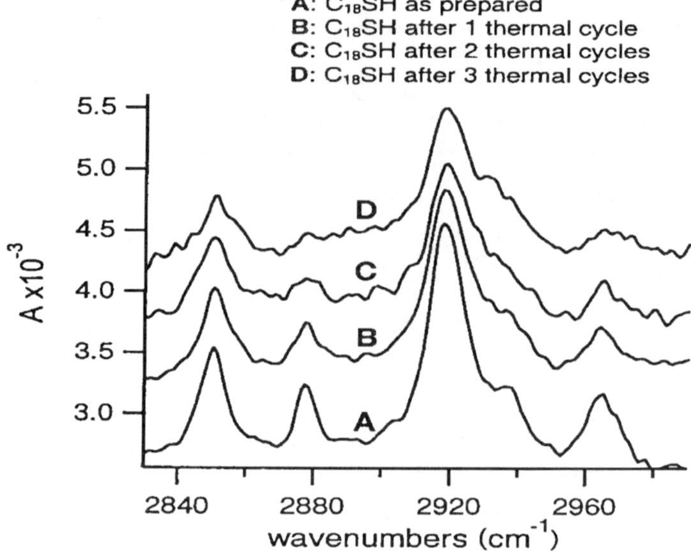



**Figure 2**. RAIRS spectra of $C_{18}SH$ on Au as a function of the number of irradiation cycles. Each cycle is equivalent to the copper evaporation process. Scan parameters: 300 scans at 84° incidence, 2 cm$^{-1}$ resolution. (A) $C_{18}SH$ as deposited; (B) $C_{18}SH$ after one thermal cycle; (C) $C_{18}SH$ after two thermal cycles; (D) $C_{18}SH$ after three thermal cycles. The spectra have been offset for clarity. [Reprinted from reference 30. With permission]

SAMs, with the bands associated with longer chain SAMs appearing at lower wavenumbers [50]. Therefore, a positive wavenumber shift of these peaks is usually interpreted as evidence of increased conformational disorder of the chains in the SAM, where the word "disorder" is intended to indicate that situation where departure from the *all-trans* conformation has taken place.

The bandwidth of the $CH_2$ stretching absorption peaks also increases upon disordering of the SAM hydrocarbon chains, since, there is in fact a distribution of *gauche* defects leading to different couplings of the C-H force constant [46, 53, 54]. An increase in the population of *gauche* conformations also leads to an increase in the intensity of the $v_{as}(CH_2)$ band [46, 47, 50]. This arises from the rotation of more of the dipoles associated to the asymmetric $CH_2$ stretching mode around the chain axis, which in turn results in a greater average component along the surface normal for this mode [46, 47, 50]. Combinations of band-shift, peak broadening and increased peak area have therefore been utilized for the detection of phase transitions in self-assembled monolayers [30, 46 and refs. therein, 51, 52, 53, 54, 55, 56].

A similar reasoning can be applied to the intensity of the peaks assigned to the methyl group, although it is usually more difficult to draw conclusions based on these peaks due to the Fermi resonance coupling [49] and because of their placement at an interface [49, 56, 57]. It has been observed that, when the substrate metal is Au, the relative intensities of symmetric and asymmetric stretching can vary according to the parity of the number of C atoms in the alkanethiol chain [57]. As shown in Figure 2, a difference of one carbon in the chain changes the orientation of the terminal $CH_3$, thus changing the components of the associated transition dipoles along the surface normal [50, 54, 57]. Contrary to the behavior of $CH_2$ associated peaks, the frequency of these modes is not as sensitive to the degree of order within alkanethiol SAMs [50, 51]. However, their absolute and relative intensities are extremely sensitive to temperature [51, 53, 56, 58], due to the methyl group being the TG and, therefore, the most mobile moiety in the molecule. In particular, it has been observed that the peak assigned to the symmetric stretching at 2878 cm$^{-1}$ progressively decreases in intensity with increasing temperature [51, 53]. On the other hand, at low temperatures (80 K), it has been observed that the peak assigned to the methyl out-of-plain asymmetric stretching, which appears as a shoulder in the spectrum at room temperature, increases in intensity and becomes well defined, resembling the peak line-shapes found in bulk polycrystalline spectra [56, 58].

## 2.3. Copper on Dodecanethiol and Octadecanethiol [30]

The studies illustrated in section 2.1 were all obtained with thin overlayer films, in the range of 0.1-20 nm, and which were deposited at very low rates, at most 1 nm/min. Most of the reported measurements were carried out without exposing the sample to the atmosphere, as it would most likely lead to the oxidation of the overlayer. As stated in the introduction, the main motivation to undertake the study of the effects of metal overlayers on SAMs in our laboratory was to correlate our findings to those obtained from molecular charge transport experiments. A high-vacuum environment, though, is not the usual environment in which molecular electronic devices have been studied and are indeed expected to function in. Additionally, many MOM junctions used to study molecular conductivity have required the presence of conventional macroscopic metallic leads. It is this necessity for macroscopic leads, that has led us to study the effect of vapor deposition of copper on molecules such as dodecanethiol ($C_{12}SH$) and octadecanethiol ($C_{18}SH$) self-assembled on polycrystalline gold, at high deposition rates (approximately 10 nm/s) which resulted in the formation of optically thick overlayers (approximately 200 nm thick) [30]. At the same time, two of the techniques used to characterize the effects of vapor deposition in these studies were carried out under normal atmospheric conditions, which is close to the conditions under which the MOM junctions will find their potential applications.

The main objectives of our experiments were to determine if the monolayer was still present after the evaporation process; to determine if the SAM had undergone any modifications; to establish if these changes are reversible; to determine whether increased electrical defects in the SAM result from evaporation; and to determine the nature of the perturbations induced by the copper coating. The SAMs were characterized both as deposited, and after the removal of the copper overlayer with nitric acid, using FT-RAIRS and electrochemistry. Control experiments were performed to confirm that the nitric acid solutions used for the stripping had no effect on either the organization or the coverage of the SAM.

The copper coating was carried out in an evaporation chamber by passing current through a Mo boat containing a Cu pellet. It was found that the heat irradiated from the Mo boat alone, which was incandescent during the deposition, was sufficient to induce the progressive degradation and ultimate desorption of the SAMs. Over multiple irradiation cycles the integrated intensity of the SAM spectroscopic signature was found

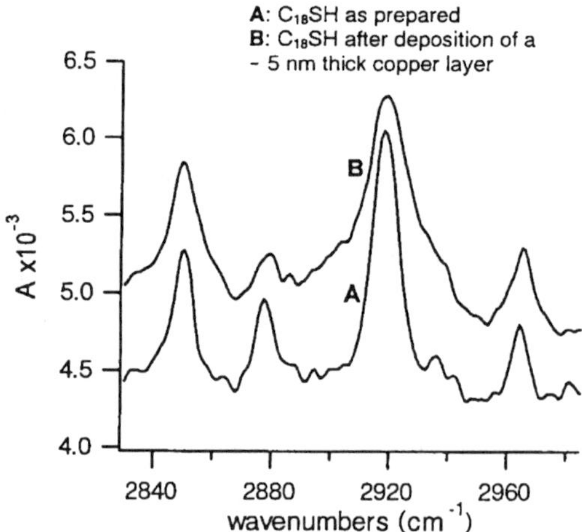


**Figure 3**. RAIRS spectra of a $C_{18}SH$ SAM with a 5 nm thick Cu overlayer (average thickness), deposited at 1nm/s. Scan parameters: 300 scans at 84° incidence, 2 cm$^{-1}$ resolution. (A) $C_{18}SH$ as deposited; (B) $C_{18}SH$ after having deposited a 5 nm thick Cu overlayer. Spectra are offset for clarity. [Reprinted from reference 30. With permission]

to decrease. Since it was the time of exposure to the incandescent boat that was responsible for deterioration of the SAMs, it was found that high deposition rates lowered the maximum temperature reached by samples during the deposition and minimized the desorption of the alkanethiols. Figure 2, shows the effect of multiple irradiation cycles on the infrared spectroscopic signature of $C_{18}SH$. $C_{12}SH$, due to its higher vapor pressure, shows an even more drastic decrease in the absorption peaks with increasing number of irradiation cycles; after only three irradiation cycles the spectral features disappeared almost completely. In the case of both $C_{12}SH$ and of $C_{18}SH$, the loss of the peak at 2878 cm$^{-1}$ was remarkably rapid, relative to the other absorption bands. This phenomenon is irreversible and is associated with the increased disordering of alkanethiol monolayers due to a phase transition to a liquid-like state above approximately 350 K (a more detailed discussion of this assertion will be given in the next paragraphs) [51, 52, 53, 55, 59, 60, 61, 62].

In the case of the actual deposition of a copper layer, the spectral changes were found to be similar. Figure 3 shows the spectra of $C_{18}SH$ after the vapor-deposition of a Cu layer 5 nm thick, and of the SAM as deposited. Such a copper layer is not thick enough to be optically opaque and the IR measurements were performed without the removal of Cu. The observed changes in the intensities and the frequency shift of the $v_{as}(CH_2)$ are consistent with the effects of the thermal treatment, as shown in Figure 2. These findings are compatible with the results obtained by Allara et al. [2] for a methyl 16-mercaptohexadecanoate $(HS(CH_2)_{15}COOH)$ SAM as a function of copper coverage. Once again, the $v_{sym}(CH_3, FR)$ peak at 2878 cm$^{-1}$ was found to decrease to a greater extent when compared to the other spectral features.

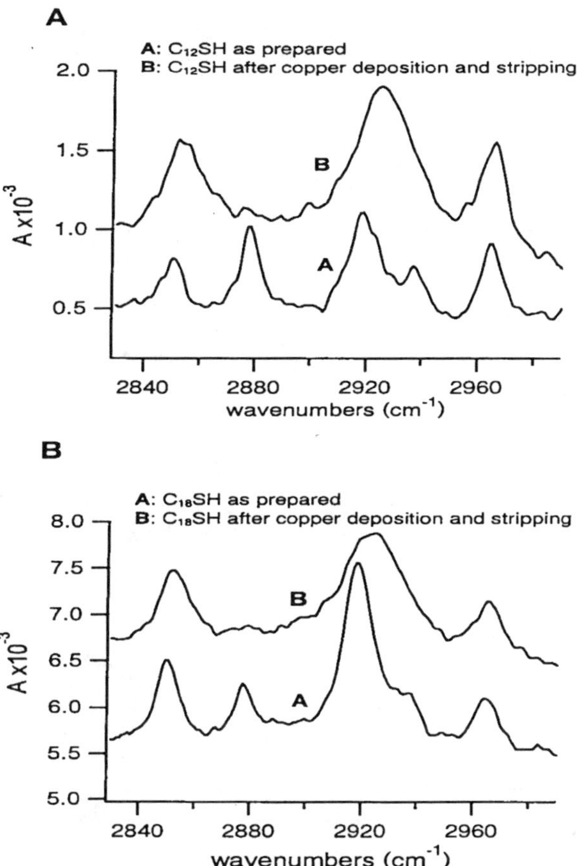


**Figure 4**. RAIRS spectra of $C_{12}SH$ (A) and $C_{18}SH$ (B) SAMs following coating with a copper overlayer 200 nm thick at the rapid deposition rate of 10 nm/s and stripping with 0.5 M nitric acid. Scan parameters: 300 scans at 84° incidence, 2 cm⁻¹ resolution. (A) $C_nSH$ as deposited; (B) $C_nSH$ after copper deposition followed by copper stripping. Spectra are offset for clarity. [Reprinted from reference 30. With permission]

When a thicker 200 nm copper layer was deposited, removal of the overlayer with nitric acid was necessary in order to observe any peaks associated with the alkanethiol SAMs. Figure 4 shows the changes induced by the evaporation process for both $C_{18}SH$ and $C_{12}SH$. As previously mentioned, the control experiments excluded the possibility that any of these changes could be ascribed either to the high vacuum in the evaporation chamber or to the exposure to nitric acid. The variations, therefore, had to have been introduced during the deposition process itself.

After the evaporation and removal of a 200 nm thick copper overlayer, the almost complete disappearance of the $\nu_{sym}(CH_3, FR)$ 2878 cm⁻¹ peak was observed [30]. This

Table 3: Position of the main peaks of the thioacetylated HS-PPB-SH molecule in the solid polycrystalline state (s) and of the SAM formed on Au substrates after deprotection (SAM/Au). Tentative assignments based on existing assignments of benzene derivatives are also given; when possible, these assignments are related to the benzene vibrational modes discussed by Varsányi [82]. $v$ = stretching; i.p. = in plane; o.p. = out of plane; asym = asymmetric; sym. = symmetric.

| Protected HS-PPB-SH (s) ($cm^{-1}$) | Deprotected HS-PPB-SH (SAM/Au) ($cm^{-1}$) | Assignment |
|---|---|---|
| 694 | - | ring skeletal vibration (mode 4)[82] |
| 825-840 | 835 | C-H o.p. bending and ring deformations[82, 83] <br> Typical C-H wag of *p*-disubstituted benzenes |
| 953-962 | - | $v[(CO)-S]$[84] <br> and <br> ring breathings[82] |
| 1012 | 1012 | C-H bending (mode 18a) |
| 1095-1117 | 1082-1099 | Several modes contributing: benzene breathings (mode 1 and mode 13)[82] <br> $CH_3$ asym. deformation[82] <br> $v[(CH_3)—(CO)S]$[84] |
| 1269 | - | C-H i.p. bending (mode 3)[82] |
| 1355 | - | $CH_3$ sym. deformation[82] |
| 1396-1437 | 1396 | several $v[C—C]$ modes[82] |
| 1480 | 1479 | $v[C—C]$[82] |
| 1514 | 1512 | $v[C—C]$ (mode 19a)[82, 75] |
| 1592 | 1587 | $v[C—C]$ (mode 8a or 8b)[82] |
| 1693 | - | $v[C=O]$[84] |
| 2200 | 2212 | $v[C≡C]$ |
| 2849-2956 | - | $v[C—H]$ of the two terminal $CH_3$ |
| 3054 | - | $v[C—H]$ of benzene rings |

mimics the behavior in the case of the vapor-deposition of a 5 nm Cu layer and of simple thermal treatments.

The most remarkable changes were observed for the $v_{as}(CH_2)$ peak at 2918 $cm^{-1}$. This peak broadened considerably both in the case of both $C_{12}SH$ and $C_{18}SH$ SAMs, and its frequency shifted towards a higher value (see Table 3). The effect on the 2850 $cm^{-1}$ $v_s(CH_2)$ mode was similar. These shifts in wavenumber are associated with an increased population of *gauche* defects, and consequently a departure from the ordered *all-trans* chain conformation, as discussed previously in section 2.2. $C_{12}SH$ and $C_{18}SH$ SAMs, however, showed differences in the total integrated intensities of these two peaks. The $C_{12}SH$ SAM showed increases in intensity of up to 50% for the symmetric stretching, and up to 300% for the asymmetric stretching. On the other hand, the $C_{18}SH$ SAM showed a decrease in intensity of approximately 12% for the symmetric stretching and no appreciable variation for the asymmetric stretching. Whereas broadening and increased peak area are features consistent with increased disorder within the monolayer and indicated by a blue-shifting of wavenumber, the reduction of the integrated area, points toward the simultaneous occurrence of another reorganization process that leads to a reduced intensity in the methylene bands. A possible explanation of this phenomenon is the untilting of chains, which, as illustrated in section 2.2, reduces the component of these modes along the surface normal.

Similar trends in the lineshape of these methylene bands, upon heating of the monolayers, have been observed by other research groups. Bensebaa et al. [51, 52, 53] have studied the effects that thermal treatment in a vacuum has on long alkanethiol SAMs ($C_nSH$ $n \geq 18$) and have observed the existence of a gradual structural transformation that leads to the irreversible transition to a more liquidlike phase of the monolayer [51, 53]. They observed a decrease in the integrated intensity of the methylene stretching bands as the temperature was increased from 150 K to 350 K, that was much larger than the common intensity drop, attributed solely to temperature, observed for $n$-alkanes [63]. The intensity drop was accompanied by a slow blue-shift of the peak maxima. These effects were found to be reversible below 350 K, however, above this threshold they observed irreversible changes with a sharp increase of these intensities and a more rapid peak blue-shift. Upon reaching 450 K, a rapid decline in intensity was observed. These experimental trends are illustrated in Figure 5. These changes were explained by a combination of chain untilting and an increase in the population of *gauche* defects, beginnning with the IR features being dominated by the untilting through the 150-350 K heat ramp and followed by the onset of high and irreversible conformational disorder between 350 K and 450 K. The untilting was considered to be the process that allowed for reduced packing density, thereby allowing for conformational changes to take place as a result of the reduced steric hindrance. At 450 K, the overall spectroscopic signature of the SAM declines, indicating the occurrence of its desorption from the gold underlying substrate [51, 53].

These thermally induced transitions have also been observed over similar temperature ranges by other experimenters. Fenter et al. [59] studied the phase transitions of $C_{12}SH$ and $C_{14}SH$ monolayers by grazing-angle X-ray diffraction. They observed a reversible order-disorder transition for $C_{12}SH$ at around 323 K. In the case of $C_{14}SH$ this transition occurred at approximately 333 K, but it was pointed out that the phase transition for this longer alkanethiol SAM displayed hysteretic behavior, i.e. it was possible for the system to remain trapped in the high temperature phase even when the temperature was lowered to room temperature. Using STM and low-energy helium-atom diffraction, Camillone et al. [60] observed a gradual disordering of the studied SAMs followed by a transition to a crystalline phase. Short monolayers ($C_nSH$, $n \leq 12$) assembled on Au(111) were observed to progressively disorder over a temperature ramp, starting at approximately 373 K and continuing up to 423 K, where they found the onset of a second crystalline phase. Using cyclic voltammetry of ferrocyanide solutions, Badia et al. [61] revealed the presence of a melting transition that occurred in the 313-338 K temperature range, which was dependent on the chain length. They also showed that kinetics play an essential role in the dynamics of the SAMs, since electrochemical measurements taken during heat-cool-reheat cycles displayed a hyteresis. Naselli et al. [55] have also reported a gradual order-disorder transition for a Langmuir-Blodgett monolayer of cadmium arachidate, as evidenced by a combination of IR and differential scanning calorimetry (DSC). They observed a reversible progressive disordering of the hydrocarbon chains as the temperature approached the melting transition located at 383

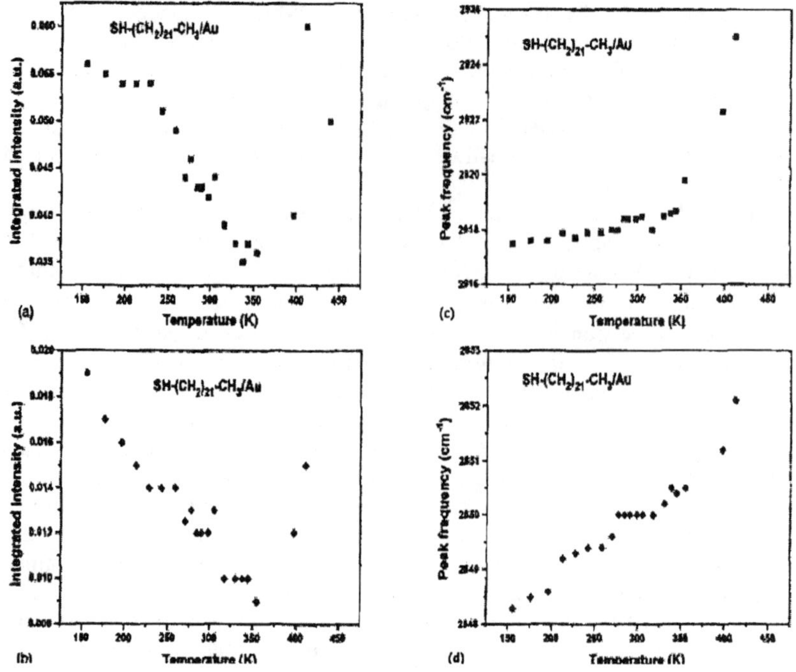

**Figure 5.** Temperature dependence of: (a) integrated intensity of the $v_a(CH_2)$ peak; (b) integrated intensity of the $v_s(CH_2)$ peak; (c) frequency shift of the $v_a(CH_2)$ peak; (d) frequency shift of the $v_s(CH_2)$ peak. [Adapted and reprinted from reference 51. With permission]

K, as measured by DSC, with the IR stretching region resembling that of an isotropic melt. Once the temperature surpassed the melting point, the spectral changes that indicated conformational disorder were irreversible.

Molecular dynamics simulations are in general agreement with the gradual melting mechanism of the SAM proposed by Bensebaa et al. [51, 52, 53]. In fact, Bhatia et al. [62] reported simulations in which alkanethiol SAMs undergo a high temperature order-disorder transition. Starting at 40 K, the chains are tilted along the Au-Au nearest-neighbor directions (azimuthal angle of 0° and 60°), constituting a crystalline phase. Upon heating the system to 275 K, the monolayer undergoes a first transition to another crystalline phase in which the chains orient along the Au-Au next-nearest-neighbor direction (90° relative to the direction of the previous phase, azimuthal angle of 30°), and their tilt angle decreases by approximately 6°. As the temperature is raised above room temperature to nearly 400 K, the hydrocarbon chains further untilt, and the angle between their axis and the surface normal broadens its statistical distribution. The behaviors of these azimuthal and tilt angle distributions are shown in Figure 6. A simultaneous increase in the population of *gauche* defects occurs, though the onset of this increase occurs at slightly higher temperatures than for the untilting. These changes indicate the

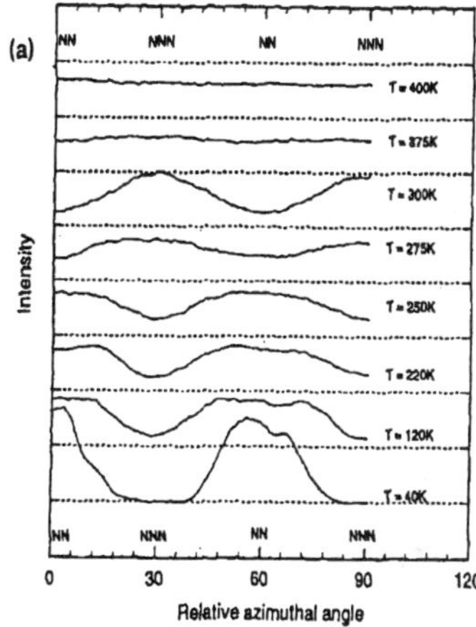
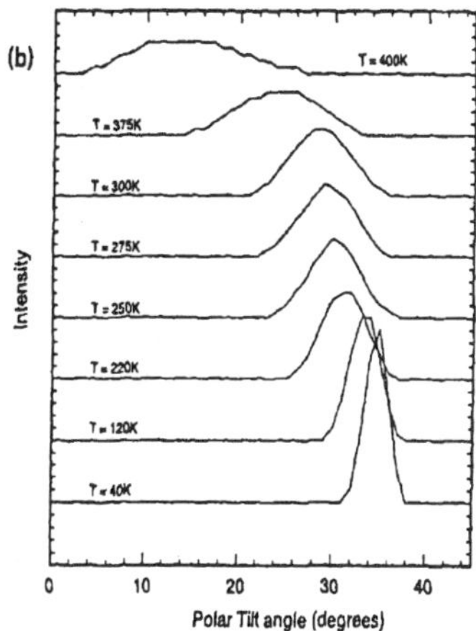

**Figure 6**. Distributions of relative azimuthal (a) and tilt (b) angles as a function of temperature in a CH$_3$(CH$_2$)SH SAM on Au(111) as obtained via molecular dynamics simulations by Bhatia et al. [Ref62]. [Adapted and reprinted from reference 62. With permission]

presence of a melting transition in the 300-400 K temperature range. They also showed that if the system was cooled down after undergoing the melting process, some of the conformational defects remained trapped within the film yielding a high-energy configuration.

The overall change in the spectra of C$_{12}$SH and of C$_{18}$SH SAMs was irreversible and the spectra did not display their original characteristics even after several hours of storage at room temperature, or after a 4 hr immersion in the SAM deposition solvent. The irreversibility of the spectral changes observed was in good agreement with the aforementioned literature regarding order-disorder transitions of alkanethiol monolayers. Therefore, all of the observations point toward a pseudo-melting phase transition induced by the copper deposition process. This molten state is probably "kinetically frozen" when the evaporation ends and the sample cools down in the deposition chamber, leaving a high degree of chain disorder in the SAM. It is not necessary to invoque any specific chemical of physical interactions between the metal and the SAM in order to explain these effects; the thermal treatments alone seem to be enough to produce them. This conclusion indicates that the temperature reached by the SAM during the evaporation process is sufficiently high to induce melting of the monolayer.

There are instances in the literature where an increase in the crystallinity of alkanethiol SAMs following annealing have been reported [59, 60, 64, 65]. These results do not necessarily contradict either our findings or those of other authors, and there are a number of reasons why this is the case. First of all, it is important to consider the role of kinetics in the phase transition behavior, especially for long hydrocarbon chains. The speed of the temperature ramp chosen for the experiment will influence the outcome. Secondly, most of the literature in which a higher degree of crystallinity at high

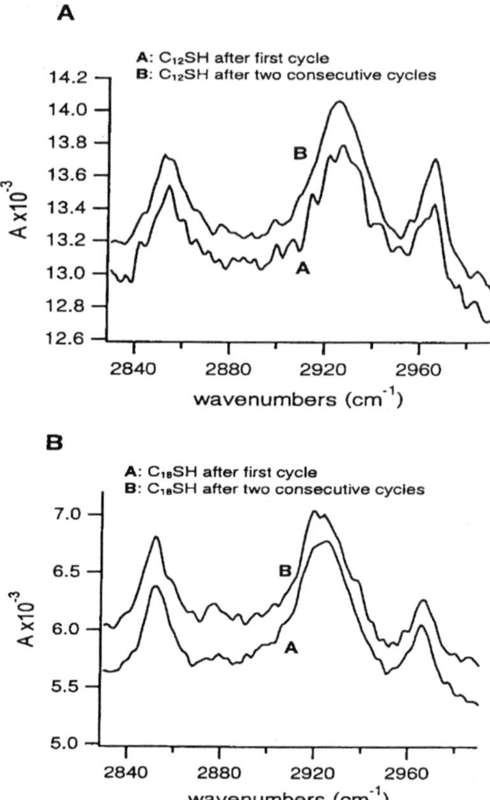


**Figure 7**. RAIRS spectra of $C_{12}SH$ (A) and $C_{18}SH$ (B) SAMs following multiple copper deposition and stripping cycles under rapid thermal evaporation rates of 10 nm/s. Scan parameters: 300 scans at 84° incidence, 2 cm⁻¹ resolution. (A) $C_nSH$ after the first cycle; (B) $C_nSH$ after two consecutinve cycles. Spectra are offset for clarity. [Reprinted from reference 30. With permission]

temperature is reported, indicates the presence of a transition temperature range in which several phases co-exist before the onset of the new highly crystalline phase. If the quenching of the SAM occurs after the highly crystalline phase has been achieved, the resulting monolayer will be more crystalline than the one freshly deposited at room temperature [59, 60]. If quenching occurs from the phase transition temperature range, however, it is likely that the annealing will not lead to a higher degree of crystallinity. Thirdly, characterization techniques can be sensitive to different types of order; for example, X-ray diffraction and STM are sensitive to long-range order, whereas IR gives information about both intra and inter molecular order.

It was suprising to find that the copper evaporation process, (at least when carried out at high deposition rates) had a protective effect, trapping the SAM and perventing its desorption. As mentioned earlier in this section, consecutive thermal treatments inside the evaporation chamber led ultimately to the desorption of the SAM from the gold substrate.

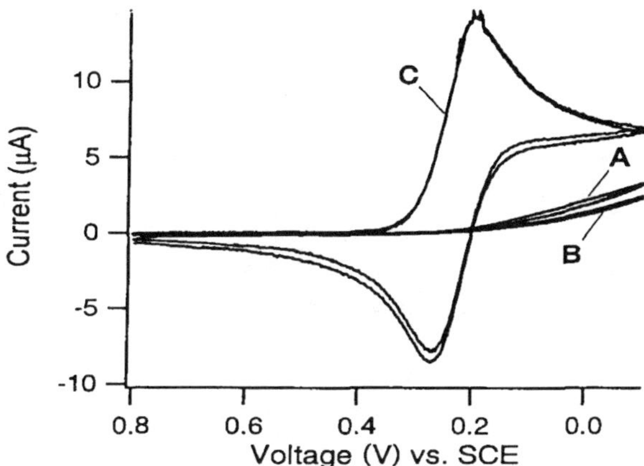


**Figure 8**. Cyclic voltammograms in ferricyanide solution, of a gold electrode passivated by a $C_{12}SH$ SAM as deposited and after being subject to copper evaporation and stripping with 3 M nitric acid. The voltammogram of a bare gold electrode is reported for comparison. (A) $C_{12}SH$ monolayer; (B) $C_{12}SH$ monolayer after undergoing copper evaporation and stripping; (C) bare gold electrode. [Reprinted from reference 30. With permission]

This was not observed when copper was deposited. In fact, multiple cycles of copper evaporation and removal yielded a fairly constant IR signature, indicating that desorption did not occur or that at least it occurred at a remarkably lower rate than for the bare SAM. Figure 7 shows how the spectra of $C_{12}SH$ and $C_{18}SH$ remain costant after two consecutive cycles of copper evaporation and subsequent removal with nitric acid.

The conclusions reached on the basis of these RAIRS measurements were supported and complemented by cyclic voltammetric and XPS experiments. Figure 8 shows the cyclic voltammograms obtained at a bare gold electrode, at a gold electrode passivated by a $C_{12}SH$ SAM, and after copper evaporation and removal with nitric acid. The bare gold electrode curve shows the forward and reverse peaks corresponding to the reaction of the $Fe(CN)_6^{4-}/Fe(CN)_6^{3-}$ redox couple. When the $C_{12}SH$ SAM is assembled on the gold electrode, the electrode becomes passivated to the redox reaction, since the SAM prevents the cyanide complexes from reaching the gold [61]. The efficiency of this passivation depends on the coverage of the monolayer and on its permeability to the chosen redox couple [61]. After the copper was deposited and removed, the passivation remained comparable to that of the SAM as deposited. This clearly indicates that the copper evaporation leaves the SAM intact in terms of its packing and coverage. It was therefore concluded that the changes in intensity and broadening previously observed using RAIRS could not be ascribed to desorption of the SAM from the substrate. Figure 9 shows the results obtained in our laboratory using XPS [30]. These XPS studies were carried out in vacuum and at a low copper deposition rate and with coverages up to only 2 ml. The conditions for this study were therefore more similar to previous studies of metal/SAM interactions. The C 1s binding region displays no changes following copper deposition, which is indicative of no changes in the oxidation state of the C atoms. This region is not very informative, given the sensitivity of this technique to the outermost

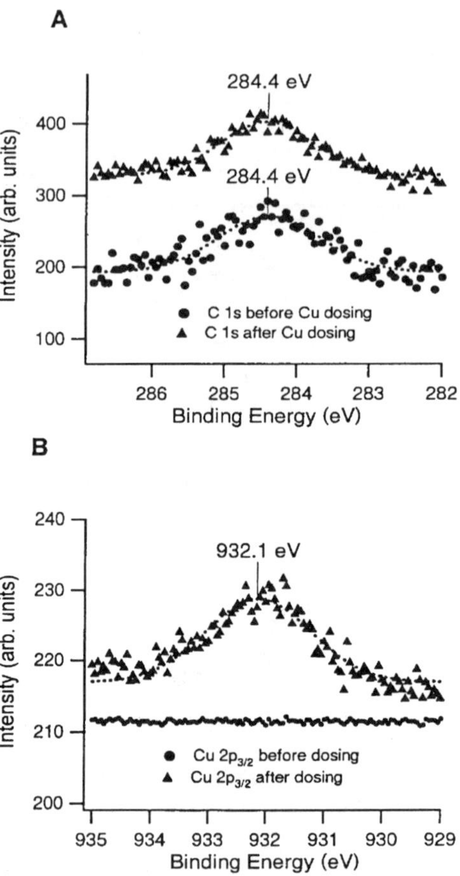



**Figure 9**. XPS of a C$_{12}$SH SAM before and after 2 ml copper coverage with best-fit gaussian curves. (A) C 1s region, before and after copper deposition; (B) Cu 2p$_{3/2}$ region before and after copper deposition. [Reprinted from reference 30. With permission]

layer of the surface. An overlayer of copper can easily obstruct the carbon atoms beneath it and obscure details about their oxidation state. The fact that the C signal intensity remains constant, though, indicates that most of the surface is free from Cu, thereby indicating that the metal aggregated, forming clusters. The Cu signal after deposition appears at marginally smaller binding energy than the literature value for Cu$^0$, indicating the presence of mostly metallic copper. Once again, however, it is possible to ascribe this signal only to the outermost copper atoms of these clusters. On the basis of this XPS data, it is not possible to rule out the formation of Cu-C chemical bonds, although it does not provide any direct evidence to support it either. The role of Cu in previous studies of metal/SAM interactions is in agreement with these findings, as can be seen from Table 1.

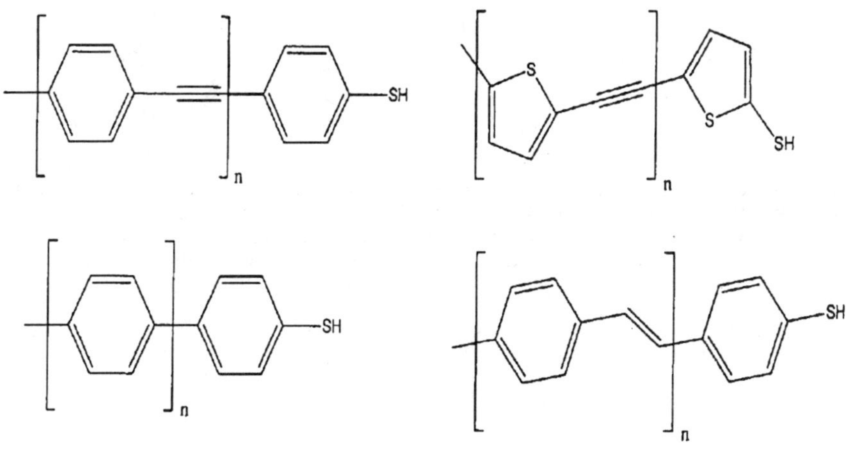

Figure 10. Typical oligomers proposed as molecular wires, with n usually in the range 0-4.

In light of these studies, it is plausible to assert that alkanethiol SAMs resist the formation of a metallic contact by vapor-deposition of a copper layer. The coverage and packing of the monolayer is preserved after the depositon process, however, its organization and ordering are altered. These measurements have not revealed the presence of partial or complete penetration of the copper layer through the SAM, as found in Cu/SAM experiments reported by other research groups. This can be explained either by a lack of sensitivity of RAIRS and electrochemistry to the presence of Cu atoms or clusters trapped among the hydrophobic chains or beneath them, or simply by the fact that the rapid evaporation rate used in our experiments favors the formation and growth of clusters as described in section 2.1.

## 3. COPPER ON CONJUGATED OGLIGOMERS

Unlike alkanethiol SAMs, there are few reports in the literature of studies on fundamental metal/SAM interactions when the monolayer is comprised of molecules with extended $\pi$-conjugation that includes aromatic rings, as is the case of prospective molecular wires. This might also be a reflection of the considerably low number of fundamental studies on arenethiol SAMs. Figure 10 shows the structure of some of the molecules proposed as suitable components of a molecular electronic device.

It is reasonable to expect that whenever conjugated oligomers are involved, the interactions of the evaporated metal with the SAM can differ dramatically from those between the same metal and alkanethiol chains. For one, the assembly and packing of these oligomers on coinage metals can be remarkably different than that of aliphatic chains. Oligomers are much more rigid than hydrocarbon chains, having a limited

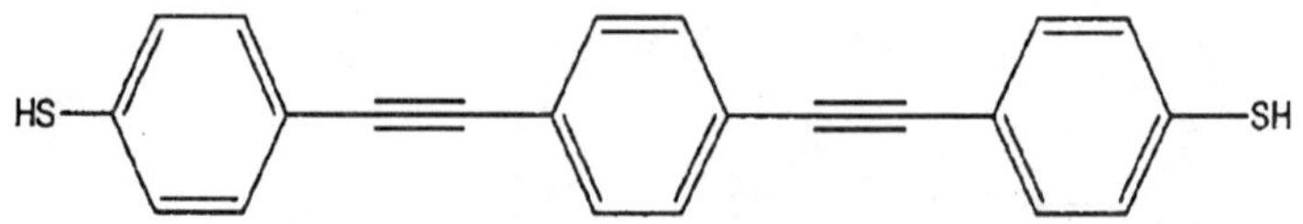

**Figure 11.** The conjugated oligomer used in the experiments described in section 3: 1-thio-4-[4'-[(4'-thio)phenylethynyl]-1'-ethynyl]-phenylbenzene (HS-PPB-SH). The molecule had the thiol group originally protected by acetyl moieties, in order to prevent polymerization through the formation of disulfide bonds between the oligomers. The protecting groups were hydrolized in the deposition solution through the addition of small quantities of ammonium hydroxide, thus restoring the thiol groups.

segmental mobility due to the high degree of conjugation, many of them, therefore, are thought of more properly as rigid rods [66, 67, 68]. The presence of aromatic rings in their structure results in stronger molecule-molecule interactions. In fact, it has been established that, while Van der Waals' interactions alone can account for the intermolecular forces that favor the packing of aliphatic thiols, the packing of aromatic rings also involves electrostatic interactions [69, 70, 71, 72, 73]. It has been consistently shown that the tilt angle of arenethiol SAMs is lower than that of alkanethiols, since the molecules do not need to tilt strongly in order to optimize their intermolecular interactions [66, 67, 73, 74, 75, 76, 77]. Finally, in the case of phases formed under low-coverage conditions, the chemisorption of alkanethiols is driven only by the formation of the thiolate bond with the substrate. It has been proposed that the chemisorption of arenethiols, on the other hand, is favored not only by the formation of the sulfur-metal bond, but also by a favorable interaction between the $\pi$-system and the metal surface. The extra stabilization would arise if the molecules adopt a flat-lying orientation in a similar manner to the adsorption of benzene on metal surfaces [78, 79]. This most likely leads to a different thermodynamic landscape of low- and high-coverage phases when compared to alkanethiols [66, 74].

A second issue to be considered is that these oligomers offer more reactive sites along their chain, as opposed to alkanethiols: electron-rich aromatic rings and olefin moieties are susceptible to chemical attacks more readily than paraffins. Furthermore, these oligomers are often substituted with electron-donor or electron withdrawing groups in order to modulate their electronic properties; these sites are also more reactive than a saturated carbon chain. Finally, although they often terminate in an aromatic C-H bond, they are sometimes capped with another thiol group at the vacuum/SAM interface. This terminal group can also react with vapor-deposited metals, this reaction being critical to understanding the nature of the coupling between the molecule and the macroscopic metallic lead.

Oligo(phenylene ethynylene)s (OPEs) are a family of molecules that have been suggested to function as molecular interconnects. They have been widely used for the construction of molecular electronic devices [21, 23] and experiments on their charge transport properties have already been conducted in this laboratory [18]. They are known

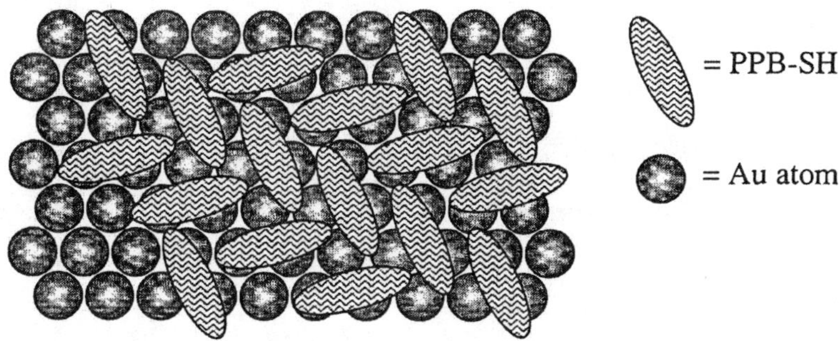


**Figure 12**. Schematic of the PPB-SH packing on Au(111) (top-down view) as measured via STM [77, 68]. The aromatic rings assemble in an "edge-on" fashion on the hexagonal arrangement of the surface.

to form well-ordered self-assembled monolayers on coinage metals (Au, Ag, Pt, etc.), that display increased packing as the number of phenylene ethynylene monomers is increased [6, 68, 75, 77, 80, 81]. Figure 11 shows the oligo(phenylene ethynylene)-thiol used in charge transport experiments in this laboratory. STM and AFM measurements on this molecule, but lacking one of the thiol terminal groups (PPB-SH), have revealed that the chains are densely packed with long-range order in an "edge-on" fashion, and oriented virtually parallel to the surface normal on Au(111) terraces (see Figure 12) [68, 77]. The lattice formed is incommensurate with Au(111) [77], and this lack of a commensurate structure is attributed to the intermolecular interactions being the driving force for the assembly, instead of the molecule-substrate interactions [6, 77].

To our knowledge, the only investigation into metal/SAM interactions for OPEs was carried out by Haynie et al. [6]. They studied the interactions between PPB-SH assembled on Au(111) and a series of vapor-deposited metals: Au, Ti, Fe, Cr, Cu, Ag and Al, via TOF-SIMS. No evidence of any chemical interactions or penetration was found in the case of Au, Fe, Cr, Ag and Al. This was attributed to the high packing density of these monolayers and the consequent steric hindrance toward penetration. In the case of Ti, the vapor-deposition destroyed the molecular array; this is important, since, in many molecular electronics experiments, Ti is often used as a "bonding" metal for the formation of molecule-metal connections [2, 23]. In the case of Cu, they found several peaks in the mass spectrum, indicating a potential reaction between the metal and the benzene rings.

In this laboratory, the study of the interactions between copper and HS-PPB-SH has recently been undertaken, and the preliminary results obtained so far will be illustrated in brief. The results will be reported in full, together with the experimental details, in a manuscript currently in preparation [35]. The same series of experiments carried out for alkanethiol SAMs have been attempted in the case of HS-PPB-SH. However, due to the aforementioned differences between alkanethiols and conjugated oligomers, the results obtained already depart from those obtained for $C_{12}SH$ and $C_{18}SH$. The monolayer of HS-PPB-SH has been characterized, via RAIRS, as deposited and after deposition of copper

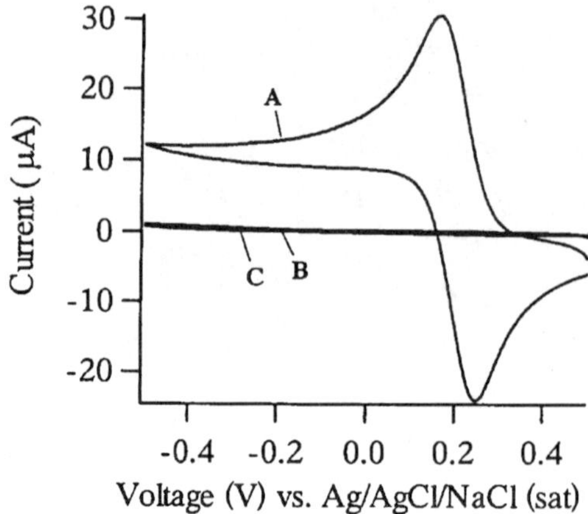


**Figure 13**. Cyclic voltammograms obtained in a 0.01 M Fe(CN)$_6^{3-}$/ 0.05 M Na$_2$SO$_4$ solution at 20 mV/s for: (A) bare gold electrode; (B) gold electrode after assembly of HS-PPB-SH SAM; (C) gold electrode with a HS-PPB-SH SAM on its surface, after having being immersed in a 0.5 M nitric acid solution for 5 min. The gold is passivated to the redox reaction when a monolayer is allowed to self-assemble on its surface. The passivation remains unaltered by the exposure to nitric acid.

and subsequent removal of the overlayer using nitric acid. The preliminary conclusions are supported by cyclic voltammetry and ellipsometry measurements.

The oligomer used for the following experiments had the terminal thiol protected by acetyl groups. In order to obtain the optimal SAM of these oligomers, it was necessary to hydrolize the TGs with ammonium hydroxide to yield the thiol anchoring groups; this procedure was carried out in the same deposition solution (*in-situ*) and in the absence of oxygen, in order to prevent the formation of disulfides. Cyclic voltammetry in ferricyanide solutions showed that excellent passivation was obtained from a solution of the molecule in THF with small quantities of hydroxide added. Figure 13 shows the cyclic voltammogram of a bare gold electrode compared to that of an electrode on which the OPE has formed a SAM. The redox peaks are almost completely inhibited in the case of the passivated electrode. This indicated that the deprotected molecules display high packing in the SAM, which is consistent with previous reports in the literature for identical or homologous OPEs [68, 75, 77, 81]. The RAIRS spectrum of the SAM as deposited was collected and compared to the spectrum of the bulk solid phase. Table 3 shows the proposed assignments for the main IR bands of the HS-PPB-SH spectrum, based on the assignments for a homologous OPE [75] and for substituted benzene rings [82]. Figure 14 shows a comparison of the spectrum obtained for the OPE SAM as deposited, and that obtained for the bulk solid. It can be seen that the bands belonging to the protecting groups do not appear in the SAM spectrum, confirming that the hydrolysis of the TGs was successful.

The first important difference between these results and those described in section 2.3 is shown in Figure 15. This figure shows the difference between the SAM as deposited and the SAM exposed to the same solution of nitric acid used in the

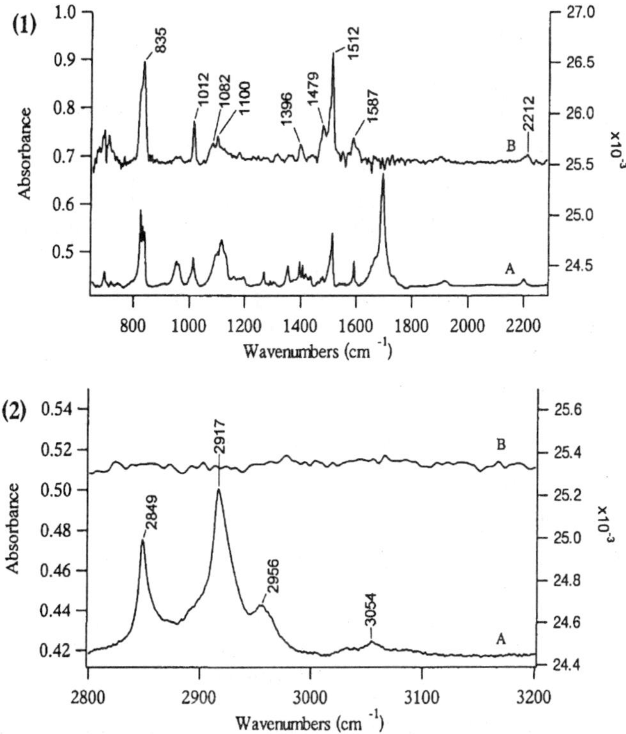



**Figure 14**. IR absorption spectra of the protected HS-PPB-SH (AcS-PPB-SAc) in the bulk solid phase (A, left axis) compared to the RAIRS spectrum of HS-PPB-SH self-assembled on gold slides (B, right axis). (1) shows the 600-2300 $cm^{-1}$ region, (2) shows the 2800-3200 $cm^{-1}$ region. The bands associated to the protecting group ($\nu$[C=O], $\nu$[CH$_3$]) disappear in the oligomer SAM, due to the *in-situ* deprotection via base-promoted hydrolysis. The $\nu$[C—H] associated with the benzene rings (see Table 3) is absent or occasionally very weak in the SAM spectrum. Scan parameters: (A) 30 scans, 1$cm^{-1}$ resolution; (B) 1000 scans at 84° incidence, 4 $cm^{-1}$ resolution. The spectra were collected in a spectrometer purged with N$_2$ to minimize the signal from water vapor and then baseline corrected.

alkanethiols experiments to dissolve the Cu overlayers. After the exposure to the acid solution, the RAIRS spectrum shows subtle changes indicating the appearance of previously absent C-H stretchings. These peaks seem to be enhanced whenever the exposure to the acid is more prolonged and the acid concentration is higher. At present, there is no conclusive explanation for this phenomenon, although it could be likely an indication of nitration of the aromatic rings or water addition to the C≡C triple bonds. These changes are not reflected in the electrochemistry at the SAM/Au electrodes, since, as shown in Figure 13, the cyclic voltammograms of the SAM-covered electrode do not differ significantly before and after the exposure to the acid.

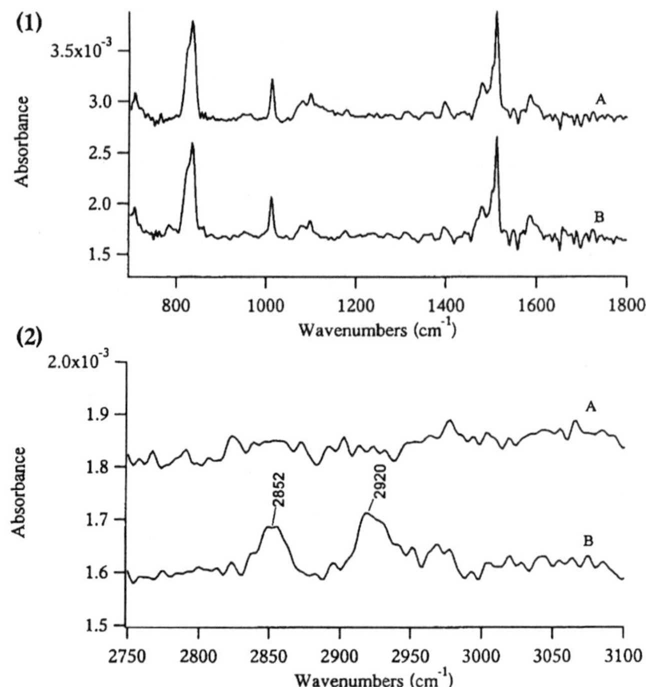

**Figure 15**. RAIRS spectra of HS-PPB-SH SAM as deposited (A) and after being immersed in a 0.5 M nitric acid solution for 5 min (B). (1) shows the 700-1800 cm$^{-1}$ region, (2) shows the 2750-3100 cm$^{-1}$ region. The C-H stretching region is the only one that shows any changes after the exposure to the acid. Scan parameters: 1000 scans at 84° incidence, 4 cm$^{-1}$ resolution. The spectra were collected in a spectrometer purged with N$_2$ to minimize the signal from water vapor and then baseline corrected.

The copper vapor deposition was carried out under the same conditions as for the experiments for the alkanethiols, with a 200 nm thick layer being deposited at high rates. It has been observed, however, that, on occasion, the copper is not easily removed by the acid solution once it is deposited on the OPE SAM, as if the SAM had a protective effect toward it. In these cases, the removal of the copper required longer exposures and higher concentrations of nitric acid resulting in difficulties distinguishing between modifications introduced by the evaporation process and those introduced by the nitric acid. Thinner copper overlayers were attempted with greater success. Figure 16 shows the effect of the vapor-deposition of 10 nm of Cu on the SAM RAIRS spectrum. The main changes observed occur for the peak around 835 cm$^{-1}$, assigned to o.p. C-H and ring deformations, and the C-C stretching peak at 1512 cm$^{-1}$ (19a mode), both of which decrease in intensity, whereas the band assigned to a C-H bending at 1012 cm$^{-1}$ (18a mode) remains comparatively constant. The dipole moment for the 19a mode lies along the long molecular axis [75] and its decrease could be either ascribed to desorption of the OPEs or to a greater tilt angle. The dipole moment of the modes contributing to the peak at 835

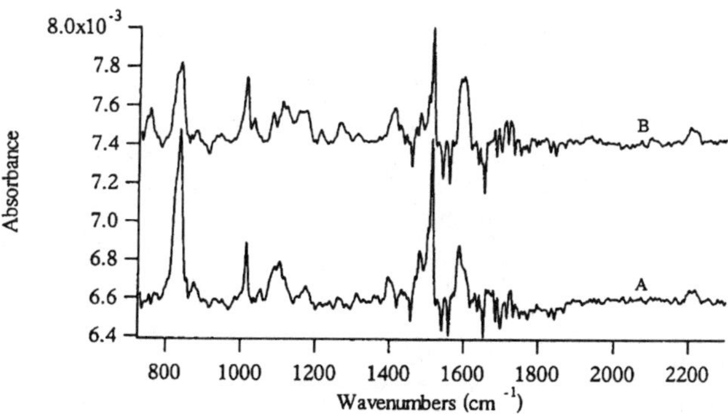

**Figure 16**. RAIRS spectra of HS-PPB-SH SAM as deposited (A) and after deposition of a 10 nm thick copper layer followed by stripping with nitric acid. Only the 750-2250 cm$^{-1}$ region is reported, higher frequencies are not informative due to the perturbing effect of nitric acid (see Figure 15). The main differences can be seen for the peaks at 835 cm$^{-1}$ and at 1512 cm$^{-1}$ (see text). The peak at 1587 cm$^{-1}$ shows an enhancement of the peak shoulder after evaporation and stripping. The peak at 1012 cm$^{-1}$ remains constant in intensity. Scan parameters: 1000 scans at 84° incidence, 4 cm$^{-1}$ resolution. The spectra were collected in a spectrometer purged with $N_2$ to minimize the signal from water vapor and then baseline corrected. Residual water peaks can still be seen in the spectrum.

cm$^{-1}$ is, instead, perpendicular to the aromatic rings and, therefore to the long molecular axis. A greater tilt angle for these molecules would lead to an increase in the 835 cm$^{-1}$ band, as long as the rings do not simultaneously rotate along the long axis resulting in a variation of the molecule average twist angle. A decrease in the intensity of this band then indicates either desorption or a departure from an assumed "edge-on" arrangement toward a more rotationally disordered organization. The peak assigned to the 18a mode at 1012 cm$^{-1}$, with a transition dipole along the molecular axis, however, does not vary in intensity. A change in tilt angle would also lead to a change in the area of this peak, but being of smaller intensity than the C-C stretching and C-H wag peaks, the change would be relatively small, probably within the error of the measurement. If the SAM desorbed, on the other hand, this peak would likely be considerably reduced instead. This indicated that the most probable assumption, until further experiments are carried out, is that the SAM undergoes a change to a more disordered organization. The hypothesis of a conformationally more disordered SAM after the vapor-deposition, appears also to be confirmed by preliminary ellipsometric results, that show no change in the average thickness of the organic film upon copper evaporation and removal.

Although these are preliminary results that need further experimental work, they are a good indication that a very different behavior should be expected when metal evaporation is carried out on a conjugated oligomer as opposed to alkanethiol SAMs. The formation of a macroscopic metallic lead on the organic surface might lead to the solvation of part of the metal within the layer and/or the introduction of conformational disorder in the layer.

## 4. ACKNOWLEDGMENTS

The authors thank the Petroleum Research Fund (ACS-PRF No. 36477-AC5) for support of this research. This material is based upon work supported under a National Science Foundation Graduate Research Fellowship. The authors thank also Dr. Timir Datta for the use of his evaporation chamber and Art Illingworth and Allen Frye for the fabrication of an evaporation chamber for this laboratory.

## 5. REFERENCES

1.  D. B. Pedersen, M. Z. Zgierski, S. Anderson, D. M. Rayner, B. Simard, S. Li and D.-S. Yang; "Bonding in Transition Metal-Ether Complexes: The Spectroscopy and Reactivity of the Zr Atom-Dimethyl Ether System", *J. Phys. Chem. A*, **2001**, *105*(51), 11462-11469.

2.  D.R. Jung, A.W. Czanderna, "Chemical and physical interactions at metal/self-assembled organic monolayer interfaces", *Crit. Rev. Solid State Mater. Sci.*, **1994**, *19*(1), 1-54

3.  A. V. Walker, T. B. Tighe, M. D. Reinard, B. C. Haynie, D. L. Allara, N. Winograd, "Solvation of zero-valent metals in organic thin-films", *Chem. Phys. Lett.*, **2003**, 369, 615-620

4.  G.L. Fisher, A. V. Walker, A.E. Hooper, T.B. Tighe, K.B. Bahnck, H.T. Skriba, M.D. Reinard, B.C. Haynie, R.L. Opila, N. Winograd and D.L. Allara, "Bond-insertion, complexation, and penetration pathways of vapor deposited aluminum atoms with HO- and $CH_3O$-terminated organic monolayers", *J. Am. Chem. Soc.*, **2002**, *124*, 5528-5541.

5.  Peter A. Willis, Hans U. Stauffer, Ryan Z. Hinrichs, and H. Floyd Davis, "Crossed beams study of C-H bond activation: $Mo(^5S_2) + CH_4MoCH_2 + H_2$", *J. Chem. Phys.*, **1998**, *108*(7), 2665-2668

6.  B.C. Haynie, A.V. Walker, T.B. Tighe, D.L. Allara, and N. Winograd, "Adventures in molecular electronics: how to attach wires to molecules", *Appl. Surf. Sci.*, **2003**, *203-204*, 433-436.

7.  A. Hartstein, J.R. Kirtley, J.C. Tsang, "Enhancement of the Infrared Absorption from Molecular Monolayers with Thin Metal Overlayers", *Phys. Rev. Lett.*, **1980**, *45*, 201-204.

8.  Z. Zhang, T. Imae, "Study of surface-enhanced infrared spectroscopy. 1. Dependence of the enhancement on thickness of metal island films and structure of chemisorbed molecules", *J. Colloid Interface Sci.*, **2001**, *233*, 99-106.

9.  Z. Zhang, T. Imae, "Study of surface-enhanced infrared spectroscopy. 2. Large enhancement achieved through metal-molecule-metal sandwich configurations", *J. Colloid Interface Sci.*, **2001**, *233*, 107-111.

10. A.E. Bjerke, P.R. Griffiths, W. Theiss, "Surface-enhanced infrared absorption of CO on platinized platinum", *Anal. Chem.*, **1999**, *71*(10), 1967-1974.

11. G.T. Merklin, P.R. Griffiths, "Effect of microscopic surface roughness in surface-enhanced infrared absorption spectrometry", *J. Phys. Chem. B*, **1997**, *101*(30), 5810-5813.

12. K. Ataka, Y. Hara, M. Osawa, "A new approach to electrode kinetics and dynamics by potential modulated Fourier transform infrared spectroscopy", *J. Electroanal. Chem.*, **1999**, *473*(1-2), 34-42.

13. A. Hooper, G.L. Fisher, K. Konstantinidis, D. Jung, H. Nguyen, R. opila, R.W. Collins, N. Winograd, and D.L. Allara, "Chemical effects of methyl and methyl ester groups on the nucleation and growth of vapor-deposited aluminum films", *J. Am. Chem. Soc.*, **1999**, *121*, 8052-8064

14. K. Konstantinidis, P. Zhang, R.L. Opila, D.L. Allara, "An in-situ X-ray photoelectron study of the interaction between vapor-deposited Ti atoms and functional groups at the surfaces of self-assembled monolayers", *Surf. Sci.*, **1995**, *338*, 300-312.

15. W.R. Salanek, S. Stafsftrom, J.-L. Bredas, Eds., in *"Conjugated polymer surfaces and interfaces: electronic and chemical structure of interfaces for polymer light-emitting devices"*; Cambridge University Press: Cambridge, UK, 1996.

16. N. L. Jeon, R.G. Nuzzo, "Physical and spectroscopic studies of the nucleation and growth of copper thin-films on polyimide surfaces by chemical-vapor deposition", *Langmuir*, **1995**, *11*, 341-355.

17. G.C. Herdt and A.W. Czanderna, "Metal overlayers on organic functional groups of self-organized molecular assemblies: VII. Ion scattering spectroscopy and X-ray photoelectron spectroscopy of $Cu/CH_3$ and $Cu/COOCH_3$", *J. Vac. Sci. Technol. A*, **1997**, *15*(3), 513-519.

18. M.S. Doescher, J.M. Tour, A.M. Rawlett and M.L. Myrick, "Stripping voltammetry of Cu overlayers deposited on self-assembled-monolayers: field emission of electrons through a phenylene ethynylene oligomer", *J. Phys. Chem. B*, **2001**, *105*, 105-110.

19. C. Zhou, M.R. Deshpande, M.A. Reed, L. Jones II and J.M. Tour, "Nanoscale metal/self-assembled monolayer/metal heterostructures", *Appl. Phys. Lett.*, **1997**, *71*(5), 611-613.

20. W. Tian, S. Datta, S. Hong, R. Reifenberger, J.I. Henderson, C.P. Kubiak, "Conductance spectra of molecular wires", *J. Chem. Phys.*, **1998**, *109*(7), 2874-2882.

21. M.A. Reed, J. Chen, A.M. Rawlett, D.W. Price and J.M. Tour, "Molecular random access memory cell", *Appl. Phys. Lett.*, **2001**, *78*(23), 3735-3737.

22. E.W. Wong, C.P. Collier, M. Behloradský, F.M. Raymo, J.F. Stoddart and J.R. Heath, "Fabrication and transport properties of single-molecule-thick electrochemical junctions", *J. Am. Chem. Soc.*, **2000**, *122*, 5837-5840.

23. J. Chen, M.A. Reed, A.M. Rawlett and J.M. Tour, "Large on-off ratios and negative differential resistance in a molecular electronic device", *Science*, **1999**, *286*, 1550-1552.

24. C.P. Collier, E.W. Wong, M. Behloradský, F.M. Raymo, J.F. Stoddart, P.J. Kuekes, R.S. Williams and J.R. Heath, "Electronically configurable molecular-based logic gates", *Science*, **1999**, *285*, 391-394.

25. D.R. Jung, A.W. Czanderna and G.C. Herdt, "Interactions and penetration at metal/self-assembled organic monolayer interfaces", *J. Vac. Sci. Technol. A*, **1996**, *14*(3), 1779-1787.

26. L.S. Dake, D.E. King and A.W. Czanderna, "Ion scattering and X-ray photoelectron spectroscopy of copper overlayers vacuum deposited onto mercaptohexadecanoic acid self-assembled monolayers", *Solid State Sciences*, **2000**, *2*, 781-789.

27. G.L. Fisher, A.E. Hooper, R.L. Opila, D.L. Allara and N. Winograd, "The interaction of vapor-deposited Al atoms with $CO_2H$ groups at the surface of a self-assembled alkanthiolate monolayer on gold", *J. Phys. Chem. B*, **2000**, *104*, 3267-3273.

28. G.C. Herdt, D.R. Jung, A.W. Czanderna, "Penetration of deposited Ag and Cu overlayers through alkanethiol self-assembled monolayers on gold", *J. Adhesion*, **1997**, *60*(1-4), 197-222.

29. G.C. Herdt, D.E. King, A.W. Czanderna, "Penetration of deposited Au, Cu, and Ag overlayers through alkanethiol self-assembled monolayers on gold or silver", *Z. Phys. Chem.*, **1997**, *202*, 163-169.

30. P.E. Colavita, M.S. Doescher, A. Molliet, U. Evans, J. Reddic, J. Zhou, D. Chen, P.G. Miney and M.L. Myrick, "Effects of metal coating on self-assembled monolayers on gold. 1. Copper on dodecanethiol and octadecanethiol", *Langmuir*, **2002**, *18*, 8503-8509.

31. J. Kattner, H. Hoffmann, "Simultaneous determination of thicknesses and refractive indices of ultrathin films by multiple incidence medium ellipsometry", *J. Phys. Chem. B*, **2002**, *106*, 9723-9729.

32. S. M. Han, W. R. Ashurst, C. Carraro, R. Maboudian, "Formation of alkanethiol monolayer on Ge(111)", *J. Am. Chem. Soc.*, **2001**, *123*, 2422-2425.

33. C. W. Meuse, "Infrared spectroscopic ellipsometry of self-assembled monolayers", *Langmuir*, **2000**, *16*, 9483-9487.

34. O. Chailapakul, L. Sun, C. Xu, R. M. Crooks, "Interactions between organized, surface-confined monolayersabd vapor-phase probe molecules. 7. Comparison of self-assembling *n*-alkanethiol monolayers deposited on gold from liquid and vapor phases", *J. Am. Chem. Soc.*, **1993**, *115*, 12459-12467.

35. P.E. Colavita, P. G. Miney, R. Priore, L. Taylor, D. Pearson, M.L. Myrick, "Effect of metal coating on self-assembled monolayers on gold. 2. Copper on a conjugated oligomer SAM", in preparation.

36. G.L. Fisher, A. Hooper, R.L. Opila, D.R. Jung, D.L. Allara, N. Winograd, "The interaction between vapor-deposited Al atoms and methylester-terminated self-assembled monolayers studied by time-of-flight secondary ion mass spectrometry, X-ray photoelectron spectroscopy and infrared reflectance spectroscopy", *J. Electron Spectrosc.*, **1999**, *99*, 139-148.

37. D. R. Jeon, Ph. D. thesis, Pennsylvania State University, 1991.

38. F. Balzer, S.D. Jett, H.-G. Rubahn, "Non-linear optically active metal clusters in nanoscaled systems including self-assembled organic films", *Thin Solid Films*, **2000**, *372*, 78-84.

39. E.L. Smith, C.A. Alves, J.W. Anderegg, M.D. Porter, L.M. Siperko, "Deposition of metal overlayers at end-group-functionalized thiolate monolayers adsorbed at Au. 1. Surface and interfacial chemical characterization of deposited Cu-overlayers at carboxylic acd-terminated structures", *Langmuir*, 1992, *8*(11), 2707-2714.

40. D. Gador, C. Buchberger, A. Soukopp, E. Sokolowski, R. Fink, E. Umbach, "Interaction of magnesium with oriented diphenyl-carbonate films", *J. Electron Spectrosc.*, **1999**, *103*, 529-537.

41. K. Bammel, J. Ellis, H.G. Rubahn, "2-Photon laser observation of diffusion of Na atoms through self-assembled monolayers on a Au surface", *Chem. Phys. Lett.*, **1993**, *201*(1-4), 101-107.

42. F. Balzer, K. Bammel, H.G. Rubahn, "Laser investigation of Na atoms deposited via inert spacer layers close to metal surfaces", *J. Chem. Phys.*, **1993**, *98*(9), 7625-7635.

43. F. Balzer, H.G. Rubahn, "Laser-excited Na atoms near metallic and dielectric surfaces", *J. Electron Spectrosc.*, **1993**, *64-5*, 321-329.

44. F. Balzer, H.G. Rubahn, "Second-harmonic generation and shielding effects of alkali clusters on ultrathin organic films", *Nanotechnology*, **2001**, *12*(2), 105-109.

45. P. Zhang, Ph. D thesis, Pennsylvania State University, 1993.

46. V.P. Tolstoy, I.V. Chernyshova, and V.A. Skryshevsky, in *Handbook of Infrared Spectroscopy of Ultrathin Films*; John Wiley & Sons, Inc., Hoboken, New Jersey, 2003.

47. A.N. Parikh, D.L. Allara, "Quantitative determination of molecular structure in multilayered thin films of biaxial and lower symmetry from photon spectroscopies. I. Reflection infrared vibrational spectroscopy", *J. Chem. Phys.*, 1992, *96*(2), 927-945.

48. D.P. Woodruff & T.A. Delchar, in *Modern Techniques of Surface Science*, 2nd Ed.; Cambridge University Press: Cambridge, UK 1994.

49. R.G. Nuzzo, L.H. Dubois, D.L. Allara, "Fundamental studies of microscopic wetting on organic surfaces. 1. Formation and structural characterization of a self-consistent series of polyfunctional organic monolayers", *J. Am. Chem. Soc.*, **1990**, *112*, 558-569.

50. M.D. Porter, T.B. Bright, D.L. Allara, C.E.D. Chidsey, "Spontaneously organized molecular assemblies. 4. Structural characterization of *n*-alkyl thiol monolayers on gold by optical ellipsometry, infrared spectroscopy, and electrochemistry", *J. Am. Chem. Soc.*, **1987**, *109*(12), 3559-3568.

51. F. Bensebaa, T.H. Ellis, A. Badia, R.B. Lennox, "Probing the different phases of self-assembled monolayers on metal surfaces: temperature dependence of the C-H stretching modes", *J. Vac. Sci. Technol. A*, **1995**, *13*(3), 1331-1336.

52. F. Bensebaa, Ch. Bakoyannis, T.H. Ellis, "Order and disorder in self-assembled monolayers probed by FT-RAIRS", *Mikrochim. Acta [Suppl.]*, **1997**, *14*, 621-623.

53. F. Bensebaa, T.H. Ellis, A. Badia, R.B. Lennox, "Thermal treatment of *n*-alkanethiolate monolayers on gold, as observed by infrared spectroscopy", *Langmuir*, **1998**, *14*, 2361-2367.

54. L.H. Dubois, R.G. Nuzzo, "Synthesis, structure, and properties of model organic surfaces", *Annu. Rev. Phys. Chem.*, **1992**, *43*, 437-463.

55. C. Naselli, J. F. Rabolt, J.D. Swalen, "Order-disorder transitions in Langmuir-Blodgett monolayers. 1. Studies of two dimensional melting by infrared spectroscopy", *J. Chem. Phys.*, **1985**, *82*(4), 2136-2140.

56. R.G. Nuzzo, E.M. Korenic, L.H. Dubois, "Studies of the temperature-dependent phase behavior of long chain *n*-alkyl thiol monolayers on gold", *J. Chem. Phys.*, **1990**, *93*(1), 767-773.

57. P.E. Laibinis, G. M. Whitesides, D.L. Allara, Y.-T. Tao, A.N. Parikh, R.G. Nuzzo, "Comparison of the structures and wetting properties of self-assembled monolayers of *n*-alkanethiols on the coinage metal surfaces, Cu, Ag, Au", *J. Am. Chem. Soc.*, **1991**, *113*, 7152-7167.

58. L.H. Dubois, B.R. Zegarski, R.G. Nuzzo, "Temperature induced reconstruction of model organic surfaces", *J. Electron Spectrosc.*, **1990**, *54*, 1143-1152.

59. P. Fenter, P. Eisenberger, K.S. Liang, "Chain-length dependence of the structures and phases of $CH_3(CH_2)_{n-1}SH$ self-assembled on Au(111)", *Phys. Rev. Lett.*, **1993**, *70*(16), 2447-2450.

60. N. Camillone, P. Eisenberger, T.Y.B. Leung, P. Schwartz, G. Scoles, G.E. Poirier, M.J. Tarlov, "New monolayer phases of *n*-alkane thiols self-assembled on Au(111): preparation, surface characterization, and imaging", *J. Chem. Phys.*, **1994**, *101*(12), 11031-11036.

61. A. Badia, R. Back, R.B. Lennox, "Phase transitions in self-assembled monolayers detected by electrochemistry", *Angew. Chem. Int. Ed. Engl.*, **1994**, *33*(22), 2332-2335.

62. R. Bhatia, B.J. Garrison, "Phase transitions in a methyl-terminated monolayer self-assembled on Au(111)", *Langmuir*, **1997**, *13*(4), 765-769.

63. R.G. Snyder, M. Maroncelli, H.L. Strauss, V.M. Hallmark, "Temperature and phase behavior of infrared intensities: the poly(methylene) chain", *J. Phys. Chem.*, **1986**, *90*, 5623-5630.

64. X. Xiao, B. Wang, C. Zhang, Z. Yang, M.M.T. Loy, "Thermal annealing effect of alkanethiol monolayers on Au(111) in air", *Surface Sci.*, **2001**, *472*, 41-50.

65. J.-P. Bucher, L. Santesson, K. Kern, "Thermal healing of self-assembled monolayers: hexane- and octadecanethiol on Au(111) and Ag(111)", *Langmuir*, **1994**, *10*(4), 979-983.

66.  F. Schreiber, "Structure and growth of self-assembling monolayers", *Progress in Surface Sci.*, **2000**, *65*, 151-256.

67.  H.-J. Himmel, A. Terfort, C. Wöll, "Fabrication of a carboxyl-terminated organic surface with self-assembly of functionalized terphenylthiols: the importance of hydrogen bond formation", *J. Am. Chem. Soc.*, **1998**, *120*, 12069-12074.

68.  A.-A. Dhirani, R.W. Zehner, R.P. Hsung, P. Guyot-Sionnest, L.R. Sita, "Self-assembly of conjugated molecular rods: a high-resolution STM study", *J. Am. Chem. Soc.*, **1996**, *118*(13), 3319-3320.

69.  J.-H. Lii, N. Allinger, "Molecular Mechanics. The MM3 force field for hydrocarbons. 3. The van del Waals' potentials and crystal data for aliphatic and aromatic hydrocarbons", *J. Am. Chem. Soc.*, **1989**, *111*, 8576-8582.

70.  S. Pérez-Casas, J. Hernández-Trujillo, M. Costas, "Experimental and theoretical study of aromatic-aromatic interactions. Association enthalpies and central and distributed multipole electric moments analysis", *J. Phys. Chem. B*, **2003**, *107*, 4167-4174.

71.  F. L. Gervasio, R. Chelli, P. Procacci, V. Schettino, "Is the T-shaped toluene dimer a stable intermolecular complex?", *J. Phys. Chem. A*, **2002**, *106*, 2945-2948.

72.  M. Zharnikov, M. Grunze, "Spectroscopic characterization of thiol-derived self-assembling monolayers", *J. Phys. Condens. Matter*, **2001**, *13*, 11333-11365.

73.  S. Frey, V. Stadler, K. Heister, W. Eck, M. Zharnikov, M. Grunze, "Structure of thioaromatic self-assembled monolayers on gold and silver", *Langmuir*, **2001**, *17*, 2408-2415.

74.  C. M. Whelan, C. J. Barnes, C.G.H. Walker, N.M.D. Brown, "Benzenthiol adsorption on Au(111) studied by synchrotron ARUPS, HREELS and XPS", *Surface Sci.*, **1999**, *425*, 195-211.

75.  J.M. Tour, L. Jones II, D.L. Pearson, J.J.S. Lamba, T.P. Burgin, G. M. Whitesides, D.L. Allara, A.N. Parikh, S.V. Atre, "Self-assembled monolayers and multilayers of conjugated thiols, a,w-dithiols, and thioacetyl-containing adsorbates. Understanding Attachments between potential molecular wires and gold surfaces", *J. Am. Chem. Soc.*, **1995**, *117*, 9529-9534.

76.  W. Geyer, V. Stadler, W. Eck, M. Zharnikov, A. Gölzhäuser, M. Grunze, "Electron-induced crosslinking of aromatic self-assembled monolayers: negative resistes for nanolithography", *Appl. Phys. Lett.*, **1999**, *75*(16), 2401-2403.

77.  G. Yang, Y. Qian, C. Engtrakul, L.R. Sita, G. Liu, "Arenethiols form ordered and incommensurate self-assembled monolayers on Au(111) surfaces", *J. Phys. Chem. B*, **2000**, *104*, 9059-9062.

78.  M. Neumann, J. U. Mack, E. Bertel and F. P. Netzer, "The molecular structure of benzene on Rh(111)", *Surface Sci.*, **1985**, *155*(2-3), 629-638.

79.  S. Haq, D.A. King, "Configurational transitions of benzene and pyridine adsorbed on Pt{111} and Cu{110} surfaces: An infrared study", *J. Phys. Chem.*, **1996**, *100*(42), 16957-16965.

80.  A. Ricca, C.W. Bauschlicher Jr., "Interaction of a conjugated phenylene ethynylene trimer with a Au(111) surface", *Chem. Phys. Lett.*, **2003**, *372*, 873-877.

81.  L. Cai, Y. Yao, J. Yang, D.W. Price Jr., J.M. Tour, "Chemical and potential-assisted assembly of thioacetyl-terminated oligo(phenylene ethynylene)s on gold surface", *Chem. Mater.*, **2002**, *14*, 2905-2909.

82.  G. Varsányi, in *Assignments for vibrational spectra of seven hundred benzene derivatives*; John Wiley & Sons, Inc., New York, 1974.

83.  D. Lin-Vien, N. B. Colthup Editor, in *The Handbook of infrared and raman characteristic frequencies of organic molecules*, 1st Ed.; Academic Press, Boston, 1991.

84.  R.A. Nyquist, W.J. Potts, "Characteristic infrared absorption frequencies of thiol esters and related compounds", *Spectrochim. Acta*, **1959**, *7*, 514-538.

# 10

# PRINCIPLES AND APPLICATIONS OF SURFACE-PLASMON FIELD-ENHANCED FLUORESCENCE TECHNIQUES

Wolfgang Knoll[*#], Fang Yu[*], Thomas Neumann[*], Lifang Niu[#], Evelyne L. Schmid[#]

## 1. INTRODUCTION

Surface plasmon resonance spectroscopy (SPR) has become a widely accepted optical technique for the characterization of interfaces and thin films[1]. The underlying physical principles have been worked out[2, 3] and are summarized in great detail in the literature[4-6]. With the availability of commercial instruments[7, 8] applications of surface plasmon spectroscopies have been reported in many diverse fields of science and engineering.

Investigations that profit in a particular way from the sensitivity of SPR are (bio)affinity studies[9-11]. Here, one of the reaction partners is chemically bound to a thin organic biofunctional layer at the surface of a solid substrate. The latter is capable of carrying a surface plasmon mode and is in contact to the analyte solution. Changes that occur at the interfacial layer upon binding of analyte molecules from solution are equivalent to a change of the effective refractive index of this layer or of its effective thickness and can be monitored in real time and in a label-free mode of operation[1]. I.e., the mere presence of the analyte molecules, or more precisely, the enrichment of the analyte at the functionalized surface (above its bulk solution concentration) via the biorecognition and binding process generates a measurable sensor signal. Detectability limits are reached if the obtainable analyte density (enrichment) at the surface (i.e., within the evanescent tail of the surface plasmon mode, cf below) is too low or if the analyte

[*] Wolfgang Knoll, Fang Yu, Thomas Neumann, Max-Planck-Institute for Polymer Research, Ackermannweg 10, 55128 Mainz, Germany

[#] Wolfgang Knoll, Lifang Niu, Evelyne L. Schmid, Departments of Chemistry and of Materials Science, National University of Singapore, 10 Science Drive 4, Singapore 11754

molecules are simply too small to generate such an effective refractive index change. One way to overcome this limitation - at least to some extent – is the use of a quasi-three dimensional surface layer of a hydrogel or a polymer brush probed by the surface plasmon wave leading to an effective increase of the binding site density available for the analyte to bind to[12]. But even then many physiologically or biomedically relevant analyte concentrations (and/or the affinities to the corresponding binding partners) are too low to allow for the application of the label-free scheme of SPR.

In an attempt to overcome this limit of detection we recently introduced surface plasmon field-enhanced fluorescence spectroscopy (SPFS) [13] following an earlier report by Attridge et al[14]. The basic principle of this approach combines the excitation of a surface plasmon mode as an interfacial light source with the well-established detection schemes of fluorescence spectroscopy: the resonantly excited surface plasmon waves excite chromophores that are attached to the analyte either chemically[15-18] or by genetic engineering techniques. The emitted fluorescence photons are then monitored and analyzed in the usual way to give information about the behavior of the analyte itself.

This approach has common features with the well-known total internal reflection fluorescence (TIRF) spectroscopy[19] that is also a surface-sensitive and surface-specific detection method, but lacks, however, the enormous enhancement of the optical fields that can be obtained at resonant excitation of a surface plasmon wave which is responsible for the substantial sensitivity enhancement in bio-affinity studies.

In this regard, SPFS has more common features with surface-enhanced Raman spectroscopies (SERS) and we will show that certain aspects worked out for SERS and related techniques[20] are directly applicable also to SPFS.

In the following chapters we first will discuss some of the basic principles underlying the excitation of surface plasmons either in the Kretschmann configuration using a prism coupler or by employing an optical diffraction grating as the coupling element, introduce the optical field enhancements associated with surface plasmon modes, and discuss some practical issues that have to be dealt with in using SPFS.

Another important aspect that needs attention is the optical behavior of a chromophore near a metal surface[21]. The loss in fluorescence intensity by Förster energy transfer mechanisms operating for chromophores in the immediate vicinity of the metal substrate[22] has to be balanced against the decrease in the excitation probability for higher separation distances governed by the exponential decay of the evanescent surface mode normal to the metal substrate[23].

We then introduce a few examples for the use of SPFS, first in surface hybridization studies and then for antigen-antibody interaction assays. We will give, in particular, examples for different versions of fluorescence spectroscopy making use of, e.g., donor-acceptor energy transfer phenomena between correspondingly labeled binding partners.

Finally, we will give an example as to how the technique can be used in a microscopic mode of operation for multiple parallel read-outs of binding events in an array format[24].

We should point out that we will not be concerned with many other issues related to fluorescence spectroscopy that are also relevant in the case of evanescent wave excitation in general and SPFS, in particular. E.g., we will not deal with the details of the distance dependence of fluorescence emission in a quantitative way[22], and are not concerned with orientational effects taking into account that surface modes have a particular polarization[2, 3, 25]. Moreover, we will ignore any lifetime effects that could be probed by time-revolved SPFS.

## 2. SURFACE PLASMONS AS INTERFACIAL LIGHT

The physical nature of surface plasmon SP waves propagating at a metal /dielectric interface, in particular, their dispersion behavior, i.e., their energy-momentum ($\omega$, $k_{sps}$) relation, does not allow for their direct excitation by simply reflecting a laser beam off that metal /dielectric interface. This is illustrated in Figure 1, which gives the dispersion curve for surface plasmons (denoted as $+PSP^0$) and the "light line" for plane waves, i.e., $a = c \cdot k$. For a given laser energy, $\omega_{ex}$, the range of possible values for the projection of the photon wave vector along the propagation direction of the surface plasmons (which has to be matched to $k_{sps}$) extends from 0 at normal incidence (Point A in Figure 1) to the full momentum $k = a /c$ (Point B in Figure 1), which would be reached at grazing incidence. However, momentum matching is not possible because the momentum of the surface plasmon wave at the same energy is substantially higher (cf. Point C in Figure 1).

Various coupling schemes between the plane waves of a laser and the evanescent surface modes have been introduced, with the prism coupler in the so-called Kretschmann configuration[26] being the most prominent one (cf. also Figure 8). According to the dispersion scheme given in Figure 1 the prism effectively "slows-down" the laser photons, i.e., for any given energy the corresponding momentum increases from $k = a /c$ to $k = (\omega/c) \cdot n_p$, with $n_p$ being the refractive index of the prism[3]. The light line has a

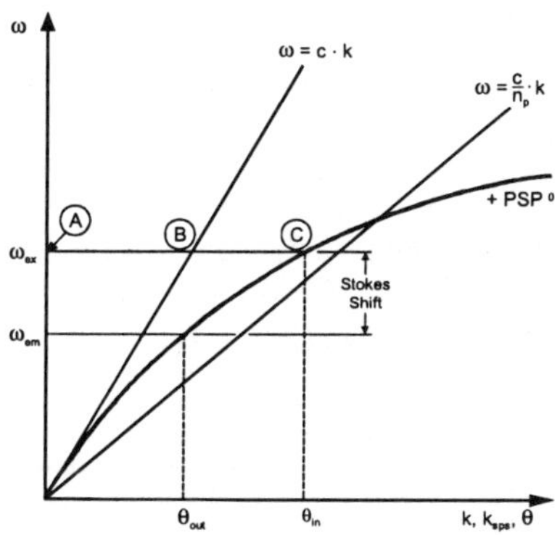


**Figure 1.** Dispersion curves ($\omega$-versus-k) of a surface plasmon mode $\left(+ PSP^0\right)$, a photon in air $\left(a = c \cdot k\right)$ or in the coupling prism ($a = c \cdot k / n_p$ with $n_p$ being the refractive index of the prism). The points A, B and C are explained in the text. $\omega_{ex}$ denotes the excitation wavelength (energy); $\omega_{em}$ is the Stokes shifted fluorescence energy.

reduced slope (equivalent to the reduced phase velocity of the photons in the prism material), thus allowing for the required matching between the energy and momentum of the photons and of the surface plasmon modes (Point C of Figure 1). (We should note that this is an approximate qualitative consideration with the correct quantitative description being given by the Fresnel algorithm which takes properly into account the finite thickness of the noble metal layer, as well as, the partial waves that are reflected and / or reradiated at the prism / metal and metal / dielectric interface, respectively).

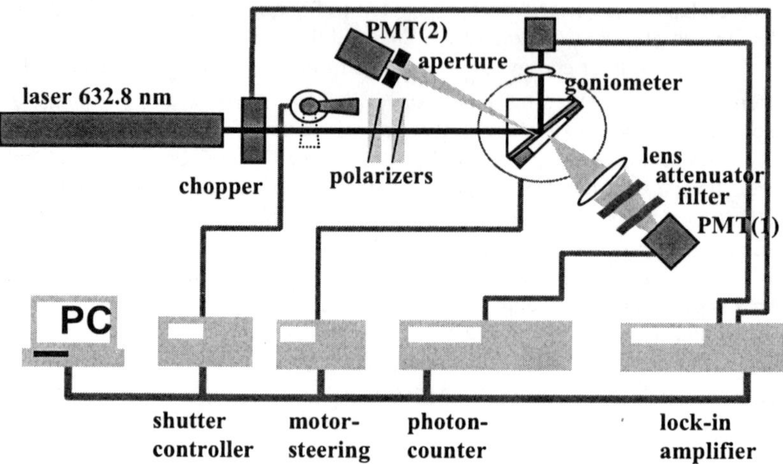


**Figure 2.** Experimental set-up for SPR measurements, as well as, for surface-plasmon fluorescence spectroscopy with the emitted fluorescence light being observed either from the base of the prism (PMT1) of re-radiating via the prism (PMT2).

A corresponding experimental setup is shown schematically in Figure 2. The beam of a p-polarized laser light source is reflected off the metal-coated (ca 50 nm in thickness) base of the coupling prism, and the reflected intensity monitored with a photodiode. The use of a light-chopper allows for lock-in detection. The coupling prism with the attached sample cell and the photodiode are mounted to a two-circle-goniometer such that angular $\theta$ - $2\theta$ scans allow for the recording of the reflectivity, R($\theta$), i.e., the reflected intensity $I_r(\theta)$, divided by the incoming laser intensity $I_0$, as a function of the angle of incidence, $\theta$:

$$R(\theta) = I_r(\theta) / I_0 \qquad (1)$$

Such an angular scan in this attenuated total internal reflection scheme is given in Figure 3(a) and (b), respectively, for the two most commonly used metal layers made from Ag and from Au, respectively. Compared to the reflectivity measured for a bare prism without the metal layer given in Figure 2(c) just as a reference situation, the reflected intensity at angles smaller than the critical angle for total internal reflection, $\theta_c$, is much higher because the thin metal film acts as a slightly transparent mirror which reflects most of the incoming laser light. However, $\theta_c$ is still clearly visible and serves a helpful practical purpose: its exact angular position depends only on the refractive index of the prism material (which is known rather precisely) and of the bulk dielectric medium (for the cases discussed in this paper an aqueous buffer solution) but is insensitive to a thin interfacial layer at the metal surface used in the bioaffinity studies. Hence, $\theta_c$ can be used as an angular calibration value for the experimental setup.

For incident angles slightly above the critical angle the reflectivity shows a sharp dip and goes through a minimum of the reflectivity that can be virtually zero. This decrease in R is a typical experimental signature for the excitation of a surface plasmon mode. However, the minimum in R is not at the angle for maximum SP excitation! This can be best seen by calculating the optical field at the surface of the metal layer in this Kretschmann setup, again as a function of the angle of incidence. This can be easily done in the Fresnel algorithm, which is generally also used to calculate reflectivity curves (and, in fact, all the R-versus-$\theta$ curves displayed in Figure 3 are Fresnel-based simulations). The resulting intensity at the metal / dielectric interface, $I_s(\theta)$, scaled to the incoming intensity, $I_0$, is also shown in Figure 3(a) for the case of a thin Ag film and in Figure 3(b) for a thin Au film, respectively, both deposited onto a high index glass prism (LaSFN9, $n_p = 1.85$ @ $\lambda = 633$ nm) in contact to an aqueous phase ($n_s = 1.33$ @ $\lambda = 633$ nm). The peak optical intensity and hence the 'true' SP resonance angle is reached at an angle somewhat smaller than the minimum angle in the angular reflectivity scan. This interfacial intensity of the excited surface plasmon mode determines the excitation of chromophores and, hence, will determine the angular dependence of the observed fluorescence intensity.

The minimum in R, however, originates from an interference phenomenon: As one sweeps in the angular scan through the SP excitation the relative phase of this optical resonance changes (as it does for any driven damped oscillator), and so does the phase of that fraction of the SP light that back-couples through the metal layer and out via the prism, where it interferes with a directly reflected part of the laser light. For a particular thickness of the metal layer this back-coupled SP light equals this directly reflected light in amplitude and with a phase, which is just different by 180 degrees and, hence, allows for destructive interference resulting in a vanishing reflectivity.

The details of this interference depend on the materials involved, in particular, on the optical properties of the metal used. For Ag with its low imaginary part $\varepsilon''$ of the complex dielectric function $\tilde{\varepsilon}_{Ag} = \varepsilon' + i\varepsilon'' = -16 + i0.6$ (@ $\lambda = 633$ nm) the resonance is rather undamped and hence the reflectivity curve relatively sharp with an angular width at R = 0.5 of only $\Delta\theta \approx 1.5$ deg. Consequently, the phase change of the SP wave re-radiating via the prism changes its phase over a relatively narrow angular range and the 180° degree out-of-phase angle is reached just shortly after the peak excitation. Hence, the angle of maximum SP intensity and the minimum reflectivity angle are separated by only $\Delta\theta = 0.2$ deg (cf. Figure 3(a)). This is significantly different in the case of Au as the metal layer with its substantially higher imaginary part: $\tilde{\varepsilon}_{Au} = -13 + i \cdot 1.3$ (@ $\lambda = 633$ nm). This higher damping leads to a broader resonance and, as a consequence, to a minimum in R which is also much more separated from the peak intensity than in the case of Ag. As can be seen in Figure 3(b) the shift for Au amounts to $\Delta\theta = 0.6$ deg.

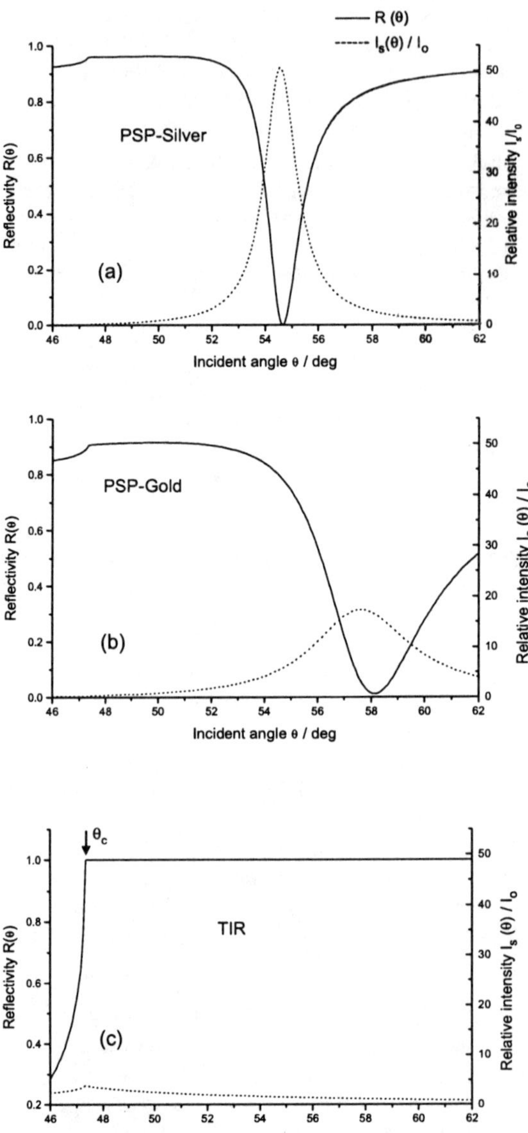



**Figure 3.** Simulated reflectivity-versus-incident angle θ (angular) scans (R(θ) full curves) for a sample configuration consisting of prism / Ag layer / $H_2O$ (a), prism / Au layer / $H_2O$ (b) and a mere dielectric interface prism / $H_2O$ (TIR). The dotted curves are the corresponding optical intensities (scaled to the incoming laser intensity $I_0$) at the metal / $H_2O$ and glass / $H_2O$ interface, respectively.

It should be pointed out that both, the reflectivity curve $R(\theta)$, as well as, the angular distribution of the optical field intensity $I_s(\theta)$ are described by Fresnel's equations, i.e., can be calculated based on Maxwell's theory of such a layered architecture. Once $R(\theta)$ is calculated by using the dielectric functions and the thicknesses of the involved materials (prism, metal layer, functional interfacial binding matrix, dielectric superstrate) the optical intensity and its angular dependence $I_s(\theta)$ can be calculated without any additional free parameter. It is important to keep in mind that for fluorescence spectroscopy with surface plasmon excitation it is the optical intensity $I_s$ (and its angular dependence) that controls the excitation process of the fluorophores.

Another feature important for SPFS can also be seen in Figure 3: the lower damping of the SP resonance for Ag results in a much higher peak intensity (again in analogy to any other resonance phenomenon: the lower the damping the higher the resonance amplitude) leading to an enhancement at resonance of about a factor of 50 compared to the incoming laser intensity. For Au this factor is only 16, owing to the higher damping described by the larger $\varepsilon''$. However, given the practical aspect that Au is chemically much more inert compared to the rather reactive Ag layer most of the experiments described below were done with Au as the substrate material in contact with the aqueous buffer phase.

The surface plasmon related optical enhancements at resonance are in sharp contrast to the maximum value one might find at the critical angle $\theta_c$ for a mere total internal reflection geometry displayed in Figure 3(c). Here, the coherent superposition of the incoming and the outgoing light beam results in an enhancement of a factor of 4 only! However, this has been utilized in the past for a surface-specific fluorescence technique, i.e., the total internal reflection fluorescence spectroscopy (TIRF)[19]!

In addition to the angular position and peak intensity (at the metal surface) of surface plasmon modes their extension into the dielectric medium is another important aspect for SPFS. Again, based on the Fresnel algorithm we have calculated the optical intensity normal to the layered architecture: prism / Au layer / aqueous buffer and display the result obtained at resonance in Figure 4 (full curve). One can see that the intensity $I_s$, indeed, peaks at the metal /dielectric interface and decays exponentially into both media.

The decay is much faster into the Au metal owing to the screening effect of the nearly free electron gas. Relevant for our spectroscopic purposes, however, is only the evanescent character of the surface mode, which leads to an extension of the optical field into the aqueous phase of about 150 nm (defined by the 1/e decay of the peak intensity). This means that only chromophores that are within this exponential decay of the excitation light will be reached for fluorescence excitation and emission. However, this field then will be much stronger than that of the in-coupling laser beam.

The other - though less frequently used - coupling scheme for surface plasmon excitation is based on a grating structure at the metal / dielectric interface. This is schematically shown in Figure 5. A substrate with a sinusoidal surface corrugation (typically fabricated holographically by lithography with photoresist and reactive ion etching) is coated with a thick (d > 150 nm) noble metal layer constituting the grating structure in contact to the dielectric medium.

As has been discussed in great detail, the grating effectively imposes a Brillouin Zone structure upon the surface plasmon dispersion, resulting in various dispersion branches originating at $k = 0$, $\pm\dfrac{2\pi}{\Lambda}$, ..., with $\Lambda$ being the grating periodicity[20, 27]. This picture is equivalent to a momentum matching scheme in which the photon wavevector projection along the x-direction (the surface plasmon propagation direction) and a multiple of the grating vector, $\vec{G}$, with $\left|\vec{G}\right| = 2\pi/\Lambda$, match the surface plasmon wavevector

$$\vec{k}_{sps} = \vec{k}_{ph}{}^{x} \pm m \cdot \vec{G} \tag{2}$$

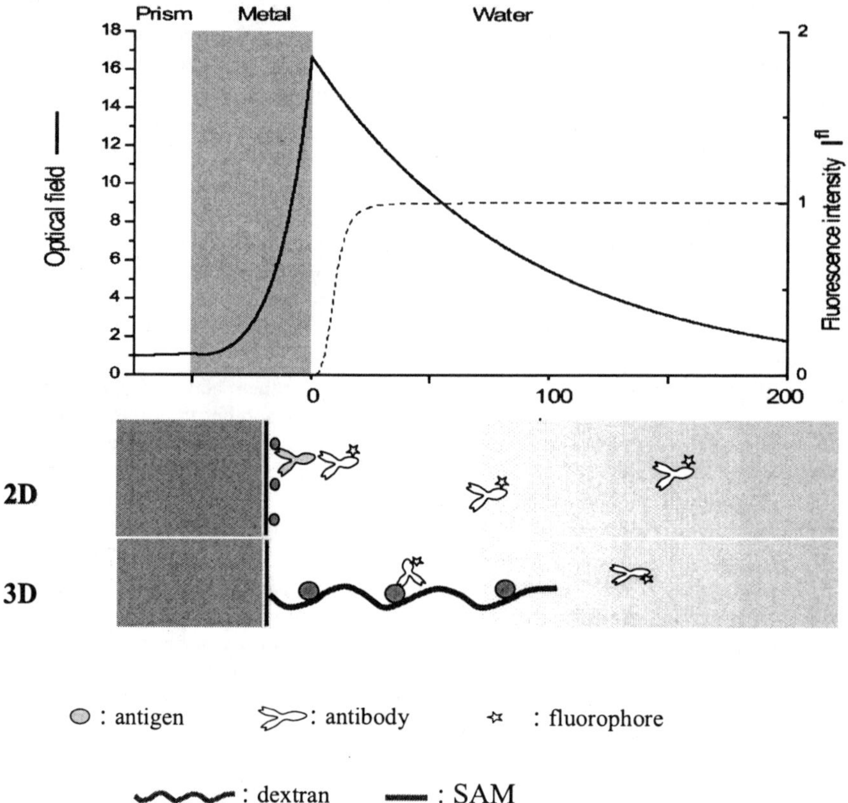


**Figure 4.** Optical Intensity (at SPR resonance) for the configuration prism / Au layer / H₂O normal to the interface (along z), full curve. The dashed curve qualitatively describes the fluorescence emitted from a chromophore that is approaching a metal surface. For comparison of distances, the schematics of biofunctional architectures in 2D and 3D, respectively, are added.

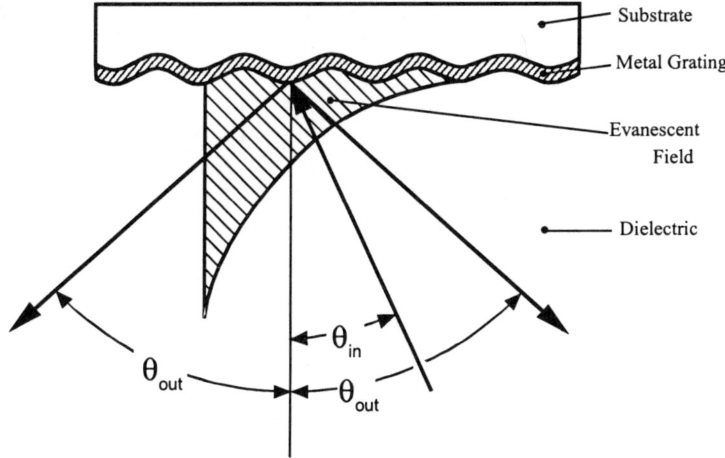


**Figure 5.** Surface plasmon excitation using a grating coupler at $\theta_{in}$, with observation of fluorescence light at $\pm\theta_{out}$.

A section of the Brillouin scheme is given in Figure 6. A dispersion branch, originating at $k = 2\pi/\Lambda$ traveling in the −x direction (marked - $PSP^{+1}$) and a forward traveling mode originating at - $2\pi/\Lambda$ (marked ⊣ $PSP^{-1}$) intersect at k = 0 (opening a gap in energy through weak interaction). Laser photons of energy $a_{ex}$ incident at $\theta_{in}$ then allow for energy and momentum matching required for resonant excitation of surface plasmon modes with their typical evanescent wave character. Note that in this coupling scheme the laser is incident from the front side thus passing through any sample cell attached to the grating coupler.

## 3. CHROMOPHORES NEAR METAL SURFACES

In addition to the considerations related to using surface plasmons as the excitation light source a second important issue that needs attention is the behavior of excited chromophores near metal surfaces. In a simple-minded approach one would argue that given the evanescent character of an SP wave (cf. Figure 4) the chromophores should be as close to the substrate as possible in order to maximize the excitation probability, which increases with the increase in optical intensity. However, already some 20 – 30 years ago the behavior of chromophores excited near metal surfaces has been investigated, both experimentally[22] and theoretically[25], in great detail.

The somewhat complex behavior is summarized in Figure 7 in a simplified scheme that contains, however, the essentials that are relevant for SPFS. Three different distance regimes are important. For dye molecules very close to the metal surface (Figure 7(a)) the 'classical' Förster energy transfer mechanism operating between an excited chromophore as the donor molecule and the metal substrate as the acceptor system leads to a dramatic loss in emission probability and, hence, decrease in fluorescence light intensity. In the semi-classical treatment given by Hans Kuhn the decay of fluorescence intensity follows a relation that depends on the 4$^{th}$ power of the dye-metal separation distance d: [22]

$$\frac{I^{fl}(d)}{I^{fl}_{\infty}} = \left[ 1 + \left( \frac{d_0}{d} \right)^4 \right]^{-1} \qquad (3)$$

Here, $I^{fl}_{\infty}$ is the fluorescence intensity far away from the metallic substrate and $d_0$ is called the Förster separation distance which indicates the separation at which the intensity decayed to 50% of its value at infinite separation. This 'quenching profile' is also given in Figure 4 (dashed curve). Typical values reported for $d_0$ are in the range of $5 - 10$ nm[22].

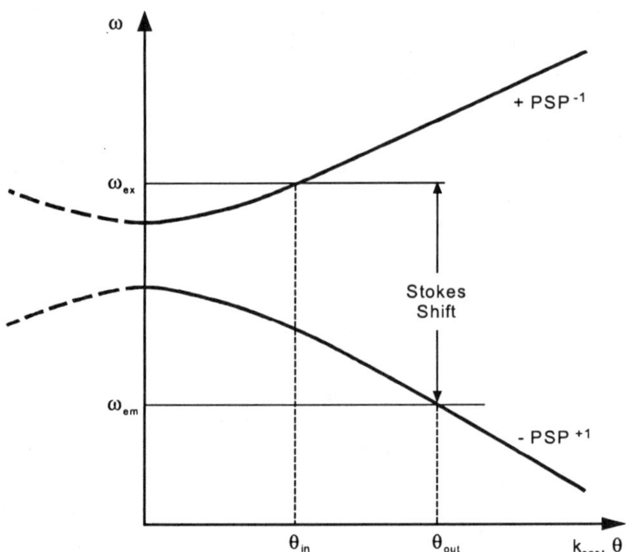


**Figure 6.** Brillouin zone scheme near $k = 0$ with a back-traveling plasmon mode, - $PSP^{+1}$ (originating from $k = 2\pi / \wedge$), and the forward-traveling mode, $\dashv PSP^{-1}$, originating from $k = -2\pi / \wedge$ intersecting at $k = 0$, however, with a gap in energy, lifting the degeneracy. Resonant excitation at $\omega_{ex}$ occurs at $\theta_{in}$, a Stokes shifted plasmon mode is emitted at $\theta_{out}$ (cf. also Figure 5).

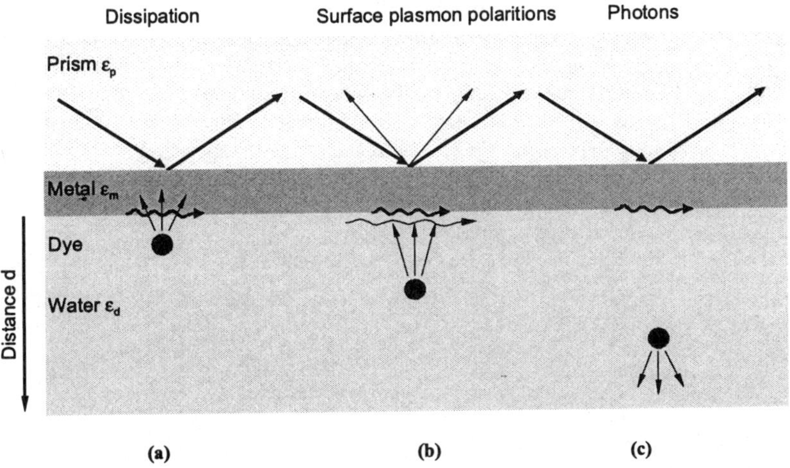


**Figure 7.** Schematic of the different electronic coupling regimes for a chromophore in water (dielectric constant $\varepsilon_d$) at different distances d to a metal film ($\varepsilon_m$) surface, upon excitation of a surface plasmon mode from the prism side ($\varepsilon_p$). (a) The chromophore is within a Förster radius (5 – 7 nm) and hence most of the fluorescence is dissipated in the metal. (b) For slightly larger separation distances, the PSP mode excites the chromophore, but some of the red-shifted fluorescence light is coupled back to the metal, exciting an energetically lower-lying surface plasmon state. This PSP-mode re-radiates via the prism in a cone at an angle corresponding to its dispersion curve. (c) If the chromophore is placed further away, but still within the evanescent tail of the PSP mode, the normal de-excitation via emission of a fluorescence photon dominates.

The most complex situation is sketched in Figure 7(b) for intermediate separation distances: the chromophores excited either by plane waves from the dielectric side or by a surface plasmon mode excited from the prism side relaxes vibronically to the bottom of the excited state level of the chromophore but then can back-couple to the metal, thereby exciting a red-shifted surface plasmon mode. This mode in turn can re-radiate via the prism (or the grating) and leads to an enhanced fluorescence emission[28]. The optimum dye-metal separation for this decay mechanism has been reported to be in the range of d = 20 nm[29].

For still larger separation distances no efficient coupling for energy transfer between the excited chromophore and the metal substrate occurs anymore and hence a nearly unperturbed emission of fluorescence light can be observed (Figure 7(c)).

Although the details of these electronic coupling schemes may be rather complicated (depending, e.g., on the chromophore orientation) it is still possible to draw a few qualitative conclusions for the practical situation in SPFS. No matter which mechanism for the dye-metal coupling may be operational leading to an undesired loss of fluorescence emission it is efficient only for a relatively short interaction distance regime of a few 10 nm. However, as one can see from Figure 4 the evanescent character of a surface plasmon mode as the excitation wave reaches out to much larger distances and, hence, allows for a compromise that offers still a substantial enhancement for SPFS. E.g., if a chromophore is placed 20 – 30 nm away form the metal surface (achieved by designing the binding matrix at the sensor surface in the corresponding way) not much fluorescence is lost by quenching into the metal, however, the optical intensity of the surface wave is still more than an order of magnitude higher than in the coupling laser beam (even in the case of a Au substrate).

## 4. RECORDING FLUORESCENCE FROM CHROMOPHORES EXCITED BY SURFACE PLASMON WAVES

The simplest way of detecting fluorescence light emitted from chromophores that are excited within the evanescent field of a surface plasmon wave is schematically given in Figure 2 (marked PMT (1)): a photomultiplier detects all photons that are emitted from the base of the prism upon SP excitation, are collected by a high NA lens and pass an interference filter set tuned to the spectral position of the peak fluorescence emission of the respective chromophore. For most situations this will be slightly red-shifted compared to the excitation wavelength (we are not dealing here with up-conversion[30] or two-photon-fluorescence experiments[31]).

For bio-affinity studies a typical experimental situation would involve a binding matrix at the sensor surface with the covalently attached ligand molecules (oligonucleotide catcher probes, antigens, antibodies, etc., cf. Figs. 4, 11 and 15) in contact to the sample cell that contains the buffer solution. The latter is schematically given in Figure 8. In a kinetic scan, at t = 0 the buffer is replaced via the liquid handling system by a solution containing the chromophore-labeled analyte which then binds to the surface-attached binding partners (cf. Figs. 12 and 16(a), respectively).

A typical angular scan in the Kretschmann format taken before and after the binding of the chromophore-labeled analyte is given in Figure 9. The analyte was the 15mer oligonucleotide T2 (cf. Table I) hybridizing to the surface attached probe oligonucleotide P2, the chromophore covalently attached to it was the cyanine dye Cy3, a much-used fluorophore. Its excitation maximum is at $\lambda = 548$ nm, hence the green HeNe laser line at $\lambda = 543$ nm had to be used for SP excitation.

However, at this wavelength a surface plasmon wave in pure Au is already highly damped due to the onset of an interband transition (from the d-band to the Fermi level) at c. $\lambda = 520$ nm with a correspondingly large $\varepsilon''$ for Au already at $\lambda = 543$ nm.

Therefore, a combination substrate was used which was composed of a 45nm thin Ag layer, covered by 5 nm of Au, evaporated subsequently. As a result, even with an excitation laser line of $\lambda = 543$ nm a reasonably narrow resonance indicates the excitation of a surface plasmon mode with a considerable optical field enhancement. However, at

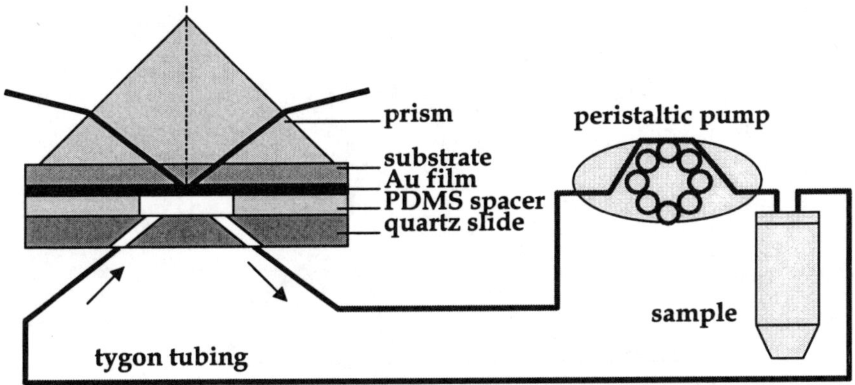


**Figure 8.** Schematics of the sample cell attached to the Au-coated coupling prism in this Kretschmann configuration, and connected to a closed liquid handling system.

**Table 1.** Oligonucleotide sequences of the probe and target strands

Probe sequences
(P1)    5'-Biotin-TTT-TTT-TTT-TTT-TTT-TGT-ACG-TCA-CAA-CTA-3'
(P2)    5'-Biotin-TTT-TTT-TTT-TTT-TTT-TGT-ACA TCA-CAA-CTA-3'
(P3)    5'-Biotin-TTT-TTT-TTT-TTT-TTT-TGT-ACG-TGA-CAA-CTA-3'

Target sequences
(T1)    3'-ACA TGC AGT GTT GAT dye-5'
(T2)    3'-ACA TGT AGT GTT GAT dye-5'
(T3)    3'-ACA TGC ACT GTT GAT dye-5'

the same time, the surface-inertness of the Au surface of the substrate gives the system the required chemical stability.

The general architecture for the hybridization studies in general and for the experiment presented in Figure 9, in particular, is shown in Figure 11(a). A binary mixed self-assembled monolayer (SAM) composed of 5 mol % of a biotinylated thiol-derivative mixed with 95 mol % of an OH-terminated thiol system is covered by a monolayer of streptavidin, a tetrameric protein with 4 binding sites of high affinity for biotin $\left(K_A = 10^{15} M^{-1}\right)$[9, 10]. Two of the binding sites are facing the buffer solution and are used to couple biotinylated 30mer oligonucleotides as catcher probes for the following hybridization experiments[32]. The charge density of the DNA oligonucleotides presumably

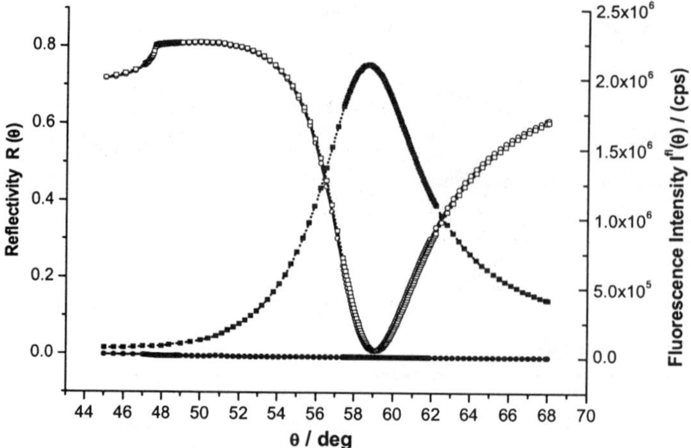

**Figure 9.** Reflectivity, $R(\theta)$, and fluorescence intensity, $I^{fl}(\theta)$, as a function of incident angle $\theta$, before (-○-, -●-) and after (---□---, ---■---) hybridization of a 15mer chromophore-labeled oligonucleotide target and a surface-attached probe oligonucleotide strand.

prevents a second strand binding to the second binding site once the first probe is attached. A corresponding angular scan of the reflectivity, as well as, of the fluorescence intensity emitted prior to the target hybridization are given in Figure 9. The reflectivity shows the relatively narrow resonance curve expected for this combination substrate with a minimum in reflectivity reaching nearly zero. The fluorescence, on the other hand, only shows the flat background level.

After the target was allowed to hybridize to the sensor surface a minimal shift of the reflectivity curve indicates the small increase of the optical thickness of the surface layer, too small to be evaluated quantitatively. However, the angular scan of the fluorescence intensity now shows the pronounced excitation profile as one sweeps in this angular scan through the resonance.

The observed fluorescence has at least 3 different contributions that pass the filters (ignoring stray light at the excitation wavelength that may not be suppressed by the filter set with sufficient efficiency): for angles below the critical angle for total internal reflection, $\theta_c$, a small fraction of the incoming laser light passes through the metal layer and is transmitted through the sample cell exciting all the chromophores that were dissolved in the bulk solution. This fluorescence light depends on the cell thickness and, in particular, on the analyte concentration in a linear way (assuming that no self-quenching in the solution occurs and no saturation effect of the photomultiplier tube has to be taken into account) but its intensity decreases as one reaches $\theta_c$.

In the angular range of the surface plasmon excitation there are still 2 contributions to the detected fluorescence light: the first one originates from chromophores in the bulk solution that are close enough to be reached by the evanescent SP wave, i.e., within a layer of c. 150 – 200 nm above the metal surface (cf. Figure 4). The second contribution

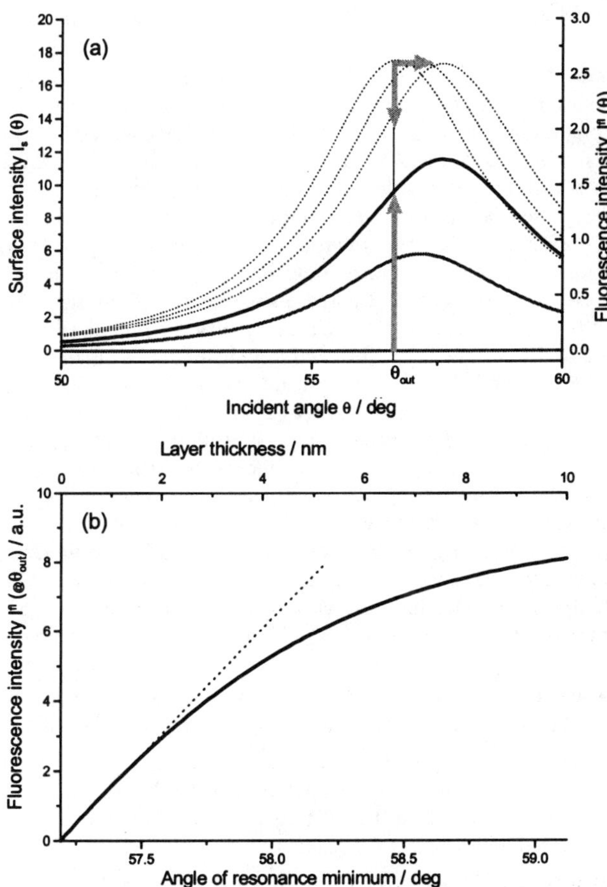



**Figure 10.** (a) Surface intensity (dotted curves) and resulting fluorescence intensity (full curve) for a bare Au / water interface, and with a thin dielectric coating of 2.5 or 5 nm, respectively, doped with a fictive fluorescent dye system. The corresponding shift of the PSP resonance angle results, at a fixed angle of observation, e.g. at $\theta_k$, in a less than linear increase of the fluorescence intensity (cf. the different shaded arrows). (b) Simulation of the deviation of a linear fluorescence intensity increase from an increase in layer thickness, equivalent to a shift of the resonance angle for PSP excitation. Note that only for very thin layers, i.e. for d < 2 nm, a linear approximation holds.

comes from chromophores that have found their binding partners at the sensor surface and now occupy the binding sites. In most cases, one is interested in only the fluorescence signal from this bound fraction of the analyte. In a general sense its relative detectability depends on the number density of binding sites at the functionalized sensor surface, compared to the weighted average number density of analyte molecules within the thin layer of the bulk solution with a thickness given by the exponentially decaying plasmon field.

Another special feature of surface plasmon fluorescence spectroscopy that needs special attention concerns an experimental situation in which the chromophore-labeled analyte, e.g., a protein, upon binding to the sensor surface not only adds to the observed fluorescence intensity but also induces – via its contribution to the optical thickness of the surface layer – to a shift in the angular SP resonance position. As it is schematically indicated in Figure 10 this can lead to a detuning of the surface plasmon resonance and, hence, to a loss of excitation probability. In order to illustrate this, Figure 10(a) displays a few simulated angular scans of the optical field intensity for 3 different analyte layer thicknesses, as well as, the expected fluorescence intensity. The first set of curves corresponds to the bare sensor surface, the two other curves simulate the situation of an analyte layer of 2.5 and 5 nm thickness, respectively, resulting in an angular shift of the peak position of the optical field. The fluorescence peak intensity shows the expected increase by a factor of 2, however, the position of the maximum has shifted according to the shift of the surface plasmon resonance conditions (cf. the horizontal gray arrow).If the fluorescence is monitored at a fixed angle of incidence this leads to a gradual loss of excitation intensity (gray arrow pointing down) and, as a consequence, to a lower increase in fluorescence intensity than would correspond to the bound analyte concentration (gray arrow pointing up). The consequence would be an underestimation of the fluorescence intensity and, hence, analyte binding as it is summarized in Figure 10(b). Here, the (simulated) fluorescence intensity is plotted as a function of the SP resonance angle position, which is – for these small shifts – a linear function of the amount of bound analyte.

One can also see from Figure 10(b) that for very small shifts corresponding to an optical thickness of the analyte layer of d < 2 nm ($@ n_f = 1.5$ in buffer of $n_s = 1.33$) this effect can be neglected. For most practical situations this, indeed, will be fulfilled: SPFS will be mostly applied in cases for which the analyte is so small or available only at such a low surface coverage that no shift in the SPR curve at all can be observed. In fact, for all the examples discussed below this is the case!

A particular detection concept, based upon the back-coupling, by which the excited chromophore upon deexcitation to the electronic ground state excites a SP mode at the metal / dielectric interface, was already introduced in Figure 7(b). This red-shifted (Stokes shifted) surface plasmon reradiates according to its dispersion curves (cf. Figs. 1 and 6, respectively) at the corresponding angle, $\theta_{out}$, indicated in Figure 2 for the case of a prism coupler (cf. also the thin arrows and the dashed ellipse in Figure 7(b) indicating the emission cone from these SP modes) and in Figure 5 for the case of a grating coupling scheme. Depending on the chromophore / metal surface separation this can be a very efficient (additional) enhancement mechanism for the sensitive recording of fluorescence light[27]. For the Kretschmann configuration this requires the recording of the fluorescence light that is emitted by back-coupling via the prism by a second detection unit (PMT(2) in Figure 2).

## 5. SURFACE HYBRIDIZATION STUDIES

In the following, a few examples for the use of surface plasmon fluorescence spectroscopy in DNA hybridization studies will be given. The general architecture of the sensor surface layer has already been introduced and is summarized in Figure 11.

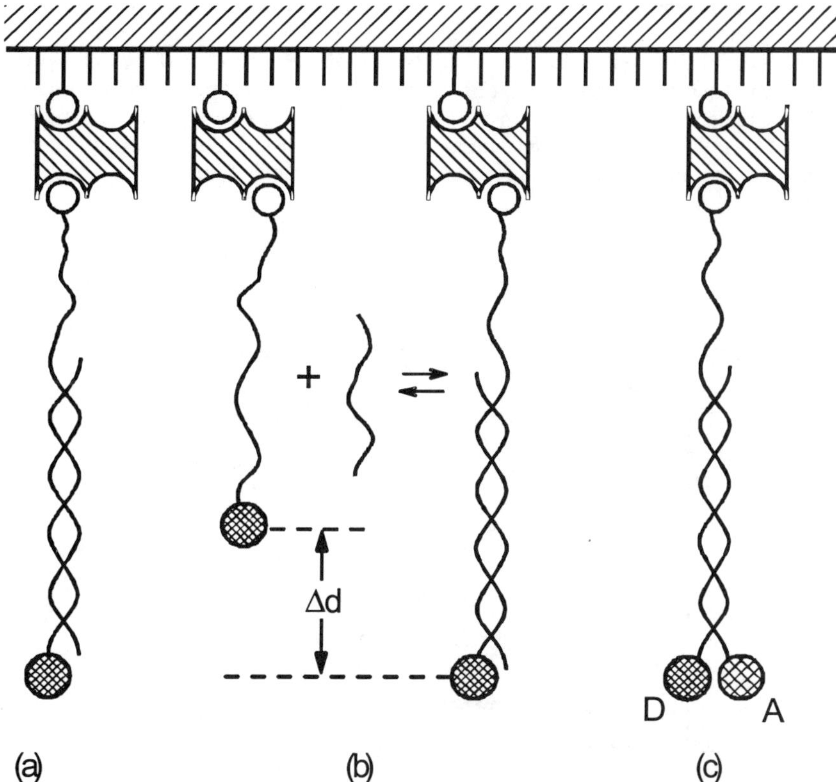


**Figure 11.** Various surface architectures and experimental schemes for surface hybridization studies: (a) a chromophore-labeled analyte hybridizes to an unlabeled surface-attached probe oligonucleotide strand; (b) a chromophore-labeled, surface-attached probe strand stretches upon hybridizing to an unlabeled target analyte strand, thus placing the chromophore further away (by Δd) from the quenching metal surface; (c) scheme for a fluorescence resonance energy transfer (FRET) experiment between a donor-dye labeled surface-attached probe oligonucleotide strand and an acceptor-dye labeled analyte target strand hybridized from solution.

Figure 11(a) schematically gives the situation in which the "ligand" – in this case a DNA oligonucleotide probe strand – is surface attached, and the analyte – the corresponding target strand – carries the chromophore. Upon binding, i.e., hybridization, the chromophores will be placed within the enhanced optical field of a resonantly excited surface plasmon mode and emit fluorescence photons. This was the experimental situation discussed in connection with Figure 9.

The excellent signal to noise of the fluorescence intensity allows then for a number of more detailed and, most importantly, quantitative studies of surface hybridization reactions. Three examples for the determination of reaction rate constants in a kinetic experiment are given in Figure 12. It should be noted that this type of experiment is possible only because SPFS can be applied on-line with the sensor surface in contact with the flow cell, hence, monitoring the binding reactions is real time. The examples given document the association- and dissociation reactions between a surface-bound catcher probe oligonucleotide strand (P2, cf. Table I) and three types of target sequences: one 15mer oligonucleotide (T2, Table I) is fully complementary (MM0) to the specific sequence of the probe, another one (T1, Table I) exhibits one mismatching base pair (MM1) by replacing a single base in the middle of the 15mer target sequence, whereas the $3^{rd}$ (T3, Table I) has 2 mismatching base pairs (MM2). The detailed molecular structures of the employed probe and target strands are given in Table I.

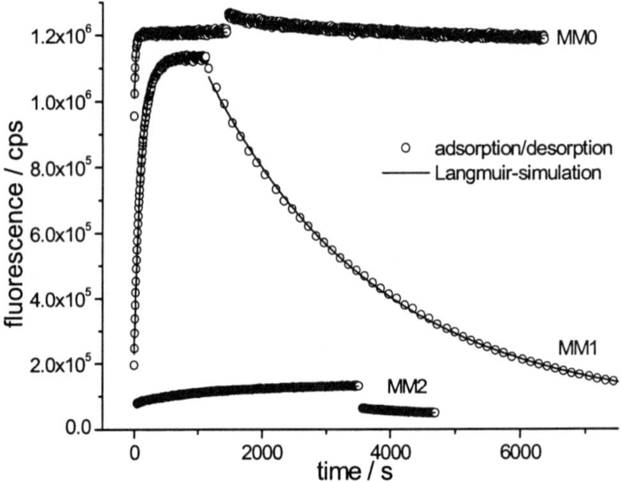


**Figure 12.** Hybridization kinetics (circles) of probe P2 with targets T1(MM1), T2(MM0) and T3 (MM2) at a target concentration of 1 µM together with the corresponding Langmuir-simulations (lines). Note, that upon rinsing with analyte-free PBS buffer MM1 can be desorbed while the MM0 duplex is nearly stable. MM2 at 1µm barely binds to the probe at all.

The experiments were done by injecting, e.g., a 1μM solution of target T1 via the liquid handling system (cf. Figure 8) into the sample cell and monitoring the increase of the fluorescence intensity at a fixed angle of incidence as a function of time. Once equilibrium between the bulk target concentration and the corresponding surface coverage was obtained (indicated by a constant plateau value of the fluorescence, cf. Figure 12), pure buffer was circulated through the flow cell inducing the dissociation process between probes and targets. As a result, the fluorescence decreases again to the background level because all bound targets leave the interfacial layer and are rinsed away.

As it was discussed in detail before, the whole association / dissociation process can be described by a simple Langmuir model with the two rate constants, $k_{on}$ and $k_{off}$, respectively, describing the complete process. This is indicated in Figure 12 by the full curve through the experimental data points. The agreement is excellent and the obtained rate constants are given in Table II.

**Table II.** Kinetic data, i.e. adsorption (hybridisation) rate constant $k_{on}$, dissociation rate constant $k_{off}$, affinity constant $K_A = k_{on}/k_{off}$ for the reaction of the probe oligonucleotide (T1)-(T3) (cf. Table I) to the target strand (P2), immobilized at the sensor surface

| | $T_2$ (MM0) | $T_1$(MM1) | $T_3$(MM2) |
|---|---|---|---|
| $k_{on}/M^{-1}s^{-1}$ | $3.7 \times 10^4$ | $8.9 \times 10^3$ | $1 \times 10^1$ |
| $k_{off}/s^{-1}$ | $7 \times 10^{-6}$ | $3.7 \times 10^{-4}$ | $7.7 \times 10^{-4}$ |
| $K_A/M^{-1}$ | $5.3 \times 10^9$ | $2.4 \times 10^7$ | $1.3 \times 10^4$ |

More important from an application point of view is the quantitative difference of the observed kinetic behavior upon injecting T2, the mismatch 0 target, which is fully complementary to the recognition sequence of the probe strand. Both the association, as well as the dissociation data is given in Figure 12. The corresponding rate constants are also presented in Table II. It is clear that the numbers of these rate constants are not as reliable as for the P2/T1 hybrid because for this concentration, i.e., $c_0 = 1\mu M$, the hybridization is too fast, whereas the stability of the MM0 duplex with its correspondingly low $k_{off}$ rate constant barely allows for a noticeable dissociation of the hybrid. However, the difference in the kinetic behavior of the two hybridization reactions is obvious. Given the strong interest in mismatch discrimination, e.g., in the context of single nucleotide polymorphisms (SNPs) detection, the observed difference in the rate constants and in the affinity constants, $K_A$, which, in the Langmuir picture, are given by

$$K_A = k_{on} / k_{off} \tag{4}$$

is quite striking: for MM0 we find $K_A \approx 5.3 \cdot 10^9 \, M^{-1}$, whereas for the MM1 case the lower affinity $K_A = 2.4 \cdot 10^7 \, M^{-1}$, indicates the substantial destabilization of the hybrid induced by a single base mismatch (out of 15 base pairs!). Figure 12 also shows that a second mismatch reduces the affinity constant by another 3 orders of magnitude with the consequence that the injection of a 1μM solution of the corresponding target barely leads to a fluorescence increase. This concentration is still below the half-saturation concentration $K_d (\equiv 1/K_A)$ and, hence, does not lead to any substantial surface coverage.

Another mode of operation is schematically sketched in Figure 11(b). This time, the probe oligonucleotides carry the chromophore. Thus, the binding of the probe strands upon injection of the corresponding solution can be observed directly as it is shown in Figure 13. Rinsing the cell removes all the non-specifically bound probe oligonucleotides. Upon injection of a 1μM unlabeled target solution a substantial increase of the fluorescence intensity is direct evidence for the hybridization event. Under these experimental conditions an increase of the fluorescence by 75% was observed (cf. Figure 13). The interpretation is schematically depicted in Figure 11(b): In the single-stranded form the surface-attached probe oligonucleotide at the relatively high ionic strength of this experiment ( $c_{ion} = 168 mM$ ) has a somewhat random configuration resulting in a separation of the chromophore from the metal surface which is less than would correspond to the contour length of the fully stretched oligonucleotide. This is very different at low ionic strength when the reduced screening of the phosphate backbone charges repel each other, resulting in a more stretched configuration. This effect is well known from other polyelectrolyte brushes[33].

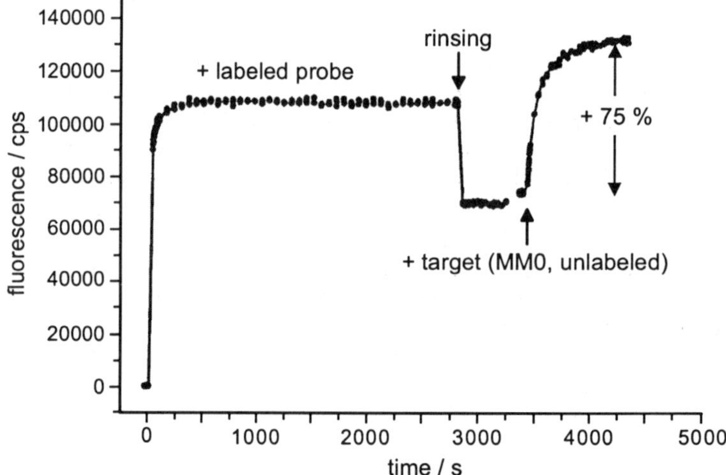


**Figure 13.** Time course of a hybridization experiment between a fluorophore-labeled, surface-attached probe strand and an unlabeled target strand of MM0. Firstly, the fluorescence increase by the binding of the probes (via their biotin moiety) to the streptavidin layer is observed. After rinsing with pure buffer the addition of unlabeled MM0 target (1 μM) results in an intensity increase by 75%.

Upon hybridization with the unlabeled target sequence the resulting double-stranded 15mer part of the surface-bound duplex is stiffened and stretches out into the buffer, thus increasing the separation distance of the chromophore from the metal substrate. As a consequence, less light is lost by quenching and the fluorescence intensity increases. In fact, the quantitative analysis of the time course of this increase, again, yields the $k_{on}$ rate constant of the hybridization step[17]. (We should point out that other effects may also contribute to the observed intensity increase, e.g., a change in the orientational distribution function of the chromophore transition dipole moment before and after hybridization.)

This experiment is a good example for a very specific feature of SPFS: being a label-detecting sensor scheme it still does not necessarily mean that the analyte has to be labeled. The surface binding reaction-induced change of chromophore properties, in this case the chromophore / metal separation distance, allows for a sensitive detection of an analyte without the need to have it chemically modified.

The last example for SPFS-based surface-hybridization studies is schematically depicted in Figure 11(c): Here, the probe strand is labeled with a chromophore that acts as a donor dye for an acceptor chromophore chemically linked to the target strand. Before hybridization the angular scan of the fluorescence shows the typical features of a SP-

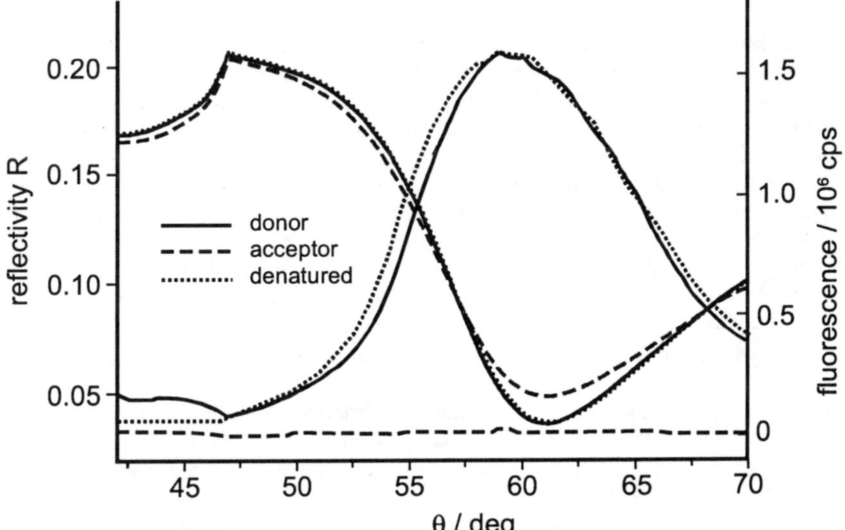


**Figure 14.** FRET (fluorescence resonance energy transfer) experiment between a surface-attached donor dye-labeled probe strand and an acceptor dye-labeled target strand from solution. Shown are the angular scans of the reflectivities and of the fluorescence emission of the donor dye before and after hybridization, as well as, after denaturing the hybrid. While the reflectivities are virtually identical the fluorescence shows a strong enhancement at surface plasmon resonance (—) which is completely quenched after hybridization (----), however, can be fully recovered upon the dissociation of the hybrid (......).

excited emission. This is shown in Figure 14. After hybridization, however, the close proximity of the donor and acceptor dyes at the end of the double-stranded part of the duplex results in a complete quenching of the donor emission via fluorescence resonant energy transfer (FRET) to the donor dye: the donor intensity is completely lost. (Unfortunately, this donor / acceptor pair is non-fluorescent and, hence, no red-shifted emission from the acceptor dye can be observed.) However, if the hybrid is denatured (by rinsing NaOH solution through the sample cell) the full intensity emitted from the donor chromophore is reestablished (cf. the dotted curve in Figure 14).This experiment is an example for yet another specific mode of operation in SPFS. Essentially all versions known from fluorescence experiments with "normal" light can be applied or adapted to the use of surface plasmon as the exciting electromagnetic mode.

## 6. PROTEIN BINDING STUDIES – THE LIMIT OF DETECTION IN SPFS

Of course, there is no reason to limit the SPFS detection schemes to surface hybridization studies: labeled proteins can be detected as sensitively.

Rather than giving a number of additional examples for the use of SPFS in proteomics we will briefly discuss only one particularly important aspect, i.e., the limit of detection (LOD) in surface binding studies.

To this end we present a series of experiments with a surface architecture that is schematically presented in Figure 15. We choose the Biacore chip CM5 with its carboxylated dextran layer as the coupling matrix. This functional polymer brush is known to extend in the swollen state about 100 nm into the buffer solution. This is schematically indicated to scale with the evanescent field of a surface plasmon mode in Figure 4. The architecture thus provides not only a quasi 3-dimensional (3D) surface architecture with an enhanced ligand density compared to a mere 2D planar surface layer, it also optimizes the overlap of the evanescent wave with the distribution of the chromophore-labeled analyte bound to the dextran-coupled ligands.Moreover, it minimizes the reduction of the fluorescence intensity via quench processes by preventing the chromophores form being placed too close to the metal surface

The system that was prepared for the LOD evaluation included the covalent coupling of mouse IgG antibodies by the usual NHS / EDC strategy to the COO⁻ groups along the dextran chains and passivating the remaining reactive ester groups by ethanolamine treatment of the surface layer. The actual experiment then involved the recording of fluorescence photons emitted from chromophore-labeled rabbit-anti-mouse RaM IgG antibodies (cf. Figure 15).

We should point out that these experiments were done in a totally different concentration range compared to the hybridization studies: for the LOD determination one operates in the mass transfer-limited regime of the binding process. I.e., the concentration of the analyte is so low that one observes only the diffusion-controlled approach of labeled analyte molecules probed by the SP field giving rise to a linear increase of the fluorescence intensity with time.

A series of such runs are shown in Figure 16(a), covering the concentration range from 67 fM to 33 pM. Each time, the base line was recorded for a few minutes, then the protein solution of a particular concentration injected and the rise of the fluorescence monitored. After a certain time, pure buffer was rinsed through the flow cell resulting in an immediate stop of the increase. Next, the sensor surface was regenerated by a pulse of glycine buffer solution (10mM, pH 1.7) resulting in a complete removal of all bound RaM antibodies.

By plotting the obtained slope of the binding curve as a function of the corresponding bulk concentration of the analyte one derives at the calibration curve displayed in Figure 16(b). It shows the expected linear relation covering almost 5 decades in concentration. With an (experimentally determined) baseline stability of c. 4 cps/min one determines a LOD in the lower femtomolar range.

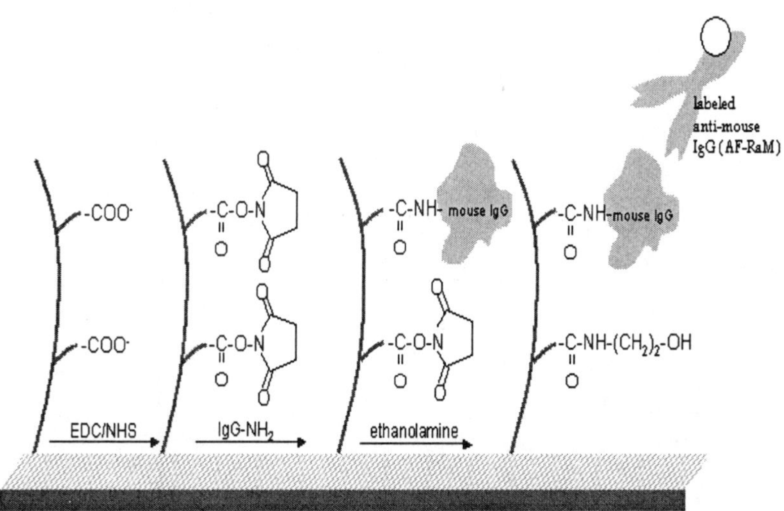


**Figure 15.** Coupling scheme for the covalent attachment of mouse IgG to the carboxy-groups of a surface-attached dextran chain via NHS / EDC reaction. The biorecognition and binding of a fluorescentlylabeled Rabbit-anti-Mouse (RaM) antibody to these sites is monitored by SPFS.

Recently, we could extend this calibration curve into the sub-femtomolar range by optimizing the set-up (e.g., by further reducing the background intensity, etc.) with the lowest concentration detected so far of $c_0 = 333$ aM $\left(3 \cdot 10^{-16} M\right)$. Knowing the relation between the number of photons detected and the number of proteins at the interface (from combined SPR / SPFS studies at higher protein concentrations) we can correlate the corresponding slope with the protein flux from solution to the sensor surface: for a 333 aM solution it corresponds to 5 protein molecules reaching every square millimeter of the sensor layer per every minute. Thus, SPFS offers a detection limit that directly connects to the single-molecule recording schemes.

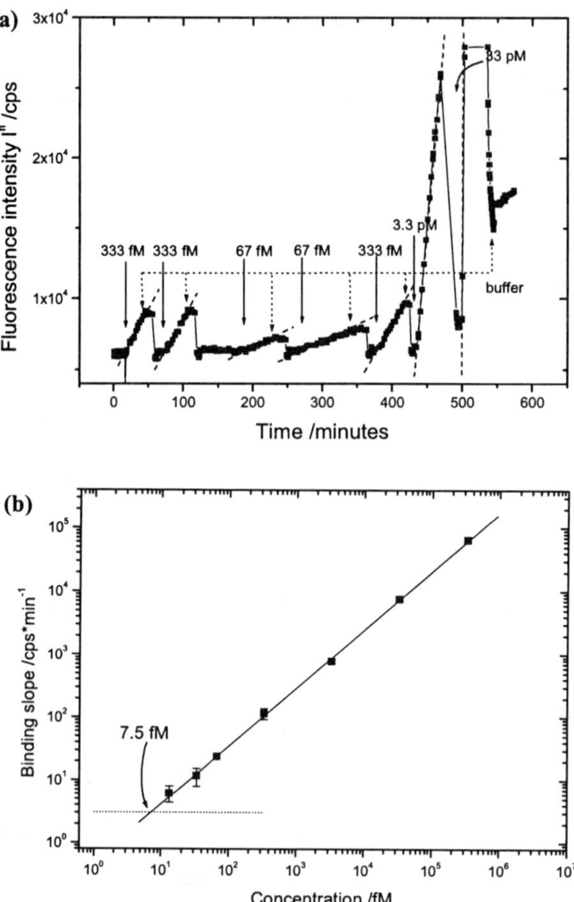


**Figure 16.**(a) Protocol of the binding (in the mass transfer limited regime) of the RaM antibodies to the surface attached IgG from solutions of different concentrations. The linear increase of the fluorescence increase is plotted in (b) as a function of the bulk concentration. Indicated is the exchange of the protein solution by pure buffer (dashed arrows, labeled: buffer), followed by regeneration via glycerine buffer pulses. The dotted line in (b) indicates the base line stability which limits the detection.

## 7. SURFACE PLASMON FLUORESCENCE MICROSCOPY

As much as the very nature of surface plasmons being surface-(bound) light has led to the introduction of surface plasmon microscopy (SPM) for the characterization of laterally structured samples surface plasmon fluorescence techniques can also be extended to microscopic formats: To this end, the emitted fluorescence light is imaged by an objective lense onto a CCD camera. This extension of a surface plasmon spectrometer to a surface plasmon fluorescence microscope (SPFM) is schematically given in Figure 17. (This figure also documents that this can be easily done while maintaining the "normal" SPM mode of operation with another CCD camera.

A series of SPFM images taken in this mode from an array of oligonucleotide sensor spots prepared by an ink-jet approach is given in Figure 18. In each frame, the left column of 3 spots each with a diameter of about 250µm was prepared with the catcher oligonucleotide probe P1, the middle column with P3 and the right column with P2. Upon injection of a 1µm solution of target T3 (cf. Table I) only the spots in the left and middle column are "decorated" by the chromophore-labeled target strands T3 upon hybridization to their MM1 and MM0 probes on the spots, respectively, and emit fluorescence light while the right column remains dark: here, the MM2 situation does not lead to any surface binding. This is in agreement with the spectroscopic data, e.g., the kinetic curves of the corresponding probe / target hybridization reactions like the ones presented in Figure 12.

By arranging the pixel intensities in each of the spots for all images taken at different times the same kinetic rate constants as determined from the spectroscopic mode of operation can be deduced, however, this time for many different hybridization reactions simultaneously. Although shown here for an array of only 3x3 sensor spots one can easily extrapolate to much larger arrays without reaching any fundamental limitations given by SPFM. A critical issue is only the photostability of the employed organic chromophores because the integration times needed per frame even with ultra-high sensitivity CCD cameras can lead to a slight distortion of the kinetic scan curves, in particular, with an overestimation of the $k_{off}$ rate constant (by the additional bleaching decay channel).

## 8. CONCLUSIONS

Surface plasmon fluorescence spectroscopy and microscopy are very young techniques. However, the results obtained so far are very promising and hold great potential both for fundamental studies as well as for practical applications, e.g., in sensor development. The obtainable signal-to-noise levels, as well as, the documented lower limit of detection are very encouraging.

Given the many different versions of general fluorescence spectroscopies one might expect still many interesting modes of operation in SPFS. One obvious extension is the use of guided optical modes as the excitation light source, with the option of using either p- or s-polarized waves thus rendering this technique also a polarization sensitive technique applicable for anisotropic samples or in depolarization studies with mobile chromophore probing molecular dynamics at different time scales. Essentially, any type

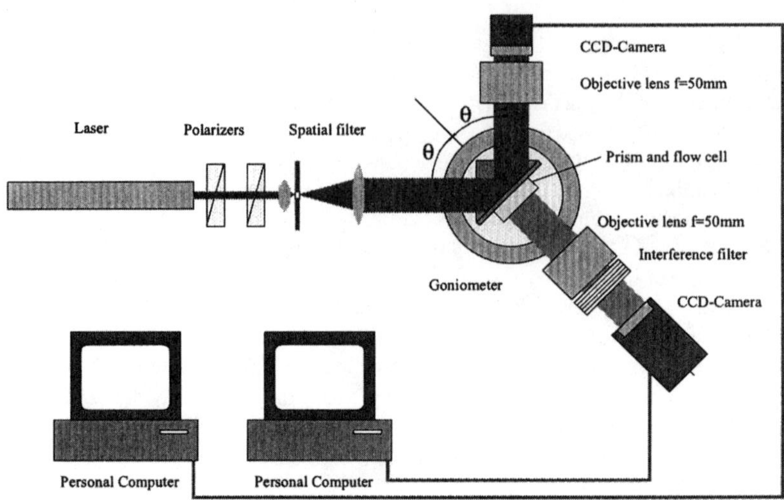



**Figure 17.** Schematic experimental setup for surface-plasmon and surface-plasmon field-enhanced fluorescence microscopy in the Kretschmann configuration.

## (a) Hybridization with target sequence T3

## (b) Desorption

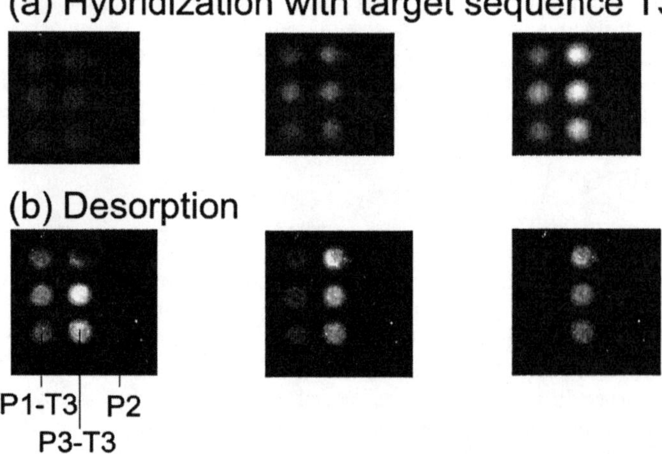



**Figure 18.** Series of time-lapse surface plasmon fluorescence microscopy images taken during hybridization (a) and dissociation (b) of the target T3, to the surface-attached probes P1 – P3 (left column of 3 spots P1 (MM1 to T3), middle column P3 (MM0), right column P2 (MM2 to T3).

of fluorescence spectroscopy or microscopy with plane waves, i.e., normal photons, can be extended to a scheme with SP excitation, e.g., two-photon fluorescence techniques[34]

From a surface plasmon optical point of view, SPFS offers many exciting opportunities for the characterization of thin films and interfaces if combined with other techniques, e.g., the quartz crystal microbalance, with electrochemical techniques or, e.g., in a high pressure cell.

There is no doubt that we will see more of these investigations in the future.

## 9. ACKNOWLEDGEMENTS

This work is based to a large extent on the dissertation of Thorsten Liebermann who introduced SPFS to our group. Thanks are also due to G. Batz, R. Herrmann, S. Lofås, B. Persson, P. Sluka and M. Zizlsperger for stimulating discussions. Financial support from Boehringer Mannheim (now Roche Diagnostics) and by the European Community (Project QLK 1-CT 2000-01658, "DNA-Track") is gratefully acknowledged. Part of this material is also based upon work supported by the Science and Research Council, Singapore, under Grant Number MCE/TP/00/001.2.

## 10. REFERENCES

1. W. Knoll, *Ann. Rev. Phys. Chem.* **49**, 565-634 (1998).
2. E. Burstein, W. P. Chen, Y. J. Chen, A. Hartstein, *J. Vac. Sci. Technol.* **11**, 1004-1016 (1972)
3. H. Raether, Springer Tracts in Modern Physics, Vol. 111, Springer Verlag, Berlin 1988
4. P. Schuck, *Ann. Rev. Biophys. Biomol. Struct.* **26**, 541-566 (1997)
5. A. G. Frutos, R. M. Corn, *Anal. Chem.* A **70**, 449A-455 (1998)
6. J. Homola, S. S. Yee, G. Gauglitz, *Sens. Actuators* **B54**, 3-15 (1999)
7. http://www.biacore.com
8. http://www.ti.com
9. A. Schmidt, J. Spinke, T. Bayerl, E. Sackmann, W. Knoll, *Biophys. J.* **63**, 1385-1392 (1992)
10. J. Spinke, M. Liley, H. J. Guder, L. Angermaier, W. Knoll, *Langmuir* **9**, 1821-1825 (1993)
11. V. H. Perez-Luna, M. J. O'Brien, K. A. Oppermann, P. D. Hampton, G. P. Lopez, L. A. Klumb, P. S. Stayton, *J. Am. Chem. Soc.* **121**, 6469-6478 (1999)
12. S. Löfås, B. J. Johnsson, *Chem. Soc., Chem. Commun.* **21**, 1526-1528, (1990)
13. T. Liebermann, W. Knoll, *Colloids Surf. A* **171**, 115-130 (2000)
14. J. P. Attridge, P. B. Daniels, J. K. Deakon, G. A. Robinson, G. P. Davidson, *Biosens. Bioelectron* **6**, 201-209 (1991)
15. T. Liebermann, W. Knoll, P. Sluka, R. Herrmann, *Colloids Surf. A* **169**, 337-350 (2000)
16. T. Neumann, D. Kambhampati, G. Stengel, W. Knoll in: *Nanotechnology Towards the Organic Photonics* (Ed: H. Sasabe), GeoTech Ltd., Chitose, Japan, 333-350 (2002)
17. T. Neumann, M.-L. Johansson, D. Kambhampati, W. Knoll, *Adv. Funct. Mat.* **12**, 575-585 (2002)
18. F. Yu, D. Yao, W. Knoll, *Anal. Chem.* **75**, 2610-2617 (2003)
19. D. Axelrod, T. P. Burghardt, N. L. Thompson, *Annu. Rev. Biophys. Bioeng.* **13**, 247-268 (1984)
20. A. Nemetz, W. Knoll, *J. Raman Spectroscopy* **27**, 587-592 (1996)
21. F. R. Aussenegg, A. Leitner, M. E. Lippitsch, H. Reinisch, Ultrafast Phenomena VI, in: T. Yajima, K. Yoshihara, C. B. Harris and S. Sionoya (Eds.), Springer Series in Chemical Physics, Springer, Berlin, 1988, p. 434

22. H. Kuhn, D. Möbius, H. Bücher, in: A. Weissberger and B. W. Rossiter (Eds.). Physical Methods of Chemistry, Wiley Interscience, New York, 1972, Part III B, Chapter 7

23. H. Knobloch, H. Brunner, A. Leitner, F. Aussenegg, W. Knoll, *J. Chem. Phys.* **98**, 10093-10095 (1993)

24. T. Liebermann, W. Knoll, *Langmuir* **19**, 1567-1572 (2003)

25. W. H. Weber, G. W. Ford, *Opt. Lett.* **6**, 122-138 (1981)

26. E. Kretschmann, H. Raether, *Z. Naturforsch. Teil A* **23**, 2135-2136 (1968)

27. W. Knoll, M. R. Philpott, J. D. Saralen, A. Girlando, *J. Chem. Phys.* **77**, 2254-2260 (1982)

28. W. Knoll, M. R. Philpott, J. D. Swalen, *J. Chem. Phys.* **75**, 4795-4799 (1981)

29. I. Pockrand, A. Brillante, D. Möbius, *Nuovo Cimento B* **63**, 350-354 (1981)

30. A. Samoc, M. Samoc, B. Luther-Davies, *Pol. J. Chem.* **76**, 345-358 (2002)

31. W. Denk, J. H. Strickler, and W. W. Webb, *Science* **248**, 73-76 (1990)

32. D. Kambhampati, P. E. Nielsen, W. Knoll, *Biosens. Bioelectron* **16**, 1109-1118 (2001)

33. M. Biesalski, J. Rühe, R. Kügler, W. Knoll, in: *Handbook of Polyelectrolytes and Their Applications,* American Scientific Publishers, 2002, Chapter 2, 39-63

34. H. Kano, S. Kawata, *Opt. Lett.* **21**, 1848-1850 (1996)

11

# OPTICALLY DETECTABLE COLLOIDAL METAL LABELS: PROPERTIES, METHODS, AND BIOMEDICAL APPLICATIONS

Steven J. Oldenburg[1] and David A. Schultz[2]

## 1. INTRODUCTION

The optical detection of sub-wavelength sized metal particles has its scientific roots nearly a hundred years ago when Zigmondy first reported the observation of individual metal colloids under a microscope[1]. However, it was not until the early 1980s that gold colloid (5 nm – 20 nm in diameter) that had antibodies attached to the particle surface were used to specially target immunogenic cellular proteins. The precise location of these electron dense markers labels (immunogold) can be visualized using Transmission Electron Microscopy (TEM) (reviewed by Hayat[2]). Alternatively, the labels can be viewed with a standard optical microscope by increasing their size with an Immuno-Gold Silver Staining (IGSS) procedure that deposits additional silver ions on the colloidal gold nucleation sites to form bulk silver [3]. Another procedure commonly used for RNA *in-situ* studies is autoradiography, a process where silver ions nucleated as a result of radioactive decay from a $^3$H or $^{35}$S labeled nucleic acid probe form colloids in a silver emulsion film and are detected using optical microscopy techniques[4].

In the 1990s, it was discovered that if larger sized colloidal metal particles were used as the biological label (40 nm – 120 nm in diameter), the particles could be easily imaged using a dark field optical microscope[5, 6]. The particles appear as bright, colored, point sources of light that do not photo-bleach. When functionalized with antibodies these highly specific labels can be used to detect low concentrations of antigens. With the recent reduction in prices of high-resolution digital cameras and the associated image processing software, the use of these labels will continue to increase. In this chapter, we

---

[1] Steven J. Oldenburg, Seashell Technology, La Jolla, California 92037.
[2] David A. Schultz, University of California, San Diego, La Jolla, California 92093-0319.

focus on recent work describing the optical detection of individual colloidal metal particles in biological applications. The fabrication, modification, and detection of different types of metal colloids will be discussed, and their use as labels in several biological and biomedical applications will be presented.


## 2. PLASMON RESONANCE

### 2.1. Experimental and Theoretical Considerations

When excited with visible light, small metallic particles (radius << wavelength) can have extremely large extinction (scattering plus absorption) cross-sections[7]. In the size regime between 40 and 100 nm diameters, the scattering cross-section can be more than 5 times larger than the geometrical cross-sectional area for silver and gold particles. At their peak scattering wavelength, silver and gold nanoparticles scatter more light than any other nanoparticle of comparable size, and are readily visualized using conventional light microscopes configured for darkfield illumination (Section 4). For example, under the same excitation conditions, an 80 nanometer diameter PRP is ~1000 times brighter than a comparable sized fluorosphere, and does not photobleach[5, 6, 8]. The peak scattering wavelength of these particles is a function of their size, shape, material properties, and local dielectric environment. Sixty nanometer diameter spherical silver particles have a peak extinction at ~400 nm (blue) whereas equivalent size spherical gold particles have a peak extinction at ~520 nm (green). Dark field images of silver and gold particles illuminated with white light are shown in Figure 1.

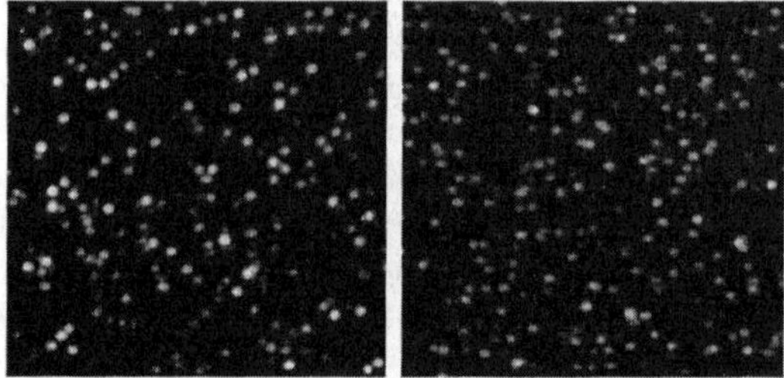

**Figure 1.** CCD images of homogeneous color populations of individual metallic particles (left side) 60 nm diameter silver particles and (right side) 60 nm diameter gold particles, as viewed using an optical microscope configured for darkfield illumination with a 50 X objective lens.


The optical properties of small metallic spheres can be explained in terms of the excitation of collective electron oscillations known as surface plasmons. Particular wavelength bands of light can excite the plasmon modes of a particle into a resonance, and hence, these particles are referred to as plasmon resonant particles (PRPs[TM]). The magnitude and frequency of this resonance is directly related to the dielectric of both the metal PRP and the surrounding media. For spherical particles in the quasi-static regime (which incorporates the time but not the spatial dependence of the electromagnetic field) the intensity of the scattered light is given by Eq. (1),

$$I_{scatter} = \left| 4\pi\varepsilon_o R^3 \frac{\varepsilon(\lambda)-\varepsilon_m}{\varepsilon(\lambda)+2\varepsilon_m} \right|^2 \qquad (1)$$

where R is the particle radius, $\varepsilon_0$ is the permitivity of free-space, $\varepsilon(\lambda)$ is the complex dielectric function of the metal, $\varepsilon_m$ is the dielectric function of the medium, and $\lambda$ is the wavelength of the incident light. When the denominator $|\varepsilon(\lambda) + 2\varepsilon_m|$ is at a minimum, a resonance is achieved resulting in a peak in the extinction spectra. Only a few materials, such as silver and gold, have both a negative dielectric value in the visible region of the spectra that results in a peak scattering wavelength in the visible, and a sufficiently low relaxation rate to permit a narrow resonance. The calculated peak scattering, absorption, and extinction cross-sections for a 40 nm diameter particle of silver and a 60 nm diameter particle of gold are shown in Figure 2.

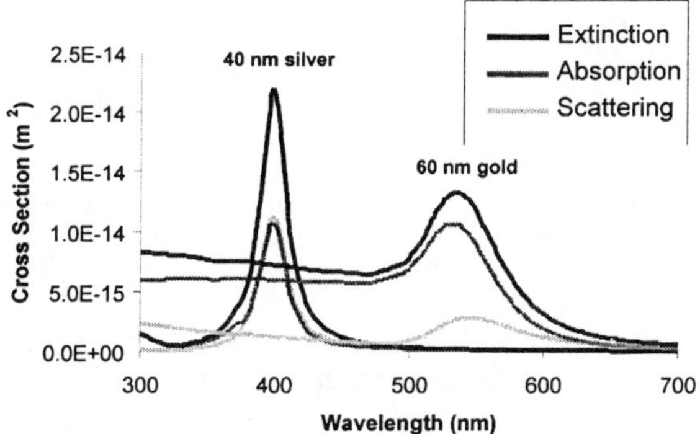


**Figure 2.** Calculated absorption, scattering, and extinction cross-section from a 40 nm diameter Ag particle and a 60 nm diameter Au particle. The spectrum from the silver particle has a sharp resonance and its extinction is equally split between scattering and absorption components, whereas, the 60 nm diameter gold particle has a broader resonance and is dominated by absorption.

A more complete description of light scattering from spherical particles solves Maxwell's equations in spherical coordinates using multi-pole expansions of the incident electric and magnetic fields. This theory was first described by Mie in 1908[9] and can be used to predict the light scattering properties of particles that have a size that is on the same order as the wavelength of incident light. Figure 3 shows the effect of increasing size on the extinction spectra of gold particles. As the particle diameter is increased, the plasmon resonance peak shifts to longer wavelengths and become broader. At diameters < 40 nm the particle extinction is due entirely to absorption while at larger diameters (> 100 nm) scattering dominates.

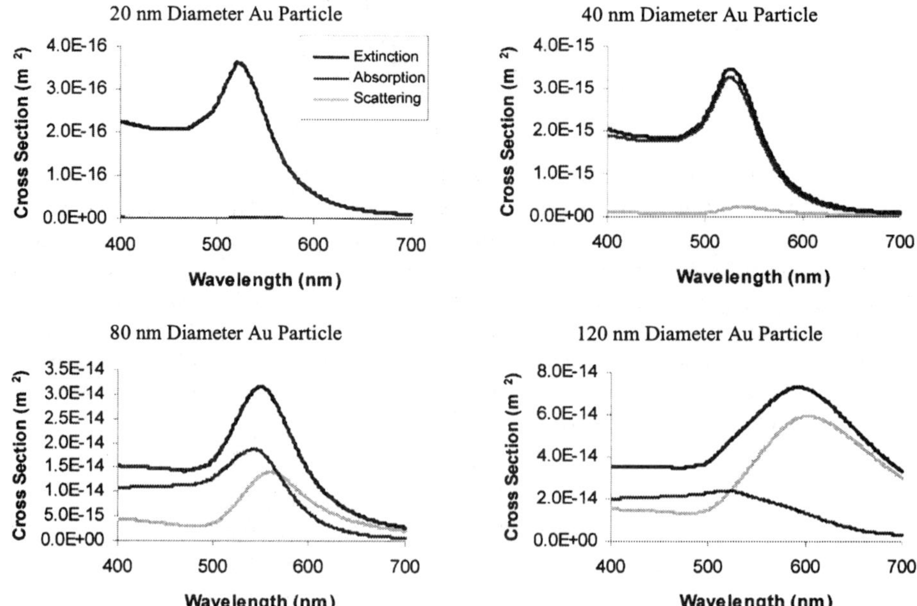


**Figure 3.** Calculated extinction (black), absorption (dark grey), and scattering (light grey) cross section as a function of wavelength for a 20, 40, 80, and 120 nm diameter gold particle in water. As the particle diameter is increased, the plasmon resonance scattering peak shifts to longer wavelengths and become broader.

More dramatic changes to the extinction spectra can be obtained by altering the shape of the metal particles. Elongation of metal particles into spheriodal or rod-like particles splits the main (dipole) resonance into multiple resonances each of which is linked to one of the geometrical axes[10-12]. For example, a gold rod with a diameter of 16 nm and a length of 63 nm has two extinction peaks, one at 525 nm that is related to a plasmon mode perpendicular to the rod length, and the other at 908 nm related to a plasmon mode along the length of the rod[12]. Thus, an sotropic particles exhibit a polarization dependent scattering dependence that clearly identifies them from isotropic particles. Other particle shapes also have a scattering spectrum that is related to their geometry (Figure 4)[13-15]. For example, spherical silver particles preferentially scatter blue light, pentagonal particles scatter primarily green light (500-600 nm), and triangular platelets or tetrahedrons principally scatter red light (600-715 nm)[14].

Yet another way to modify the particle's extinction spectra is to fabricate particles with a core-shell geometry. For a dielectric core that is surrounded by a thin shell of gold or silver, the peak plasmon resonant scattering wavelength is a function of both the size of the core and the thickness of the shell[16, 17]. Figure 5 shows that the the peak extinction of a 120 nm silica core that has been shelled with gold can be shifted from 740 nm to 1030 nm by reducing the shell thickness from 20 nm to 5 nm. The smaller, red shifted peak at 580 nm on the extinction of the 20 nm thick shell is due to a second oscillation mode of the conduction electrons in the shelled particle. The placement and location of the primary and secondary peaks act as a unique "fingerprint" that is directly related to the size of the core and the thickness of the shell. Depending on the choice of metal shell

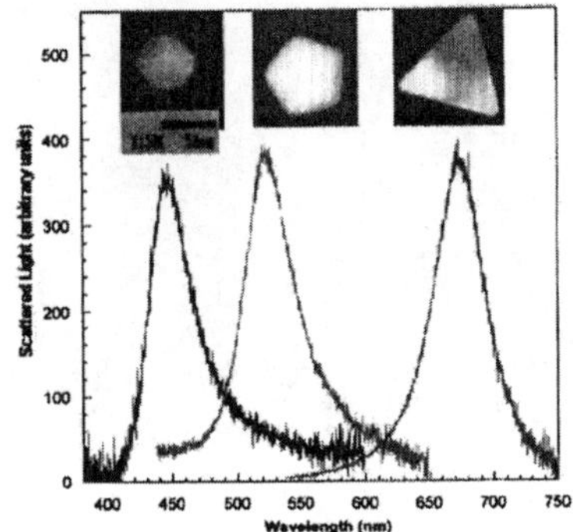

**Figure 4.** Effect of silver particle's shape on the peak plasmon resonance scattering wavelength. Reproduced from [14] with permission from the publisher.

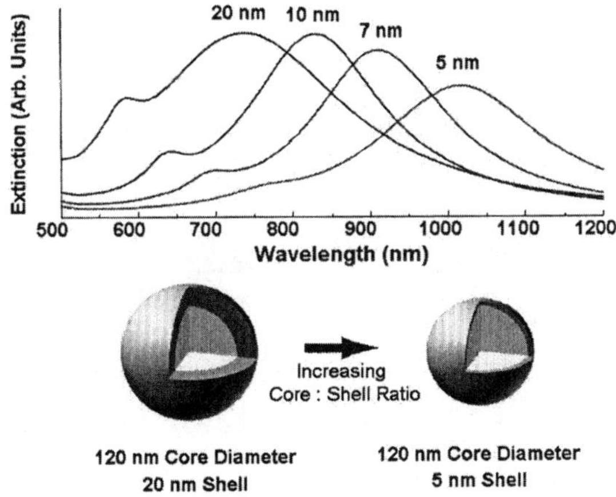


**Figure 5.** Calculated spectra from a 120 nm diameter silica particle coated with gold shell thickness that ranges from 5 nm to 20 nm. As the thickness of the shell is decreased, the peak extinction location shifts from 730 nm out to 1030 nm. Reproduced from [18] with permission from the author.

the peak plasmon resonant scattering wavelength can be placed anywhere between 400 and 2000 nanometers. Thus, particles can be fabricated in a large number of distinguishable colors and are well suited for assays where a high degree of multiplexing is desired (Figure 6).

## 2.2. Particle Fabrication

### 2.2.1. Spherical Particles

There are a large number of methods for producing metallic colloids in solution[2].

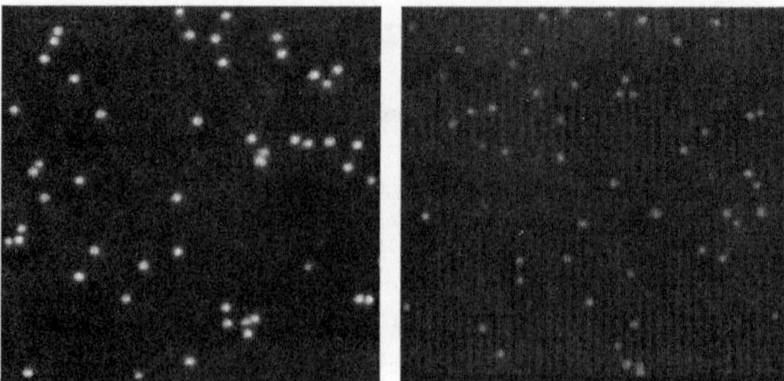

**Figure 6.** Darkfield image of gold-shelled particles with either 130 nm (left) or 200 nm (right) diameter silica cores.

The selection of recipes that produce particle populations with narrow size and shape distributions is critical for developing labels with uniform brightness and color.

Fabrication methods that utilize a small amount of smaller colloid as a "seed" for subsequent particle growth typically produce narrow size distributions of particles[18, 19]. As an example, fabrication procedures for the production of homogeneous populations of gold and silver spheres in the 10 – 100 nm size range are listed in Table 1. The final particle size can be adjusted by changing the initial concentration of the 5 nm seed gold.

**Table 1.** Recipes for colloid fabrication

| Fabrication of 80 nm diameter gold colloid | Fabrication of 60 nm silver colloid |
| --- | --- |
| Add 25 mg KCarbonate and 1.5 mL of HAuCl4 (aged at least 3 days in a dark glass container) to 100 mL of water. | Add 50 µL of 5 nm gold to 80 mL of water |
| Add 30 µL of 5 nm gold (British Biocell International, $5^{13}$ part/mL) to 100 mL of aged KCarbonate solution. | Add 1.5 µL of silver initiator* <br> Add 400 µL of silver enhancer* <br><br> Wait 5 minutes for color to change to a yellow/brown solution. |
| Add 50 µL of formaldehyde and stir until color has stabilized to a dark red (1-5 hours). | * Silver enhancement kit was obtained from British Biocell International |
| Collect and concentrate by centrifugation. | |

### 2.2.2. Anisotropic, Elliptical or Rod Shaped Particles

Anisotropic particles, such as rods, are fabricated using two different methods. The first are solution based methods where a surfactant assembles "soft" templates in solution and promotes growth of particles in one direction[11, 20]. A variety of particles with extremely large aspect ratios can be produced with this method. An alternative method is to fabricate particles within the pores of a template (e.g. nanopore alumina or polycarbonate filters) that have well defined pore sizes[21, 22]. Metal is deposited into the pores using electrochemical procedures followed by the dissolution of the supporting template. By varying the time and type of metal deposition, composite materials with interesting optical properties can be generated[23].

### 2.2.3. Core: Shell Particles

Metal coated particles are typically fabricated on the surface of commercially available monodisperse cationic cores such as amine terminated silica or polystyrene (Bangs Labs, Polysciences, Duke Scientific). When mixed with a solution of small gold colloid (typically 2-4 nm), the negatively charged gold particles bind to the surface of the positively charged core. The colloidal fabrication chemistry described in Table 1 is used to increase the size of the "seed" particles until the gold coalesces into a complete shell as shown in Figure 7[17, 24]. Depending on the ratio of the size of the core to the thickness of the shell, the color of solution will span the visible spectrum (blue, purple, green, yellow, red) (Figures 5 & 6).

## 2.3. Particle Characterization

Since the plasmon resonance gives rise to a distinct optical spectrum, UV-VIS spectra of PRP solutions yields information on particle size (position of the plasmon resonant peak), homogeneity (width of the peak), and concentration (peak intensity). For the case of shell particles, the magnitude and location of the various peaks in the spectrum is directly related to the size and completeness of the metal shell. At early time points, when the shell is incomplete, the spectrum is relatively flat and featureless. At coverage of 80% or more, distinct peaks appear in the spectra indicating successful shell growth. Dark field imaging of individual particles (see Section 4) is useful for measuring the brightness, color, and homogeneity of particle populations. Other useful characterization methods include transmission electron microscopy for high

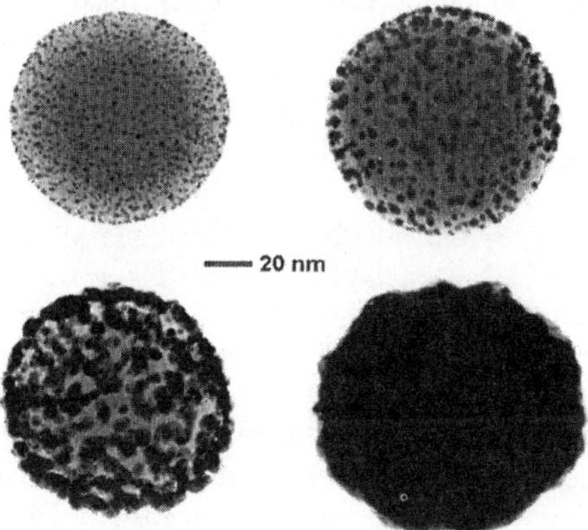


**Figure 7.** TEM micrographs of gold shell growth on silica substrates. The shell is complete at a thickness of ~8 nm. Figure reproduced from [17] with permission from the publisher.

magnification images of individual particles, dynamic light scattering, for rapid quantification of particle size and concentration, and zeta potential for determining the PRP's surface charge.


## 3. PARTICLE SURFACE MODIFICATION TO PRODUCE BIOLOGICAL LABELS

### 3.1. Particle Surface Modification

There are several methods available for preparing surface functionalized metallic colloids for use as biological labels. The three most widely used fabrication methods include non-covalent absorption of biomolecules to the particle surface[2], binding to the metal surface via a thiol functionalized cross linker[25-27], and covalent binding to a derivatized silica or polystyrene shell that encompasses the metal particle[28-31].

Non-covalent absorption of proteins to metal colloids is typically achieved by mixing the metal colloid and the protein in a buffered solution at the pI of the protein. Excess, non-absorbed, protein is removed by several successive low speed centrifugation steps with redispersion of the particles in low salt, neutral pH buffer containing a blocking reagent such as bovine serum albumin (BSA). Alternatively, by direct or indirect chemical modification of the surface with positively charged molecules, negatively charged DNA can be electrostatically bound to the positively charged particle surface[32].

Thiol-gold binding chemistry has been used to modify gold surfaces with functional groups that immobilize nucleic acids and proteins. The thiol-gold bond forms spontaneously in both organic and aqueous solutions and is stable to disassociation[25-27]. Further increases in binding stability have been obtained with trithiol-capped biomolecules that have successfully stabilized PRPs with diameters greater than 100 nm while retaining bio-functionality[25]. The use of thiol linkage chemistry facilitates molecular orientation of bound molecules, which increases binding efficiencies.

A third strategy is to encapsulate the particle with a functionalized polymer shell that is between 1 and 20 nm thick. A representative example is shown in Figure 8 where an 8-nanometer thick silica shell is present on a 100-nanometer diameter gold particle. Fabrication procedures for the production of a silica shell on gold particles[33] are listed in Table 2. The chemical properties of the particle's polymer surface can then be functionalized with any of the hundreds of available silane derivatives (Gilest, UCT) using standard recipies[34]. For example, composite silica-gold nanoparticles that contain reactive epoxide, and aldehyde functional groups can be prepared, and proteins can be directly bound using standard coupling protocols. Alternatively, primary amines or carboxylic acid groups can be incorporated at the particle surface and used for subsequent biomolecule immobilization using bi-functional cross-linking reagents. In addition, the polymer coating protects the metal cores and results in improved stability of the particles.

Covalent linkages are advantageous because they increase the stability of particle/biomolecule conjugates, and allow for molecular orientation of the attached biomolecules so that the ligand recognition site is fully exposed. Attachment conditions should be adjusted so that the biomolecules (for example, an antibody or nucleic acid) will bind to the particle at a surface density sufficient to allow for efficient target binding yet not so dense as to impede target molecules recognition. Utilization of linker

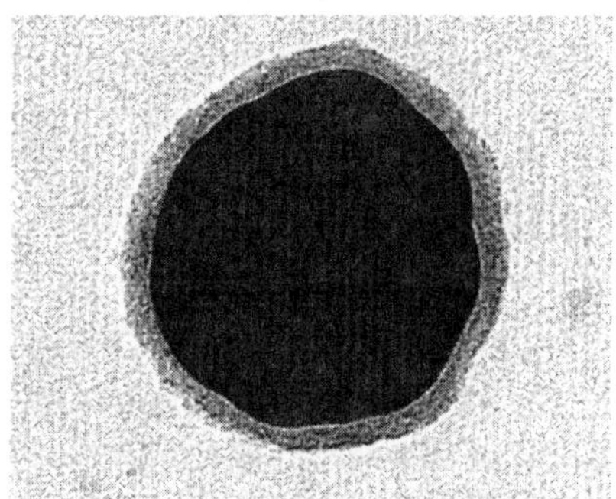

Figure 8. ~8 nm thick silica shell on a 100 nanometer diameter gold particle.

**Table 2.** Fabrication of silica shell on gold colloid

---

Shift the pH of a 100 mL batch of gold colloid to ~3.5 using 0.1N HCl
Dilute 10 uL of 100% aminopropyltriethoxy silane (APTES) in 1 mL of ethanol
Add 30 uL of 1% APTES to 1 mL of water.  Vortex and add to gold colloid mixture.
Mix 60 uL of 28% NaSilicate in 3 mL of water.
After 15 minutes add the diluted NaSilicate to the gold colloid solution.
After 3-7 days, a thin (2-4 nm) silica shell will form.

molecules to increase the separation between the bound biomolecules and the particle surface are useful for improving the accessibility of the bound biomolecules for their targets.

After the attachment of biomolecules to the PRPs, the particle's surface must be "blocked" to minimize particle-particle aggregation and non-specific binding of the colloids to the target sample which can result from a combination of electrostatic, hydrophobic, van der Waals, and other interactions. Typically, colloidal particles are surface stabilized by blocking reagents such as bovine serum albumin (BSA), or casein present in the reaction solution.  Stability and storage lifetime are maximized by adding polymers such as PVA (0.1% by volume) and by storing the colloids at 4 °C.

## 4. DARK FIELD OPTICAL MICROSCOPE DESIGNS FOR PLASMON RESONANT PARTICLE (PRP) DETECTION

### 4.1. Microscope Configuration

Under the proper illumination, PRPs can be individually visualized as bright, colored, point source scatters that can be readily identified with a properly configured microscope.  Unlike fluorophores, PRPs scatter light elastically (there is no wavelength shift of the scattered light) and thus microscope illumination must be in a "dark field" configuration, similar in this respect to epi-fluorescence where the optical field of view is black in the absence of fluorescent molecules.  The sample is illuminated at an angle such

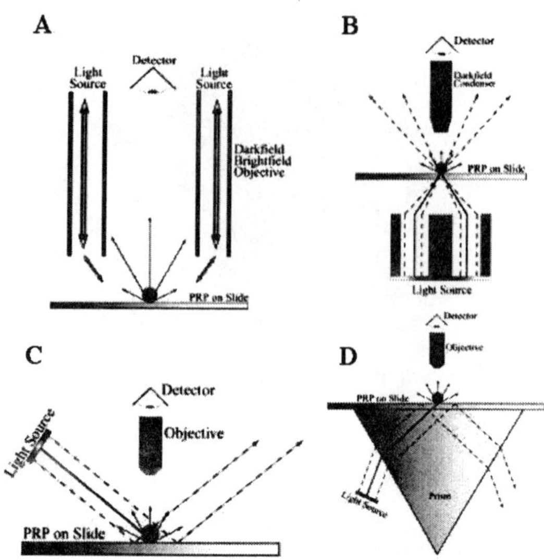


**Figure 9.** Four different dark field illumination and detection configurations. (A) Dark field objective, (B) Dark field condenser, (C) Glancing angle illumination, and (D) Total internal reflection. Reproduced from [35].

that only light that is scattered from the sample is directed into the objective lens. Figure 9 shows a schematic of several dark field illumination and detection configurations that include (A) dark field (DF) objectives, (B) a standard dark field condenser, (C) glancing angle illumination, (D) total internal reflection (T.I.R.). The choice of which dark field design to use is dependent upon the sample characteristics, the inherent brightness of the scattering objects, the background scattering from non-PRP entities, the use of polarization, and the choice of illumination source. The instrument can either be in a standard upright or in an inverted microscope configuration.

The simplest apparatus uses a microscope that is equipped with a dark field cube that selectively blocks the center portion of the epi-illumination beam so that a cylindrical tube of light is directed into a DF objective (Figure 9A). Only the light scattered from the particles is collected in a lens that is present in the center of the ring lens. Typically, 20X, 40X, 50X, 100X or 200X objectives are used, all of which are available from the major microscope manufacturers. The dark field cube and dark field objectives are convenient since they can be incorporated entirely within a standard optical microscope, but have the drawback that a large portion (typically 70%) of the illuminating light intensity is eliminated prior to reaching the sample. An additional drawback is that the integration of glancing angle illumination into the objective lens may limit the working distance of some objectives. Alternatively, an external light source, such as a 1 mm diameter multimode optical fiber, can be used to illuminate the sample (Figure 9C).

In the Total Internal Reflection (T.I.R.) configuration, the sample is illuminated from below at an angle sufficiently large so that only evanescent waves interact with the sample (Figure 9D). The PRPs scatter the evanescent waves and this light is collected by the objective lens. For T.I.R., the illumination light can be introduced into a glass or quartz right angle prism at an angle resulting in total internal reflection. Alternatively, the light can be introduced into the side of a microscope slide (Microvideo) or a fluid filled glass capillary. Because the PRPs are excited by evanescent waves, very efficient rejection of the incident illumination is observed (less background scattering), and scattering from non-PRP entities is minimized. This method of illumination has the additional advantage that intense (brighter) illumination can be introduced, which allows for smaller PRPs to be used as labels. The disadvantage of this design is that the substrate containing the PRPs must be index matched to the prism with index matching oil so that the evanescent field reaches the PRP. Also, the evanescent field falls off exponentially from the surface of the slide, which requires that the label be in close proximity to the substrate.

Another alternative microscope configuration for PRP observation is epi-illumination using an oil immersion bright field lens. To reduce non-PRP scattering from the sample interface, a microscope slide or cover slip is typically set on top of a quartz right angle prism with a drop of index matching oil. The PRPs appear bright and this configuration has the advantage that polarized illumination can be used to detect the polarization dependent light scattering properties of anisotropic PRPs, for example plasmonic rods or ellipses. The disadvantage of this approach is that index-matching oil is required at both the slide-prism and objective-slide interface, which makes sample storage and subsequent examinations more difficult.

## 4.2. Illumination Light Sources

Gold spheres > 80 nm and silver spheres > 60 nm in diameter have large scattering cross-sections, and they are readily observed either visually or with a CCD camera using standard microscope light sources (e.g. a 100-Watt halogen white light source). To visualize smaller diameter particles, such as 50 nm diameter gold or 35 nm silver particles, a more intense light source (e.g. a 75-Watt Xenon source) is required. Still smaller diameter particles can be imaged with laser illumination, however, since the particles only scatter the incident light, multiple laser wavelengths are necessary to discern different colored PRPs, and constructive and destructive interference produce speckled illumination patterns.

The choice of illumination light source depends on several factors that include the peak scattering wavelength of the PRPs, the brightness of the source, and the frequency dependent output of the lamp. For example, the frequency of the illumination light should be close to the plasmon resonance peak of the PRPs to be detected. For solid spherical gold and silver colloids in the 40-100 nanometer diameter size range, a Xenon or Halogen light source that has strong output in the visible, between 400 and 700 nm is recommended. For spectroscopic studies, or to clearly see differences between different colored PRPs, the frequency dependent output of the lamp, as well as the frequency dependence of the apparatus including the CCD camera is corrected for by normalizing all acquired spectra to a control spectra obtained using a white-light scatterer (Labsphere).

An alternative to the use of a broadband white light source is the use of laser illumination that can be tightly focused on the area of interest. In contrast to organic fluorophores, PRPs do not photobleach and even at very high light intensities, the scattered light is directly proportional to the laser power.  Thus, short exposure times (< 100 milliseconds) can be used to acquire images at a high frame rate allowing for real-time tracking of PRP movements, and may be an important consideration for biological and pharmacological applications that require high-throughput screening. However, since lasers are monochromatic, multiple lasers are necessary for distinguishing PRPs that have different colors and thus laser illumination is not typically used for analysis of PRPs with multiple colors.

## 4.3. Apparatus for Individual Plasmon Resonant Particle Spectral Determination

To obtain the optical scattering spectrum of individual PRPs a spatially distributed population of PRPs is imaged using a 50x or 100x magnification objective, and individual particles are selected using an image plane aperture and monocular.  In our laboratory, the spectrum of the scattered light from the PRP is obtained using a dark field microscope (Nikon Labophot), modified as described in Section 4.1 by the addition of a spectrometer (Spex 270M) equipped with two precision controlled gratings (1200g/mm and 150g/mm), a fixed mirror, re-focusing lenses, and a CCD camera (Figure 10).  Figure 11 shows spectra of an individual blue, green and red PRP that have been background subtracted and normalized for the spectral output/response of the light source and CCD.

## 4.4. Single Particle Counting

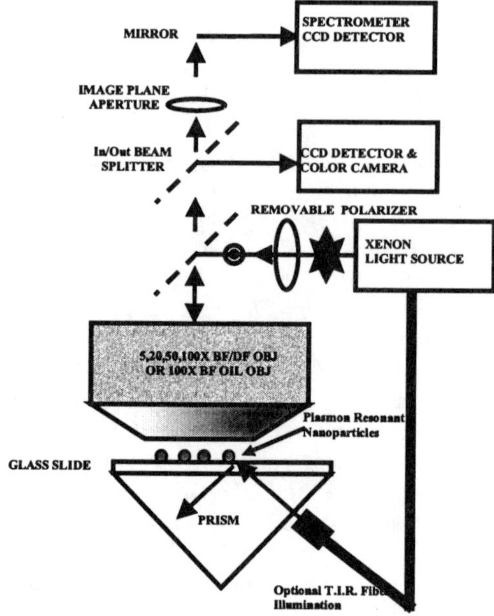


**Figure 10.** Apparatus for individual plasmon resonant particle spectral determination[35].  Reproduced from [35].

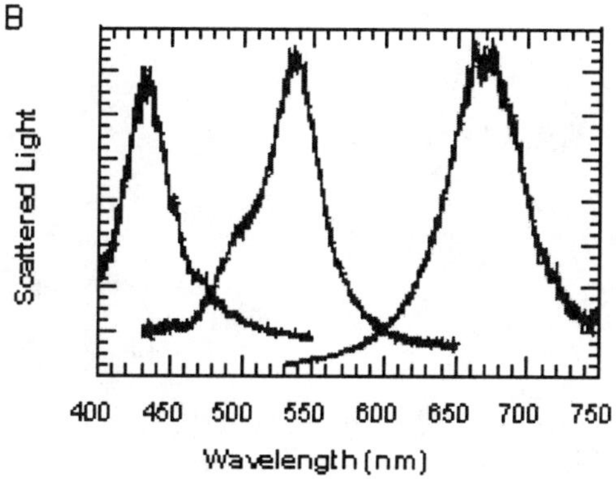


**Figure 11.** Spectrum of an individual blue, green and red plasmon resonant particle (PRP). Reproduced from [5] with permission from the publisher. (Copyright 2000 National Academy of Sciences, U.S.A.)

With objective magnifications greater than 20, individual PRPs can be identified and classified with respect to their size, brightness, and color. Figure 12 shows the identification and classification of PRPs on a typical "clean" glass substrate. The left panel is a picture of a 50:50 mix of silver (blue) and gold (yellow-red) PRPs. The right panel is an example where software (Image-Pro, Media Cybernetics, Baltimore MD) was used to classify and count the gold PRPs. Also, since light is directly scattered from the substrate, captured images not only include light scattered from bound PRPs, but also light scattered from dust, scratches, and coating imperfections on the slide. Fortunately, the distinct optical spectra of PRPs (size, shape, color, and brightness) allow all background scattering objects that are not PRPs to be easily eliminated for quantitative analysis.

## 5. BIOLOGICAL APPLICATIONS

There are a number of formats for the use of PRPs in biological diagnostic applications (reviewed in[36]). In direct analogy to fluorescent labels, bio-functionalized PRPs can be used to directly target biological samples and have yielded high sensitivity in a variety of different formats[5, 6, 8, 37-42]. Alternatively, PRP aggregation assays monitor changes in the plasmon resonant optical spectrum (i.e. shifts in plasmon resonant peak, or in the intensity of scattered or transmitted light) from a population of particles due to inter-particle interaction[5, 6, 8, 37-43]. Since the plasmon resonant peak is sensitive to the external environment, small shifts in the scattering spectrum can also be used to detect molecules binding to the particle's surface[44-49]. Finally, ultrasensitive assays have been developed that utilize the ability to localize and identify individual PRPs[5, 6, 50, 51]. Two specific examples of PRP detection assays are described below.

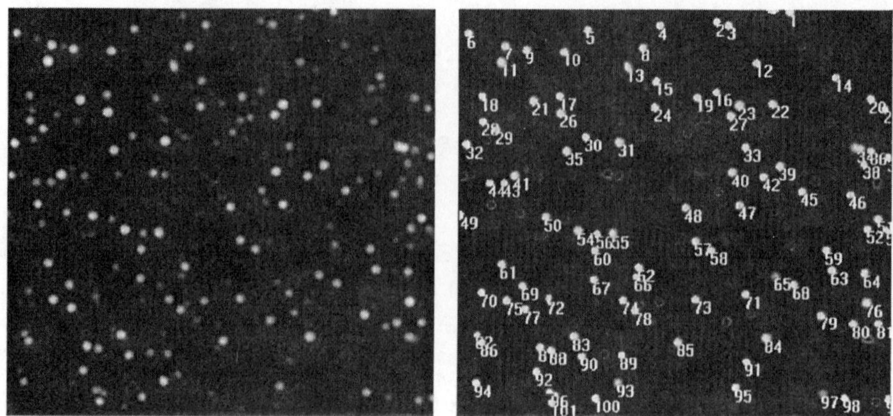

**Figure 12.** (Left) A typical image of a 50:50 mix of silver (blue) and gold (yellow-red) PRPs. (Right) Same image after applying automated PRP identification and counting algorithms. The identified silver particles are colored blue and the identified gold particles are colored yellow and numbered. See Fig. 11.12 in the color insert at the end of this volume.

## 5.1. Individual PRP Detection and Counting in a Protein or DNA Microarray Format

The simultaneous optical imaging of multiple individual PRPs (40-100 nanometers in diameter) that are bound to target sites via specific molecular recognition allows for sensitive detection of biomolecules[5, 6]. Due to their small physical dimensions, PRPs are ideally suited for nucleic acid and protein microarray and microfluidic based applications. In a three-component base pair mismatch recognition DNA hybridization assay (Figure 13) the measured signal is the total number of individual PRPs bound to the complementary capture spot of a microarray minus those that bind to a non-complementary capture spot. Figure 14 displays the result of a DNA hybridization experiment between a biotinylated oligonucleotide (the product of an oligonucleotide ligation assay – see Figure 13), and a complementary capture oligonucleotide. A robust optical signal, originating from a large number of bound mouse anti-biotin antibody-conjugated PRP labels, is observed[51].

A comparison of the relative sensitivity of PRP based detection to that of fluorescence labels as a function of concentration of a labeled oligonucleotide hybridizing to a microarray of complementary capture oligonucleotides is shown in Figure 15. A detection sensitivity limit of $1 \times 10^6$ molecules achieved with PRP labels is a 60-fold improvement when compared to that of fluorescent labels[51].

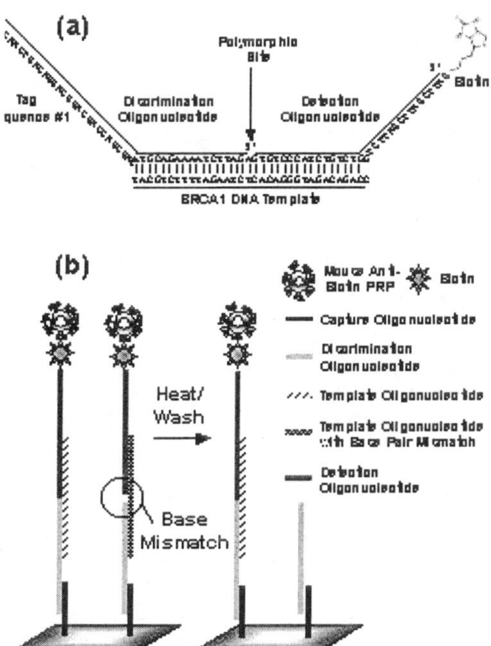



**Figure 13.** (a) A schematic representation of how SNP detection is performed using an oligonucleotide ligation assay (OLA). A template DNA acts as a scaffold to hold the WT discrimination and detection oligonucleotides in place, thereby allowing the ligase to link the two oligonucleotides together if their 5' and 3' bases are adjacent. (b) The ligated product can bind to a microarray spot containing capture oligonucleotides complementary to the DNA extension sequence, Tag 1. A wash step at an elevated temperature removes the non-ligated detection oligonucleotide. Mouse anti-biotin PRPs that bind to the ligated biotinylated product are then counted. (Reproduced from [51] with permission from Academic Press.)

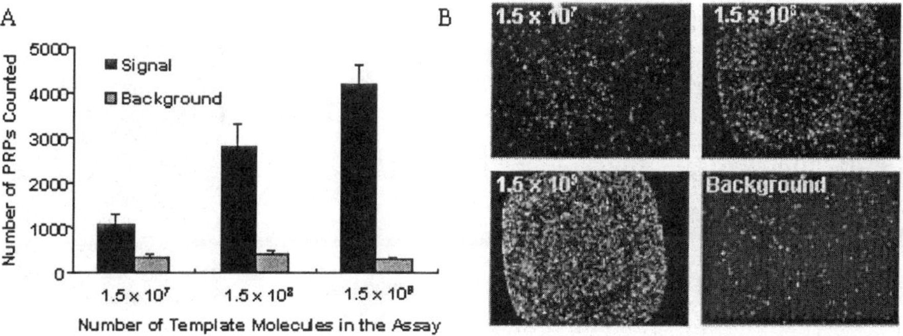



**Figure 14.** An OLA detection assay with decreasing amounts of synthetic WT template oligonucleotide. (A) A plot of the number of PRPs counted versus the number of template molecules in the assay. A detection limit of $1.5 \times 10^7$ molecules was obtained with a signal-to-noise ratio of 8.2. (B) Representative dark field images of PRP labeled 190-um-diameter microarray spots including an image of the non-specific background observed on a non-complementary capture spot. (Reproduced from [51] with permission from Academic Press.)

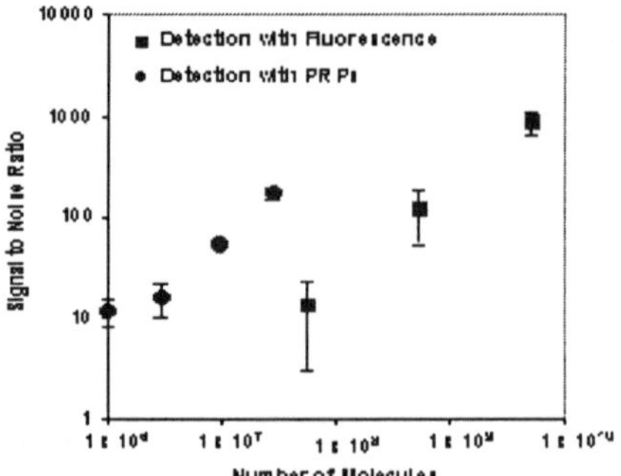



**Figure 15.** A comparison of the relative detection sensitivity of PRP and fluorescent labels. The mouse anti-biotin PRPs detect a smaller number of biotinylated oligonucleotides than can be observed by fluorescence detection of Cy5-labeled oligonucleotides. (Reproduced from [51] with permission from Academic Press.)

## 5.2. Individual PRP Detection of Immuno-Labeled Tissue

Individual PRPs can also be used as optically detectable multicolor labels for *in situ* applications. Analogous to the use of nanometer size immunogold as electron dense markers for electron microscopy studies of tissue sections, antibody modified PRPs are target molecule specific labels that are viewed using a dark field optical microscope. An example is shown in Figure 16A, where ryanodine receptors present in a section of chicken skeletal muscle are detected with PRP labels[5]. The ryanodine receptors are arranged spatially in a parallel series of lines, with a periodic separation distance of 0.6 microns (Figure 16B). The same spatial distribution of the ryanodine receptors is observed by TEM measurement (Figure 16C). Optical observation of PRPs, when compared to TEM detection of immunogold has the benefits of inexpensive and user-friendly instrumentation, amenable to either aqueous or dry environments, real-time monitoring of motion, and multi-color labeling.

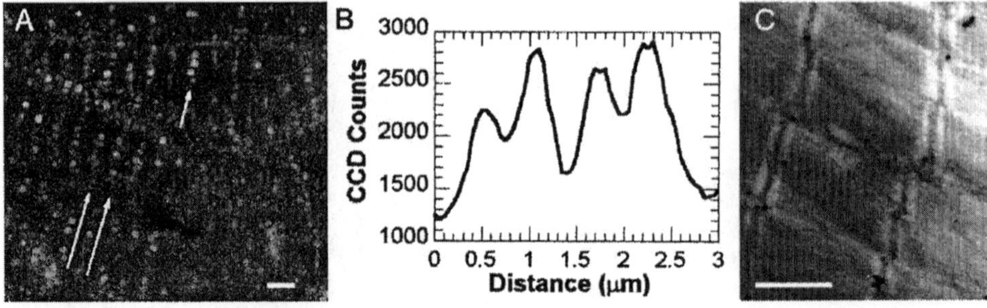



**Figure 16.** (*A*) A b+w photograph of a chicken muscle tissue section in which the ryanodine receptors have been immunogold labeled, and then silver enhanced until individual PRPs have been formed. The optical microscope configuration and detection system is similar to that used for the data of Fig. 2. The white arrows indicate the direction of the parallel set of *z* lines located between the lines of PRPs. (Bar = 2 μm.) (*B*) An intensity (count) line scan along the direction indicated by the single arrow in *A*, demonstrating that the individual PRP image peaks are readily resolved. (*C*) A transmission electron micrograph of a tissue section similar to that used for *A*, confirming that the PRPs are found along the *z* lines in a lattice like fashion characteristic of the known spatial distribution of ryanodine receptors at the sarcoplasmic reticulum and transverse tubule system junctions (Bar = 1 μm). Reproduced from [5] with permission from the publisher. Copyright 2000 National Academy of Sciences, U.S.A.)

## 5.3. Plasmon Resonant Particles and other Nanoparticles as Labels for Biomedical Applications

Recently, there is a growing interest in the use of nanometer size particles as optically detectable labels for use in single particle detection assays. Potential nanoparticles include fluorospheres, quantum dots[52], selenium particles[53], europium nanoparticles[54], upconverting phosphors[55], and PRPs. A major advantage of using particles as labels is that they are readily detected using standard microscopy techniques. This is in contrast to the observation of individual organic fluorophores that are susceptible to photobleaching and require ultrasensitive detectors. PRP labels are ideal for applications that require multi-color, bright, photostable particles that can be imaged with white light illumination.

## 6. REFERENCES


1. Zsigmondy, R.A., *Colloids and the Ultramicroscope - A Manual of Colloid Chemistry and Ultramicroscopy.* 1914, NY: John Wiley and Sons, Inc.
2. Hayat, *Colloidal Gold: Principles, Methods, and Applications*, ed. M.A. Hayat. Vol. 1. 1989, San Diego, CA: Academic Press.
3. Hayat, M.A., *Immunogold-Silver Staining.* 1995, New York: CRC Press.
4. Wilkinson, D.G., *In Situ Hybridization.* 1995, New York: Oxford University Press.
5. Schultz, S., et al., *Single target molecule detection with non-bleaching multicolor optical immunolabels.* Proc. Natl. Acad. Sciences, 2000. **97**: p. 996-1001.
6. Schultz, S., et al., *Nanoparticle based biological assays.* Journal of Clinical Ligand Assay, 1999. **22**(2): p. 214-216.
7. Kreibig, U., Wollmer, M, *Optical Properties of Metal Clusters.* 1995: Springer-Verlag.
8. Yguerabide, J. and E. Yguerabide, *Resonance Light Scattering Particles as Ultrasensitive Labels for Detection of Analytes in a Wide Range of Applications.* J. of Cell. Biochem. Supp., 2001. **37**: p. 71-81.
9. Mie, G., *Beiträge zur optik trüber Medien speziell kolloidaler Metallösungen.* Ann. Phys., 1908. **24**: p. 377.
10. Yu, Y., et al., *Gold Nanorods: Electrochemical Synthesis and Optical Properties.* Phys. Chem. B., 1997. **101**(34): p. 6661-6664.
11. Jana, N.R., L. Gearheart, and C.J. Murphy, *Wet Chemical Synthesis of High Aspect Ratio Cylindrical Gold Nanorods.* J. Phys. Chem. B., 2001. **105**: p. 4065-4067.
12. van der Zande, B.M.I., et al., *Colloidal Dispersions of Gold Rods: Synthesis and Optical Properties.* Langmuir, 2000. **16**: p. 451-458.
13. Jin, R.C., et al., *Photoinduced conversion of silver nanospheres to nanoprisms.* Science, 2001. **294**: p. 1901-1903.
14. Mock, J.J., et al., *Shape effects in plasmon resonance of individual colloidal silver nanoparticles.* J. Chem. Phys., 2002. **116**(15): p. 6755-6759.
15. Sun, Y. and Y. Xia, *Gold and Silver Nanoparticles: A Class of Chromophores with Colors Tunable in the Range from 400 nm to 700 nm.* Analyst, 2003. **128**: p. 686-691.
16. Jackson, J.B. and N.J. Halas, *Silver Nanoshells: Variations in Morphologies and Optical Properties.* J. Phys. Chem. B., 2001. **105**: p. 2743-2746.
17. Oldenburg, S.J., et al., *Nanoengineering of Optical Resonances.* Chem. Phys. Lett., 1998. **288**: p. 243-247.
18. Duff, D.G. and A. Baiker, *A new hydrosol of gold clusters. 1. Formation and particle size variation.* Langmuir, 1993. **9**: p. 2301.
19. Brown, K.R. and M.J. Natan, *Hydroxylamine Seeding of Colloidal Au Nanoparticles in Solution and on Surfaces.* Langmuir, 1998. **14**: p. 726-728.
20. Sun, Y., et al., *Crystalline Silver Nanowires by Soft Solution Processing.* Nano Letters, 2002. **2**(2): p. 165-168.

21. Colby, A.F., et al., *Template-Synthesized Nanoscopic Gold Particles: Optical Spectra and the Effects of Particle Size and Shape.* J. Phys. Chem., 1994. **98**: p. 2963-2971.
22. Brumlik, C. and C.R. Martin, *Template Synthesis of Metal Microtubules.* J. Am. Chem. Soc., 1991. **113**: p. 3174-75.
23. Mock, J.J., et al., *Composite Plasmon Resonant Nanowires.* Nano Letters, 2002. **2**(5): p. 465-469.
24. Ji, T., et al., *Preparation, Characterization, and Application of Au-Shell/Polystyrene Beads and Au-Shell/Magnetic Beads.* Adv. Mater., 2001. **13**: p. 1253-1256.
25. Li, Z., et al., *Multiple thiol-anchor capped DNA-gold nanoparticle conjugates.* Nucleic Acids Research, 2002. **30**(7): p. 1558-1562.
26. Letsinger, R.L., et al., *Use of a steroid cyclic disulfide anchor in constructing gold nanoparticle-oligonucleotide conjugates.* Bioconjugate Chemistry, 2000. **11**(2): p. 289-291.
27. Elghanian, R., et al., *Selective colorimetric detection of polynucleotides based on the distance-dependent optical properties of gold nanoparticles.* Science, 1997. **277**(5329): p. 1078-1081.
28. Mandal, T.K., M.S. Fleming, and D.R. Walt, *Production of hollow polymeric microspheres by surface-confined living radical polymerization on silica templates.* Chemistry of Materials, 2000. **12**(11): p. 3481-3487.
29. Caruso, F., *Nanoengineering of particle surfaces.* Advanced Materials, 2001. **13**(1): p. 11-22,3.
30. Obare, S.O., N.R. Jana, and C.J. Murphy, *Preparation of polystyrene- and silica-coated gold nanorods and their use as templates for the synthesis of hollow nanotubes.* Nano Letters, 2001. **1**(11): p. 601-603.
31. Quaroni, L. and G. Chumanov, *Preparation of polymer-coated functionalized silver nanoparticles.* Journal of the American Chemical Society, 1999. **121**(45): p. 10642-10643.
32. Kneuer, C., et al., *A Nonviral DNA Delivery System Based on Surface Modified Silica-Nanoparticles Can Efficiently Transfect Cell in Vitro.* Bioconjugate Chemistry, 2000. **11**(926-932).
33. Liz-Marzan, L.M., M. Giersig, and P. Mulvaney, *Synthesis of Nanosized Gold-Silica Core-Shell Particles.* Langmuir, 1996. **12**: p. 4329-4335.
34. Stober, W., A. Fink, and E. Bohn, *Controlled Growth of Monodisperse Silica Spheres in the Micron Size Range.* J. Coll. Int. Sci., 1968. **26**: p. 62.
35. Mock, J., et al., *Spectroscopic studies of individual plasmon resonant nanoparticles.* SPIE, 2003. ***In press.***
36. Schultz, D.A., *Plasmon resonant particles for biological detection.* Current Opinion in Biotechnology, 2003. **14**: p. 13-22.
37. Reichert, J., et al., *Chip-based optical detection of DNA hybridization by means of nanobead labeling.* Analytical Chemistry, 2000. **72**(24): p. 6025-6029.
38. Kohler, J.M., et al., *Selective labeling of oligonucleotide monolayers by metallic nanobeads for fast optical readout of DNA-chips.* Sensors and Actuators B-Chemical, 2001. **76**(1-3): p. 166-172.
39. Taton, T.A., C.A. Mirkin, and R.L. Letsinger, *Scanometric DNA Array Detection with Nanoparticle Probes.* Science, 2000. **289**: p. 1757-1760.
40. Bao, P., et al., *High-sensitivity detection of DNA hybridization on microarrays using resonance light scattering.* Analytical Chemistry, 2002. **74**(8): p. 1792-1797.
41. Yguerabide, J. and E.E. Yguerabide, *Light-scattering submicroscopic particles as highly fluorescent analogs and their use as tracer labels in clinical and biological applications.* Anal. Biochem, 1998. **262**: p. 157-176.
42. Reynolds, R.A., C.A. Mirkin, and R.L. Letsinger, *A gold nanoparticle/latex microsphere-based colorimetric oligonucleotide detection method.* Pure Applied Chemistry, 2000. **72**(1-2): p. 229-235.
43. Taton, T.A., C.A. Mirkin, and R.L. Letsinger, *Two-color labeling of oligonucleotide arrays via size-selective scattering of nanoparticle probes.* J. Am. Chem. Soc, 2001. **123**: p. 5164-5165.
44. Silva, T.J., *A Scanning Near-Field Optical Microscope with Magneto-Optic Kerr Effect Contrast for the Imaging of Magnetic Domains with 200 A Resolution.* 1994, University of California, San Diego.
45. Haynes, C.L. and R.P. Van Duyne, *Nanosphere Lithography: A Versatile Nanofabrication Tool for Studies of Size-Dependent Nanoparticle Optics.* J. Phys. Chem. B, 2001. **105**: p. 5599-5611.
46. Englebienne, P., A. Van Hoonacker, and M. Verhas, *High-throughput screening using the surface plasmon resonance effect of colloidal gold nanoparticles.* Analyst, 2001. **126**(10): p. 1645-1651.
47. Englebienne, P., *Use of colloidal gold surface plasmon resonance peak shift to infer affinity constants from the interactions between protein antigens and antibodies specific for single or multiple epitopes.* Analyst, 1998. **123**(7): p. 1599-1603.
48. Haes, A.J. and R.P. Van Duyne, *A nanoscale optical biosensor: Sensitivity and selectivity of an approach based on the localized surface plasmon resonance spectroscopy of triangular silver nanoparticles.* Journal of the American Chemical Society, 2002. **124**(35): p. 10596-10604.
49. McFarland, A.D. and R.P. Van Duyne, *Single Silver Nanoparticles as Real-Time Optical Sensors with Zeptomole Sensitivity.* Nano Letters, 2003. **3**: p. 1057-1059.

50.  West, J.L. and N.J. Halas, *Applications of nanotechnology to biotechnology - Commentary.* Current Opinion in Biotechnology, 2000. **11**(2): p. 215-217.
51.  Oldenburg, S.J., Genick, C., Clark, K., Schultz, D.A., *Base Pair Mismatch Recognition Using Plasmon Resonant Particle Labels.* Analytical Biochemistry, 2002.
52.  Chan, W.C.W., et al., *Luminescent quantum dots for multiplexed biological detection and imaging.* Current Opinion in Biotechnology, 2002. **13**(1): p. 40-46.
53.  Stimpson, D.I., et al., *Real-time detection of DNA hybridization and melting on oligonucleotide arrays by using wave guides.* Proc. Natl. Acad. Sci., 1995. **92**: p. 6379-6383.
54.  Harma, H., T. Soukka, and T. Lovgren, *Europium Nanoparticles and Time-resolved Fluorescence for Ultrasensitive Detection of Prostate-specific Antigen.* Clinical Chemistry, 2001. **47**(3): p. 561-568.
55.  Van de Rijke, F., et al., *Up-converting phosphor reporters for nucleic acid microarrays.* Nature Biotechnology, 2001. **19**: p. 273-276.

# 12

# NOBLE METAL NANOPARTICLE BIOSENSORS

Nidhi Nath and Ashutosh Chilkoti[*]

## 1.  INTRODUCTION

Receptor-ligand binding assays are central to medical diagnostics, proteomics, drug discovery, environmental monitoring and food processing.  A typical binding assay uses a "capture molecule" (often loosely termed as the "receptor") which is typically an antibody, DNA, peptide, or protein, that binds to a target analyte (ligand) of interest in a sample with high affinity and specificity.  Binding of the analyte to the "receptor" is coupled to a transduction step to enable detection of the binding event and quantification of the analyte concentration in the sample.  Based on their mode of detection, most binding assays can be divided into two categories.  The first category includes assays that require a label or tracer–a radioisotope, chromophore or fluorophore–to transduce the binding event into a quantifiable signal.  The second category of analyte binding assays is label-free assays that do not require the addition of extrinsic reagents or labels.  Instead, a change in a physical parameter upon analyte binding such as the mass, thickness, or refractive index, is directly transduced into a measurable signal.  These two categories are broad, and some overlap exists between them.  For example, recent protein engineering approaches have yielded direct fluorescence biosensors, in which binding is coupled to a change in the fluorescence of the receptor through allostery.[1]

A radio-immunoassay (RIA) to quantify insulin concentration in plasma via binding of an insulin-specific antibody to $^{131}$I–labeled insulin was the first labeled assay to be introduced in the early 1960's.[2]  In RIA, a known amount of a radiolabeled analyte (typically $^{125}$I-labeled) is added to the sample containing an analyte of interest and compete with the unlabeled analyte for binding to the receptor.  After separation of the unbound analyte, the amount of radioactivity from the radiolabeled analyte-receptor complex is quantified and the amount of bound radioactivity is inversely proportional to the concentration of the unlabeled analyte in the sample.

Since their introduction, immunoassays have undergone several advances, both in terms of the assay configuration and detection methods, with the goal of increasing sensitivity, specificity and resolution of the assay. Due to their high cost and regulatory and safety issues, RIAs have now been largely supplanted by enzyme immunoassays

[*] Nidhi Nath and Ashutosh Chilkoti, Department of Biomedical Engineering, Duke University, Durham, North Carolina 27708

(EIA) and fluorescence immunoassays (FIA). EIA and FIA use colorimetric or fluorescence indicators, respectively, to optically transduce the primary binding event into a measurable signal. In particular, the introduction of new chromogenic, fluorescent, and luminescent substrates as well as the development of new assay configurations that use avidin–biotin or liposome complexes to amplify the signal have significantly enhanced the sensitivity of immunoassays.[3] Fluorescence detection has gained popularity because of the higher sensitivity of fluorescence as compared to absorbance. In addition, homogeneous fluorescence assays that can independently detect the concentration of bound and unbound fluorescently labeled analyte by the change in fluorescence polarization of the two species have gained popularity because they enable simultaneous, real-time monitoring of biomolecular binding without the need to separate the two species. The demand for high-throughput ligand binding assays has also resulted in automation of the washing, dispensing, and reading of 384 and 1536 well plates in EIA and FIA, leading to a large increase in sample throughput.

DNA and protein arrays are a recent and exciting development that are conceptually derived from plate-based binding assays.[4-6] Instead of immobilizing the receptors within plastic wells, proteins and DNA are "printed" on chemically activated glass plates at a high density. Fluorescence is exclusively used for detection of ligand binding at the surface of these arrays. The high density of arrays reduces sample volume and increases throughput over 1536 well plate assays. Protein arrays using fluorescence labels have also been developed for functional assays on thousands of spots.[4, 7]

Despite their extensive use, assays employing labeled analytes suffer from several disadvantages. First, analytes need to be conjugated to an enzyme, fluorophore or a radioactive moiety, which increases the cost of the assay. Second, assays that employ reporters typically require multiple washing and incubation steps and in most cases are end-point assays. Finally, in many instances, conjugation of reporters near the active site of the "receptor" can interfere with receptor-analyte binding.

Label-free assays offer an alternative and competing technology to RIA, EIA and FIA. The two major categories of label-free biosensors are based on acoustic and optical transducers. In acoustic biosensors, such as the quartz crystal microbalance (QCM) and surface acoustic wave (SAW) devices, biomolecular binding at the surface of the transducer causes a mass change resulting in a proportional modulation of the acoustic frequency[8]. Apart from the mass change, acoustic biosensors can also measure the viscoelastic properties of immobilized molecules; this information, uniquely available from this class of sensors, is valuable in quantifying the morphological changes in cells or changes in protein configuration when they bind to or attach at the solid-liquid interface.[9, 10]

Surface plasmon resonance (SPR) spectroscopy is an alternative label-free method, which detects the change in the local refractive index at the sensor surface due to the binding of a biomolecule.[11, 12] In a typical SPR sensor, a laser beam that is incident on a glass prism coated with a thin layer of gold (~50 nm) undergoes total internal reflection at an angle greater than a specific critical angle. The evanescent wave generated at the point of incidence can couple into the metal and excites surface plasmons (SPR)–longitudinal electron oscillations at the surface of metal–that decay exponentially from the surface. The excitation of surface plasmons occurs at a specific angle of incidence and can be observed as a minimum in the reflected light at that specific angle. Any change in the local refractive index at the interface of the gold layer and the surrounding medium alters the angle of incidence at which SPR is observed and is correlated with the refractive index change at the surface.  In a SPR biosensor, a "receptor" is immobilized onto the metal surface, and binding of its analyte to the receptor is quantified by the shift in the angle of minimum reflection due to the altered refractive index in the vicinity of the metal surface.

Although label-free methods for detection of biomolecular binding are attractive, they are limited by complex and expensive instrumentation and by their limited ability to perform high-throughput assays, especially in high density array format. Motivated by these limitations of current label-free sensors, we are developing a label free biosensor—which we call nanoSPR sensor—that exploits the change in the color of immobilized noble metal nanoparticles (i.e. gold and silver) due to the change in the local refractive index upon a receptor-ligand binding at the nanostructure-liquid interface.[13] The fabrication of the "chip-based' nanoSPR sensor is simple and flexible and can be easily implemented in an array format.  Similar approaches to SPR biosensing using immobilized metal nanostructures have also been demonstrated by Van Duyne and colleagues.[14,15] In the remainder of this chapter, we briefly trace the history of silver and gold nanoparticles and their application in bioanalytical assays, followed by a discussion of nanoSPR.

## 2. NOBLE METAL NANOPARTICLE

### 2.1. Optical Properties of Noble Metal Nanoparticles

Gold and silver belong to a family of "free" electron metals that have a filled valence shell but an unfilled conduction band.  When nanoparticles of these metals are irradiated with incident light, the "free" electrons move under the influence of the electromagnetic field and are displaced relative to their positive core, which creates an oscillating dipole (Figure 1A).[16, 17] Oscillating dipoles absorb maximum energy at their resonance frequency, which lies in the visible range of the electromagnetic spectrum for gold and silver nanoparticles.

In 1908, Mie proposed a theoretical model to explain the optical extinction (sum of the absorption and scattering properties) of noble metal nanoparticles. For nanoparticles with a radius (r) much smaller than the wavelength of light (2r << $\lambda$), the extinction profile can be adequately explained by the simplified Mie formula (Eq. 1).[18, 19]

$$\sigma_{ext}(\omega) = 9\frac{\omega}{c}\varepsilon_m^{3/2}VN\frac{\varepsilon_2(\omega)}{[\varepsilon_1(\omega)+2\varepsilon_m]^2 + \varepsilon_2(\omega)^2} \quad \{1\}$$

where $\sigma_{ext}$ is the extinction coefficient of the nanoparticles and is the sum of the absorption and scattering contributions, $\omega$ (= $2\pi\nu$) is the angular frequency of the incident light, $\varepsilon_m$ ($=n_{med}^2$) is the wavelength independent dielectric constant of the medium surrounding the nanoparticles, $\varepsilon(\omega) = \varepsilon_1(\omega) + i\varepsilon_2(\omega)$ is the wavelength dependent dielectric constant of the metal nanoparticles, and is assumed to be the same as bulk material. $\varepsilon_1(\omega) = n^2 - k^2$ and $\varepsilon_2(\omega) = 2nk$, where n is the refractive index and k is the extinction constant of bulk gold or silver. V (= $4/3\pi r^3$) is the nanoparticle volume and N is the number density of the nanoparticles. The resonance condition is fulfilled at a wavelength for which $\varepsilon_1(\omega)$ = $-2\varepsilon_m$ if $\varepsilon_2$ is weakly dependent on $\omega$, and the nanoparticles display maximum extinction at that wavelength. For gold and silver particles with a diameter of 20 nm, the maximum extinction is at ~530 nm and 400 nm, respectively. Hence, gold nanoparticles appear red while silver nanoparticles appear yellow.

However, for larger particles (approximately $2r \geq \lambda/20$) the peak extinction wavelength shifts to longer wavelengths, and their optical properties cannot be accurately modeled by Eq. (1). Instead, a complete solution of Mie theory is needed to describe their optical properties. Although the mathematical details of the complete solution of Mie theory are beyond the scope of this chapter, we simply describe the physical concepts behind the deviation from Eq. (1). As shown in Figure 1B, for small nanoparticles with a diameter much smaller than the wavelength of the incident light (r $<<\lambda$), all the surface electrons in a nanoparticle "see" the same phase of the electromagnetic field and are displaced simultaneously with respect to the positively charged core, which creates a dipole (dipolar approximation). When the dipole approximation is valid, the optical properties of metal nanoparticles can be determined from Eq. (1) with a reasonable degree of accuracy. However, for larger nanoparticles, the phase of the electromagnetic field is not homogeneous over the particles and electrons in different parts of the particle undergo multipolar oscillations so that the dipole approximation is no longer valid. Because multipolar oscillations exhibit a peak at lower energy compared to dipoles, the color of larger nanoparticles exhibit a red shift that is not accounted for by Eq. (1). Mie theory also fails to adequately explain the optical properties of anisotropic particles. Several new theoretical models have been proposed for anisotropic particles, and readers interested in theoretical modeling of the optical properties of noble metal nanoparticles are referred to several excellent papers on this subject.[20-22]

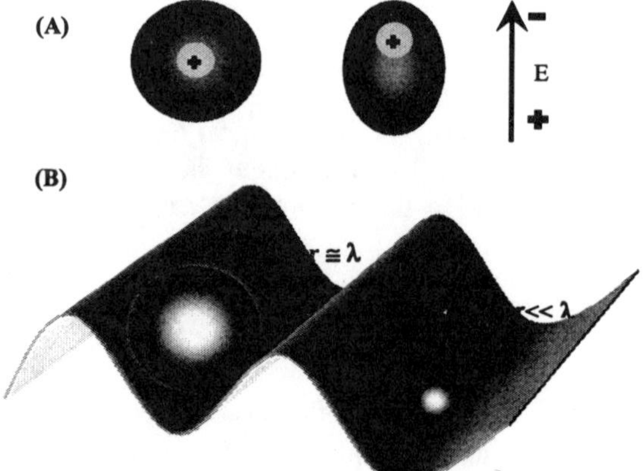


**Figure 1.** (A) Surface 'free' electrons of metal nanoparticles respond to the electromagnetic field of the incident light and oscillate as a dipole. (B) If the particle size is much smaller than the wavelength of light all the surface electrons experience the same phase of the incident light and hence oscillate as a dipole. For large particles different regions of the nanoparticles experience a different phase, leading to multipolar oscillations.

## 2.2. Noble Metal Nanoparticles: Historical Perspective

The brilliant red and yellow colors of noble metal nanoparticles have fascinated humans since ancient times. The Lycurgus Cup, a Roman era artifact, is the first evidence of the use of gold nanoparticles to impart red color to glass. The synthesis of noble metal nanoparticle was rediscovered in the 17th century in Europe and their use since then can be broadly divided into three historical phases. The first phase, starting 17th century, was when gold colloids–called "Purple of Cassius" because the protocol for their synthesis was first published in 1685 by Andreas Cassius–were used to make stained ruby glass.[23, 24] The synthesis involved dissolving fine gold powder in aqua regia, adding water, and then a piece of pure tin. After an hour or two, a brilliant purple precipitate, the Purple of Cassius, was formed. Though this pigment was extensively used, not much was known about its chemistry or the origin of its color. First scientific study of the gold nanoparticle was performed by Faraday who developed a new method for the synthesis of gold colloids by reducing a gold salt using phosphorous.[25] In 1903 Zsigmondy developed ultramicroscopy and used it to systematically analyze gold colloids.[26] He also invented the nucleus method of growing very small colloids into larger colloids and was awarded the Noble Prize in 1925 for these advancements. On the theoretical front, Mie formulated his eponymous theory in 1908 to explain the physical origins of the brilliant colors of gold sols, a theory that remains largely valid to this day.

The second phase of research in gold colloids was their application in biology, beginning with the work of Faulk and Taylor in 1970, who first described a method to label polyclonal antibodies with colloidal gold.[27] This was immediately followed by a landmark paper by Frens in 1971, in which he described a simple protocol for the solution synthesis of monodisperse gold colloids in the range of 10–100 nm by reduction of tetrachloroauric acid ($HAuCl_4$) with sodium citrate.[28] These developments led to an explosion in the application of gold nanoparticles as an electron dense marker to label lectins, protein A, protein G and antibodies to visualize specific antigens and carbohydrates on cells using scanning electron microscopy (SEM) and transmission electron microscopy (TEM).[29] In the 1980's, gold colloids were also tested as visual markers for rapid immunoassay based diagnostics.[30] The home pregnancy test is an example of a commercially available immunoassay that uses gold colloids for rapid detection of human gonadotropin.

The third and current phase started in the mid 1990's, and focused on exploiting the modulation of the optical and electromagnetic properties of the nanoparticles as a function of their size, shape and interparticle distance. The first significant application of metal nanoparticles, in this phase, was initiated by the demonstration that gold nanoparticles can be self-assembled on substrates that present amine or thiol groups.[31] Although it was well known by this time that roughened metal surfaces are ideal substrates for surface enhanced Raman spectroscopy (SERS) because they provide a huge amplification of the spectroscopic signal, methods to reproducibly create such substrates were somewhat limited. Self-assembled monolayers of metal nanoparticles provided a technically simple and reproducible method to fabricate SERS substrates with nanometer scale resolution. Subsequently using self-assembled silver nanoparticles, Nie and colleagues demonstrated the feasibility of single molecule detection, using SERS.[32] Using the enhanced Raman scattering signature from silver nanoparticles, Mirkin and

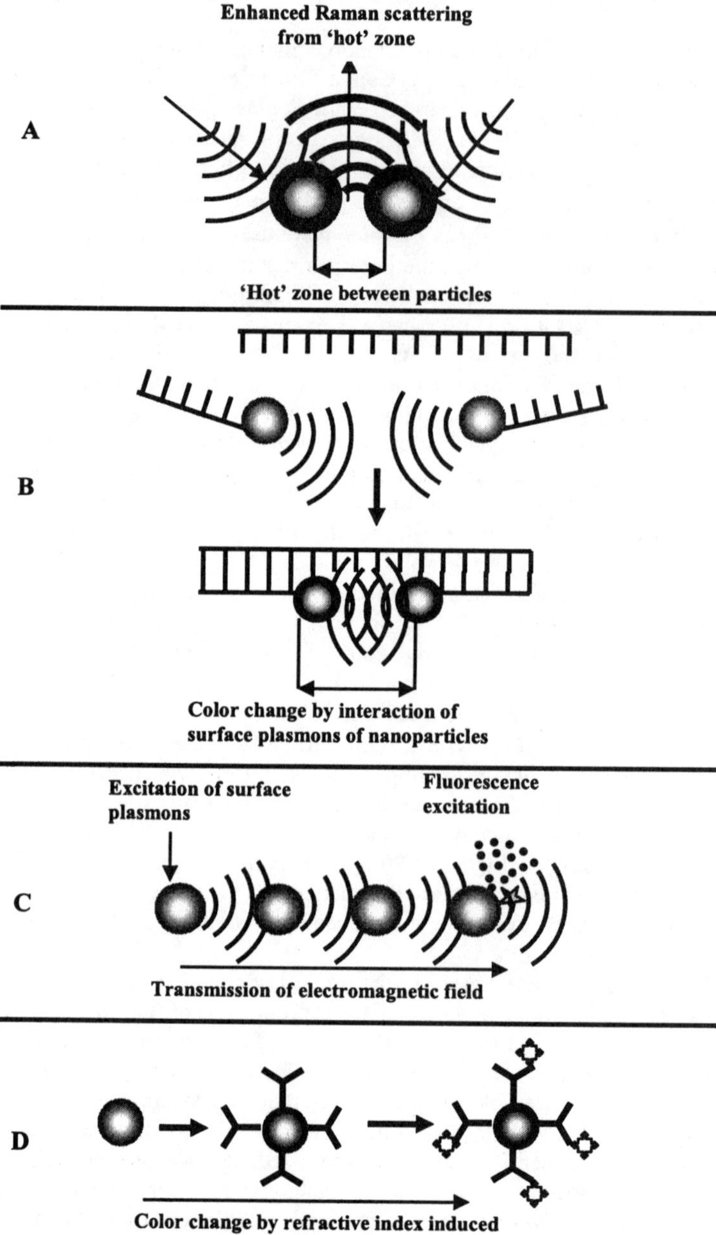


**Figure 2.** Schematic representation of four different sensing configurations using metal nanoparticles. A) SERS due to electromagnetic enhancement from fractal aggregates of metal nanoparticles, B) Detection of DNA hybridization by color change in gold nanoparticles, C) Excitation of fluorescence by transporting electromagnetic field by surface plasmon coupling in an array of nanoparticles, and D) Color change in individual nanoparticles on binding of an analyte

colleagues also independently demonstrated the detection of RNA and DNA at the femtomolar level.[33]

The phenomenon that made these studies possible is the electromagnetic field enhancement that occurs close to the surface of metal nanoparticles when they are excited at a wavelength close to their surface plasmon resonance (SPR) wavelength. Although the electromagnetic field enhancement from a single nanoparticle is typically modest (3 to 6 orders of magnitude), the enhancement increases dramatically to 10–15 orders of magnitude for fractal aggregates of single nanoparticles.[34] The dramatic increase in the electromagnetic field strength in these aggregates has been attributed to the presence of 'hot' zone due interparticle coupling of the electromagnetic field associated with nanoSPR of the individual particles (Figure 2A).

The second significant application of nanoSPR, is their use as probes of oligonucleotide hybridization. The interparticle coupling of surface plasmons triggers a change in the color of nanoparticles, a property that has been used for the development of a sensor capable of detecting femtomolar concentration of oligonucleotides.[35] In this application, a nanoparticle probe is prepared by attaching a short strand of capture DNA (30 mer) to the gold nanoparticles. When the probe is added to a solution containing a complementary oligonucleotide–the "target"–the probe hybridizes with the target, resulting in the aggregation of the gold nanoparticles. Aggregation triggers a visible shift in the color of the gold nanoparticles from red to blue (Figure 2B). In addition to acting as a visible marker for hybridization, double-stranded DNA modified with gold nanoparticles exhibit a sharp melting temperature that allows a single base-pair mismatch in the DNA to be easily distinguished from the wild-type (no mismatch) DNA.[36] It has also been demonstrated that when combined with signal amplification by silver reduction, the sensitivity of the gold nanoparticle-based colorimetric assay exceeds that of fluorescence based methods by two-orders of magnitude.[37]

Recently, the change in the optical properties of gold nanoshells, upon aggregation, has been used to devise an immunoassay in complex biological media such as blood and serum.[38] Nanoparticle probe for the immunoassay was prepared by conjugating antibodies to gold nanoshells (96 nm silica core coated with 22 nm gold shell). Incubation of the probe with the solution containing a multivalent antigen specific to the antibody results in the formation of large aggregates causing a shift in optical spectrum proportional to the antigen concentration. An attractive feature of gold nanoshells for immunoassays is the fact that their plasmon extinction occurs in near-IR region where the interference from biological component in the serum or blood is less, leading to higher sensitivity.[38, 39]

In the third notable application of nanoparticle SPR, a periodic array of silver nanoparticles has been used as a waveguide to transport electromagnetic energy below the diffraction limit (Figure 2C).[40] As proof-of-principle of this concept, a near field-scanning optical microscope (NSOM) was used to excite surface plasmons in a single nanoparticle, which are coupled into neighboring particles, resulting in propagation of the incident light along the nanoparticle array. A fluorescent particle, with an excitation wavelength that overlapped the surface plasmon wavelength, could be excited up to 500 nm away from the NSOM source by the propagated surface plasmons.

Finally, the dependence of the position and peak intensity of the extinction spectrum of the noble metal nanoparticles, due to nanoSPR, on the refractive index of the surrounding medium has also been exploited for biosensing. Gold or silver nanostructures can be functionalized with a receptor, so that binding of an analyte causes a change in the local refractive index at the nanoparticle-solution interface, which optically transduces the binding event as a small but visible change in the peak intensity and wavelength of extinction (Figure 2D). The refractive index-dependent color change in gold nanoparticles was first used to develop a solution phase immunoassay.[41, 42] Gold nanoparticles were coated with monoclonal antibodies, and binding of the antibody-coated nanoparticles to their antigen in solution caused a red shift and an increase in the absorbance of the nanoparticles. The increase in absorbance at 600 nm was used to quantify the affinity of an antibody-antigen pair. The assay was performed in an automated clinical analyzer, which also demonstrated the feasibility of high sample throughput.

Subsequently, Okamoto and colleagues demonstrated that gold nanoparticles, self-assembled on a functionalized glass surface exhibit an optical response to change in refractive index upon spin casting of polymers of different thickness onto the immobilized gold nanoparticles.[43] We further extended this concept by demonstrating proof-of-principle of a colorimetric biosensor in a chip format that is capable of real-time, label-free detection of biomolecular binding.[13] The sensor was fabricated by functionalizing immobilized gold nanoparticles on an optically transparent substrate nanoparticles with an receptor. The binding of ligand to the receptor changed the refractive index in the vicinity of the nanoparticles that results in a colorimetric response. We have named this transmission mode sensor, nanoSPR sensor for brevity and to distinguish it from conventional SPR reflectometry. Van Duyne and colleagues have also developed a nanoSPR sensor using silver nanostructures that are directly fabricated on glass by nanosphere lithography.[14, 15] The use of anisotropic silver nanostructures for biosensing is attractive for two reasons: first, the surface plasmon response of silver nanostructures to the refractive index of the surrounding medium, which determines the sensitivity of nanoparticle SPR is greater than that of gold; second, the sensitivity is further amplified by the anisotropy of the nanostructures. The importance of the structural anisotropy of nanoparticles on the sensitivity of nanoSPR sensors was further demonstrated by Xia and colleagues who showed that gold nanoshells that are self-assembled on glass have a much greater sensitivity (~7 times) to the change in refractive index of the surrounding medium when compared to gold spheres of a similar size.[44]

## 3.   NANOPARTICLE SPR BIOSENSOR

In the remainder of this review, we discuss our implementation of the two components that together constitute a surface-based nanoSPR biosensor–the optical transducer and biological detector (Figure 3). We first discuss the fabrication and characterization of the optical transducer, which involves: (a) solution phase synthesis of metal nanoparticles, (b) activation of a glass substrate and self-assembly of metal nanoparticles on the functionalized glass substrate, (c) characterization of the sensing volume and sensitivity of the sensor to refractive index changes. We then describe coupling of the biomolecular detector to the optical transducer to create a functional nanoSPR sensor, which involves: (a) activation of the nanoparticle surface and immobilization of the receptor and (b) binding of target biomolecules and

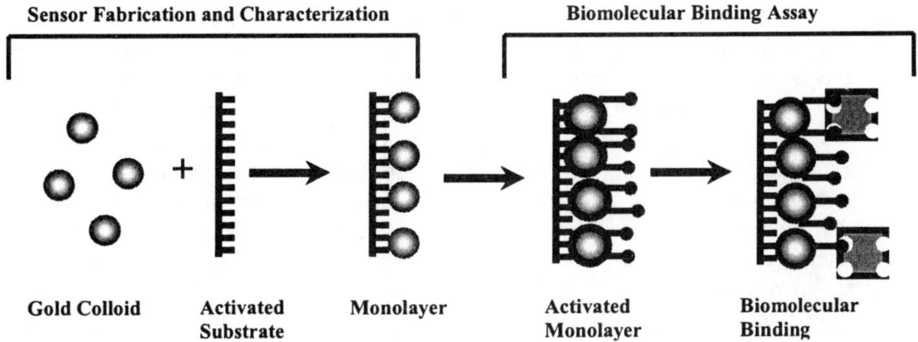


**Figure 3.** Various steps involved in the fabrication of a nanoSPR biosensor and its use for quantification of biomolecular interactions.

characterization of the performance of the sensor as defined by its detection limit, dynamic range, and signal-to-noise (S/N) ratio.

## 3.1. Noble Metal Nanoparticles: Synthesis

A typical synthesis of spherical metal nanoparticles involves reduction of its metal salt in solution, and proceeds in three stages.[45-47] In the first stage, metal ions are reduced to metal atoms, which then undergo rapid collisions to form stable icosahedral nuclei of 1-2 nm in size. The initial concentration of nuclei depends on the concentration of the reducing agent, solvent, temperature and reduction potential of the reaction. This stage, which is typically complete in a few seconds, is important because it determines the heterogeneity in the size and shape of the nanoparticles. Further growth of the nanoparticles occurs, in the second stage, by reduction of metal ions on the surface of the nuclei, until all the metal ions are consumed. The third and final stage involves prevention of nanoparticle aggregation, which is typically achieved by the addition of stabilizing agents. For example, gold nanoparticles are stabilized by electrostatic repulsion due to adsorbed citrate ions on their surface that impart negative charge to the nanoparticles. Gold nanoparticles with diameters in the range of 2-5 nm have been stabilized by thiol-capping agents.[48] Polyvinylpyrolidone (PVP) and bis(p-sulfonatophenyl)-phenylphosphine (BSPP) have been used to stabilize silver nanoparticles.[49, 50] Adsorbed proteins can also stabilize nanoparticles, and prevent their aggregation at high salt concentrations.

The size of nanoparticles is controlled by varying the ratio of the reducing agent to metal salt. A high concentration of reducing agent causes rapid formation of a large number of nuclei and leads to smaller, monodisperse metal nanoparticles. In contrast, a low concentration of reducing agent leads to slow formation of a few nuclei, resulting in larger nanoparticles with a greater heterogeneity in size. Another method to control the size of nanoparticles is by using different combinations of reductant and stabilizing agent. For gold nanoparticles, the use of sodium citrate as the reductant typically results in particles with a diameter in the range of 12–100 nm, whereas the use of white phosphorus produces nanoparticles in a size range of 5–12 nm.[51] Even smaller nanoparticles of a few nanometers in diameter can be synthesized by reduction of $HAuCl_4$ in organic solvent in the presence of a thiol-capping stabilizing agent.[48] For silver nanoparticles, $NaBH_4$ reduction of silver nitrate in the presence of BSPP[50] has been shown to result in the formation of ~8 nm diameter particles whereas the use of PVP as the reducing and protective agent results in larger particles with diameters that range from 15–36 nm.[49]

Due to their unique shape-dependent optical properties, substantial efforts have been devoted to the synthesis of anisotropic structures such as high-aspect ratio nanorods,[52] multimetal micrometer rods,[53] silver nanoprisms,[54] and silver and gold nanocubes.[55] The different growth rates of various faces of the nuclei can be exploited to synthesize anisotropic nanostructures. In one synthetic approach, nanoparticles are physically entrapped inside rod-like micelles or nanotubes so as to force growth in a specific direction leading to the formation of high-aspect ratio nanorods and nanowires. Polymers can also reduce or enhance the growth in a specific crystallographic direction resulting in structurally anisotropic nanoparticles. For example, PVP is believed to reduce the growth rate in the <100> directions and/or increase growth rate in the <111> direction, to produce nanocubes.[55] Core-shell nanoparticles are another class of nanostructures of interest for biosensing, and their optical properties can be independently controlled by either changing the thickness of the shell or by the overall diameter of the core-shell nanoparticle. Core-shell nanoparticles can be synthesized either by silica nanotemplate-assisted electroless deposition[39] or by using silver nanoparticles as sacrificial templates.[56]

## 3.2. Self-Assembly of Noble Metal Nanoparticles on Substrate

Noble metal nanostructures have been directly fabricated on surfaces by numerous methods including photolithography,[40] nanosphere lithography,[57] protein or DNA mediated metal organization,[58-60] and electrochemical deposition of metals in nanoporous aluminum.[61] We have taken a different approach, in which metal nanoparticles are first synthesized in solution, and are then self-assembled on functionalized glass. We chose this approach because it is technically simple and yields reproducible and stable nanostructures on glass. From a technological perspective, this method is also attractive, because it can be scaled up for production of nanoSPR chips far more easily and at significantly less cost than methods involving lithography. An added advantage of solution-based assembly of preformed nanoparticles is the ability to control the nanoparticle density on the surface by changing the solution conditions during colloidal self-assembly. A significant limitation of this method is that nanoparticle assembly is random and fabrication of a periodic array of nanostructures on the surface with long-range order is not possible.

The first step in the self-assembly of nanoparticles on the surface is functionalization of the substrate to introduce the necessary functional groups to enable chemisorption of the metal nanoparticles. Functionalization of the substrate, typically involves introduction of amine or thiol groups at the surface, moieties that are known to form strong bonds with gold and silver nanoparticles.[31] Because glass is the preferred substrate due to its optical transparency and ubiquitous use in laboratories, the preferred route for introduction of these moieties is by silanization of the glass surface using amine or thiol-terminated alkoxysilanes. We note, parenthetically, that, the quality of the silane monolayer is critical for the reproducible formation of nanoparticle monolayers and their long-term stability on the substrate. Several critical steps determine the quality of silane monolayers on glass: these include the use of clean glass substrate, silane concentration, reaction time, water content and baking of glass after silanization. The most common problem encountered in silanization is uncontrolled polymerization of the silane, leading to the formation of multilayers on the surface, which can compromise the assembly of the nanoparticles on the surface. A number of excellent papers on silanization are available, which provide greater experimental details on the fabrication of high-quality silane monolayers.[62-64]

The assembly of gold nanoparticles on an amine-functionalized surface is driven by two forces: diffusion of negatively charged nanoparticles towards the positively charged amine-functionalized surface and, interparticle repulsion between negatively charged nanoparticles on the surface. The solution concentration of nanoparticles, time of incubation, temperature, pH and ionic strength of the solution are variables that can be used to control the density of nanoparticles on the surface. In the initial stages of assembly when a significant fraction of the surface is available for chemisorption, the assembly is controlled by diffusion of gold nanoparticles to the surface. According to Park $et$ $al.$, the rate of diffusion of nanoparticles from solution to the surface is given by:[65]

$$q = 0.163\,pn(kTt\,/\,yr)^{1/2} \quad \{2\}$$

where q is the number of particles that reach the surface per unit time, p is the probability that a particle reaching the surface will adsorb to the surface, n is the solution concentration of particles per unit volume (particles cm$^{-3}$ ), k is the Boltzman constant ($1.38 \times 10^{-16}$ g cm$^2$ sec$^{-2}$ K$^{-1}$), T is the absolute temperature in Kelvin, t is the time in seconds, y is the viscosity (g.cm$^{-1}$.sec$^{-1}$) and r is the radius of the particle (cm). We assume that initially every particle that reaches the surface binds to the surface (p =1), an assumption that is reasonable due to the chemisorptive bond of amine and thiol groups with gold, so that Eq. (2) can be simplified to describe the initial kinetics of nanoparticle assembly at room temperature, as shown below:

$$\Gamma = 3.5x10^{-7}\,n(t\,/\,r)^{1/2} \quad \{3\}$$

where $\Gamma$ is the particle density (number of particles per unit area) at the glass surface at time t. According to this equation, the nanoparticle density in the initial phase of adsorption will increase with time as $t^{1/2}$, linearly with particle concentration in solution (n), and decrease with particle radius as $r^{1/2}$. Indeed, Natan and colleagues experimentally showed that the assembly of gold nanoparticles on a thiol-functionalized glass surface exhibits a $t^{1/2}$ dependence in the initial phase of assembly.[66] However, the kinetics of self-assembly deviate from a $t^{1/2}$ dependence at longer times, as interparticle repulsion becomes significant with increasing surface density of the nanoparticles.

We have examined the assembly of ~12 nm diameter gold nanoparticles (solution concentration of 11.6 nM) and of ~39 nm diameter particles (0.36 nM) on amine-functionalized glass by monitoring the increase in optical extinction from the surface at the peak resonance wavelength ($\lambda^{max}$) as a function of time. Figure 4A shows the

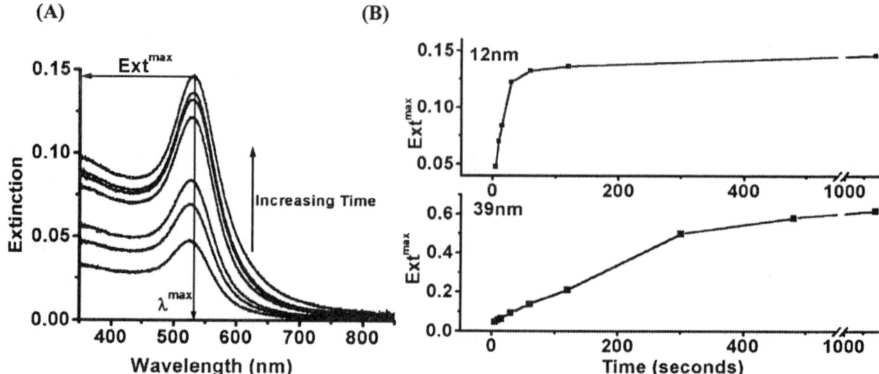

**Figure 4.** (A) Assembly of 12nm nanoparticle on glass monitored by optical extinction. Spectrum was collected at 5, 10, 15, 30, 60, 120, and 1080minutes. (B) Kinetics of gold nanoparticle adsorption on glass measured by the change in $Ext^{max}$ as a function of time.

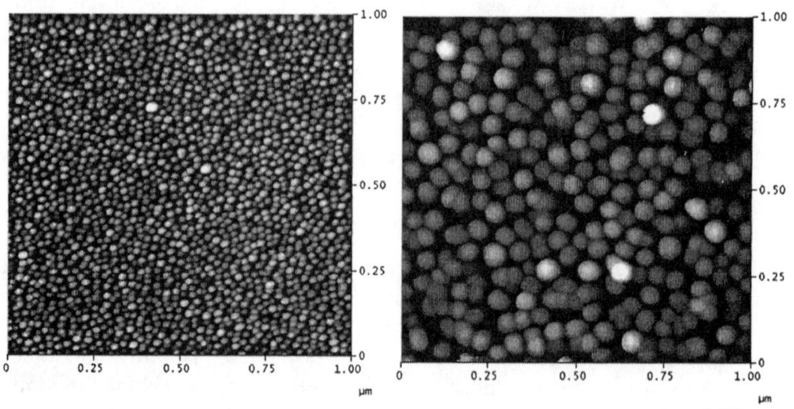

**Figure 5.** AFM image of a monolayer of 12 nm (left) and 39 nm(right) diameter gold nanoparticles immobilized on amine-functionalized glass

evolution of extinction spectra as the 12nm nanoparticles assemble on the functionalized glass. Kinetics of nanoparticle assembly can be studied by plotting $Ext^{max}$ as a function of time, as shown in Figure 4B for 12 nm and 39 nm particle sizes. $Ext^{max}$ increases linearly with time for both types of nanoparticles and then reaches a steady state density. A typical AFM image of a randomly packed layer of gold nanoparticles of 12 nm and 39 nm diameter, on an amine-functionalized glass, is shown in Figure 5. We calculated the particle density on the surface, in the initial, linear phase, from Eq. (3) as well as by direct measurement using atomic force microscopy (AFM). The particle density calculated by the two methods show good agreement for both particle sizes, indicating that adsorption in the initial phase is primarily a diffusion-controlled process and that the particle density in this phase can be accurately determined from Eq. (3) for nanoparticles of different sizes.

With time, as the available sites on the surface are occupied by chemisorbed nanoparticles, the lateral repulsion between nanoparticles becomes significant, and the experimental results deviate from the linear diffusion model. The final, steady state density of nanoparticles at the surface is observed to be dependent both upon the solution concentration of the nanoparticles and the magnitude of lateral repulsion between the nanoparticles. Given the importance of lateral repulsion in controlling the assembly process, the particle density at the surface can be increased at a constant solution concentration by decreasing the lateral repulsion between nanoparticles. The screening of surface charges by adsorption of counterions reduces the repulsion between nanoparticles and increases the particle density at the surface.[66] An alternative is to increase the ionic strength of the solution, which reduces the Debye screening length of the nanoparticles, and thereby allows adjacent nanoparticles to approach each other on the surface.[67-69] High ionic strength however, can be deleterious to nanoparticle assembly, because it also drives aggregation of the nanoparticles in solution, which can be visually discerned by a change in color of the gold hydrosol from red to blue.

## 3.3. Optical Properties of Self-Assembled Gold Nanoparticle on Glass

In order for the self-assembled monolayer of gold nanoparticles to act as an optical transducer of receptor analyte binding, two important properties need to be characterized; (1) optical response of the nanoparticle assembly to a refractive index change in the surrounding medium and (2) the sensing volume of the nanoparticle, defined as the region around the nanoparticle that is optically responsive to the refractive index change. The sensing volume and sensitivity to refractive index of the surrounding medium is controlled by material and geometric properties of the nanoparticles, which include the nanoparticle size, shape, and composition. These material and geometric parameters determine the strength and the range of the electromagnetic field associated with the plasmon extinction of the nanoparticles and their modulation by the refractive index of the surrounding medium.

### 3.3.1. Refractive index response of metal nanoparticles

Theoretically, the sensitivity of the optical properties of metal nanoparticles to the refractive index of the surrounding medium is described by simplified Mie theory. Plasmons resonate in a nanoparticle when the denominator in Eq. (1) is at a minimum, as shown below

$$\varepsilon_1(a) + 2\varepsilon_m = \text{minimum} \qquad\qquad \{4\}$$

For gold nanoparticles, $\varepsilon_1$ ($=n^2-k^2$) is negative and decreases with increasing wavelength, so that as $\varepsilon_m$ increases the resonance condition is satisfied at higher wavelengths and the peak red-shifts with an increase in refractive index of the solution. According to Eq. (4), the resonance condition is independent of particle size and the red shift in the peak resonance wavelength is not expected to change with size but will depend on $\varepsilon_1$, which is determined by the composition of the nanoparticle. Apart from the red shift in the peak resonance wavelength, the extinction of nanoparticles increases with the refractive index of the medium as $\varepsilon_m^{3/2}$ (or $n_m^3$) as well as with the volume of the nanoparticle (Eq. 1).

Experimentally, the sensor optical response (R), defined as the change in extinction or shift in peak resonance wavelength as a function of the refractive index of the surrounding medium can be represented by Eq. (5), as follows:[70]

$$R = m(R.I_{\text{final}} - R.I_{\text{initial}}) \qquad\qquad \{5\}$$

where m is the sensitivity of the sensor, and $R.I_{\text{initial}}$ and $R.I_{\text{final}}$ are the initial and final refractive index at the nanoparticle surface, respectively. The sensitivity (m) of a nano SPR sensor is an intrinsic property of the nanoparticle size, shape and composition and can be determined experimentally by measuring the sensor response as a function of a known refractive index change. In a typical biomolecular interaction, the change in refractive index is caused by binding of the analyte to the receptor on the surface. Proteins have a refractive index of 1.6 in their dry crystalline form, but in aqueous medium the refractive index of proteins is much lower due to the large amount of water associated with the protein. Hence, upon binding of a protein to the surface of a nanoparticle in an aqueous solution, the maximum change in refractive index ($R_{\text{max}}$) that is theoretically expected is $R.I_{\text{protein}} - R.I_{\text{water}} = 1.6 - 1.33 = 0.27$. In actual practice, the change in refractive index is much smaller, due to the formation of a spatially heterogeneous and incomplete layer of protein on the surface and also because most proteins will only partially occupy the sensing volume of the nanoparticle.

We undertook a systematic experimental study to determine the sensitivity of immobilized spherical gold nanoparticles (m) as a function of nanoparticle size to the refractive index of the surrounding environment. The aim of this study was to compare the magnitude of the optical change as a function of nanoparticle size to determine the optimum particle size for maximum sensitivity. Monolayers of spherical gold nanoparticles were prepared on glass with seven different sizes in the range of 12 nm to 50 nm. The immobilized nanoparticles were immersed in different solvents with refractive indices ranging from 1.33 to 1.495 and their extinction spectra were measured in the range of 350–850 nm.

Typical extinction spectra for immobilized nanoparticles of 12 nm and 39 nm diameter show that the sensor response to an increase in the solution refractive index is manifested as a red shift of the resonance wavelength and increase in the extinction (Figure 6). The change in maximum extinction ($\text{Ext}^{\text{max}}$) or extinction at 575 nm ($\text{Ext}_{575\text{nm}}$) as a function of refractive index was linear for both the 12 nm and 39 nm diameter gold nanoparticles (Figure 6). The measurement at off peak resonance wavelength was

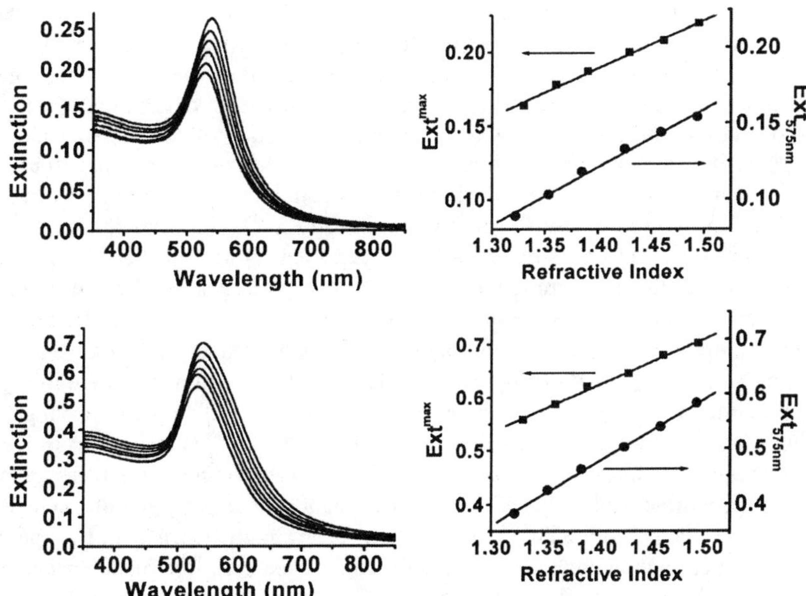

**Figure 6.** Extinction spectra and plot of $Ext^{max}$ and $Ext_{575nm}$ as a function of the refractive index of the surrounding medium for 12 nm (top panels) and 39 nm (bottom panels) diameter gold particles. The solvents were: water (n = 1.33), ethanol (n = 1.36); 3:1 (v/v) ethanol:toluene (n = 1.39), 1:1 (v/v) ethanol:toluene (n = 1.429), 1:3 (v/v) ethanol:toluene (n = 1.462), and toluene (n = 1.495).

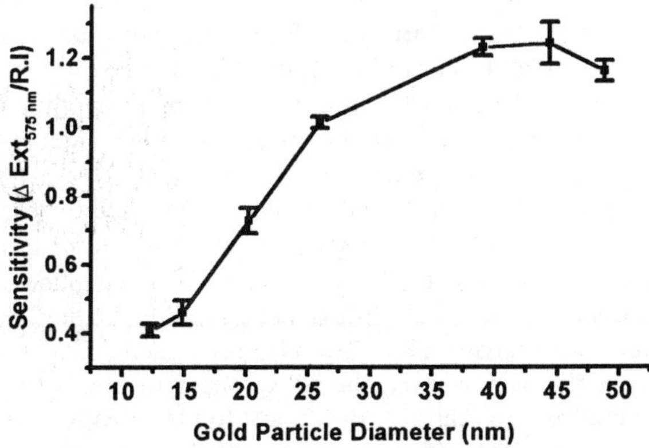

**Figure 7.** Sensitivity as a function of nanoparticle diameter.

chosen because it is experimentally easier to implement as is explained later and also since extinction at 575nm changes linearly with refractive index in the range of 1.33–1.495. The sensitivity of gold nanoSPR sensor, defined as change in $Ext_{575nm}$ per unit change in refractive index of the solution, is determined by the slope of the linear fit. The sensitivity of all the nanoparticle sizes were similarly determined (results not shown). A plot of sensitivity of nanoSPR sensor as a function of particle size shows that the sensitivity increases with size and reaches a plateau at a size of ~ 39 nm (Figure 7).

The shift in $\lambda^{max}$ ($\Delta\lambda^{max}$) was 10-13 nm for a change in the solvent refractive index from 1.33 to 1.5, but there was no consistent trend in $\Delta\lambda^{max}$ with particle size. A review of the literature showed that a similar $\Delta\lambda^{max}$ was reported for spherical gold nanoparticles ranging in diameter from 5.2 nm to 50 nm.[44, 71, 72] A much larger $\Delta\lambda^{max}$ has been reported for non-spherical nanoparticles, indicating that the shape of the nanoparticles rather than their size is more important in modulating the $\Delta\lambda^{max}$ with refractive index. This is best exemplified by the work of Van Duyne and colleagues, who used nanosphere lithography to prepare a series of truncated trapezoid nanostructures of silver with varying height, width, and shape on the surface and demonstrated a large $\Delta\lambda^{max}$ (i.e. 191nm) in response to refractive index change of the environment.[73, 74] We have found that silver nanoprisms prepared by photoinduced conversion of silver nanospheres are equally sensitive as truncated tetrahedrons, with a $\Delta\lambda^{max}$ of ~270 nm per refractive index unit (unpublished results). Gold core-shell nanoparticles are the other class of noble metal structures that have been studied, and were shown to have a $\Delta\lambda^{max}$ per unit change in refractive index of 409 nm for a gold core-shell nanoparticle with a core radius of 25 nm and a shell thickness of 4.5 nm.[44]

### 3.3.2.    Spatial Sensitivity of Immobilized Gold Nanoparticles on Glass

The electromagnetic field strength due to the excitation of surface plasmons and its penetration depth–the distance from the surface where the intensity falls to 1/e or ~37% of its initial value–from the particle surface into the surrounding medium determines the "sensing volume" around a nanostructure. The significance of the sensing volume in a biomolecular-binding assay is shown in Figure 8A. For detection of a biomolecular binding event at the surface of a nanostructure, the penetration depth of the surface plasmon should be similar to the sum of the size of the capture agent and the target. For example, the dimensions of an antibody are 23.5 x 4.5 x 4.5 nm,[75] so that the penetration depth must be greater than 23.5 nm to detect the binding of an antigen to the antibody. The optimal penetration depth will obviously depend on the size of the antigen, which can vary greatly depending upon whether it is a small molecule, peptide or a large protein.

The penetration depth of the electromagnetic field due to surface plasmons has been extensively studied for planar SPR.[70] The electromagnetic field in planar SPR decays exponentially from the surface, and the SPR sensor responds to the refractive index change upto 200 nm from the surface. In contrast to planar SPR, there is less information available on the sensing depth of metal nanoparticles. The electromagnetic field due to SPR in nanoparticles depends on the metal composition, size and shape of the nanoparticles.[20] A few experimental results available for 14 and 15 nm diameter gold nanoparticles indicate that the sensing distance is ~20–25 nm from the surface of the nanoparticles.[43, 76] Theoretical studies by Schatz and co-workers have shown that the

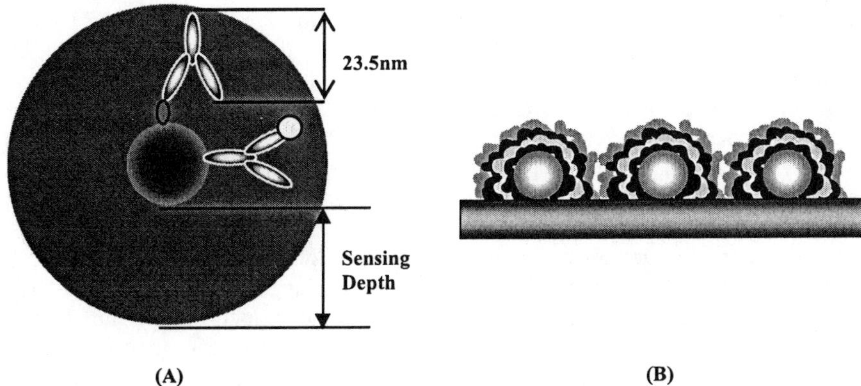



**Figure 8.** A) Schematic showing the significance of the relationship between the sensing volume of a nanoparticle to the size of the biomolecular complex for detection in nanoSPR.   B) Schematic showing sequential, "layer-by-layer" adsorption of polyelectrolytes at the nanoparticle surface to experimentally measure the sensing volume.

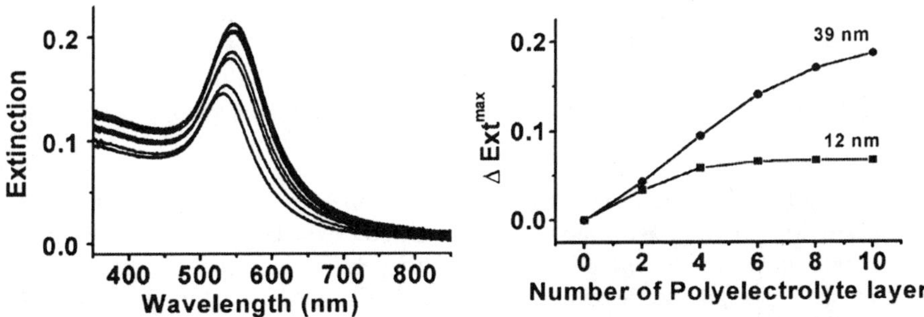



**Figure 9.**  A) Change in extinction spectrum of 12 nm diameter gold nanoparticle monolayer with layer-by-layer deposition of polyelctrolytes. B) Plot of peak extinction as a function of the number of deposited polyelectrolyte layers for immobilized 12 nm and 39 nm diameter gold nanoparticles on amine-functionalized glass.

electromagnetic field intensity decays to its asymptotic value within ~ 60 nm from the surface of a 30 nm diameter silver sphere.

We have taken an experimental approach to measure the spatial range of the electromagnetic field–the sensing volume–as a function of size of the gold nanoparticles. We used layer-by-layer deposition to sequentially deposit polyelectrolyte layers of opposite charge on the colloidal gold surface to systematically vary the thickness (Figure 8B).[76,77] Positively charged poly(allylamine) hydrochloride (PAH; MW = 70,000 Da) and negatively charged polystyrene sulfonate (PSS; MW = 70,000 Da) were used for layer-by-layer deposition.  A monolayer of gold nanoparticles on glass is negatively charged due to adsorbed citrate and residual $AuCl_2$ anions. Incubation of a glass surface with immobilized gold nanoparticles with 0.003 monomol/L solution of PAH in 1M NaCl (where monomol is defined as moles of monomer/Liter) for 30 min cause the deposition of a monolayer of positively charged PAH on the negatively gold nanoparticles. The adsorption of PAH is evident from the red shift in the extinction spectrum of the surface (Figure 9A). The surface was subsequently incubated in 0.003 monomol/L solution of PSS in 1 M NaCl for 30 min. Negatively charged PSS adsorbed to the positively charged PAH on gold colloid, as is evident from a further red-shift in the extinction spectrum of the surface. Ten polyelectrolyte layers were deposited in a similar manner and the extinction spectrum was measured after each step, and is shown for nanoSPR chip fabricated with 12 nm diameter gold particles (Figure 9A). The $Ext^{max}$ for 12 nm and 39 nm particles is plotted for layers 0, 2, 4, 6, 8, 10 (Figure 9B). There are two important observations that are relevant to the performance of these immobilzed nanoparticles as nanoSPR transducers. First, the magnitude of extinction change ($\Delta Ext^{max}$) in response to polyelectrolyte deposition increases with particle size and second, $\Delta Ext^{max}$ saturates at a smaller number of layers for smaller nanoparticles, indicating that smaller particles have a smaller sensing volume compared to larger particles.

These studies provide important insights into the design of nanoSPR biosensors. Assuming a 4 nm thickness for each polyelectrolyte layer,[78] a 12 nm diameter gold nanoparticle has a penetration depth of 24 nm whereas a a monolayer fabricated with 39 nm diameter gold particles has a penetration depth of 40 nm. Because an antibody has dimensions of 23.5 x 4.5 x 4.5 nm,[75] binding of an antibody to a small analyte can be detected using both nanoparticles, although the smaller nanoparticle is close to the limit of its sensitivity.  In contrast, binding of an antibody to a larger biomolecule such as albumin  with dimensions of 14 x 4 x 4 nm[75] will result in a marginal optical signal from immobilized 12 nm gold particles  but should be easily detected by  immobilized 39 nm diameter gold particles.

## 3.4. Biosensing Using Noble Metal Nanoparticles

The previous discussion has focused on the fabrication, characterization and optimization of the transducer. Here, we discuss, coupling of the biomolecular detector to the optical transducer to create a functional nanoSPR biosensor. The two important steps in coupling the biological detector to the optical transducer are presentation of the receptor on the nanoparticle surface and biomolecular binding between receptor and analyte. It has been shown that the density and spatial distribution of the receptor on the surface influences the effective binding affinity of the biomolecular recognition interaction as well as the surface density of the biomolecular complex, so that optimization of the receptor density at the surface is critical to maximize the refractive index change at the interface in response to receptor-analyte binding.[79, 80] Apart from the surface density of the biomolecular complex, the nature of the ligand binding to the receptor will play an important role in optimizing the sensor design. Planar SPR studies, done on four different proteins, have concluded that sensor response is dependent only on the amount of the protein bound to the surface and is independent on the nature of the protein. The quantitative correlation between the amount of the surface bound protein and sensor response will also apply to the nanoSPR biosensor.[81] Finally, as described in the previous section the nanoparticles have limited sensing volume on the order of 50nm compared to 200nm for planar SPR, hence the dimensions of the receptor-ligand binding partners will dictate the optimal particle size required to maximize the sensitivity and dynamic range for detection of a particular analyte by a specific receptor.

### 3.4.1. Receptor presentation

The fabrication of all biosensors typically involves the incorporation of a receptor or "capture agent"–a small molecule, peptide, DNA or protein- to bind an analyte from the sample of interest. Typically, in surface-based sensors the receptor is immobilized onto the surface of the transducer. A successful immobilization method should create a stable bond between the receptor and the surface, should minimize loss of functional activity of the receptor and prevent non-specific binding of the receptor to the surface during immobilization. The two most common methods to immobilize receptors onto the gold nanoparticles are by physical adsorption and covalent conjugation. Faulk and Taylor first demonstrated that proteins can be physically adsorbed onto gold nanoparticles and used as a tracer for electron microscopy.[27] Proteins remain active on adsorption to nanoparticles and can react with their corresponding ligands or substrates.[82] Maximum adsorption of proteins to gold and silver nanoparticles is observed at a pH at or slightly above the isoelectric point of the protein and is concentration dependent.[83] Although adsorption is the simplest method to immobilize protein "receptors" on nanoparticles, adsorption provides little control over the orientation of the immobilized receptor, needs to be optimized for every protein and may not be applicable for small receptors. In addition, adsorbed proteins do not always remain bound to the surface because adsorbed proteins can be desorbed during reaction or displaced by other proteins when exposed to a complex mixture such as serum or plasma. Adsorption is therefore not the preferred method to couple proteins to nanoparticles for biosensing applications.

A versatile approach to immobilize different types of receptors is by their covalent immobilization to self-assembled monolayers (SAMs) of alkanethiols that present a suitable reactive functional group. SAMs of alkanethiols are formed spontaneously when a gold or silver surface is incubated with a dilute solution of an alkanethiol, and are most versatile method to tailor the physical and chemical properties of gold and silver surfaces, including nanoparticles.[84-86] SAMs offer two important features for the presentation of receptors that are important for label-free sensors such as nanoSPR. First, SAMs containing short oligoethyleneglycol chains [$(EG)_n$-SH SAMs] resist the adsorption of proteins, so that non-specific adsorption, the bane of label-free biosensors can be minimized. Second, derivatives of these SAMs that present a terminal reactive group ($NH_2$, COOH, CHO etc.) are available so that the receptor can be covalently immobilized to the transducer against a nonfouling background.

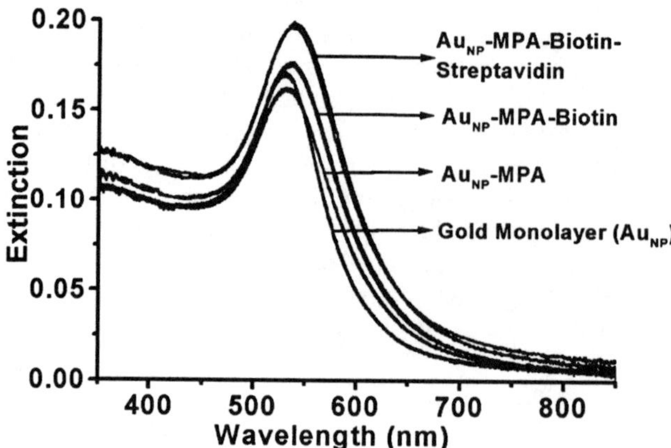

Figure 10 Change in extinction spectrum of immobilized gold nanoparticles on glass at each step in the biotin-streptavidin binding assay. Data from two separate experiments are plotted on same graph to show the reproducibility of the assay.

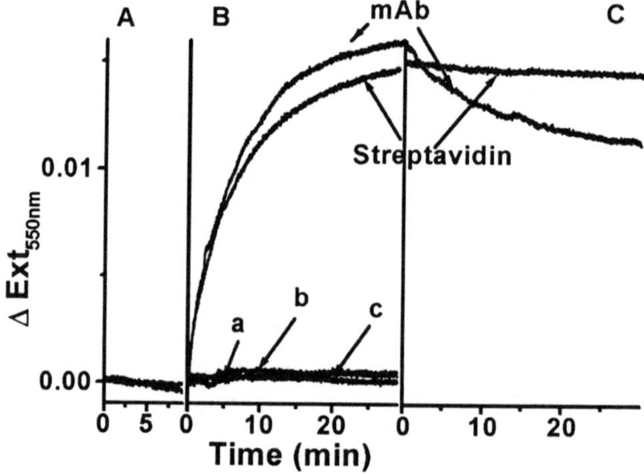

Figure 11 Binding of streptavidin and anti-biotin mAb to a biotin-functionalized gold nanoparticle monolayer studied by surface plasmon absorbance of gold nanoparticles at 550 nm. (A) Baseline extinction in PBS-Tween as a function of time. (B) Incubation of the biotin-functionalized surface with streptavidin (10 μg/ml) or antibiotin mAb (50 μg/ml) results in an increase in the absorbance due to protein-ligand binding. No increase in extinction was observed on incubation of biotin-functionalized surface with BSA(10 μg/ml) (a), human IgG (50 μg/ml) (b) or streptavidin (30 μg/ml) pre-incubated with 1.0 mM biotin (c). (C) Incubation of the protein-ligand complex on the surface with 1 mM biotin in solution causes decrease in signal due to dissociation of biotin-mAb complex. No dissociation was observed for biotin-streptavidin complex due to its slow off-rate constant.

Control of the receptor density on the surface is another critical variable in optimization of the sensor. A very high receptor density can sterically hinder the binding of large analytes to the immobilized receptor on the surface.[79, 80] In addition, in protein arrays that utilize small sample volumes, the surface area to volume ratio is high so that a high receptor density can rapidly deplete the target analyte from the sample leading to slow binding kinetics as well as a low S/N ratio.[6] Mixed SAMs of $(EG)_n$-SH, prepared from binary mixtures, in which one thiol has a terminal reactive group for receptor immobilization (e.g. COOH), and the second thiol presents a non-reactive terminal group (e.g. OH or $CH_3$) and acts as a diluent, are very useful in this regard. This is because the surface concentration of the reactive group, which determines the maximum density of the receptor on the surface, can be systematically controlled by the solution molar ratio of the reactive alkanethiol in the binary mixture.

### 3.4.2. Biomolecular Binding

We selected the streptavidin-biotin binding interaction for proof-of-concept studies of the immobilized nanoSPR sensor[13] for the following reasons: first, because of its very high association constant, streptavidin-biotin binding is insensitive to washing steps, which simplifies the experimental setup. Second, the wide use of this model system for validation of other biosensors enables comparison of the nanoSPR biosensor with other platforms under development.[4, 15, 87, 88]

A monolayer of 13.4 nm gold nanoparticles on glass was functionalized by formation of a SAM of 3-mercaptopropionic acid (MPA) to present terminal COOH groups at the surface of the gold nanoparticles. The COOH groups were then activated with a 1:1 mixture of 1-ethyl-3-(dimethylamino)propyl carbodiimide (EDAC) and pentafluorophenol (PFP) in ethanol and then reacted with an amine-terminated biotin derivative to covalently tether the biotin to the gold surface. The biotin functionalized nanoSPR chip was subsequently incubated with 5.0 µg/ml of streptavidin for 1 h, and the extinction spectrum of the nanoparticle monolayer was measured in a conventional UV-visible spectrophotometer. A shift in $\lambda^{max}$ and an increase in $Ext^{max}$ was observed to accompany each functionalization step (Figure 10), which is consistent with an increase in refractive index in the vicinity of the gold surface. We also observed that the extinction spectra from two different sets of experiments were virtually identical, which demonstrates the reproducibility with which the nanoSPR chips can be fabricated.

In nanoSPR, measurement of the change in extinction at a single wavelength (e.g. 550 nm) allows binding events to be monitored in real time. Figure 11 shows results from a typical experiment in which biotin functionalized nanoSPR chips were sequentially immersed in PBS–Tween20 buffer (to characterize baseline stability), a solution of streptavidin or anti-biotin mAb (to initiate binding), followed by a 1 mM solution of biotin (to initiate dissociation of the protein-ligand complex). A small negative baseline drift of $3 \times 10^{-4}$ extinction unit/min was observed for all sensor chips in buffer, but this drift was extremely reproducible as seen by the overlaid plots for two different chips (Figure 11A). Incubation of the biotin functionalized nanoSPR chip with a 10 µg/ml solution of streptavidin or 50 µg/ml antibiotin mAb resulted in a dramatic, time-dependent increase in extinction (Figure 11B). In control experiments, incubation of sensor chips with BSA or with anti-human IgG or with streptavidin whose biotin binding sites were blocked by pre-incubation with 1.0 mM biotin did not result in an extinction

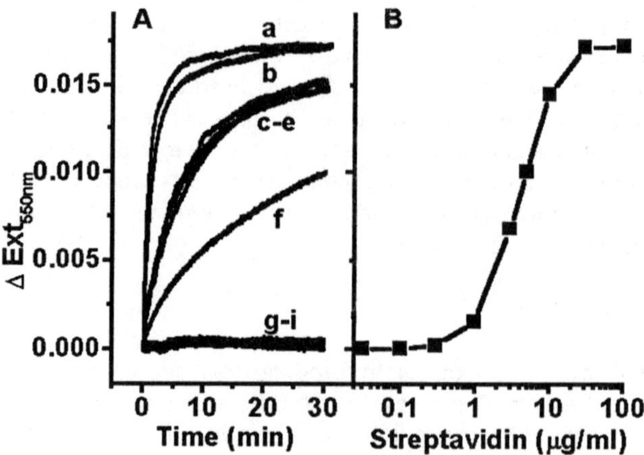

**Figure 12.** (A) Time-dependent change in surface plasmon extinction at 550 nm as a result of specific binding of streptavidin to biotinylated gold nanoparticles surface. (a) 100 μg/ml, (b) 30 μg/ml, (c-e) 10 μg/ml, (f) 5 μg/ml, (g) BSA, (h) biotin saturated streptavidin (i) human IgG. **(B)** Absorbance change at 550 nm as a function of streptavidin concentration.

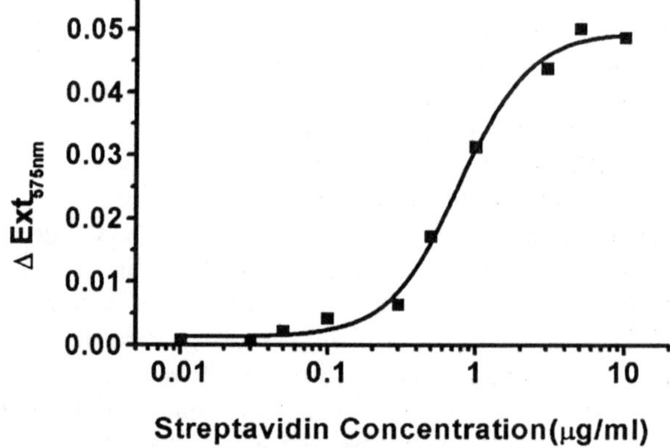

**Figure 13.** Extinction change at 575 nm for 39 nm biotin functionalized gold nanoparticle monolayer as a function of streptavidin concentration.

change (Figure 11B), which confirmed that the increase in extinction observed upon incubation with streptavidin and antibiotin mAb is due to molecular recognition between the protein and immobilized ligand on the surface. Incubating the biotin-mAb complex in an aqueous solution of 1 mM biotin resulted in a decrease in extinction as a function of time, due to dissociation of the mAb from the surface (Figure 11C). The lack of dissociation of the streptavidin-biotin complex upon incubation in a biotin solution is consistent with the extremely slow off-rate constant of the complex ($\sim10^{-6}$ s$^{-1}$).

We further investigated the concentration-dependent extinction change to determine the dynamic range and sensitivity that can be achieved for streptavidin-biotin binding using this sensor. Biotin functionalized sensor chips were incubated in streptavidin as a function of solution concentrations ranging from 0.03 µg/ml to 30 µg/ml, and the extinction change at 550 nm was measured as a function of time. Figure 12 shows representative plots of the sensor response as a function of time for different streptavidin concentrations. Both the kinetic and steady-state response of the sensor are highly reproducible, as shown for three different replicates at the same solution concentration of streptavidin (10 µg/ml) in Figure 12A (plots c-e). A calibration plot of the absorbance change at 550 nm, after 30 min incubation, as a function of streptavidin concentration yielded a detection limit of 1 µg/ml streptavidin (16.6 nM streptavidin tetramer) (Figure 12B) and dynamic range was 1.0–30 µg/ml. These results indicate that a self-assembled monolayer of gold nanoparticles on glass can be used to transduce ligand-receptor binding at a surface into an extinction change with a sensitivity that is useful for biosensor applications.

More recently, we have shown that a much lower detection limit can be achieved by a nanoSPR sensor fabricated from 39 nm gold nanoparticles to the streptavidin-biotin interaction. The extinction change was measured at 575 nm for this nanoSPR sensor, due to non-linear behavior at 550 nm, and incubation time was increased to 3 h. The detection limit of this sensor is 0.05 µg/ml streptavidin, which amounts to a 20-fold improvement over the nanoSPR sensor fabricated from 12 nm diameter gold nanoparticles, and can be attributed to optimization of the particle size and increased incubation time(Figure 13). Several additional improvements can substantially increase the sensitivity of the sensor. First, receptor-ligand binding assay at solid-liquid interface is dependent on the ligand density at the surface and optimization of the biotin density at the gold particle surface will increase the sensitivity.[79, 80, 89] Furthermore, implementation of this assay in a microfluidic flow cell will substantially reduce mass transfer effects, decrease the assay time, and improve the kinetic response of the sensor. The primary advantage of this biosensor is its simplicity and flexibility at several different levels. First, gold nanoparticles are easily synthesized, and can be easily and reproducibly deposited on glass (or other optically transparent substrate) by solution self-assembly. Second, the spontaneous self-assembly of alkanethiols on gold allows convenient fabrication of surfaces with well-defined interfacial properties and reactive groups, which allows the chemistry at the interface to be easily tailored for a specific application of interest, an advantage this sensor shares with conventional SPR on gold or silver films. Third, this sensor enables label free detection of biomolecular interactions. The biosensor can also be easily multiplexed to enable high-throughput screening in an array-based format for applications in genomics, proteomics and drug discovery. Finally, we note that the fabrication of this nanoSPR biosensor is not restricted to the use of gold nanoparticles, but can be easily extended to anisotropic metal nanostructures, which are attractive because of their enhanced optical sensitivity.


## 4.  FUTURE DIRECTIONS

The design of biosensors using metal nanoparticles is an exciting area of research, with many opportunities to translate fundamental research findings into new sensing technology. An important area of fundamental research– and one that we believe has not received sufficient attention is the development of theoretical models that can enable in silico design of nanoparticles with the direct goal of optimizing sensor response.  At the simplest level, such a theoretical model would allow input of nanoparticle composition, size and shape into the computational model, with outputs that are directly relevant to biosensing, such as the sensing volume and sensitivity to the refractive index of the surrounding medium. At an increased level of sophistication that would allow reverse engineering of the biosensor from first principles, such a model would provide the necessary design parameters of nanoparticles that would maximize the sensitivity and dynamic range for target-receptor pairs of known sizes and binding constants. It will also be important to couple these theoretical efforts to experimental studies of the optical properties of metal nanostructures with the goal of validating these theoretical models. These studies will develop a fundamental understanding of how sensor performance is controlled by nanostructure composition, shape, size, chemical binding, and packing density on surface. Together, we believe that these fundamental studies will provide the design rules for nanoparticle sensors optimized for specific interactions. that are characterized by a known binding constant, size of interaction partners and the likely concentration that will be encountered in actual use.

Another area of research that is of great importance to nanoparticle biosensors are new methods to *reproducibly synthesize* anisotropic nanostructures.  This is an active area of research as seen by new methods for synthesis of anisotropic nanostructures such as nanoprisms, cubes, wires, and belts that have recently appeared in the literature.[56] However, the optimization of these synthetic procedures to control the shape and size dispersity, scale up of the synthesis, and long-term stability of the nanostructures– issues that, to date, have received significantly less attention– will need to be addressed as these nanostructures begin to make the transition from basic science to technology.


## 5.  ACKNOWLEDGEMENTS

This work was supported in part by grants to A.C. from the National Science Foundation (NSF-BES-99-86477-NANOSCALE), the National Institutes of Health (R01-GM-61232-01), and the Center for Disease Control (NCD-1-R01-CI00097-01).   The AFM instrumentation used in these studies was acquired through multi-investigator research instrumentation awards from the National Science Foundation (NSF-DBI-96-04785) and the North Carolina Biotechnology Center (9703-IDG-1002) to Duke University.

## 6. REFERENCES


1    R. M. De Lorimier, J. J. Smith, M. A. Dwyer, L. L. Looger, K. M. Sali, C. D. Paavola, S. S. Rizk, S. Sadigov, D. W. Conrad, L. Loew and H. W. Hellinga. Construction of a fluorescent biosensor family. *Protein Science* 11 (11), 2655-2675 (2002).

2    R. S. Yalow and S. A. Berson. Assay of plasma insulin in human subjects by immunological methods. *Nature* 184, 1648-1649 (1959).

3    J. P. Gosling. Enzyme Immunoassay. In *Immunoassay* (Diamandis, E.P. and Christopoulos, T.K., eds.), pp. 287-308, Academic Press, San Diego (1996).

4    G. MacBeath and S. L. Schreiber. Printing proteins as microarrays for high-throughput function determination. *Science* 289 (5485), 1760-1763. (2000).

5    M. C. Pirrung. How to make a DNA chip. *Angewandte Chemie-International Edition* 41 (8), 1277-+ (2002).

6    M. F. Templin, D. Stoll, M. Schrenk, P. C. Traub, C. F. Vohringer and T. O. Joos. Protein microarray technology. *Trends in Biotechnology* 20 (4), 160-166 (2002).

7    H. Zhu, M. Bilgin, R. Bangham, D. Hall, A. Casamayor, P. Bertone, N. Lan, R. Jansen, S. Bidlingmaier, T. Houfek, T. Mitchell, P. Miller, R. A. Dean, M. Gerstein and M. Snyder. Global analysis of protein activities using proteome chips. *Science* 293 (5537), 2101-2105 (2001).

8    A. Janshoff, H. J. Galla and C. Steinem. Piezoelectric mass-sensing devices as biosensors - An alternative to optical biosensors? *Angewandte Chemie-International Edition* 39 (22), 4004-4032 (2000).

9    F. Hook, B. Kasemo, T. Nylander, C. Fant, K. Sott and H. Elwing. Variations in coupled water, viscoelastic properties, and film thickness of a Mefp-1 protein film during adsorption and cross-linking: A quartz crystal microbalance with dissipation monitoring, ellipsometry, and surface plasmon resonance study. *Analytical Chemistry* 73 (24), 5796-5804 (2001).

10   F. Hook, M. Rodahl, B. Kasemo and P. Brzezinski. Structural changes in hemoglobin during adsorption to solid surfaces: Effects of pH, ionic strength, and ligand binding. *Proceedings of the National Academy of Sciences of the United States of America* 95 (21), 12271-12276 (1998).

11   W. Knoll. Interfaces and thin films as seen by bound electromagnetic waves. *Annual Review of Physical Chemistry* 49, 569-638 (1998).

12   J. Homola, S. S. Yee and G. Gauglitz. Surface plasmon resonance sensors: review. *Sensors and Actuators B-Chemical* 54 (1-2), 3-15 (1999).

13   N. Nath and A. Chilkoti. A colorimetric gold nanoparticle sensor to interrogate biomolecular interactions in real time on a surface. *Analytical Chemistry* 74 (3), 504-509 (2002).

14   J. C. Riboh, A. J. Haes, A. D. McFarland, C. R. Yonzon and R. P. Van Duyne. A nanoscale optical biosensor: Real-time immunoassay in physiological buffer enabled by improved nanoparticle adhesion. *Journal of Physical Chemistry B* 107 (8), 1772-1780 (2003).

15   A. J. Haes and R. P. Van Duyne. A nanoscale optical biosensor: sensitivity and selectivity of an approach based on the localized surface plasmon resonance spectroscopy of triangular silver nanoparticles. *J Am Chem Soc* 124 (35), 10596-10604 (2002).

16   J. Yguerabide and E. E. Yguerabide. Light-scattering submicroscopic particles as highly fluorescent analogs and their use as tracer labels in clinical and biological applications - I. Theory. *Analytical Biochemistry* 262 (2), 137-156 (1998).

17   J. Yguerabide and E. E. Yguerabide. Light-scattering submicroscopic particles as highly fluorescent analogs and their use as tracer labels in clinical and biological applications - II. Experimental characterization. *Analytical Biochemistry* 262 (2), 157-176 (1998).

18   U. Kreibig and M. Vollmer. (1995) *Optical properties of metal clusters*, Springer-Verlag, Berlin ; New York

19   S. Link and M. A. El-Sayed. Shape and size dependence of radiative, non-radiative and photothermal properties of gold nanocrystals. *International Reviews in Physical Chemistry* 19 (3), 409-453 (2000).

20   E. J. Zeman and G. C. Schatz. An Accurate Electromagnetic Theory Study of Surface Enhancement Factors for Ag, Au, Cu, Li, Na, Al, Ga, in, Zn, and Cd. *Journal of Physical Chemistry* 91 (3), 634-643 (1987).

21   W. H. Yang, G. C. Schatz and R. P. Vanduyne. Discrete Dipole Approximation for Calculating Extinction and Raman Intensities for Small Particles with Arbitrary Shapes. *Journal of Chemical Physics* 103 (3), 869-875 (1995).

22   D. L. Feldheim and C. A. Foss. (2002) *Metal nanoparticles : synthesis, characterization, and applications*, Marcel Dekker, New York

23   F. E. Wagner, S. Haslbeck, L. Stievano, S. Calogero, Q. A. Pankhurst and P. Martinek. Before striking gold in gold-ruby glass. *Nature* 407 (6805), 691-692 (2000).

24   J. Roth. The silver anniversary of gold: 25 years of the colloidal gold marker system for immunocytochemistry and histochemistry. *Histochemistry and Cell Biology* 106 (1), 1-8 (1996).

25   M. Faraday. Experimental Relations of gold (and other metals) to light. *Philosophical Transactions of Royal Society of London* 147, 145 (1857).

26   R. Zsigmondy and J. Alexander. (1909) *Colloids and the ultramicroscope; a manual of colloid chemistry and ultramicroscopy*, Wiley Chapman & Hall, New York, London,

27   W. P. Faulk and G. M. Taylor. An immunocolloid method for the electron microscope. *Immunochemistry* 8 (11), 1081-1083 (1971).

28   G. Frens. Controlled nucleation for the regulation of the particle size in monodisperse gold solutions. *Nature* 241, 20 (1973).

29   M. A. Hayat. (1989) *Colloidal gold : principles, methods, and applications*, Academic Press, San Diego

30   J. H. W. Leuvering, B. C. Goverde, P. Thal and A. Schuurs. A Homogeneous Sol Particle Immunoassay for Human Chorionic- Gonadotropin Using Monoclonal-Antibodies. *Journal of Immunological Methods* 60 (1-2), 9-23 (1983).

31   R. G. Freeman, K. C. Grabar, K. J. Allison, R. M. Bright, J. A. Davis, A. P. Guthrie, M. B. Hommer, M. A. Jackson, P. C. Smith, D. G. Walter and M. J. Natan. Self-Assembled Metal Colloid Monolayers - an Approach to Sers Substrates. *Science* 267 (5204), 1629-1632 (1995).

32   S. M. Nie and S. R. Emery. Probing single molecules and single nanoparticles by surface-enhanced Raman scattering. *Science* 275 (5303), 1102-1106 (1997).

33   Y. W. C. Cao, R. C. Jin and C. A. Mirkin. Nanoparticles with Raman spectroscopic fingerprints for DNA and RNA detection. *Science* 297 (5586), 1536-1540 (2002).

34   M. Moskovits, L. L. Tay, J. Yang and T. Haslett. SERS and the single molecule. *Optical Properties of Nanostructured Random Media* 82, 215-226 (2002).

35   R. Elghanian, J. J. Storhoff, R. C. Mucic, R. L. Letsinger and C. A. Mirkin. Selective colorimetric detection of polynucleotides based on the distance-dependent optical properties of gold nanoparticles. *Science* 277 (5329), 1078-1081 (1997).

36   J. J. Storhoff, R. Elghanian, R. C. Mucic, C. A. Mirkin and R. L. Letsinger. One-pot colorimetric differentiation of polynucleotides with single base imperfections using gold nanoparticle probes. *Journal of the American Chemical Society* 120 (9), 1959-1964 (1998).

37   T. A. Taton, C. A. Mirkin and R. L. Letsinger. Scanometric DNA array detection with nanoparticle probes. *Science* 289 (5485), 1757-1760 (2000).

38   L. R. Hirsch, J. B. Jackson, A. Lee, N. J. Halas and J. L. West. A whole blood immunoassay using gold nanoshells. *Anal Chem* 75 (10), 2377-2381 (2003).

39   S. J. Oldenburg, R. D. Averitt, S. L. Westcott and N. J. Halas. Nanoengineering of optical resonances. *Chemical Physics Letters* 288 (2-4), 243-247 (1998).

40   S. A. Maier, P. G. Kik, H. A. Atwater, S. Meltzer, E. Harel, B. E. Koel and A. A. G. Requicha. Local detection of electromagnetic energy transport below the diffraction limit in metal nanoparticle plasmon waveguides. *Nature Materials* 2 (4), 229-232 (2003).

41   P. Englebienne. Use of colloidal gold surface plasmon resonance peak shift to infer affinity constants from the interactions between protein antigens and antibodies specific for single or multiple epitopes. *Analyst* 123 (7), 1599-1603 (1998).

42   P. Englebienne, A. Van Hoonacker and J. Valsamis. Rapid homogeneous immunoassay for human ferritin in the cobas mira using colloidal gold as the reporter reagent. *Clinical Chemistry* 46 (12), 2000-2003 (2000).

43   T. Okamoto, I. Yamaguchi and T. Kobayashi. Local plasmon sensor with gold colloid monolayers deposited upon glass substrates. *Optics Letters* 25 (6), 372-374 (2000).

44   Y. G. Sun and Y. N. Xia. Increased sensitivity of surface plasmon resonance of gold nanoshells compared to that of gold solid colloids in response to environmental changes. *Analytical Chemistry* 74 (20), 5297-5305 (2002).

45   J. Turkevich. Colloidal Gold. Part 1. *Gold Bulletin* 18 (3), 86-91 (1985).

46   D. V. Goia and E. Matijevic. Preparation of monodispersed metal particles. *New Journal of Chemistry* 22 (11), 1203-1215 (1998).

47   D. V. Goia and E. Matijevic. Tailoring the particle size of monodispersed colloidal gold. *Colloids and Surfaces a-Physicochemical and Engineering Aspects* 146 (1-3), 139-152 (1999).

48   M. Brust, M. Walker, D. Bethell, D. J. Schiffrin and R. Whyman. Synthesis of Thiol-Derivatized Gold Nanoparticles in a 2-Phase Liquid-Liquid System. *Journal of the Chemical Society-Chemical Communications* (7), 801-802 (1994).

49      P. Y. Silvert, R. HerreraUrbina, N. Duvauchelle, V. Vijayakrishnan and K. T. Elhsissen. Preparation of colloidal silver dispersions by the polyol process .1. Synthesis and characterization. *Journal of Materials Chemistry* 6 (4), 573-577 (1996).

50      Y. Cao, R. Jin and C. A. Mirkin. DNA-modified core-shell Ag/Au nanoparticles. *J Am Chem Soc* 123 (32), 7961-7962 (2001).

51      D. A. Handley. Methods for synthesis of colloidal gold. In *Colloidal gold : principles, methods, and applications* (Vol. 1) (Hayat, M.A., ed.), pp. 13-32, Academic Press, San Diego (1989).

52      C. J. Murphy and N. R. Jana. Controlling the aspect ratio of inorganic nanorods and nanowires. *Advanced Materials* 14 (1), 80-82 (2002).

53      S. R. Nicewarner-Pena, R. G. Freeman, B. D. Reiss, L. He, D. J. Pena, I. D. Walton, R. Cromer, C. D. Keating and M. J. Natan. Submicrometer metallic barcodes. *Science* 294 (5540), 137-141 (2001).

54      R. C. Jin, Y. W. Cao, C. A. Mirkin, K. L. Kelly, G. C. Schatz and J. G. Zheng. Photoinduced conversion of silver nanospheres to nanoprisms. *Science* 294 (5548), 1901-1903 (2001).

55      Y. Sun and Y. Xia. Shape-controlled synthesis of gold and silver nanoparticles. *Science* 298 (5601), 2176-2179 (2002).

56      Y. G. Sun and Y. N. Xia. Gold and silver nanoparticles: A class of chromophores with colors tunable in the range from 400 to 750 nm. *Analyst* 128 (6), 686-691 (2003).

57      T. R. Jensen, M. D. Malinsky, C. L. Haynes and R. P. Van Duyne. Nanosphere lithography: Tunable localized surface plasmon resonance spectra of silver nanoparticles. *Journal of Physical Chemistry B* 104 (45), 10549-10556 (2000).

58      Z. Li, S. W. Chung, J. M. Nam, D. S. Ginger and C. A. Mirkin. Living templates for the merarchical assembly of gold nanoparticles. *Angewandte Chemie-International Edition* 42 (20), 2306-2309 (2003).

59      M. Reches and E. Gazit. Casting metal nanowires within discrete self-assembled peptide nanotubes. *Science* 300 (5619), 625-627 (2003).

60      M. G. Warner and J. E. Hutchison. Linear assemblies of nanoparticles electrostatically organized on DNA scaffolds. *Nature Materials* 2 (4), 272-277 (2003).

61      J. S. Choi, G. Sauer, P. Goring, K. Nielsch, R. B. Wehrspohn and U. Gosele. Monodisperse metal nanowire arrays on Si by integration of template synthesis with silicon technology. *Journal of Materials Chemistry* 13 (5), 1100-1103 (2003).

62      P. VanDerVoort and E. F. Vansant. Silylation of the silica surface a review. *Journal of Liquid Chromatography & Related Technologies* 19 (17-18), 2723-2752 (1996).

63      J. J. Cras, C. A. Rowe-Taitt, D. A. Nivens and F. S. Ligler. Comparison of chemical cleaning methods of glass in preparation for silanization. *Biosensors & Bioelectronics* 14 (8-9), 683-688 (1999).

64      C. M. Halliwell and A. E. G. Cass. A factorial analysis of silanization conditions for the immobilization of oligonucleotides on glass surfaces. *Analytical Chemistry* 73 (11), 2476-2483 (2001).

65      K. Park, H. Park and R. M. Albrecht. Factors affecting the staining with colloidal gold. In *Colloidal gold : principles, methods, and applications* (Vol. 1) (Hayat, M.A., ed.), pp. 489-518, Academic Press, San Diego (1989).

66      K. C. Grabar, P. C. Smith, M. D. Musick, J. A. Davis, D. G. Walter, M. A. Jackson, A. P. Guthrie and M. J. Natan. Kinetic control of interparticle spacing in Au colloid-based surfaces: Rational nanometer-scale architecture. *Journal of the American Chemical Society* 118 (5), 1148-1153 (1996).

67      E. S. Kooij, E. A. M. Brouwer, H. Wormeester and B. Poelsema. Ionic strength mediated self-organization of gold nanocrystals: An AFM study. *Langmuir* 18 (20), 7677-7682 (2002).

68      M. Semmler, E. K. Mann, J. Ricka and M. Borkovec. Diffusional deposition of charged latex particles on water-solid interfaces at low ionic strength. *Langmuir* 14 (18), 5127-5132 (1998).

69      M. Semmler, J. Ricka and M. Borkovec. Diffusional deposition of colloidal particles: electrostatic interaction and size polydispersity effects. *Colloids and Surfaces a-Physicochemical and Engineering Aspects* 165 (1-3), 79-93 (2000).

70      L. S. Jung, C. T. Campbell, T. M. Chinowsky, M. N. Mar and S. S. Yee. Quantitative interpretation of the response of surface plasmon resonance sensors to adsorbed films. *Langmuir* 14 (19), 5636-5648 (1998).

71      S. Underwood and P. Mulvaney. Effect of the Solution Refractive-Index on the Color of Gold Colloids. *Langmuir* 10 (10), 3427-3430 (1994).

72      A. C. Templeton, J. J. Pietron, R. W. Murray and P. Mulvaney. Solvent refractive index and core charge influences on the surface plasmon absorbance of alkanethiolate monolayer- protected gold clusters. *Journal of Physical Chemistry B* 104 (3), 564-570 (2000).

73    M. D. Malinsky, K. L. Kelly, G. C. Schatz and R. P. Van Duyne. Chain length dependence and sensing capabilities of the localized surface plasmon resonance of silver nanoparticles chemically modified with alkanethiol self-assembled monolayers. *Journal of the American Chemical Society* 123 (7), 1471-1482 (2001).

74    T. R. Jensen, M. L. Duval, K. L. Kelly, A. A. Lazarides, G. C. Schatz and R. P. Van Duyne. Nanosphere lithography: Effect of the external dielectric medium on the surface plasmon resonance spectrum of a periodic array of sliver nanoparticles. *Journal of Physical Chemistry B* 103 (45), 9846-9853 (1999).

75    M. Malmsten. Ellipsometry Studies of Protein Layers Adsorbed at Hydrophobic Surfaces. *Journal of Colloid and Interface Science* 166 (2), 333-342 (1994).

76    J. Schmitt, P. Machtle, D. Eck, H. Mohwald and C. A. Helm. Preparation and optical properties of colloidal gold monolayers. *Langmuir* 15 (9), 3256-3266 (1999).

77    G. Decher. Fuzzy Nanoassemblies: Toward Layered Polymeric Multicomposites. *Science* 277, 1232-1237 (1997).

78    K. Buscher, K. Graf, H. Ahrens and C. A. Helm. Influence of adsorption conditions on the structure of polyelectrolyte multilayers. *Langmuir* 18 (9), 3585-3591 (2002).

79    L. S. Jung, K. E. Nelson, P. S. Stayton and C. T. Campbell. Binding and dissociation kinetics of wild-type and mutant streptavidins on mixed biotin-containing alkylthiolate monolayers. *Langmuir* 16 (24), 9421-9432 (2000).

80    L. Haussling, H. Ringsdorf, F. J. Schmitt and W. Knoll. Biotin-Functionalized Self-Assembled Monolayers on Gold - Surface-Plasmon Optical Studies of Specific Recognition Reactions. *Langmuir* 7 (9), 1837-1840 (1991).

81    E. Stenberg, B. Persson, H. Roos and C. Urbaniczky. Quantitative-Determination of Surface Concentration of Protein with Surface-Plasmon Resonance Using Radiolabeled Proteins. *Journal of Colloid and Interface Science* 143 (2), 513-526 (1991).

82    M. Horisberger, J. Rosset and H. Bauer. Colloidal gold granules as markers for cell surface receptors in the scanning electron microscope. *Experientia* 31 (10), 1147-1149 (1975).

83    W. D. Geoghegan and G. A. Ackerman. Adsorption of horseradish peroxidase, ovomucoid and anti-immunoglobulin to colloidal gold for the indirect detection of concanavalin A, wheat germ agglutinin and goat anti-human immunoglobulin G on cell surfaces at the electron microscopic level: a new method, theory and application. *J Histochem Cytochem* 25 (11), 1187-1200 (1977).

84    C. D. Bain and G. M. Whitesides. Modeling Organic-Surfaces with Self-Assembled Monolayers. *Angewandte Chemie-International Edition in English* 28 (4), 506-512 (1989).

85    M. Mrksich and G. M. Whitesides. Patterning Self-Assembled Monolayers Using Microcontact Printing - a New Technology for Biosensors. *Trends in Biotechnology* 13 (6), 228-235 (1995).

86    E. Ostuni, L. Yan and G. M. Whitesides. The interaction of proteins and cells with self-assembled monolayers of alkanethiolates on gold and silver. *Colloids and Surfaces B-Biointerfaces* 15 (1), 3-30 (1999).

87    L. Movileanu, S. Howorka, O. Braha and H. Bayley. Detecting protein analytes that modulate transmembrane movement of a polymer chain within a single protein pore. *Nature Biotechnology* 18 (10), 1091-1095 (2000).

88    Y. Cui, Q. Wei, H. Park and C. M. Lieber. Nanowire nanosensors for highly sensitive and selective detection of biological and chemical species. *Science* 293 (5533), 1289-1292 (2001).

89    Z. P. Yang, W. Frey, T. Oliver and A. Chilkoti. Light-activated affinity micropatterning of proteins on self-assembled monolayers on gold. *Langmuir* 16 (4), 1751-1758 (2000).

# SURFACE PLASMON-COUPLED EMISSION: A NEW METHOD FOR SENSITIVE FLUORESCENCE DETECTION

Ignacy Gryczynski[*], Joanna Malicka, Zygmunt Gryczynski and Joseph R. Lakowicz

## 1. INTRODUCTION

In recent years, there has been a growing interest among the scientific community in the modification of the fluorescence properties of dyes near metallic surfaces. It has been well established that the brightness of fluorophores significantly increases at close proximity to silvered surfaces (deposited colloids or islands) [1-6]. This effect is accompanied by dramatically decreased lifetimes, indicating a significant increase in radiative decay rates [3, 6-8]. Shorter lifetimes enable higher photostability of fluorophores [8-10]. The strongest enhancements were observed for fractal silver structures [11] and for overlabeled macromolecules where the self-quenching was released [12-13]. The effects of noble metal particles on fluorescence are described in several chapters in this volume.

In this chapter we will focus on fluorescence which couples to plasmon resonances in continuous metal surfaces, resulting in directional and wavelength-resolved emission. For the metallic surface, we used a layer with a thickness of only a few tens of nanometers rather than deposited particles. Such a metallic slide is nearly opaque, with a transmission of about 10 - 30 %. The most common preparation of these mirrors is done by vapor-deposition. Surface plasmon-coupled emission (SPCE) appears to be a reverse process of surface plasmon resonance (SPR) as seen in the angle-dependent absorption of thin metal films.

## 2. SURFACE PLASMON RESONANCE ANALYSIS

---

[*] *Center for Fluorescence Spectroscopy, University of Maryland, School of Medicine, 725 W. Lombard Street, Baltimore, MD 21201*

SPR technology has already been adapted to surface plasmon resonance analysis (SPRA). This method is now widely used in the biosciences and provides a general approach to the measurement of biomolecular interactions on surfaces [14-18]. A schematic description of SPRA is shown in Figure 1. The measurement is based on the interaction of light with thin metal films on a glass substrate. The films are typically made of gold 40-50 nm thick. The analysis surface consists of a capture biomolecule which has affinity for the analyte of interest. The capture biomolecule is typically covalently bound to the gold surface. The analysis substrate is optically coupled to a hemispherical or hemicylindrical prism by an index matching fluid. Light impinges on the gold film through the prism, which is called the Kretschmann configuration. The instrument measures the reflectivity of the gold film at various angles of incidence ($\theta$), with the same angle used for observation ($\theta$). Other configurations can be used, such as a triangular prism or more complex optical geometry and a position-sensitive detector. In any event the measurement is the same, reflectivity of the gold surface versus the angle of incidence.

The usefulness of SPRA is due to the large dependence of the gold film reflectivity on the refractive index and the thickness of the dielectric layer immediately above the gold film. Binding of macromolecules above the gold film causes small changes in the refractive index which result in changes in reflectivity. It is instructive to visualize these changes with simulations. The dependence of the reflectivity on the thickness of the dielectric layer- bound proteins with an assumed refractive index of $n_s = 1.50$ ($\epsilon_s = n_s^2 = 2.25$) is presented in Figure 2. These dependencies are shown for the BK7 glass prism with a refractive index $n_p = 1.52$ ($\epsilon_p = n_p^2 = 2.31$), coated with a 50 nm gold mirror and air as the exit medium. The reflectivity of such four-phase systems (Figure 2, top) can be easily calculated either from derived equations [19-21] or by using a web-based program developed by the R. M. Corn group [22]. The reflectivity minimum shifts significantly with increased thickness of the sample and can be detected even without sophisticated equipment. The detection becomes more difficult if the exit medium is water (buffer) as shown in Figure 3 (top). The reflectivity minima are shifted to high angles and become much wider. Experiments with angles larger than 80° are highly unreliable. The use of a prism with a higher refractive index is preferred (Figure 3, bottom). There are a number of high refractive index materials such as SF-11 (n = 1.78), LaSFN9 (n = 1.85), or sapphire (n = 1.76).

The theory of SPR involves the solution of the Maxwell equation with proper boundary conditions [23-25]. The optical and reflective properties of silver and gold depend on the interplay of incident frequency and electron mobility, as well as underlying absorption bands not related to electron oscillations. In this chapter, we will present only an intuitive approach to the SPR phenomenon. For a more detailed description, we refer readers to [19-21, 26].

The interplay of electron mobility and incident frequency results in complex and imaginary optical constants. The refractive index and dielectric constant of a metal are given by $n_m = n_r + n_{im} i$ and $\epsilon_m = \epsilon_r + \epsilon_{im} i$ , respectively. Subscripts indicate the real (r) and imaginary (im) components. These constants are wavelength (frequency) dependent. The imaginary part is related to light absorption, which can be seen by the larger values of $\epsilon_{im}$ of gold for wavelengths below 500 nm (Figure 4) [27-28]. The real part of $\epsilon_m$ becomes increasingly more negative as the wavelength increases. This effect can be interpreted as electron oscillations with the charge opposite to the incident field. As the incident frequency

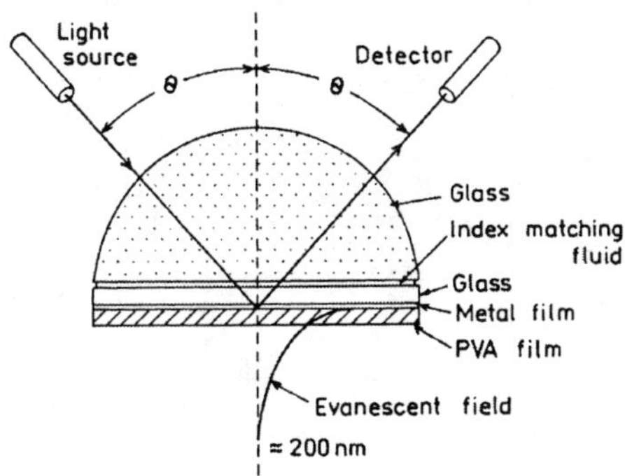


Figure 1. Typical configuration for surface-plasmon resonance analysis. The incident beam is p-polarized (adopted from [27]).

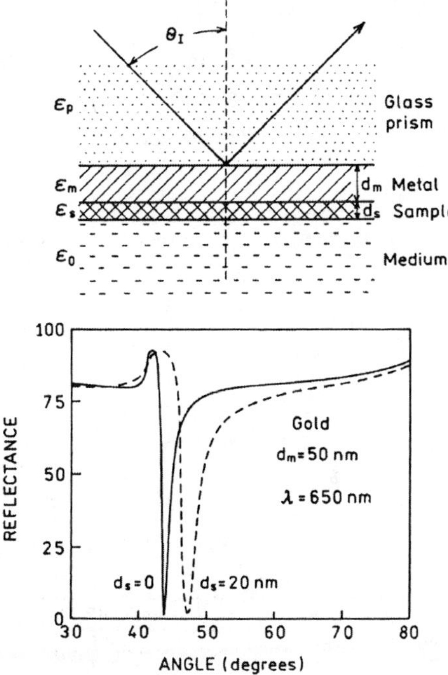


Figure 2. Bottom: Dependence of the reflectance of a 650nm light beam on the incidence angle for a 50nm gold film on glass ($n_p$=1.52) without (——) and with (------) a dielectric layer ($n_s$=1.50) The dielectric constant for gold was taken from Figure 4. The exit medium was air ($\epsilon_0$=1). Top: A four-phase system used for reflectance calculations.

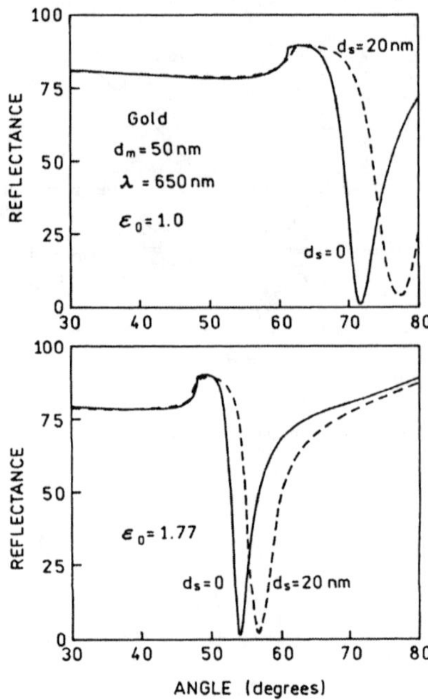

Figure 3. Top: The reflectances of a 50 nm gold film calculated for the parameters as described in Figure 2 with water as the exit medium ($\epsilon_0=n_0^2=1.77$). Bottom: Calculations as above with a higher refractive index (SF-11) prism ($\epsilon_p=n_p^2=3.19$).

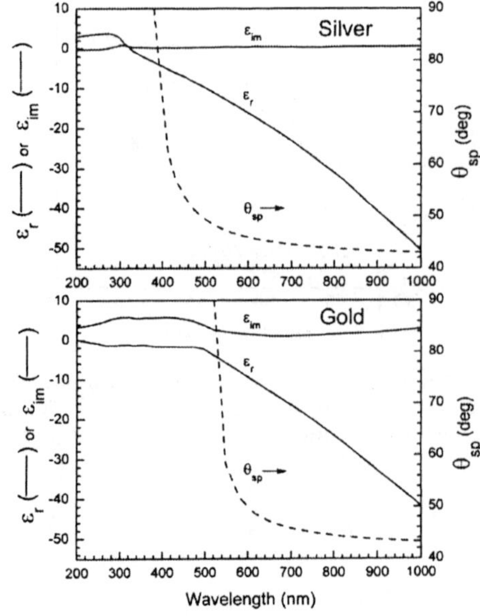

Figure 4. Complex dielectrical constants for silver (top) and gold (bottom), (from [27-28]).

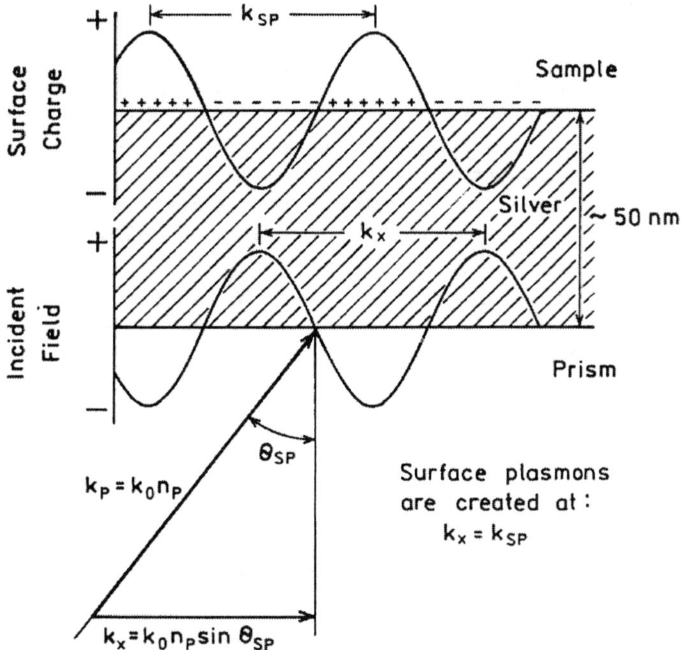


Figure 5. Schematic showing propagation constants in a prism and a thin film (adopted from [27]).

decreases, $\epsilon_r$ becomes more negative, reflecting a more complete response of the electrons to the lower frequency. For a perfect conductor $\epsilon_r$ approaches minus infinity.

The incident light can excite a surface plasmon when its wavevector x-axis component $(k_x)$ equals the propagation constant for the surface plasmon, $k_{SP}$. However, $k_{SP}$ is always greater than $k_x$ in free space, i.e. surface plasmons cannot be excited with light incident from the medium with a lower dielectric constant. In order to obtain SPR the magnitude of the light propagation vector $k_x$ must be increased to equal or exceed the surface plasmon propagation vector $k_{SP}$. This can be accomplished using the configuration shown in Figure 1 where light is incident on the metal film from the prism side. This approach increases the wavevector to $k_p = n_p k_0$. The refractive index of the prism reduces the wavelength to $\lambda = \lambda_0 / n_p$. This results in the maxima and minima of the electric field being more closely spaced, but now $k_{SP}$ is less than $k_p$. To obtain resonance the x-components of the electric field distribution are matched by adjustment of the x-component of $k_p$ by a factor of $\sin \theta_p$ (Figure 5).

For SPR measurement the incident beam should be p-polarized (the electric vector is parallel to the plane of incidence) since s-polarized light will not excite the surface plasmon and will not decrease reflectivity at some angles of incidence. Similarly to a total internal reflection (TIR), the incident beam which excites surface plasmons creates an evanescent field. This evanescent field penetrates the medium next to the metal up to a few hundred nanometers [27-29] and can provide effective excitation of fluorophores.

## 3.  SURFACE PLASMON-COUPLED EMISSION

The concept of SPCE has its ground in the fact that excited fluorophores can excite surface plasmons and create a radiative beam (Figure 6). The proximity of a fluorophore to a metallic film should result in a directional emission with sharply defined angles ($\theta_F$). These angles would be the same as $\theta_{SP}$ obtained for the emission wavelength. The predictions for $\theta_{SP}$ dependence on wavelength apply to $\theta_F$ by analogy. Figure 7 shows the reflectivity calculated for several wavelengths with parameters for the silver taken from Figure 4. From these calculations it is immediately evident that fluorescence emission should occur at a smaller angle than the incident excitation since excitation wavelengths are shorter than the emission (two-photon excitation is a different case). It should be noted that the depth and width of the reflectance profile depends on excitation/emission wavelengths as well as other parameters involved in reflectance calculation. It is wise to calculate the expected reflectances before performing SPCE experiments.  Also, the metal and its thickness should be carefully chosen. Figure 8 shows the reflectance profiles obtained for various silver thicknesses at 600nm. The 50nm thickness of silver seems to be ideal for SPCE experiments. At longer wavelengths, a slightly thicker layer of metal should be used in order to better define minimum reflectance.

There is no need to employ surface plasmon excitation in the creation of SPCE, as excited fluorophores couple to the surface plasmon irrespective of the origin of the excitation. Figure 9 shows excitation both through a glass prism with a higher refractive index at $\theta_{SP}$ (Kretschmann configuration)(top) and directly from the sample site (reverse Kretschmann configuration)(bottom). In both configurations, SPCE emission appears through the prism at the same angle $\theta_F$. In the fluorescence experiments described below, we used four possible excitation/emission configurations as shown in Figure 10.  KR and RK refer to Kretschmann and reverse Kretschmann configurations for excitation, respectively. SPCE and FS refer to directional SPCE and free-space fluorescence observation, respectively.

In order to study SPCE, it is important to prepare the appropriate samples and sample holders, as well as proper excitation and observation systems. We found it convenient to deposit the fluorescence samples on glass, quartz, or sapphire slides which were previously metalized by vapor deposition. The slides can be attached to the prism with matching index fluid, such as that used in fiber connections or microscopy. For precise angle-dependent measurements we used hemicylindrical or hemispherical prisms. The prism with attached slide was positioned on a rotary stage (Figure 11). This simple construction provides the possibility of fluorescence detection at any angle. It can also change the vertical position of the sample. The fiber bundle can be positioned at any distance from the sample up to 150mm. The observations were usually done through a 200µm slit positioned vertically at the fibre mount which results in an acceptance angle of about 0.1°. The other end of the fibre goes to a fluorometer or spectrofluorometer as needed.

### 3.1. Properties of SPCE with Reverse Kretschmann Excitation

<u>Sulforhodamine 101 in Polyvinyl Alcohol Films</u>

We felt that the most convincing demonstration of SPCE would be obtained with RK excitation where the incident light cannot induce surface plasmons. The sample was a Sulforhodamine 101 (S101) in polyvinyl alcohol (PVA), spin coated on the silvered site of the slide. Figure 12 (top) shows the dependence of emission intensity on observation angle with RK excitation. The emission is sharply distributed at ±47° on the prism side (back side,

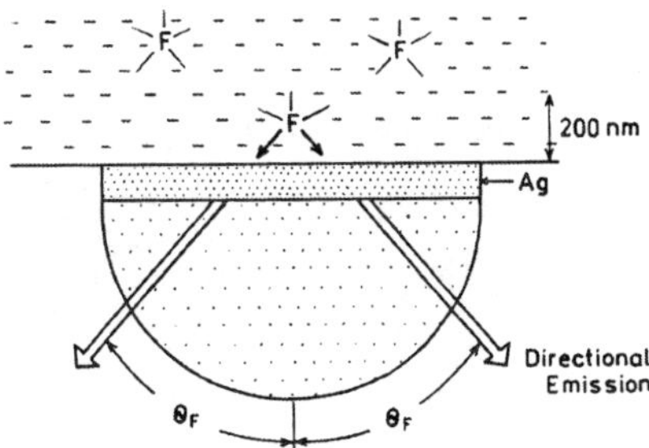


Figure 6. Surface plasmon-coupled emission. F is a fluorophore.

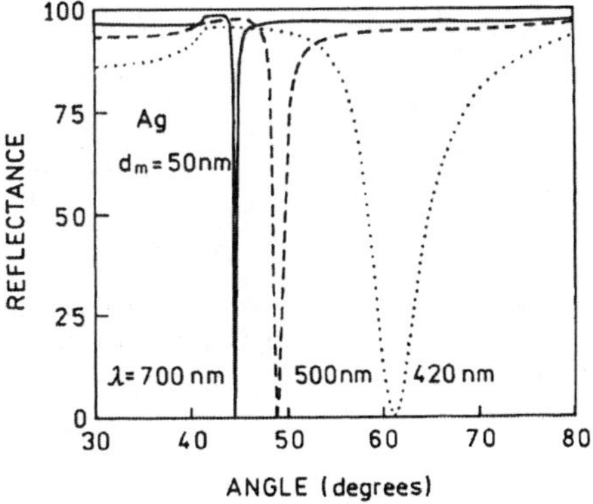


Figure 7. Dependence of reflectivity on the wavelength of the incidence light beam for a 50nm silver film. Calculations were done (see Figure 2, top) for a BK7 glass prism, 20nm dielectric layer ($n_s$=1.50) and air as an exit medium. The wavelength-dependent dielectric constants for silver were taken from Figure 4. The wavelength-dependent indexes of BK7 glass were taken from the Melles Griot catalog.

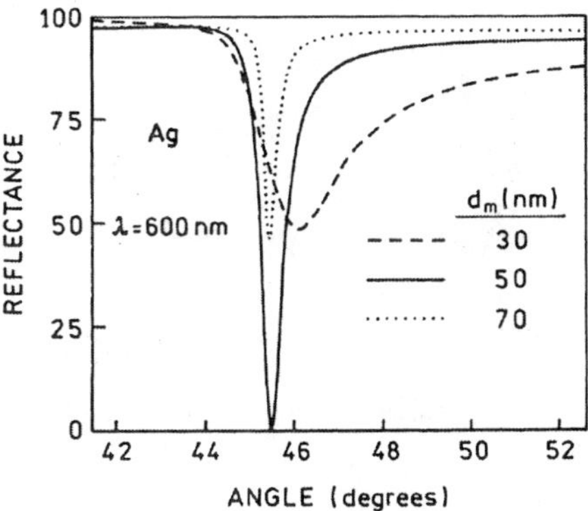


Figure 8. Dependence of reflectance on the thickness of silver film. The parameters used for calculations (see Figure 2, top) were: $\lambda$=600nm, $\epsilon_p$=2.31 (BK7), $d_m$=50nm, $d_s$=20nm, $\epsilon_s$=2.25, $\epsilon_0$=1.0. The dielectric constant for silver was taken from Figure 4.

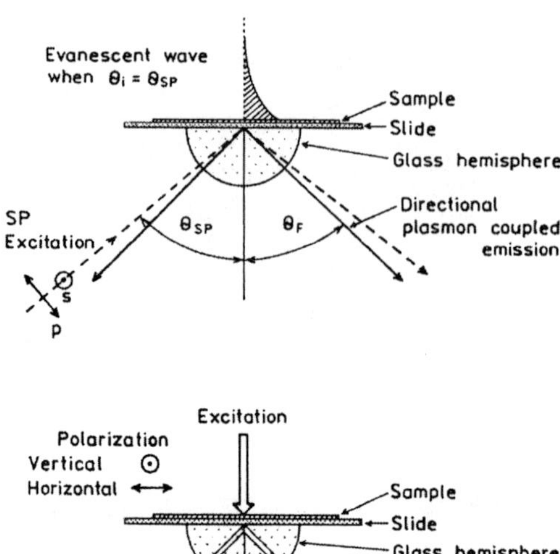



Figure 9. Geometries used for SPCE measurements. Top: For SPE (Kretschmann configuration, KR). The excitation enters through the coupling prism. Bottom: In reverse Kretschmann configuration, RK, the excitation directly reaches the sample and does not excite surface plasmons.

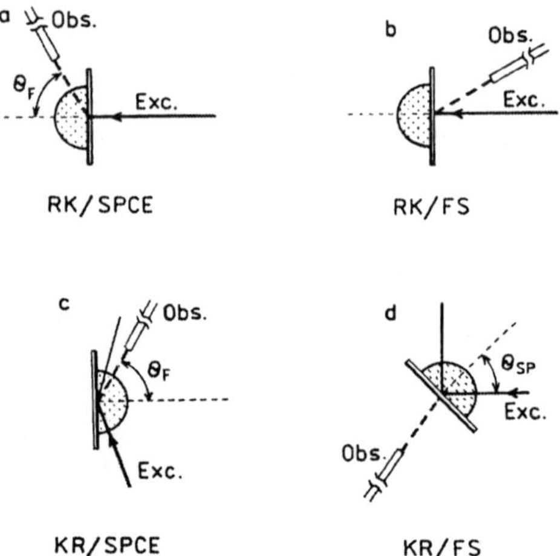



Figure 10. Optical configurations used to study SPCE.

B) of the sample. The intensity observed on the front (F) side of the sample was much lower and not sharply distributed at any particular angle. To determine the relative intensities of the coupled and free-space emissions, we integrated the intensities observed at all angles on the front side ($180° \pm 90°$) and the back side ($0° \pm 90°$). The total back-to-front intensity ratio ($I_B$

/ $I_F$) was 0.96. This suggests a coupling efficiency of about 49 %. Slightly lower coupling efficiencies were found for thicker samples [30]. Figure 12 (bottom) shows the reflectance profile calculated for the parameters corresponding to the SPCE experiment. There is a good agreement between a narrow reflectance minimum and the angular distribution of a 600nm emission.

Figure 11. Rotation stage used for SPCE measurements. Rotation of the sample and observation fiber are independent over 360°. The height of the sample can be precisely controlled.

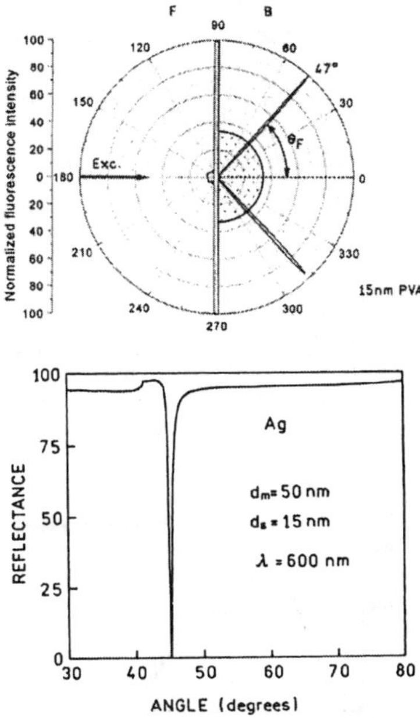


Figure 12. Top: Angular distribution of S101 emission excited using the reverse Kretschmann configuration. The PVA thickness was approximately 15nm. The ratio of the angle-integrated intensities was $I_B$ / $I_F$ = 0.96. Bottom: The reflectivity profile calculated for the parameters corresponding to the experiment: $\lambda$=600nm, $n_p$=1.52, $d_m$=50nm, $d_s$=15nm, $n_s$=1.50, $\epsilon_0$=1.0. The dielectric constant for the silver was taken from Figure 4.

We considered the possibility that the strong emission near 47° was due to scattered light or a reflection. We proceeded to measure the emission spectra of the directional SPCE at 47° (——) and the free-space emission at 149° (-------, Figure 13). We found that the notch filter eliminated scattered incident light and the emission spectra were consistent with S101. For the 15nm thick sample, the SPCE at the maximum angle is about 10-fold more intense than the free-space emission.

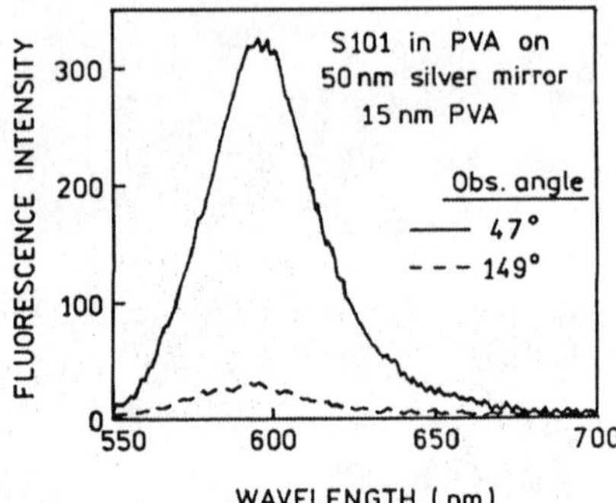


Figure 13. Emission spectra of S101 in PVA. The spectra are for the SPCE (——) and the free-space emission (----) measured at the indicated angle (adopted from [30]).

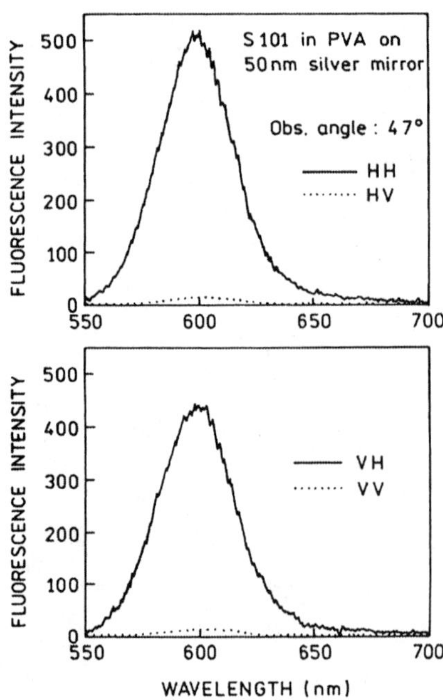


Figure 14. Polarized emission spectra with polarized excitation.

SPR only occurs for p-polarized incident light. Similarly, we expected the SPCE to be p-polarized even with RK excitation. Upon examining the polarization of the SPCE, we found that when the excitation was horizontally (H) polarized in the lab x-y plane, the emission was strongly polarized in the same horizontal direction (Figure 14, top). The intensity seen through a vertically-oriented polarizer (V) is 30-fold less, corresponding to a polarization of 0.94. This is already an unusual value, exceeding the limit of 0.5 for an isotropic sample with co-linear transition moments [31]. Examination of Figure 14 (bottom) shows that the emission is polarized in the plane of incidence (p-polarized) irrespective of the polarization of the excitation. In contrast to the SPCE, the free-space emission (RK/FS) displayed similar intensities for both polarized components of the emission [30].

The polarization of the directional emission proves that it is due to coupling with the surface plasmons and that the polarization of the SPCE is independent of the polarization of the normal incidence excitation. Similarly, SPCE is also p-polarized, suggesting that the emission dipoles perpendicular to the plane of incidence do not result in SPCE or at least display less efficient coupling. For accuracy we note that the relationship between fluorophore orientation and coupling to the metal surface has been studied theoretically (as summarized in [27]), but not experimentally. Additionally, we refer to the fluorophore as emitting into the substrate, but one can also consider the surface plasmons to be the source of emission into the substrate.

Since the emission displays the same p-polarization with vertically or horizontally polarized light, one can reason that the emission should be the same for all azimuthal angles $(\theta_A)$ around an axis normal to the metal. The symmetry of the SPCE can be preserved using a hemispherical rather than a hemicylindrical prism. In this case we expect the emission to form a cone central around the normal axis (Figure 15). The SPCE is expected to be p-polarized, which means in this case the electric vector points radially from the central axis (Figure 16).

We looked for a cone of emission using a hemispherical prism. The emission was visualized by incidence on a sheet of tracing paper (Figure 17). The cone remained unchanged upon rotation of the excitation polarization. When view through a polarizer the two sides or the top and bottom of the cone was extinguished, depending on the orientation of the polarizer. This observation, and the polarized intensities shown in Figure 14 demonstrate the SPCE is p-polarized, the electric vector pointing away from the central axis at all azimuthal angles.

In previous studies of the effects of silver particles on fluorophores we observed substantial decreases in lifetime as the intensities increased [6-8]. We interpreted this effect as an increase in the radiative decay rates near the metal particles [32-33]. In an analogous way we expected the radiative decay rate to be increased in the direction of the surface plasmon angle. Stated alternatively, we imagined that a large fraction of the emission appeared as SPCE because the rate of transfer to the surface plasmon was larger than the rate of spontaneous free-space emission. Hence, we expected the lifetime of SPCE to be shorter than the free space emission.

Figure 18 shows frequency-domain intensity decays for the free-space emission (top) and the surface plasmon-coupled emission (bottom). Overall, the lifetimes of SPCE (bottom) and free-space do not differ significantly. This was an unexpected result which we do not fully understand. We carefully considered possible artifacts and the effects of sample geometry, but can only conclude that our experiments indicate that the component of SPCE that we observe occurs without a substantial change in lifetime. At present we do not understand the origin of this discrepancy.

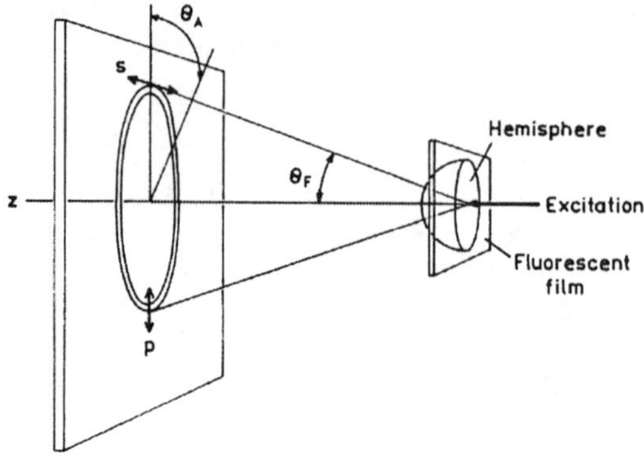



Figure 15. Cone of emission with a hemispherical prism.

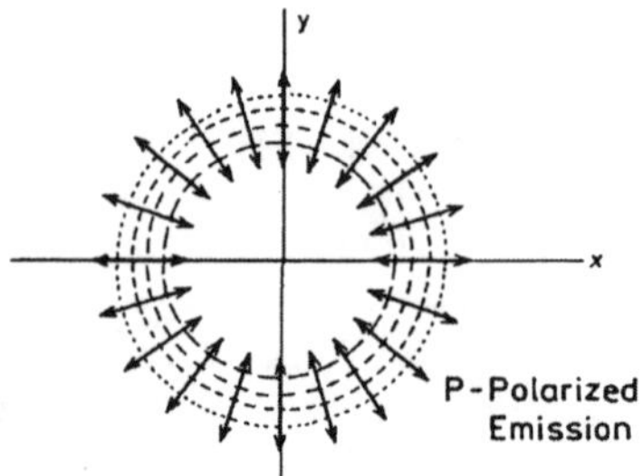



Figure 16. Cone of SPCE as seen from its central z axis.

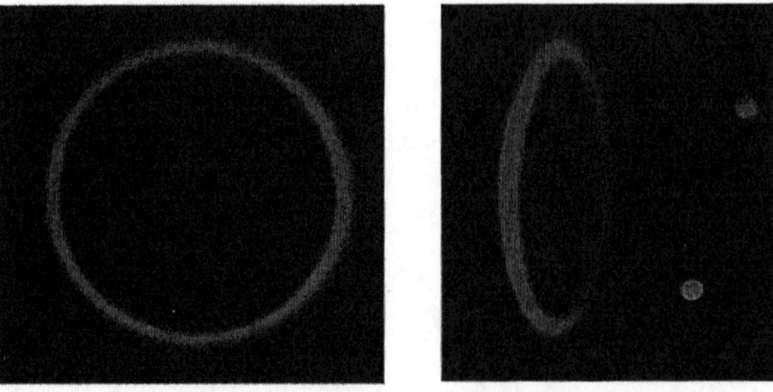

Figure 17. Cone of emission for S101 in PVA observed with a hemispherical prism and RK excitation. The emission was incident on tracing paper and photographed through a LWP 550 filter, without a notch filter (adopted from [30]). See Fig. 13.17 in the color insert at the end of this volume.

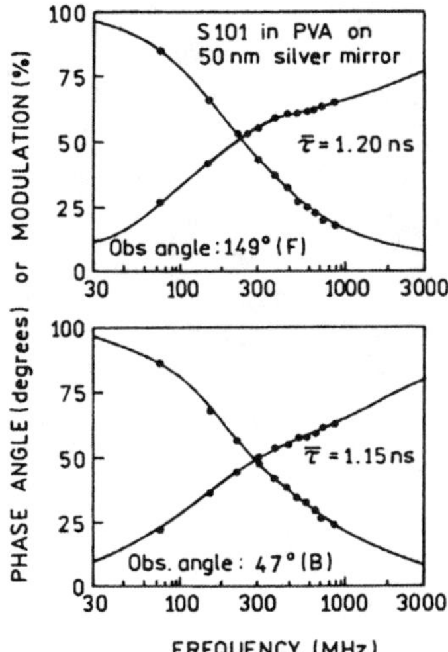


Figure 18. Frequency-domain intensity decays of S101 in PVA. Top, free-space emission. Bottom, surface plasmon-coupled emission. Adopted from [30].

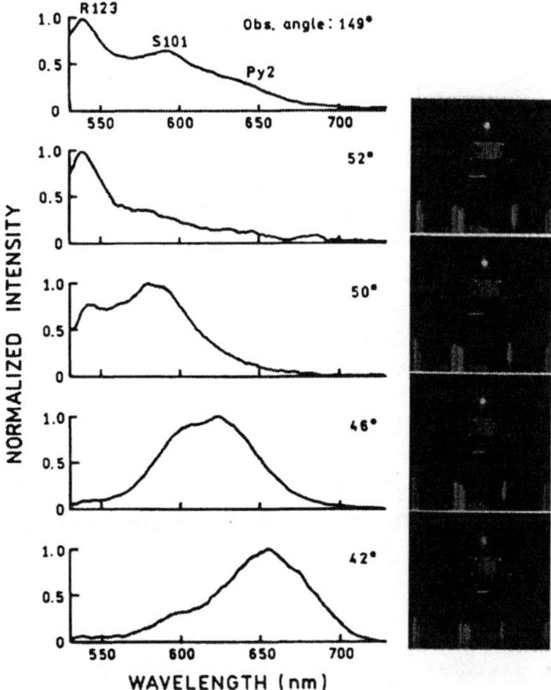


Figure 19. Emission spectra and photographs of SPCE for a three fluorophore mixture observed at different angles from the normal axis using the hemi-cylindrical prism. All concentrations of dyes in the evaporated films were about 10mM (adopted from [30]). See Fig. 30.19 in the color insert at the end of this volume.

Figure 7 indicates that the surface plasmon angle is dependent on wavelength. This suggests that fluorophores with different emission maxima will display SPCE at different angles. We tested this possibility using a mixture of three fluorophores (rhodamine 123, R123, sulforhodamine 101, S101, and pyridine 2, Py2) in a 30 nm thick PVA film. The concentrations were adjusted to obtain comparable emission from each fluorophore when exciting at 514 nm and observing the free-space emission (Figure 19, top panel). We recorded the emission spectra at different observation angles. These spectra are clearly distinct at each angle, with shorter wavelengths occurring for larger angles. This effect can be understood by considering the component of the p-polarized incident wavevector in the metal plane. This component increases with an increase in $\theta$. For a larger wavevector (smaller wavelength) the angle has to be smaller to decrease the in-plane component to match $k_{SP}$. The separation of wavelengths can be seen visually in the horizontal plane. Minor movements in the position of your eye shifts the color of the SPCE from red to green (Figure 19, right panels). Only a small movement is required, as can be seen from the apparent constant position of the optical hardware.

We then examined the SPCE from the fluorophore mixture using the hemi-spherical prism (Figure 20). The SPCE was imaged on white tracing paper. A circular rainbow pattern was observed with the shorter wavelength of emission appearing at the outer edges of the rainbow. This result demonstrated that SPCE can provide wavelength resolution without additional dispersive optics.

## 3.2 Properties of SPCE with Kretschmann Excitation

It is of interest to examine SPCE using the Kretschmann (KR) configuration (Figure 9, top), which provides some important benefits. With Kretschmann illumination the fluorophores are excited by the evanescent field which occurs when the incident angle equals the surface plasmon angle for the excitation wavelength ($\theta_{SP}$). At this angle the incident intensity is amplified about 20-fold due to the resonance interaction [34-35]. Additionally, the evanescent field is localized near the metal film, providing localized excitation near the metal. For a multi-photon processes the increased local intensity would result in a quadratic, 400-fold increase in the rate of excitation. Figure 21 shows the angular distribution of the S101 emission with surface plasmon excitation (SPE). In this case emission is observed on the same side of the metal film as the excitation. The emission appears to be more sharply distributed than for RK excitation (Figure 12), but this observation requires further study. In contrast to RK excitation, emission from S101 is only observed when the angle of incidence equals $\theta_{SP}$ and when the excitation is p-polarized. There appears to be less free-space emission with RK excitation (Figure 21) than with surface plasmon KR excitation (Figure 12). This difference is consistent with our expectation that KR excitation is localized near the metal, which excites those fluorophores which are more strongly coupled with the surface plasmon. This coupling is evidenced by the strongly p-polarized emission at 47° (not shown). We compared the relative SPCE intensities of the same S101 sample with surface plasmon (KR) or RK excitation (Figure 22). The emission intensity is about 10-fold larger with SPE, consistent with a resonance-enhanced incident field. We note that the actual increase in SPCE with SPE is likely to be larger than 10-fold because SPE is localized near the metal film and RK excitation is uniform across the film, which will increases the amount of free-space emission. The fluorescence observed with the KR/SPCE configuration shows similar wavelength dependence to the RK/SPCE configuration presented in Figure 19 [30].

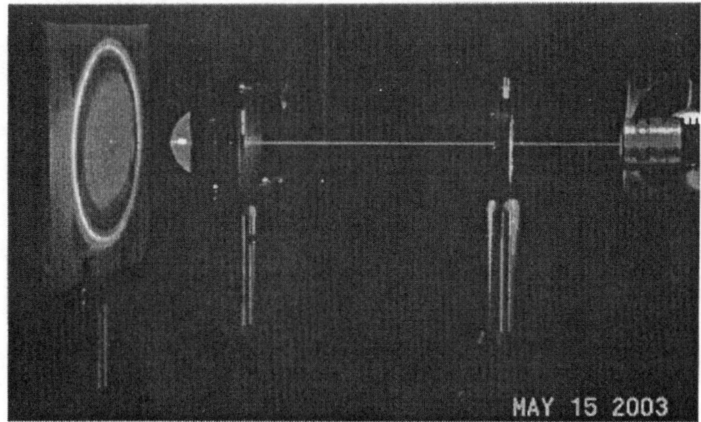


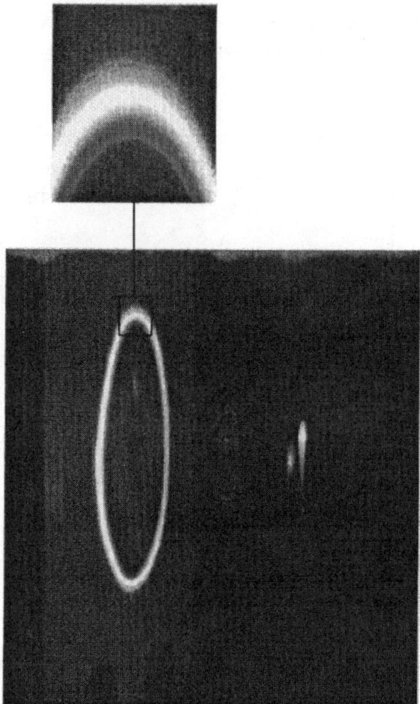

Figure 20. Photograph of SPCE from the mixture of fluorophores using RK excitation and a hemi-spherical prism, 532nm excitation. Top, no emission filter. Bottom, through a long pass filter but no notch filter (adopted from [30]). See Fig. 13.20 in the color insert at the end of this volume.

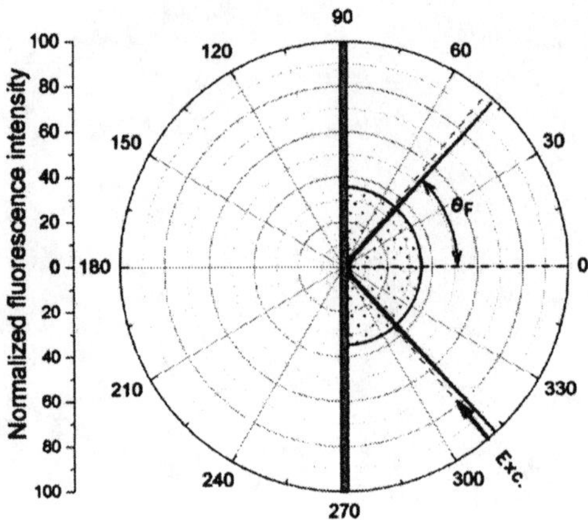


Figure 21. Angular distribution of the emission of S101 in 15nm thick PVA with Kretschmann excitation of $\theta_{SP}$=50°. The emission maximum was about 47° (adopted from [30]).

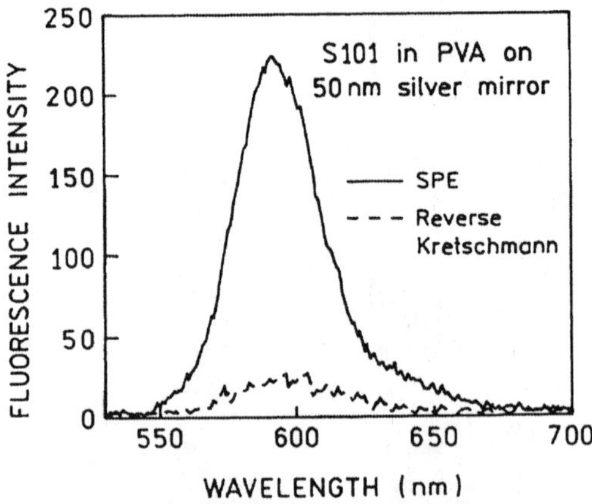


Figure 22. Comparison of the emission intensities of S101 in PVA with surface plasmon excitation (KR/SPCE) and reverse Kretschmann configurations (RK/SPCE), (adopted from [30]).

## 3.3 Background Rejection with SPCE

SPCE offers the possibility of strong background suppression. We reasoned that background rejection should be more effective with SPE because of its localization near the metal. This concept was tested using S101 in PVA as the desired emission, and a 1 mm thick ethanol solution of pyridine (Py2) as the background (Figure 23). In this figure the concentration of Py2 was increased progressively from top to bottom. For the free-space spectra (RK/FS) the emission is dominated by Py2 at all concentrations. For the lowest concentration of Py2 (Figure 23, top) the surface plasmon-coupled emission with RK excitation (RK/SPCE) is dominated by S101, which is closer to the metal film. If the concentration of Py2 is increased, the SPCE of Py2 becomes substantial with RK excitation (middle and lower panel). However, with surface plasmon KR excitation (KR/SPCE) the emission is always dominated by S101, even at the highest concentration of Py2 (———). These spectra demonstrate that the background from fluorophores distal from the metal can be suppressed by observing the SPCE, and suppressed further by observation of the SPCE with surface plasmon KR excitation. The fact that only fluorophores in proximity to the metal couple to surface plasmons opens up new possibilities for assays. The kinetics of binding and conformational changes on the surface can be observed without a change in the quantum yield of the fluorescent probe.

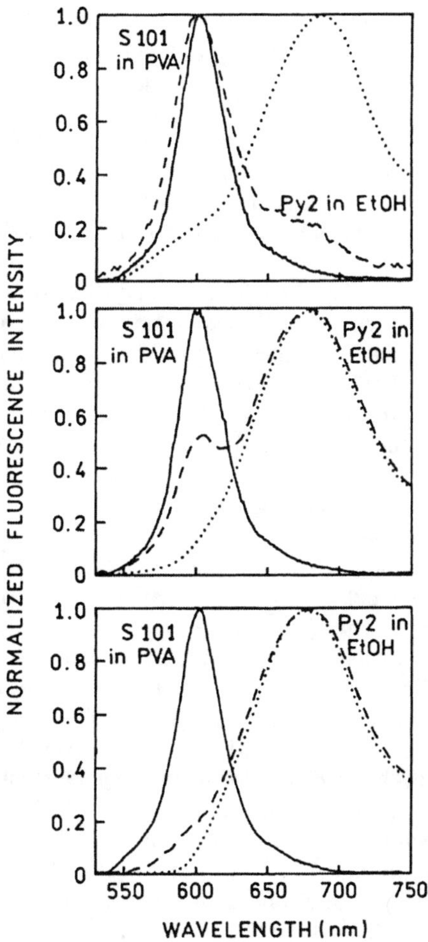


Figure 23. Emission spectra and background rejection as seen with free-space (RK/FS, ·······), reverse Kretschmann (RK/SPCE, ----), and Kretschmann (KR/SPCE, ———). The "background" was Py2 in ethanol in a 1mm thick layer adjacent to the PVA film with S101. The concentration of S101 in PVA was near 10mM. The concentrations of Py2 in ethanol from top to bottom were 7.5, 15, and 30μM (adopted from [30]).

### 3.4 DNA Hybridization Using SPCE

Measurement of DNA hybridization is now a central component of biotechnology and medical diagnostics. A variety of approaches are available to detect DNA hybridization including the use of intercalating fluorophores [36-37], dyes which bind to double stranded (ds) DNA [38-39], fluorescence resonance energy transfer [40-41], and excimer formation [42]. In all these methods hybridization is detected by a change in the emission spectral properties of the probe which occurs upon formation of double-stranded DNA, typically an increased quantum yield of the fluorophore.

We now describe a new approach to measurement of nucleic acid hybridization which does not depend on a spectral change in the fluorophores. The intensity change is due to localization near a thin metal surface by the binding reaction. We examined SPCE for fluorophore-labeled DNA oligomers, which were complementary (ssDNA-Cy3, Figure 24) or not complementary (ssDNA-Cy5) to a surface-bound capture oligomer [43].

The surface-bound capture oligomer was labeled with biotin. The oligomer complementary to the capture oligomer was labeled with Cy3. A shorter oligomer which was not complementary to the capture oligomer was labeled with Cy5. The concept of the experiment is shown in Figure 25. The ssDNA-biotin is bound to the silver surface by a layer of biotinylated BSA covered with streptavidin. ssDNA-biotin binds to this surface. The bathing solution can contain ssDNA-Cy3 and/or ssDNA-Cy5. We expect some of the ssDNA-Cy3 to bind to the surface and any excess to remain unbound. ssDNA-Cy5 will be unbound and more distant from the silver. The silvered slide was used as a cover slip for a 1mm path length demountable cuvette with silver inside. This mounted cuvette was attached to a hemicylindrical prism using index matching fluid between the prism and the cover slip.

Upon injection of ssCy3-DNA there was a time-dependent increase in the emission of Cy3 (Figure 26). Upon injection of ssCy3-DNA to a sample which contained protein (BSA and streptavidin) but no capture oligomers, there was no increase in Cy3 emission. The latter result shows that there was no significant non-specific binding of ssCy3-DNA to the protein surface which lacked the complementary oligomer. This result also shows there is little observable emission from ssCy3-DNA which was in the sample but not bound near the silver surface. In total, the data in Figure 26 demonstrated that DNA hybridization can be detected from the SPCE. Furthermore, detection of hybridization depends on proximity to the silver surface and does not require a change in quantum yield of the fluorophore. The bound and free-in-solution Cy3-DNA displays similar fluorescence quantum yields [7-8].

We tested the possibility of background rejection by examining a sample containing both surface bound Cy3-DNA and non-complementary ssCy5-DNA. We measured both the SPCE (KR/SPCE) and the free-space emission (RK/FS). At the excitation wavelength of 514nm Cy5 absorbs light more weakly than Cy3. To obtain comparable intensities in the free space emission of Cy3 and Cy5 we used an approximate 30-fold higher concentration of ssCy5-DNA than Cy3-DNA, resulting in the free space emission spectrum shown in Figure 27 (——). We then changed the optical configuration to use surface plasmon (KR) excitation and to observe the SPCE. Using these conditions the emission was almost completely due to Cy3 (Figure 27, -----). The emission from Cy5 was suppressed 20-fold or more. Hence SPCE can be used with samples containing multiple fluorophores or autofluorescence, and only fluorophores close to the metal will result in SPCE.

Cy3-5'-GAA GAT GGC CAG TGG TGT GTG GA-3'

**ss Cy3-DNA**

3'-CTT CTA CCG GTC ACC ACA CAC CT-5'-**biotin**

**ss DNA-biotin**

Cy3-5'-GAA GAT GGC CAG TGG TGT GTG GA-3'
3'-CTT CTA CCG GTC ACC ACA CAC CT-5'-**biotin**

**ds Cy3-DNA-biotin**

Cy5-5'-ATT GTT AT-3'

**ss Cy5-DNA**

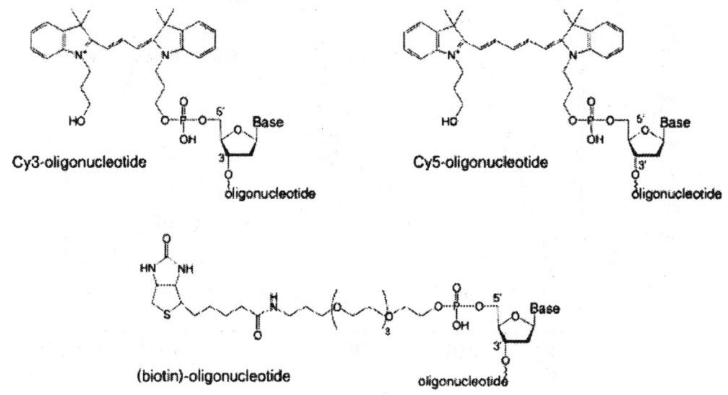

Figure 24. Structures of Cy3-DNA and Cy5-DNA.

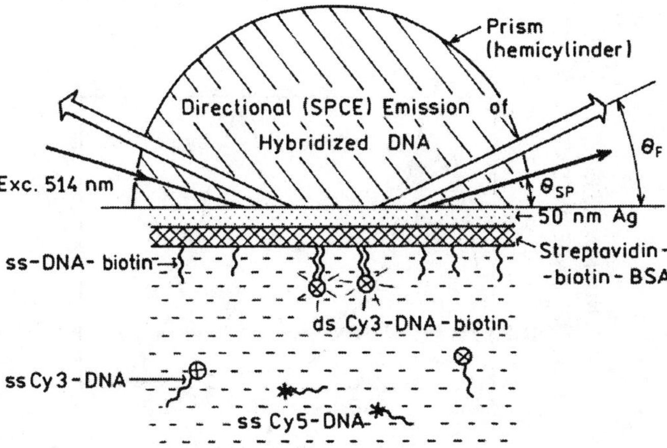


Figure 25. Intuitive description of directional fluorescence emission from hybridized DNA. Figure not drawn to scale, BSA-streptavidin ≈90Å, Cy3-DNA ≈ 70Å (adopted from [43]).

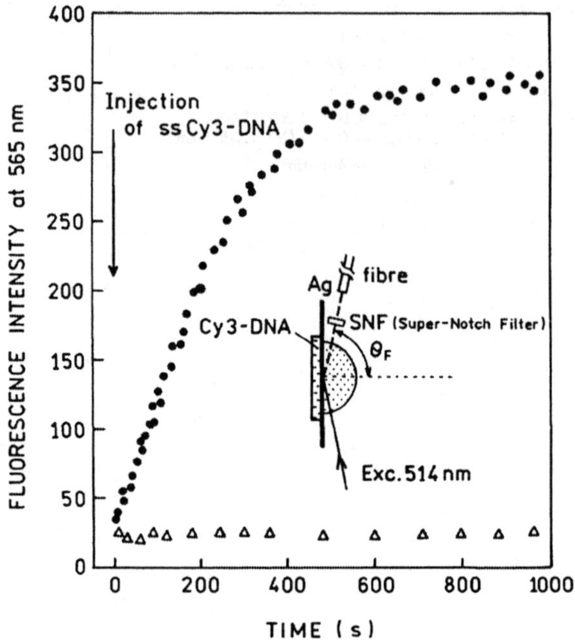



Figure 26. SPCE fluorescence observed at 565nm (Cy3-DNA emission) upon injection of a ssCy3-DNA in presence (•) and absence (∆) of a complementary ssDNA-biotin deposited on the protein coated Ag 50nm mirror (adopted from [43]).

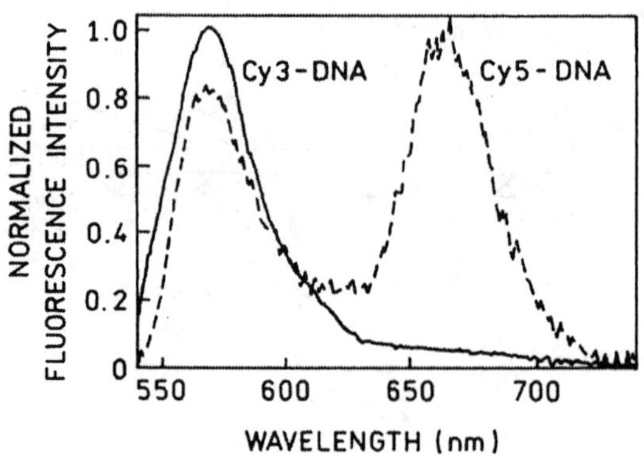



Figure 27. SPCE spectrum of dsCy3-DNA in presence of excess of ssCy5-DNA with surface plasmon (KR) excitation (——). Also shown is a free space emission observed in RK configuration (------). [dsCy3-DNA] = 5.4 x $10^{-9}$ M and [ssCy5-DNA] = 150 x $10^{-9}$ M (adopted from [43]).

## 4. DISCUSSION

In summary, SPCE appears to offer several advantages for measurement of DNA hybridization and other binding interactions. The increased intensity seen at the surface plasmon angle is due to localization near the metal surface. Hence, binding can be detected without a change in probe intensity due to the binding event. Since the intensity change is due to surface localization, an intensity change can be observed for any association reaction. Additionally, SPCE occurs over moderately large distances, typically up to several hundred nanometers from the metal surface. Since the biomolecules are typically much smaller, several layers of the capture molecules can be placed on the metal surface for increased sensitivity.

Another advantage of SPCE is effective rejection of the emission from fluorophores more distant from the metal. This suppression occurs by two mechanisms, the decreased efficiency of coupling at larger distances from the metal and selective excitation near the metal when using the Kretschmann configuration.

Another important characteristic of SPCE is high sensitivity because plasmon coupling can result in light collection efficiency near 50%, much higher than efficiencies of a few percent with more typical optics. In our experiments we utilized only a small fraction of directional SPCE. The entire SPCE emission can be collected to the detector as shown in [27] improving the sensitivity by about two orders of magnitude. The use of SPCE is technically simple, requiring only an easily prepared thin silver film. And finally, we note that SPCE can be observed with thin gold films [44] which are chemically stable and for which the surface modification chemistry is well developed. These attributes suggest SPCE will find numerous applications for nucleic acid and protein binding reactions. Finally, the intrinsic wavelength resolution in SPCE can be used in miniaturized spectrofluorometers. Also, the polarized emission can be combined with polarization-based sensing [45-47] in smart and simple sensing devices.

## 5. ACKNOWLEDGMENT

This work was supported by the NIH National Center for Research Resources, RR-08119, HG-002655, EB-000682, and EB-00981. Zygmunt Gryczynski thanks Philip Morris USA, Inc. for financial support.

## 6. REFERENCES

1. Sokolov, K., Chumanov, G., and Cotton, T.M. (1998). Enhancement of molecular fluorescence near the surface of colloidal metal films, *Anal. Chem.,* **70**:3898-3905.
2. Selvan, S. T., Hayakawa, T., and Nogami, M. (1999). Remarkable influence of silver islands on the enhancement of fluorescence from $Eu^{3+}$ ion-doped silica gels, *J. Phys. Chem. B,* **103**:7064-7067.
3. Weitz, D. A., Garoff, S., Hanson, C. D., and Gramila, T. J. (1982). Fluorescent lifetimes of molecules on silver-island films, *Optics Letts.,* **7**(2):89-91.
4. Lakowicz, J. R., Gryczynski, I., Shen, Y., Malicka, J., and Gryczynski, Z. (2001). Intensified fluorescence, *Photonics Spectra,* 96-104.
5. Lakowicz, J. R., Malicka, J., Gryczynski, I. (2003). Silver particles enhance the emission of fluorescent DNA oligomers, *BioTechniques,* **34**:62-68.
6. Lakowicz, J. R., Shen, Y., D'Auria, S., Malicka, J., Gryczynski, Z. and Gryczynski, I. (2002). Radiative decay engineering 2: Effects of silver island films on fluorescence intensity, lifetimes and resonance energy transfer, *Anal. Biochem.,* **301**:261 277.

7. Malicka, J., Gryczynski, I., Gryczynski, Z., and Lakowicz, J. R.(2003). Effects of fluorophore-to-silver distance on the emission of cyanine dye-labeled oligonucleotides, *Anal. Biochem.,* **315**:57-66.
8. Malicka, J., Gryczynski, I., Fang, J., and Lakowicz, J. R. (2003). Fluorescence spectral properties of cyanine dye-labeled DNA oligomers on surfaces coated with silver particles, *Anal. Biochem.,* **31** 7:136-146.
9. Geddes, C. D., Cao, H., Gryczynski, I., Gryczynski, Z., Fang, J., and Lakowicz, J. R.(2003). Metal-enhanced fluorescence (MEF) due to silver colloids on a planar surface: Potential applications of indocyanine green to in vivo imaging, *J. Phys. Chem. A,* **107**:3443-3449.
10. Malicka, J., Gryczynski, I., Fang, J., Kusba, J., and Lakowicz, J. R.(2002). Photostability of Cy3 and Cy5-labeled DNA in the presence of metallic silver particles, *J. Fluorescence,* **12**(3/4):439-447.
11. Parfenov, A., Gryczynski, I., Malicka, J., Geddes, C. D., and Lakowicz, J. R.(2003). Enhanced fluorescence from fluorophores on fractal silver surfaces. *J. Phys. Chem. B.,* **107**:8829-8833.
12. Lakowicz, J. R., Malicka, J., D'Auria, S., and Gryczynski, I. (2003). Release of the self-quenching of fluorescence near silver metallic surfaces, *Anal. Biochem.,* **320**:13-20.
13. Malicka, J., Gryczynski, I., and Lakowicz, J. R. (2003). Enhanced emission of highly labeled DNA oligomers near silver metallic surfaces, *Anal. Chem.,* **75**:4408-4414.
14. Salamon, Z., Macleod, H. A., and Tollin, G. (1997). Surface plasmon resonance spectroscopy as a tool for investigating the biochemical and biophysical properties of membrane protein systems. I: Theoretical principles, *Biochim. et Biophy. Acta.,* **1331**:117-129.
15. Melendez, J., Carr, R., Bartholomew, D.U., Kukanskis, K., Elkind J., Yee, S., Furlong, C., and Woodbury, R. (1996). A commercial solution for surface plasmon sensing, *Sensors and Actuators B,* **35-36**:212-216.
16. Liedberg, B., and Lundstrom, I. (1993). Principles of biosensing with an extended coupling matrix and surface plasmon resonance, *Sensors and Actuators B,* **11**:63-72.
17. Cooper, M. A. (2002). Optical biosensors in drug discovery, *Nature Reviews,* **1**:515-28.
18. Wegner, G. J., Lee, H. J., and Corn, R. M. (2002). Characterization and optimization of peptide arrays for the study of epitope-antibody interactions using surface plasmon resonance imaging, *Anal. Chem.,* **74**:5161-5168.
19. Raether, H. (1977). Surface plasma oscillations and their applications, in *Physics of Thin Films, Advances in Research and Development,* (Hass, G., Francombe, M. H., and Hoffman, R. W., Eds.), Academic Press, New York, Vol.9, pp. 145-261.
20. Pockrand, I. (1978). Surface plasma oscillations at silver surfaces with thin transparent and absorbing coatings, *Surface Sci.,* **72**:577-588.
21. Nelson, B. P., Frutos, A. G., Brockman, J. M., and Corn, R. M. (1999). Near-infrared surface plasmon resonance measurements of ultrathin films. 1. Angle shift and SPR imaging experiments, *Anal. Chem.,* **71**:3928-3934.
22. www.corninfo.chem.wisc.edu
23. Sambles, J. R., Bradbery G. W., and Yang, F. (1991). Optical excitation of surface plasmons: an introduction, *Contemporary Physics,* **32**(3):173-183.
24. Levi, L. (1968). *Applied Optics. A Guide to Optical System Design/Volume 1,* John Wiley & Sons, New York, 620 pp.
25. Born, M., and Wolf, E. (1980). Principles of Optics. Electromagnetic theory of propagation, interference and diffraction of light, Pergamon Press, New York, pp. 808.
26. Natan, M. J. and Lyon, L. A.(2002). Surface plasmon resonance biosensing with colloidal Au amplification, in *Metal Nanoparticles* (D. L. Feldheim and C. A. Foss, Jr., eds.) Marcel Decker, Inc., Bassel, New York. pp.183-205.
27. Lakowicz, J. R. (2003). Radiative decay engineering 3. Surface plasmon-coupled directional emission, *Anal. Biochem.,* **324**:153-169.
28. Feldhelm, D. L., and Foss, C. A. Jr. (Eds.) (2002). Overview. In *Synthesis, Characterization, and Applications, Metal Nanoparticles,* Marcel Dekker, Inc, New York. pp. 1-15.
29. Ford, 0. W., and Weber, W. B. (1984). *Electromagnetic Interactions of Molecules With Metal Surfaces,* North-Holland Physics Publishing Amsterdam, **113**:195-287.
30. Gryczynski, I., Malicka, J., Gryczynski, Z., and Lakowicz, J. R. (2003). Radiative decay engineering 4. Experimental studies of surface plasmon-coupled directional emission, *Anal. Biochem.,***324**:170-182.
31. Lakowicz, J. R. (1999). *Principles of Fluorescence Spectroscopy, 2nd Edition.* Kluwer Academic/Plenum Press, New York, 698 pp.
32. Barnes, W. L. (1998). Fluorescence near interfaces: the role of photonic mode density, *J. Modern Optics,* **45**(4):661-699.
33. Lakowicz, J. R., Malicka, J., Gryczynski, I., Gryczynski, Z., and Geddes, C. D. (2003). Radiative Decay Engineering: The role of photonic mode density in biotechnology, *J. Physics., D. Applied Physics,* **36**:R240-R249
34. Neumann, T., Johansson, M. L., Kambhampati, D., and Knoll, W. (2002). Surface plasmon fluorescence spectroscopy, *Adv. Funct. Mater,* **12**(9):575-586.

35. Liebermann, T., and Knoll, W. (2000). Surface-plasmon field-enhanced fluorescence spectroscopy, *Colloids and Surfaces,* **171**:115-130.
36. Le Pecq, J.-B., Le Bret, M., Barbet, J., and Roques, B. (1975). DNA polyintercalating drugs: DNA binding of diacridine derivatives, *Proc. Natl. Acad. Sci.,* **72**(8):2915-2919.
37. Markovits, J., Roques, B. P., and Le Pecq, J.-B. (1979). Ethidium dimer: A new reagent for the fluorimetric determination of nucleic acids, *Anal. Biochem.,* **94**:259-264.
38. Glazer, A. N., Peck, K., and Mathies, R. A.(1990). A stable double-stranded DNA ethidium homodimer complex: Application to picogram fluorescence detection of DNA in agarose gels, *Proc. Natl. Acad. Sci, USA,* **87**:3851-3855.
39. Rye, H. S., Yue, S., Wemmer, D. E., Quesada, M. A., Haugland, R. P., Mathies, R. A., and Glazer, A. N. (1992). Stable fluorescent complexes of double-stranded DNA with bisintercalating asymmetric cyanine dyes: Properties and applications, *Nucleic Acids Res.,* **20**(11):2803-2812.
40. Parkhurst, K. M. and Parkhurst, L. J.(1996). Detection of point mutations in DNA by fluorescence energy transfer, *J. Biomed. Optics,* **1**(4):435-441.
41. Morrison, L. E. and Stols, L. M.(1993). Sensitive fluorescence-based thermodynamic and kinetic measurements of DNA hybridization in solution, *Biochem.,* **32**:3095-3104.
42. Ebata, K., Masuko, M., Ohtani, H., Kashiwasake-Jibu, M. (1995). Nucleic acid hybridization accompanied with excimer formation from two pyrene-labeled probes, *Photochem. Photobiol.,* **62**(5):836-839.
43. Malicka, J., Gryczynski, I., Gryczynski, Z., and Lakowicz, J. R. (2003). DNA hybridization using surface plasmon-coupled emission, *Anal. Chem.,* **75**:6629-6633.
44. Gryczynski, I., Malicka, J., Gryczynski, Z., and Lakowicz, J. R. (2003). Surface plasmon-coupled emission using gold films, *submitted.*
45. Lakowicz, J. R., Gryczynski, I., Gryczynski, Z., Tolosa, L., Dattelbaum, J. D., and Rao, G. (1999). Polarization-based sensing with a self-referenced sample, *Appl. Spectroscopy,* **53**:1149-1157.
46. Gryczynski, I., Gryczynski, Z., and Lakowicz, J. R. (1999). Polarization sensing with visual detection, *Anal. Chem.,* **71**:1241-1251.
47. Gryczynski, Z., Gryczynski, I., and Lakowicz, J. R. (2000). Simple apparatus for polarization sensing of analytes, *Optical Engineering,* **39**:2351-2358.

# RADIATIVE DECAY ENGINEERING (RDE)

Chris D. Geddes*[1,2], Kadir Aslan[1], Ignacy Gryczynski[2], Joanna Malicka[2], and Joseph R. Lakowicz*[2]

## 1. INTRODUCTION

Fluorescence experiments are typically performed in sample geometries that are large relative to the size of the fluorophores and relative to the absorption and emission wavelengths. In this arrangement the fluorophores radiate into free space. Most of our knowledge and intuition about fluorescence is derived from the spectral properties observed in these free-space conditions. However, the presence of nearby metallic surfaces or particles can alter the free-space condition, which can result in dramatic spectral changes which are distinct from those observable in the absence of metal surfaces. Remarkably, metal surfaces can increase or decrease the radiative decay rates of fluorophores and increase the extent of resonance energy transfer (RET) (Figure 1). These effects are due to interactions of the excited-state fluorophores with free electrons in the metal, the so-called surface plasmon electrons, which polarize the metal and produce favorable effects on the fluorophore. The effects of metallic surfaces are complex and include quenching at short distances, spatial variation of the incident light field, and changes in the radiative decay rates (Figure 2). We refer to the use of fluorophore-metal interactions as radiative decay engineering (RDE) or metal-enhanced fluorescence (MEF).

The concept of modifying the radiative decay rate is unfamiliar to the fluorescence spectroscopists. It is therefore informative to consider the novel spectral effects expected by increasing the radiative rate. Assume the presence of a nearby metal (m) surface increases the radiative rate by addition of a new rate $\Gamma_m$ (Figure 1, Right). In this case the quantum yield and lifetime of the fluorophore near the metal surface are given by:

---

[1] Institute of Fluorescence, University of Maryland Biotechnology Institute, [2] Center for Fluorescence Spectroscopy, 725 W. Lombard St., Baltimore, MD 21201 USA,
* Corresponding authors, cfs@cfs.umbi.umd.edu

*Topics in Fluorescence Spectroscopy, Volume 8: Radiative Decay Engineering*
Edited by Geddes and Lakowicz, Springer Science+Business Media, Inc., New York, 2005

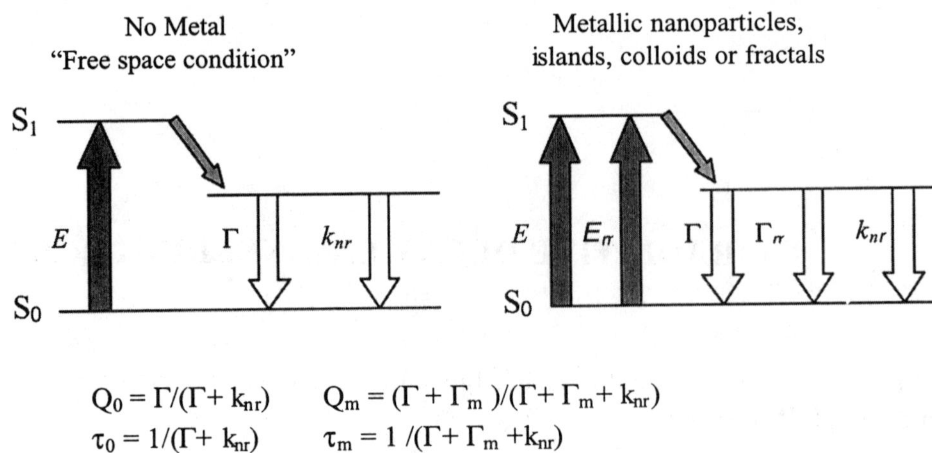



**Figure 1.** Classical Jablonski diagram for the *free space condition* and the modified form in the presence of metallic particles, islands, colloids or silver nanostructures. $E$-excitation, $E_m$, Metal-enhanced excitation rate; $\Gamma_m$, radiative rate in the presence of metal. For our studies, we do not consider the effects of metals on $k_{nr}$. Adapted from reference 59.

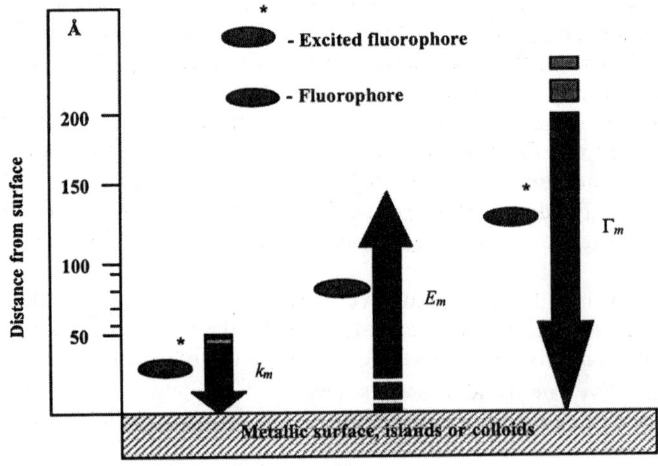



**Figure 2.** Predicted distance dependencies for a metallic surface on the transitions of a fluorophore. The metallic surface can cause Förster-like quenching with a rate ($k_m$) can concentrate the incident field ($E_m$) and can increase the radiative decay rate ($\Gamma_m$). Adapted from reference 59.

$$Q_m = \frac{\Gamma + \Gamma_m}{\Gamma + \Gamma_m + k_{nr}}$$

$$\tau_m = \frac{1}{\Gamma + \Gamma_m + k_{nr}}$$

These equations result in unusual predictions for a fluorophore near a metal surface. As the value of $\Gamma_m$ increases, the quantum yield increases while the lifetime decreases. To illustrate this point we calculated the lifetime and quantum yield for fluorophores with an assumed natural lifetime $\tau_N = 10$ ns, $\Gamma = 10^8$ s$^{-1}$ and various values for the non-radiative decay rates and quantum yields. The values of $k_{nr}$ varied from 0 to 9.9 x 10$^7$ s$^{-1}$, resulting in quantum yields from 1.0 to 0.01. Suppose the metal results in increasing values of $\Gamma_m$. Since $\Gamma_m$ is a rate process returning the fluorophore to the ground state, the lifetime decreases as $\Gamma_m$ becomes comparable and larger than $\Gamma$ (Figure 3, Left).

As a result of these calculations, we predicted that the metallic surfaces can create unique fluorophores with increased quantum yields and shorter lifetimes. Figure 4 illustrates that the presence of a metal surface within close proximity of a fluorophore with low quantum yield ($Q_0 = 0.01$) increases its quantum yield ~10-fold resulting in brighter emission, while reducing its lifetime 10-fold, resulting in an enhanced photostability of the fluorophore due to spending less time in an excited state.

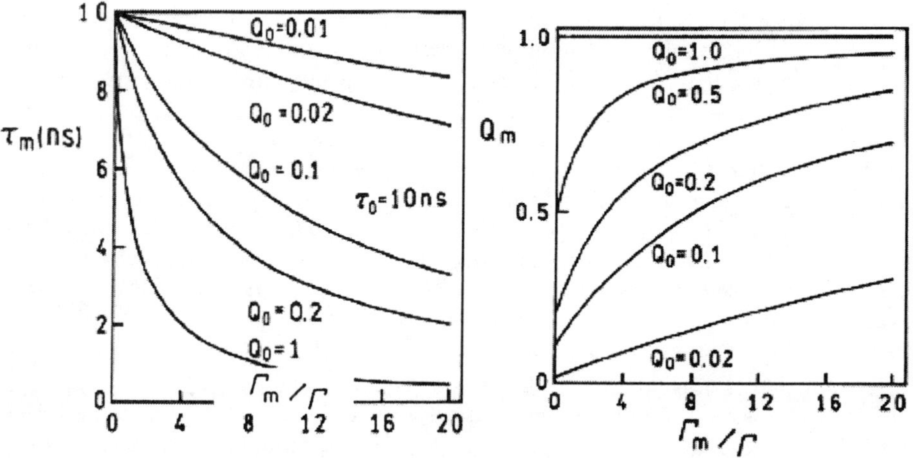


**Figure 3.** The effect of an increase in radiative decay rate on the lifetime and quantum yield. Adapted from reference 5.

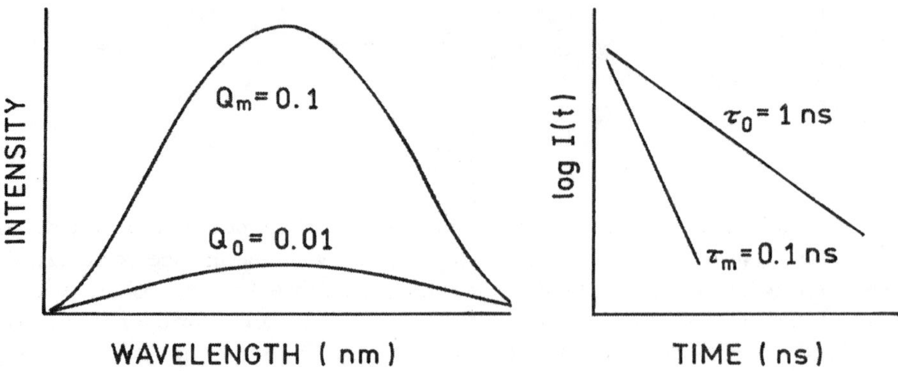



**Figure 4.** Metallic surfaces can create unique fluorophores with high quantum yields and short lifetimes. Adapted from reference 5.

The possibility of altering the radiative decay rates was demonstrated by measurements of the decay times of europium ($Eu^{3+}$) complex positioned at various distances from a planar silver mirror.[1-4] In a mirror the metal layer is thicker than the optical wavelength. The lifetimes oscillate with distance but remain a single exponential at each distance (Figure 5). This effect can be explained by changes in the phase of the reflected field with distance and the effects of this reflected field on the fluorophore. A decrease in lifetime is found when the reflected field is in-phase with the fluorophores' oscillating dipole (i.e., a $\Gamma_m$ modification). An increase in the lifetime is found if the reflected field is out-of-phase with the oscillating dipole. As the distance increases, the amplitude of the oscillations decreases. The effects of a plane mirror occur over distances comparable to the excitation and emission wavelengths. At short distances below 20 nm the emission is quenched c.f. Figure 2, $k_m$. This effect is due to coupling of the dipole to oscillating surface charges on the surface of the metal, which are called surface plasmons resonances (SPR). One of the most dramatic effects of silver islands on fluorescence is shown in Figure 6. Silver islands were coated with a thin film of $Eu(ETA)_3$,[6] where ETA is a ligand which chelates europium. This chelate displayed a quantum yield near 0.4. The sample contained an inert coating between the islands so $Eu^{3-}$ chelates positioned between the islands were not emissive. When the $Eu(ETA)_3$ chelate was deposited on the silica substrate, without the silver islands, it displayed a single exponential decay time of 280 μs and a quantum yield near 0.4. However, when deposited on silver island films, the intensity increases about 5-fold and the lifetime decreases by about 100-fold to near 2 μs (Figure 5). Also, the decay is no longer a single exponential on the silver island films.[6] The silver islands had the remarkable effect of increasing the intensity 5-fold while decreasing the lifetime 100-fold. Such an effect can only be explained by an increase in the radiative rate, $\Gamma_m$.

The five-fold increase in the quantum yield of $Eu(ETA)_3$ results in an apparent quantum yield of 2.0, which is obviously impossible. This high quantum yield is probably due to a cumulative increase in the local excitation field near the metal particle. For this reason, it is important to recognize the intensities measured on surfaces represent "apparent" quantum yields which can include an unknown factor due to incident field enhancement. This increase in the local intensity of the incident light cannot explain the decreased lifetime because an unperturbed $Eu^{3-}$ chelate, excited by an enhanced field, would still decay with a 280-$\mu$s lifetime. According to the authors[6] the decreased lifetime is due to electromagnetic coupling between the $Eu^{3-}$ and the silver islands.

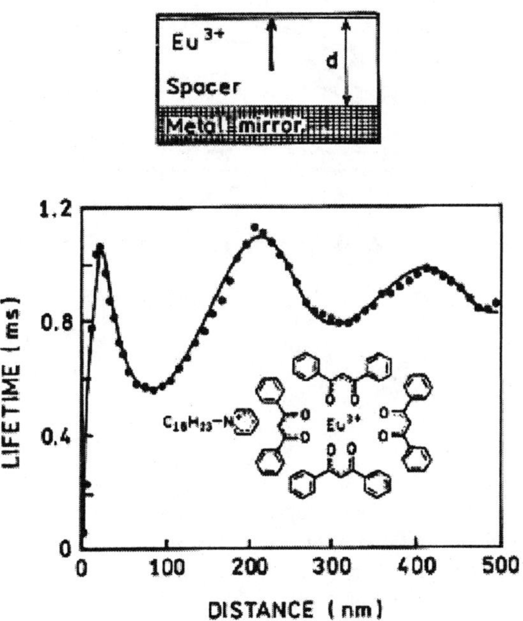


**Figure 5.** The effect of a plane metal surface on fluorescence lifetimes. Adapted from reference 5.

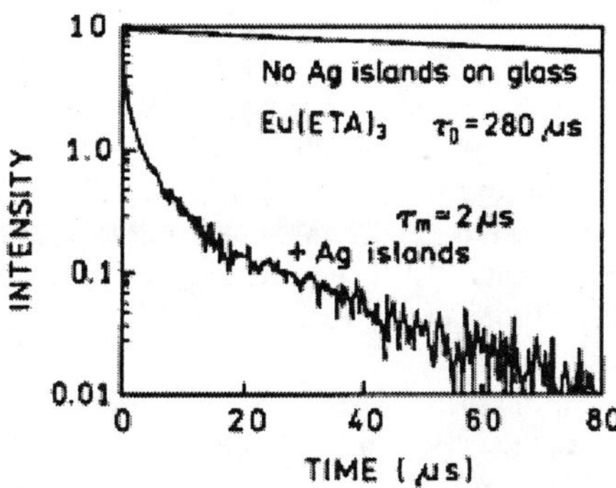


**Figure 6.** The effect of silver islands on fluorescence lifetimes. Adapted from reference 5.

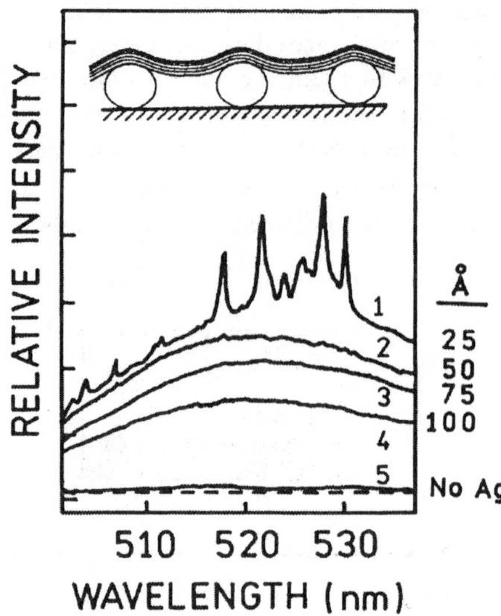

**Figure 7.** Enhanced fluorescence emission is a through space interaction as compared to SERS. Adapted from reference 5.

One can compare MEF with Surface-enhanced Raman scattering (SERS). While MEF and SERS depend on metal particles or a rough metal surface, it seems that these are different phenomena, at least in part. MEF and SERS show different dependencies on the distance to the surface. The enhancement of Raman scatter and fluorescence by metal colloids is illustrated in Figure 7. Silver colloids were prepared in suspension by known procedures. The particles were reacted with a glass surface which was modified to contain sulfohydryl groups. This colloidal metal film was then covered with one or more layers of octadecanoic acid (ODA) as a Langmuir–Blodgett film. Each layer of ODA was thought to be 25 Å thick. A fluorescein-labeled lipid Fl-DPPE was positioned in the layer immediately adjacent to a silver-colloidal film (Figure 7, line 1) and in layers more distant from the metal (lines 2–4). In the absence of metal on the bare glass slide there is little detectable emission (line 5). When Fl-DPPE is immediately adjacent to the metal, the emission is enhanced, as are the Raman peaks (line 1). When Fl-DPPE is present in the more distant layers, the emission is still enhanced, but the shorter range SERS effect is no longer present. It appears that SERS requires molecular contact of the fluorophore with the metal particles. The fluorescence enhancement occurs at larger distances from the metal, up to 100 Å (lines 2 to 5).

In the present review, we describe the effects of different silver nanostructures that were prepared by various methods in our laboratories on the emission intensity of fluorophores with various quantum yields and on biochemical fluorophores. The silver nanostructures consist of subwavelength size nanoparticles of silver deposited on inert substrates. These particles display a surface plasmon absorption, which in the small particle limit can be calculated from the real and imaginary dielectric constants of the metal[7-9] We show that proximity to silver nanostructures results in a preferential increase in intensity of low-quantum-yield fluorophores, and that the lifetimes decrease as the intensities increase, we subsequently discuss the use of RDE and some of the possible Bio/technological applications.

## 2. ENHANCED EMISSION FROM LOW AND HIGH QUANTUM YIELD SPECIES USING SILVER ISLAND FILMS (SiFs)

The presence of a nearby metallic surface can modify the radiative rate of an excited fluorophore. This results in an increase in the fluorescence intensity, a reduction in the lifetime and an in increase in quantum yield of the fluorophores. It is expected that the properties of the fluorophores with various quantum yields will be affected differently by the presence of a nearby metal due to the difference in radiative rate modifications. This hypothesis was tested by investigating the changes in emission properties of two fluorophores with similar absorption/emission spectra but with different quantum yield (Rhodamine B, quantum yield, $Q_0 = 0.48$ and Rose Bengal, $Q_0 = 0.02$). Figure 8 summarizes our observations with these fluorophores. The emission from Rhodamine B on SiFs was 20 % higher as compared to the emissions from the glass side. On the other hand, emissions from Rose Bengal on SiFs were ˜ 5-fold higher than those on the glass side. Figure 8 also shows the sample geometry used, which is important for the interpretation of these results; the fluorophores were placed in between two quartz plates (approximately half of the plates were coated with silver island films). We estimate the distance between the plates to be ˜ 1 μm, and such a configuration of the samples results in only small fraction of the fluorophores to be present within the distance over which metallic surfaces can exert effects. The region of varying photonic mode density is expected to extend about 200 Å into the solution (c.f. Figure 2). Hence only about 4% of the liquid volume between the plates is within the active volume. This suggests that the fluorescence intensity of Rose Bengal within 200 Å of the islands is indeed increased 125-fold.

We also investigated the emission spectral properties of the fluorophore labeled-DNA using the same geometry as shown in Figure 8. Emission spectra of Cy3-DNA and Cy5-DNA are shown in Figure 9. The emission intensity is increased 2- to 3- fold between SiFs as compared to between the quartz slides for Cy3-DNA and Cy5-DNA, respectively. The slightly larger increase in emission intensity for Cy5-DNA compared to Cy3-DNA is consistent with the results where larger enhancements are observed with low quantum yield fluorophores. Figure 9 also shows the photographs of the labeled oligomers on quartz and on SiFs. The emission from the labeled-DNA on quartz is almost invisible and is brightly visible on the SiFs. This difference intensity is due to an increase in the photonic mode density near the fluorophore, which in turn results in an increase in the radiative decay rate and quantum yield of the fluorophores. We note that the photographs are taken through emission filters and the increase in emission intensity is not due to an increased excitation scatter from the silvered plates.

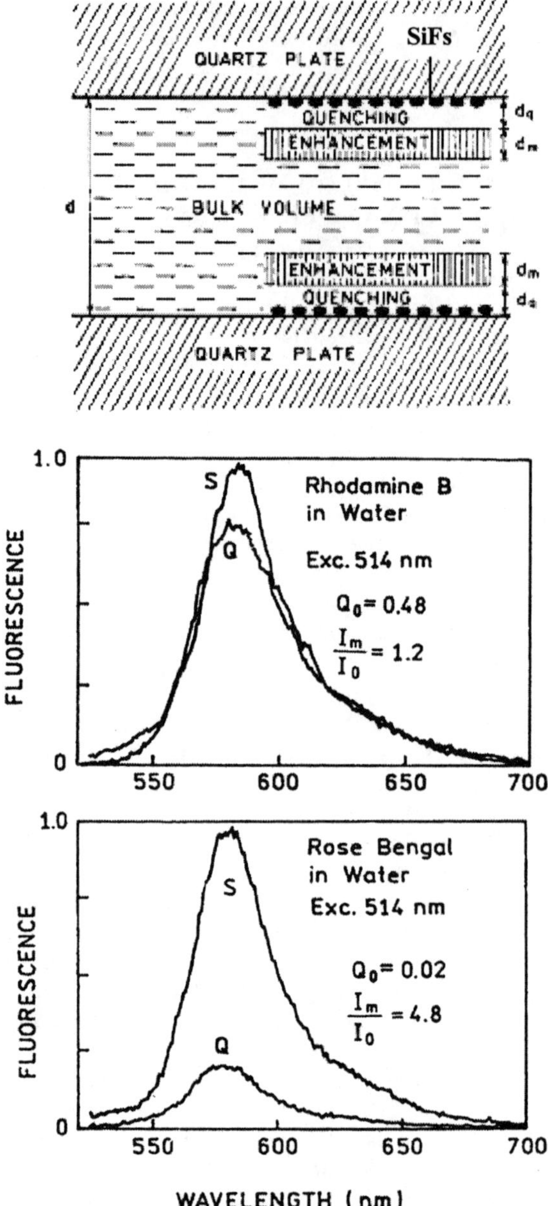



**Figure 8.** Schematic for a fluorophore solution between two silver island films. The solid ellipsoids represent the silver island films (Top). The effect of silver island films on the emission spectra of Rhodamine B (Middle) and Rose Bengal (Bottom). Adapted from reference 17.

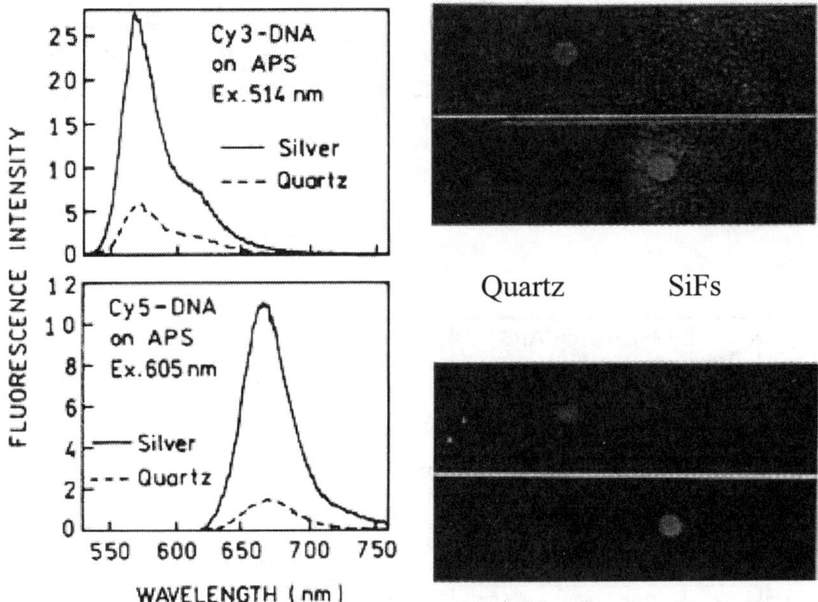


**Figure 9.** See Fig. 14.9 in color insert. Emission spectra of Cy3-DNA (Top-Left) and Cy5-DNA (Bottom-Left) between quartz plates with and without SiFs. Photographs of corresponding fluorophores. Adapted from ref. 19.

Photostability of the labeled-DNA was studied by measuring the emission intensity during continuous illumination at a laser power of 20 mW (Figure 10). The intensity initially dropped rapidly, but became more constant at longer illumination times. Although not a quantitative result, examination of these plots visually suggests slower photobleaching at longer times in the presence of silver particles compared with quartz slides. We also questioned the nature of emission from the fluorophores remaining after 300 s illumination. The emission spectra of Cy3-DNA and Cy5-DNA were identical before and after illumination, both in the absence and presence of silver islands (data not shown). This result indicates that the detected emission, even after intense illumination, is still due to Cy3 and Cy5, and not a photolysis byproduct.

The results from Rhodamine B and Rose Bengal, Figure 8, were consistent with our expectations that the presence of metal increases the emission intensity and quantum yields, and decreases the lifetime of the fluorophores. Nonetheless, one could be concerned with possible artifacts due to dye binding to the surfaces or other unknown effects. For this reason, we examined a number of additional fluorophores between uncoated quartz plates and between silver island films. In all cases, the emission was more intense for the solution between the silver islands. For example, [Ru(bpy)$_3$] and [Ru(phen)$_2$dppz] have quantum yields near 0.02 and 0.001, respectively. A larger enhancement was found for [Ru(phen)$_2$dppz] than for [Ru(bpy)$_3$] (data not shown). The enhancements for 10 different fluorophore solutions are shown in Figure 11. In all cases, lower bulk-phase quantum yields result in larger enhancements for samples between silver island films.

The results in Figures 8–11 *provide strong support* for our assertion that proximity of the fluorophore to the metal islands resulted in increased quantum yields (i.e., a modification in $\Gamma_m$). It is unlikely that these diverse fluorophores would all bind to the silver islands or display other unknown effect results that resulted in enhancements that increased monotonically with decreased quantum yields.

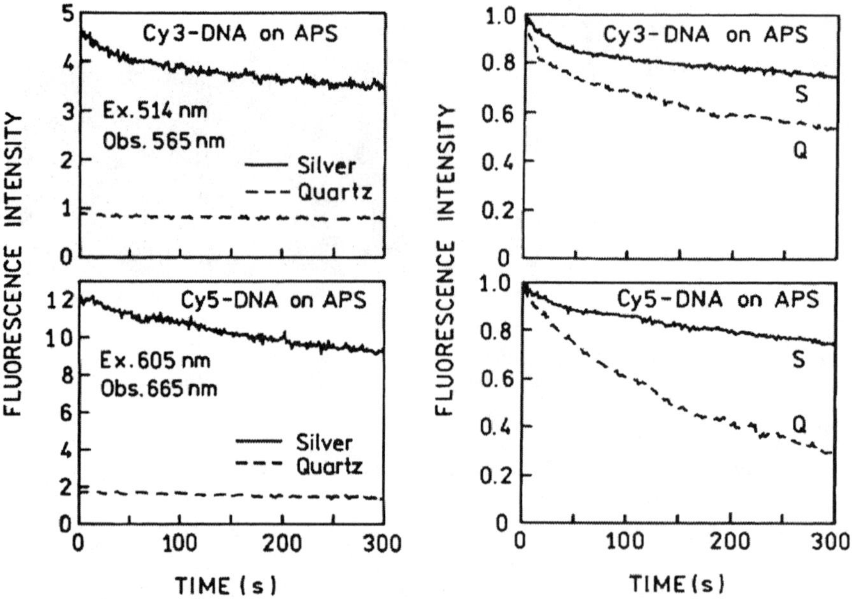


**Figure 10.** Photostability of Cy3-DNA and Cy5-DNA between quartz plates with and without silver island films. The laser power was 20 mW. Adapted from reference 19.

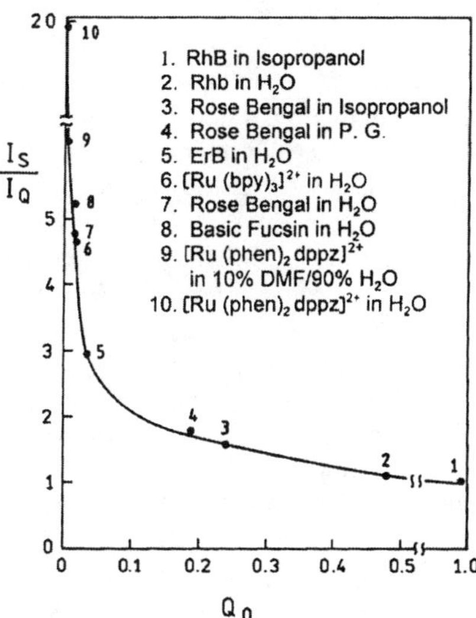


**Figure 11.** The effect of silver island films on the quantum yields of fluorophores. Adapted from reference 17.

## 3. ENHANCED INTRINSIC FLUORESCENCE USING SiFs

Intrinsic fluorophores are those which occur naturally in biomolecules. These molecules include fluorescent tryptophan residues in proteins, enzyme cofactors like flavin and NADH, collagen and green fluorescent protein.[10] In DNA, each nucleotide residue contains a UV absorbing (˜ 260 nm) base which might be expected to display fluorescence. However, the intrinsic fluorescence of the bases is exceedingly weak in DNA or in the isolated bases.[11,12] The emission is so weak that there are no practical uses of intrinsic DNA emission. For this reason a vast array of fluorophores have been developed which bind to DNA and display UV, visible to NIR fluorescence.[13-16] In fact the renaissance of fluorescence some 15 years ago is attributed to the widespread use of fluorescent probes developed for DNA sequencing as compared to the now outlawed use of radiolabels. The difficulty with DNA is not that the radiative rates are slow, but that the non-radiative rates are exceedingly fast, so that the excited bases return to the ground state prior to emission. That is, fluorescence is a competitive process between radiative and non-radiative decay. In the case of DNA, the low quantum yields are due to non-radiative rates which are much larger than the radiative rate. Suppose that the photonic mode density (PMD) near the DNA can be increased so that the radiative rate increases. This may be accomplished by bringing the DNA into proximity of silver particles which display plasmon resonance suggesting usefully high intrinsic fluorescence from DNA.

We examined the emission of DNA in micron thick samples between quartz plates and between silver island films (SiFs). When a solution of DNA is examined between two quartz slides the emission is barely detectable (Figure 12). There is a dramatic increase in intrinsic emission for the DNA between SiFs.[18] In this experiment, the information is only qualitative because the DNA is not bound to the surfaces and only a small fraction is near the silver particles. Hence, the increase in quantum yield for those molecules near the silver particles is likely to be larger than seen in Figure 12. The emission spectra in Figure 12 do not demonstrate an increase in the radiative rate. However, such an increase can be demonstrated by the additional measurement of the intensity decays. These results show that the intensity decay is more rapid near the SiFs (Figure 12). We interpret this decrease in lifetime as due to the increased PMD near the DNA bases.

While the human genome and other organisms have been sequenced [20,21] there is still a need for faster, cheaper, and more sensitive DNA sequences. Some groups are attempting to sequence a single DNA strand using a single strand of DNA.[22,23] The basic idea is to allow an exonuclease to sequentially cleave single nucleotides from the strand, label the nucleotide with a fluorophore, and detect and identify the labeled nucleotide. This goal is more difficult than single molecule detection because every nucleotide must be detected and identified, not the simpler task of finding one fluorophore among many. Also, the labeling of nucleotides is likely to result in a larger number of fluorophores which have not reacted. The use of RDE could allow base detection and identification without labeling. Suppose the released nucleotides pass through a specially designed flow chamber (Figure 14). The size and shape of the chamber could be such that the bases displayed intrinsic emission. Also, the surface would be shaped and periodic in a manner which directs the emission toward a detector. The design would be such that the directed emission occurs wherever the unlabeled nucleotides flow through the laser beam. SERS of DNA bases has been observed using silver and gold colloids,[24] which suggests to us that metal-enhanced fluorescence will also be observed for DNA and its bases. Additionally, it has now become possible to create silver particles on surfaces with broadly tunable SPR spectra using nanolithography, [25] and a variety of methods are appearing for self-assembly of metallic nanoparticles.[26-28] If successful, such an apparatus would allow DNA sequencing using intrinsic nucleotide fluorescence.

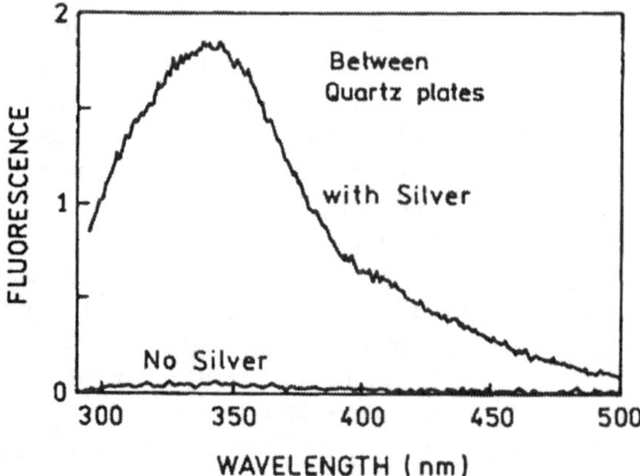


**Figure 12.** Emission spectra of DNA solution between quartz plates (no silver) and between silver island films. Adapted from reference 18.

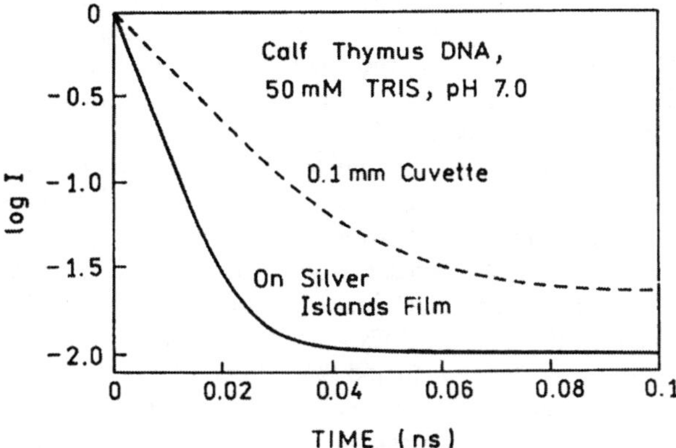


**Figure 13.** Time-dependent intensity decays of calf thymus DNA without metal (- -) and between silver island films (–).Adapted from reference 18.

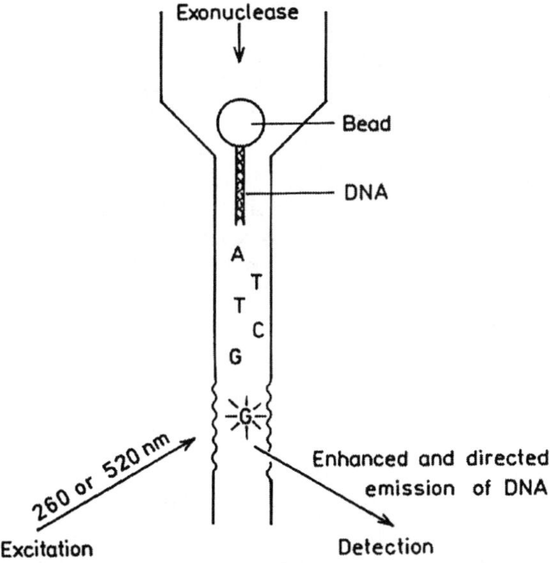


**Figure 14.** Notion of DNA sequencing using intrinsic base fluorescence. Adapted from reference 5.

## 4. DISTANCE DEPENDENCE OF ENHANCED FLUORESCENCE USING SiFs

The interactions of fluorophores with metallic particles have been the subject of a number of publications.[29,30-39] There is reasonable agreement that the maximal enhancements occur within about 100Å from the surface, but some publications report the optimal distance to be as large as 600 Å.[38] In our opinion the use of metallic particles to enhance fluorescence has great potential for advances in medical diagnostics and biotechnological methodology.[5] For this reason, we examined the effects of metal-to-fluorophores distance on enhanced fluorescence. To investigate this distance dependence we used alternating monolayers of biotinylated bovine serum albumin (BSA) and avidin (Figure 15). It is known that BSA adsorbs as a monolayer onto glass or silver surfaces, and that subsequent exposure to avidin and then biotinylated BSA results in additional monolayers of these proteins.[38] Because of the relevance to genomic analysis, we examined dsDNA oligomers labeled with Cy3 or Cy5. An unlabeled biotinylated oligomer was used to bind to the outermost layer of avidin. This biotinylated oligomer was previously hybridized to a complementary oligomer labeled with Cy3 or Cy5.

Emission spectra of the Cy3- and Cy5-labeled oligomers are shown in Figure 16. The largest intensity is seen for the labeled DNA bound to a single layer of BSA-biotin-avidin. The intensity, relative to uncoated quartz, decreases progressively with the number of protein layers to an enhancement near 2 for six layers. We note that the intensities were measured relative to that found for the labeled oligomers on quartz with the same number of protein layers. We noticed a progressive increase in the intensity of both probes on quartz with increasing numbers of protein layers, with the overall increase being about 3-fold for six layers. We interpret this increase as due to penetration of the labeled DNA to lower layers of BSA-biotin/avidin. Assuming this interpretation is correct then the 2-fold enhancements seen for six layers may be an overestimate of the actual enhancement for six protein layers. This penetration may explain the difference between our results and that of Sokolov *et al.*[38] who found increasing enhancements of fluorescein out to six layers. It was not clear from the article whether they considered the penetration of deeper protein layers. An increasing amount of bound fluorescein-biotin, and normalization of the signal to that with a single layer of protein, would explain the apparent increase in intensity out to six layers reported in reference 38.

In summary, by the use of protein layers we found that the maximum increase in intensity and maximum decrease in lifetime were found for a single BSA-avidin layer which positions the fluorophore about 90 Å from the surface. Hence the optimal distance for metal-enhanced fluorescence (MEF) can be readily studied using protein monolayers.

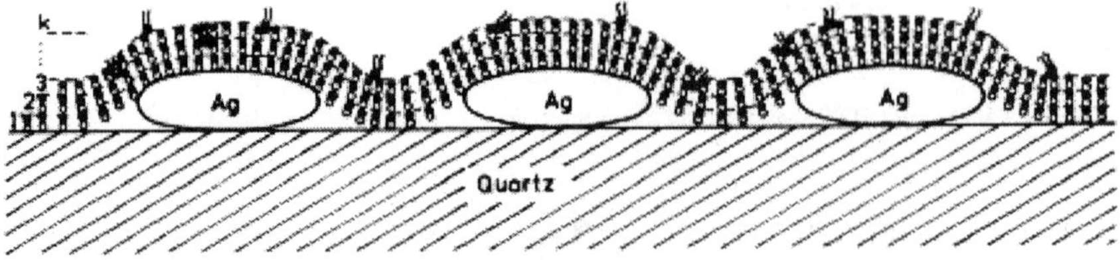


**Figure 15.** Schematic of BSA-avidin monolayers with labeled DNA. Adapted from reference 29.

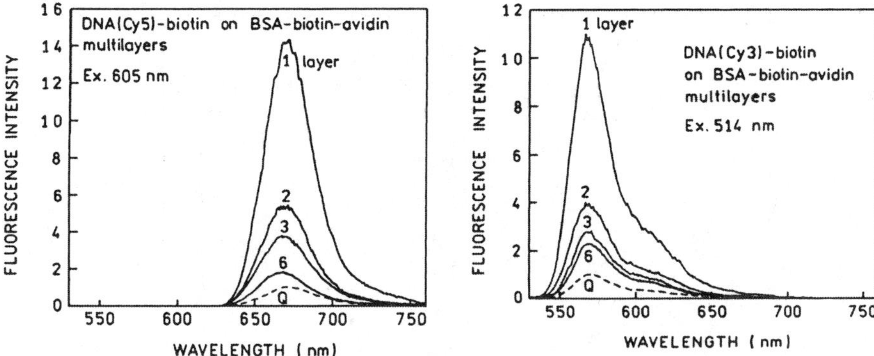

**Figure 16.** Emission spectra of DNA(Cy5)-biotin (Left) and DNA(Cy3)-biotin (Right) on BSA-biotin/avidin layers. Spectra are normalized to the spectrum on quartz (Q) with same number of protein layers. Adapted from reference 29.


## 5. RELEASE OF SELF-QUENCHING USING SiFs

Proteins covalently labeled with fluorophores are widely used as reagents, such as immunoassays or immuno-staining of biological specimens with specific antibodies. In these applications, fluorescein is one of the most widely used probes. An unfortunate property of fluorescein is self-quenching, which is due to Forster resonance energy transfer between nearby fluorescein molecules (homo-transfer).[40] As a result, the intensity of labeled protein does not increase with increased extents of labeling, but actually decreases. Figure 17 shows the spectral properties of FITC–HSA with molar labeling ratios (L) ranging from 1-to-1 (L=1) to 1-to-9 (L=9) (the samples had the same optical density at 490 nm). The relative intensity decreased progressively with increased labeling. The insert in Figure 17 shows the intensities normalized to the same amount of protein, so that the relative fluorescein concentration increases nine-fold along the x axis. It is important to note that the intensity per labeled protein molecule does not increase and in fact decreases, as the labeling ratio is increased from 1 to 9. We found that the self-quenching could be largely eliminated by the close proximity to SiFs,[41] as can be seen from the emission spectra for labeling ratios of 1 and 7 (Figure 18) and from the dependence of the intensity on the extent of labeling (Figure 19). We speculate that the *decrease* in self-quenching is due to an increase in the rate of radiative decay, $\Gamma_m$.

The dramatic difference in the intensity of heavily labeled HSA on glass and on SiFs is shown pictorially in Figure 20. The effect is dramatic as seen from the nearby invisible intensity on quartz (left side) and the bright image on the SiFs (right side) in this unmodified photograph. These results suggest the possibility of ultra bright labeled proteins based on high labeling ratios and metal-enhanced fluorescence. We conclude that SiFs, and most probably colloidal silver, can be utilized to obtain dramatically increased intensities of fluorescein-labeled macromolecules.

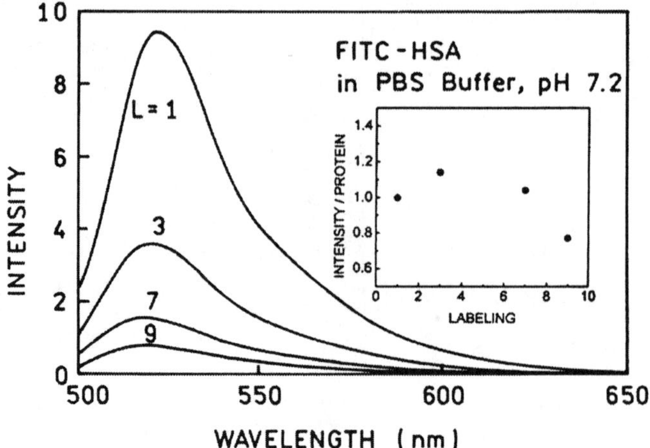

**Figure 17.** Dependence of emission intensity on the degree of labeling. Adapted from reference 41.

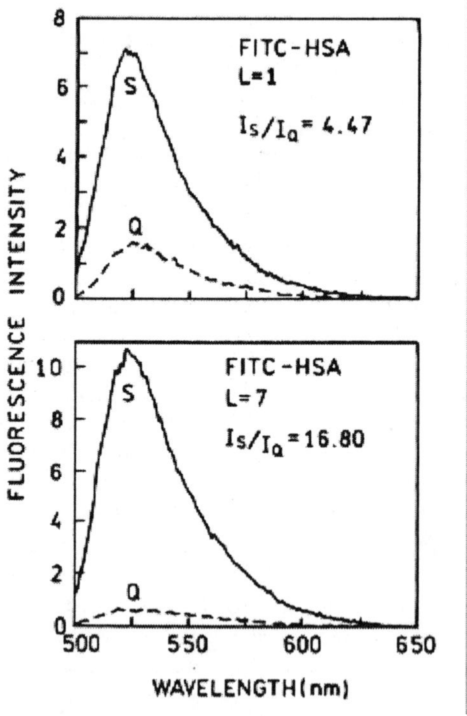

**Figure 18.** Emission spectra of FITC-HSA with different degrees of labeling on quartz (Q) and on SiFs (S). Adapted from reference 41.

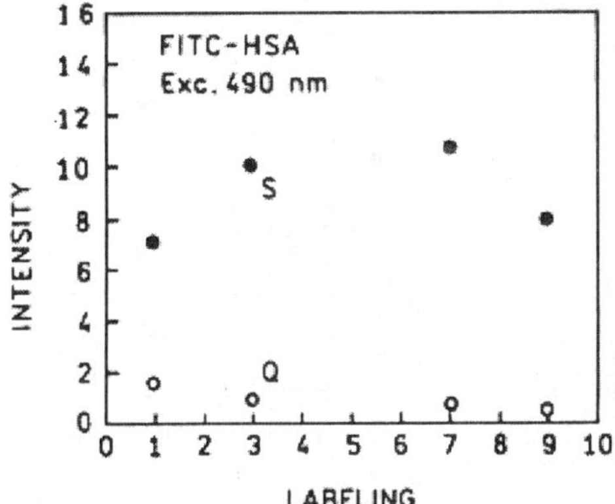


**Figure 19.** Emission intensity of FITG HSA at 520 nm vs. different degrees of labeling on quartz and on SiFs (S). Adapted from reference 41.

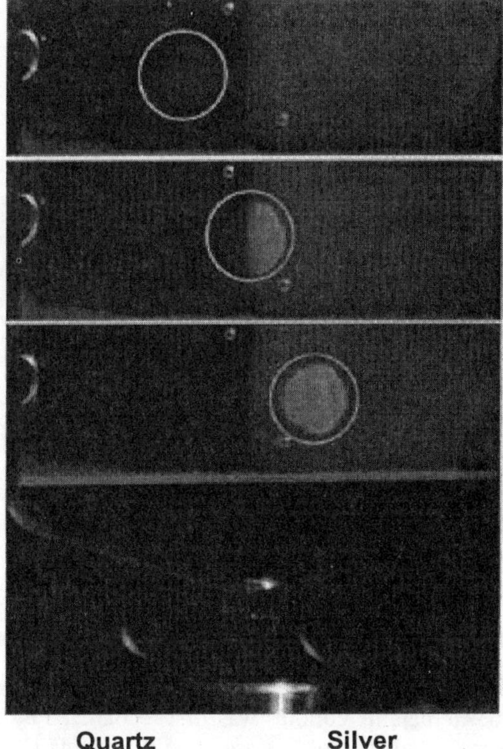


**Figure 20.** See Fig. 14.20 in color insert at the end of this volume. Photograph of fluorescein-labeled HSA (molar ratio of fluorescein/HSA =7) on quartz (left) and on Si Fs (right) as observed with 430-nm excitation and a 480-nm long-pass filter. The excitation was progressively moved from the quartz side to the silver, Top to Bottom, respectively. [41]

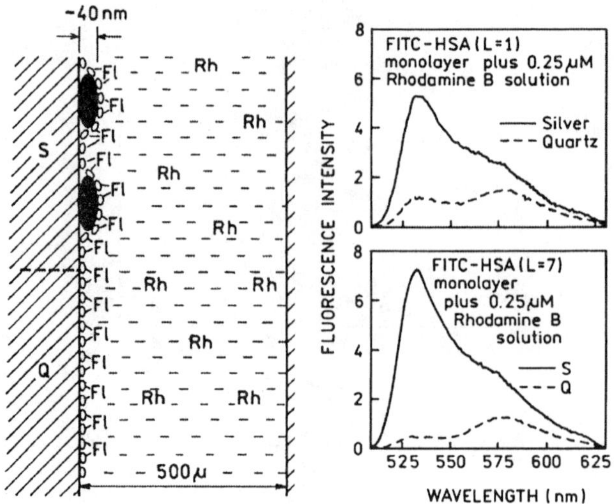


**Figure 21.**   Emission spectra of a monolayer of FITC–HSA L=1 (Right-Top) and L=7 (Right-Bottom) containing 0.25 µM Rhodamine B between the quartz plates (Q) or one SIF (S). Schematic of the sample with bound fluorescein and free Rhodamine B (Left). Adapted from reference 41.

It is informative to consider how silver-enhanced fluorescence, particularly of a heavily labeled sample, can be used for improved assays. Figure 21 shows emission spectra of a quartz plate coated with FITC–HSA to which we adjusted the concentration of Rhodamine B (0.25 µM) to result in an approximate 1.5-fold larger Rhodamine B intensity. One can consider the Rhodamine B to be simple auto-fluorescence or any other interference signal. When the same conditions are used for FITC–HSA on silver with L=1 the fluorescein emission is now two- to three-fold higher than that of Rhodamine B (Figure 21, Right-top). When using the heavily labeled sample (L=7) the fluorescein emission becomes more dominant (Figure 21, Right-bottom).

In conclusion, silver particles or colloids, when bound to a protein heavily labeled with fluorophores, can provide significantly higher intensities due to a decrease in the extent of self-quenching.

## 6. OTHER METAL NANOSTRUCTURES FOR MEF USING INDOCYANINE GREEN (ICG)

Metal colloids have been used for centuries to make some colored glasses.[42] The origin of the color as due to metallic colloids was first recognized by Faraday in 1857.[43] Typical absorption spectra of silver colloids are shown in Figure 22. The long wavelength absorption is called the surface plasmon absorption, which is due to electron oscillations[8,9] on the metal surface. These spectra can be calculated for the small particle limit ($r \ll \lambda$) from the complex dielectric constant of the metal.[8, 9] Larger particles display longer wavelength absorption. The absorption spectra are also dependent on the shape of the particles, with prolate spheroids displaying longer absorption wavelengths. Most studies of surface effects on fluorescence have been performed using silver particles to avoid the longer wavelength absorption of gold.

Several groups have considered the effects of metallic spheroids on the spectral properties of nearby fluorophores.[31, 32, 44-46] A typical model is shown in Figure 23 (Left), for a prolate spheroid with an aspect ratio of $a/b$. The particle is assumed to be a metallic ellipsoid with a fluorophore positioned near the particle. The fluorophore is located outside the particle at a distance $r$ from the center of the spheroid and a distance $d$ from the surface. The fluorophore is located on the major axis and can be oriented parallel or perpendicular to the metallic surface. The presence of a metallic particle can have dramatic effects on the radiative decay rate of a nearby fluorophore. Figure 23 (Right) shows the radiative rates expected for a fluorophore at various distances from the surface of a silver particle and for different orientations of the fluorophore transition moment. The most remarkable effect is for a fluorophore perpendicular to the surface of a spheroid with $a/b = 1.75$. In this case the radiative rate can be enhanced by a factor of 1000-fold or greater. The effect is much smaller for a sphere ($a/b = 1.0$), and much smaller for a more elongated spheroid ($a/b = 3.0$) when the optical transition is not in resonance.

In our work we utilized the long-wavelength dye indocyanine green (ICG) (Figure 24), which is widely used in a variety of *in vivo* medical applications.[47, 51-54] ICG is not toxic and is approved by the U.S. Food and Drug Administration for use in humans, typically by injection. ICG displays a low quantum yield in solution, ~ 0.016,[55] and a somewhat higher quantum yield when bound to serum albumin.[56-58] Albumin adsorbs to form a monolayer and ICG spontaneously binds to albumin. ICG is chemically and photochemically unstable, and thus provided us with an ideal opportunity to test deposited silver for both metal-enhanced emission and increased ICG photochemical stability.

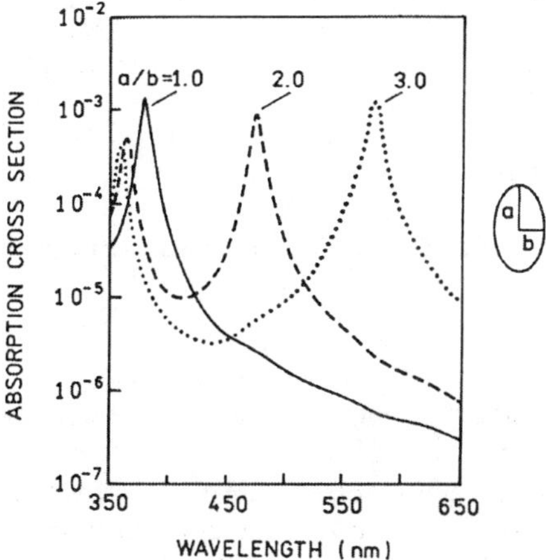


**Figure 22.** Calculated absorption cross-section of silver nanostructures. Adapted from reference 5.

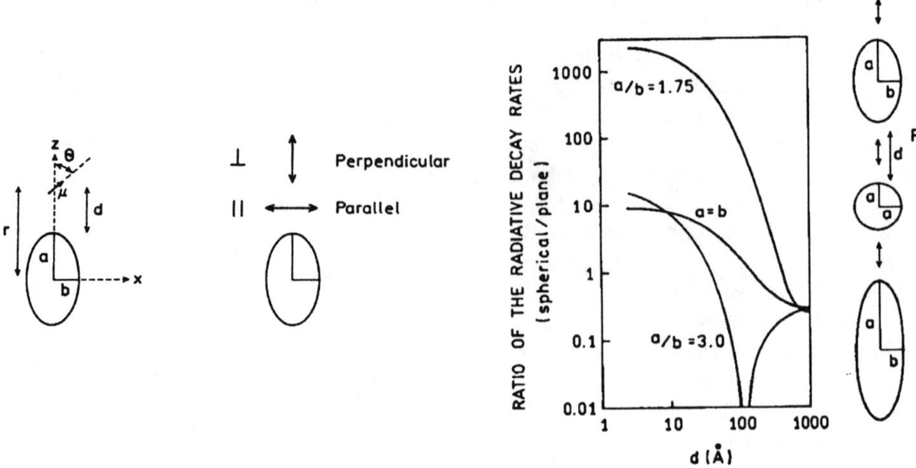



**Figure 23.** Effect of a metal spheroid on the radiative decay rate Adapted from reference 5.

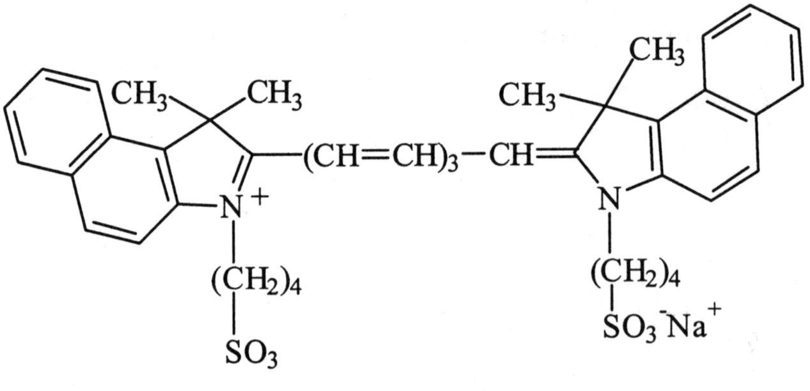

**Figure 24.** Chemical structure of indocyanine green

## 6.1. Silver Island Films (SiFs)

In recent years we have been reporting our observations on the favorable effects of silver nanoparticles deposited randomly on glass substrates (silver island films) for increasing the intensities and photostability of fluorophores, particularly those with low quantum yields. These reports described fluorophores with visible excitation and emission wavelengths. Since the fluorophores interact with metal through the surface plasmon resonance, for silver the absorption maximum is near 430 nm, we did not know if silver particles would enhance the emission of ICG with absorption and emission maxima of 795 and 810 nm, respectively. In this regard, we have investigated the effects of SiFs on the properties of ICG, which is bound to HSA.

The emission spectra of ICG-HSA bound to quartz and SiFs are shown in Figure 25. The intensity of ICG is increased approximately 10-fold on the SiFs as compared with the quartz. The emission spectrum is similar both on quartz and SiFs. We found the same amount of increase in the emission of ICG whether the surfaces were coated with HSA, which already contained bound ICG, or if the surfaces were first coated with HSA followed by exposure to a dilute solution of ICG. From on-going studies of albumin-coated surfaces, we estimated that the same amount of HSA binds to each surface, with the difference in binding being less than a factor of two. Hence the observed increase in the intensity on SiFs is not due to the increased ICG-HSA binding but rather to a change in the quantum yield and rate of excitation of ICG near particles.

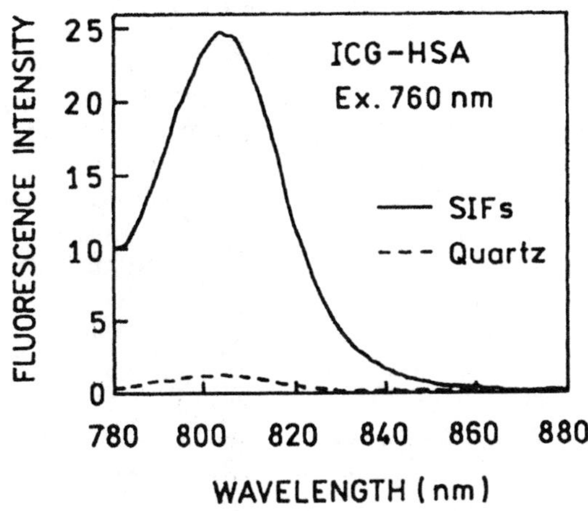


**Figure 25.** Fluorescence spectra of ICG-HSA on SiFs and on quartz. Adapted from reference 73.

## 6.2. Immobilized Silver Colloids

As shown in the previous section, SiFs (random shapes of silver) increase the emission intensity of ICG bound to HSA in agreement with the theoretical predictions that elongated silver particles should have significantly larger effects on fluorescence than for spheres.[31] Silver island films are formed by chemical reduction of silver with direct deposition onto a glass substrate, which is difficult to control. In contrast, preparation of colloidal suspensions of silver is rather standard and easily controlled to yield homogeneously sized spherical silver particles. An advantage of a colloidal suspension is the possibility of injection for medical imaging. Thus, we also investigated the effects of spherical silver particles on the emissions of ICG.

The sample geometry was similar to that of SiFs, except the spherical silver particles were prepared separately and were immobilized to an APS-coated glass slides by immersing the glass in a solution of the particles (Figure 26-Top). Figure 26 shows an absorption spectrum typical of our-colloid coated APS glass slides. The absorption peak centered near 430 nm is typical of colloidal silver particles with sub-wavelength dimensions but not completely at the small particle limit. An AFM image of silver colloid coated glass slides shows that the size of the silver colloids was smaller than 50 nm with partly aggregated sections. The surfaces were incubated with ICG-HSA to obtain a monolayer surface coating. The emission spectra showed a ˜ 30-fold larger intensity on the surfaces coated with silver colloids (Figure 27). We also measured the lifetimes of ICG on both surfaces and observed a significant reduction on the lifetimes on the silver colloids (data not shown) providing additional evidence that the increases in intensity is in fact due to modification of the radiative decay rate, $\Gamma_m$, by silver colloids.

## 6.3. Photo-Deposition of Silver onto Glass

It would be useful to obtain metal-enhanced fluorescence (MEF) at desired locations in the measurement device for use in medical and biotechnology applications, such as diagnostic or micro-fluidic devices. While a variety of methods could be used, we reasoned that the light-directed deposition of silver would be widely applicable. In recent years, a number of laboratories have reported light-induced reduction of silver salts to metallic silver.[68-71] Typically, a solution of silver nitrate is used that contains a mild potential reducing agent such as a surfactant[70] or dimethylformamide.[71] Exposure of such solutions to ambient or laser light typically results in the formation of silver colloids in suspension or on the glass surfaces. These results suggested the use of light-induced silver deposition for locally enhanced fluorescence.

In a typical preparation of light-induced deposition of silver on glass slides; the silver-colloid-forming solution was prepared by adding 4 mL of 1% trisodium citrate solution to a warmed 200 mL $10^{-3}$ M $AgNO_3$ solution. This warmed solution already contains some silver colloids as seen from a surface plasmon absorption optical density near 0.1. A 180 mL aliquot of this solution was syringed between the glass microscope slide and the plastic cover slip, which created a micro-sample chamber 0.5 mm thick (Figure 28). For all experiments a constant volume of 180 mL was used. Irradiation of the sample chamber was undertaken using a HeCd laser, with a power of ~8 mW, which was collimated and defocused using a 10X microscope objective, numerical aperture (NA) 0.40, to provide illumination over a 0.5 mm diameter spot.[48]

We examined the emission spectrum of ICG–HSA when bound to illuminated or non-illuminated regions of the APS treated slides. For APS treated slides (Figure 30) the intensity of ICG was increased about 7-fold in the regions with laser-deposited silver.

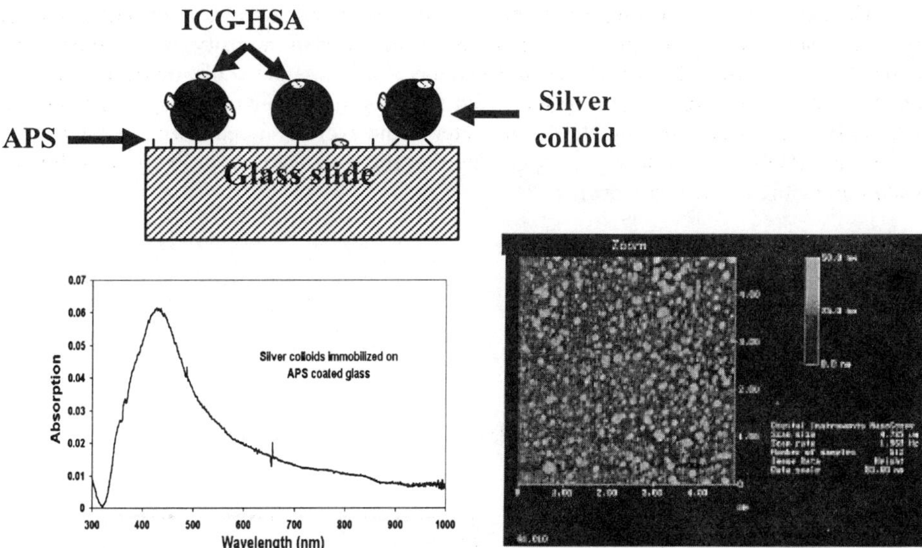



**Figure 26.** Glass surface geometry (Top); APS is used to functionalize the surface of the glass with amine groups which readily bind silver colloids, absorption spectrum of silver colloids on APS-coated glass (Bottom-Left), and AFM image of a silver colloid coated glass (Bottom-Right). APS; aminopropylethoxy silane.

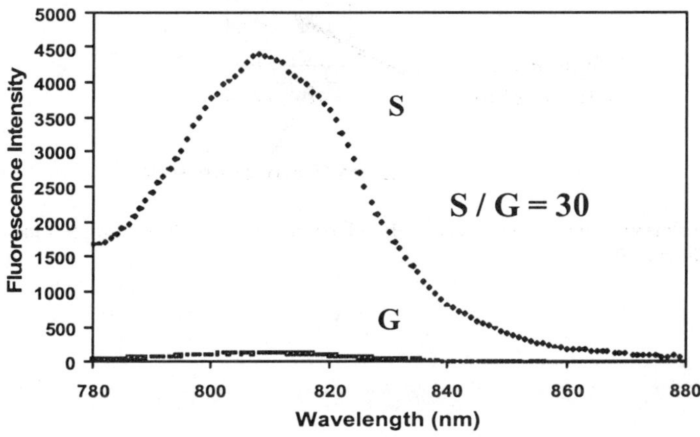



**Figure 27.** Fluorescence spectra of ICG-HSA on silver colloids and on quartz. S = Silver, G = Glass/Quartz.

The extent of the ICG enhancement was variable from spot to spot, or within a single spot, with some regions displaying much greater increases in intensity, which appeared to depend on the optical density of the spot (Figure 31). It was generally observed, within a single spot, that the ICG intensity increased towards the center of the spot, although no qualitative data is available due to the diameter of the excitation beam with respect to that of the spot itself. We do, however, speculate that the enhancement is likely to follow the Gaussian nature of the beam profile.

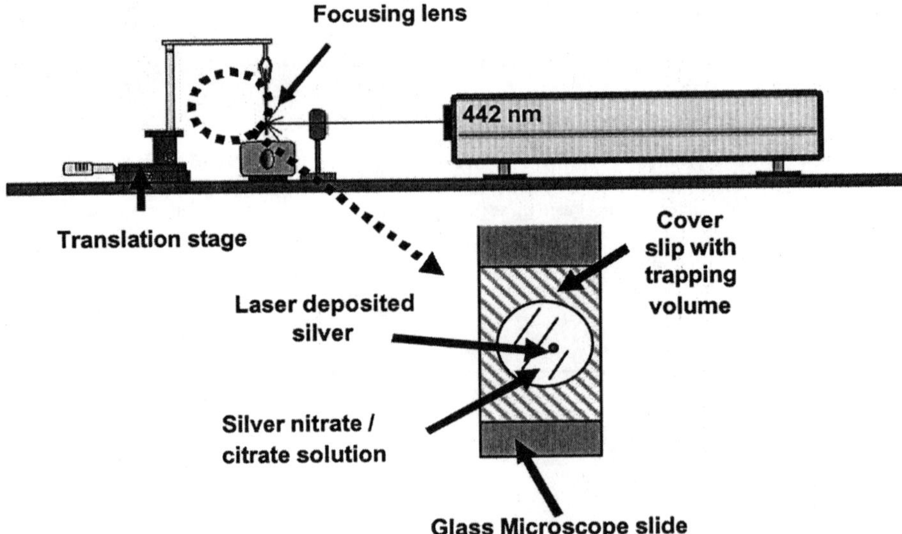


**Figure 28.** Experimental setup for light-induced deposition of silver on APS-coated glass microscope slides. Adapted from reference 48.

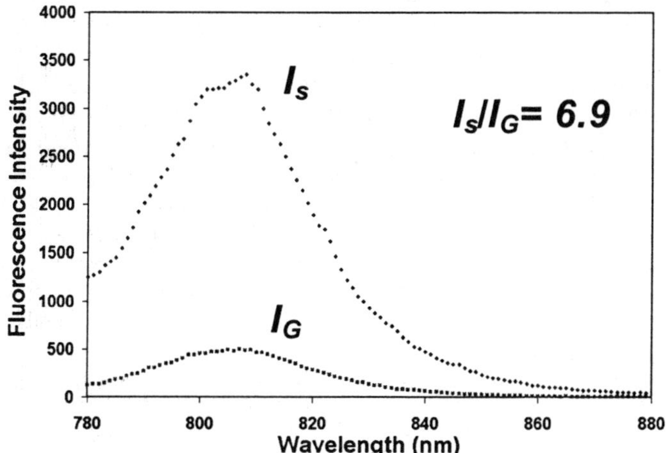


**Figure 30.** Fluorescence spectra of ICG-HSA on glass and on light-deposited silver. $I_S$-intensity on silver, $I_G$-intensity on glass.

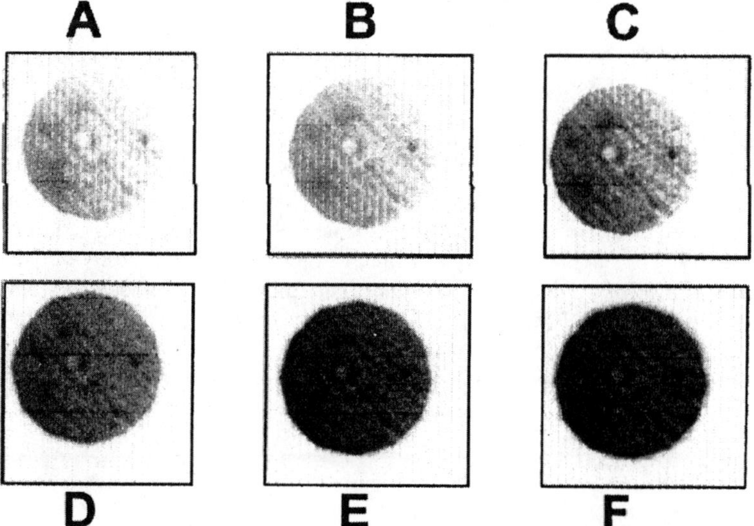


**Figure 31.** Image of silver spots, produced by laser deposition, using transmitted light illumination. The diameter of a spot is typically 50 mm. The irradiance was 560 W/cm$^2$ with a 40X, NA 1.2 objective. Images (A–F) were taken 5 s apart. Adapted from reference 48.

We examined the photostability of ICG–HSA when bound to glass or laser-deposited silver. We reasoned that ICG molecules with shortened lifetimes should be more photostable because there is less time for photochemical processes to occur. The intensity of ICG–HSA was recorded with continuous illumination at 760 nm. When excited with the same incident power, the fluorescence intensities, when considered on the same intensity scale, decreased somewhat more rapidly on the silver (Figure 32, top). However, the difference is minor. Since the observable intensity of the ICG molecules prior to photobleaching is given by the area under these curves, it is evident that at least 10-fold more signal can be observed from ICG near silver as compared to glass. Alternatively, one can consider the photostability of ICG when the incident intensity is adjusted to result in the same signal intensities on silver and glass. In this case (Figure 32, bottom) photobleaching is slower on the silver surfaces. The fact that the photobleaching is not accelerated for ICG or silver indicates that the increased intensities on silver *are not due* to an increased rate of excitation.

We investigated the use of other possible methods and substrates for producing localized silver surfaces for the applications of metal-enhanced fluorescence. One of our methods of silver deposition was to pass a controlled current between two electrodes in pure water (Figure 33, top). The two silver electrodes were mounted in a quartz cuvette containing deionized (Millipore) water. The silver electrodes had dimensions of 9 x 35 x 0.1 mm separated by a distance of 10 mm. For the production of silver colloids, a simple constant current generator circuit (60 μA) was constructed and used. After 30 min of current flow, a clear glass microscope slide was positioned within the cuvette (no chemical glass surface modifications) and was illuminated (HeCd, 442 nm). We observed silver deposition on the glass microscope slide, the amount depending on the illumination time. Simultaneous electrolysis and 442-nm laser illumination resulted in the deposition of metallic silver in the targeted illuminated region, 5-mm-focused spot size (Figure 33, Top). Absorption spectrum of the deposited silver is shown in Figure 33 (Bottom). A single absorption band is present on glass indicating that the silver particles are somewhat spherical. We examined ICG-HSA when coated on glass (G) or silver particles (S). The emission intensity was increased about 18-fold on the silver particles (Figure 34).

We examined the photostability of ICG-HSA when near silver particles on glass. We found a dramatic increase in the photostability near the silver particles (Figure 35). This very encouraging result indicates that a much higher signal can be obtained from each fluorophore prior to photodestruction and that more photons can be obtained per fluorophore before the ICG on silver eventually degrades. Our photostability data presented here are very encouraging and suggest the use of metal-enhanced fluorescence in fluorescence surface assays and lab-on-a-chip-type technologies, which are inherently prone to fluorophore instability and inadequate fluorescence signal intensity.

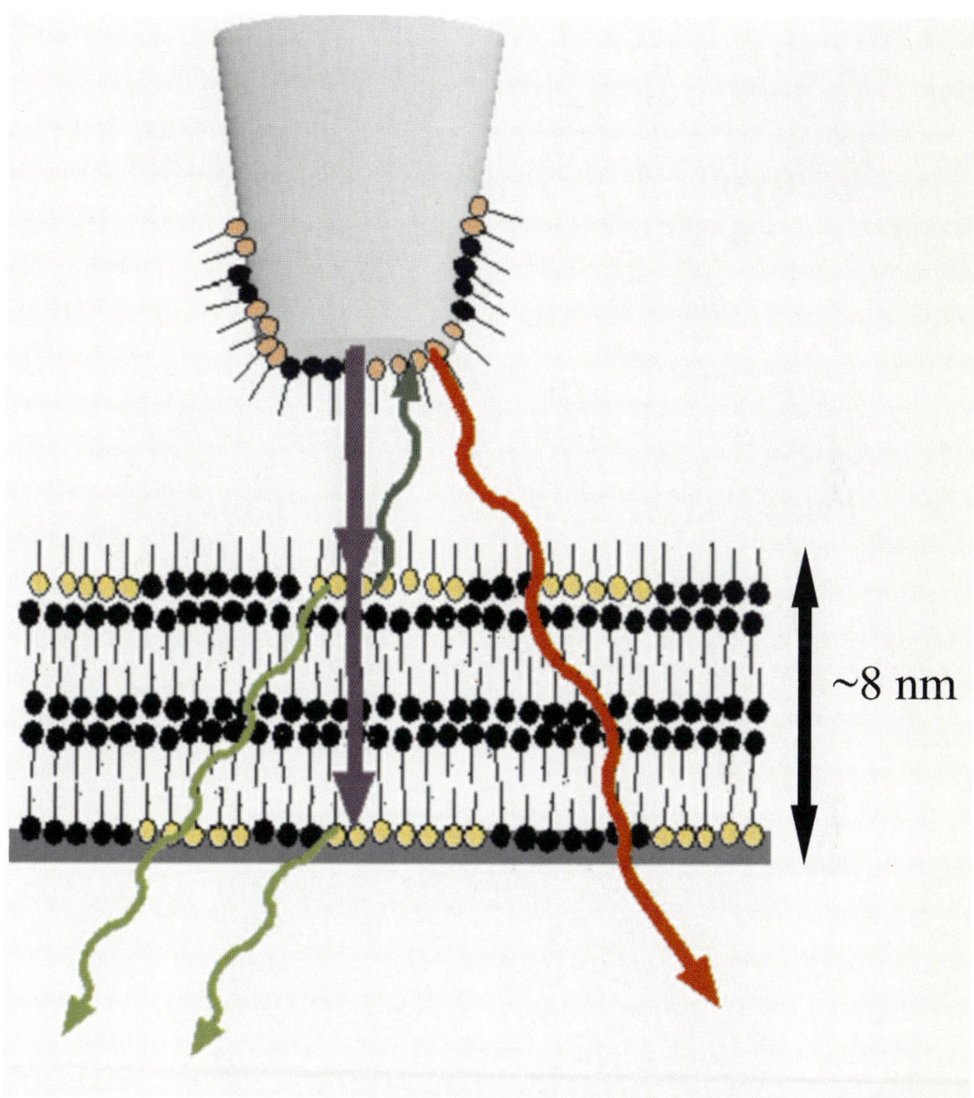

**Figure 2.2.** The above illustration represents the configuration of tip and sample used in FRET/NSOM. The acceptor dye of the FRET pair is rhodamine. Rhodamine was incorporated into a DPPC monolayer at 0.5 mol % and used to coat an NSOM probe lacking any metal. Fluorescein was used as the donor dye in the sample, and was incorporated into two DPPC/0.5 mol % fluorescein layers separated by a spacer of three arachidic acid layers. Light exiting the tip, (*blue arrow*) excites the donor dye in the sample but does not directly excite the acceptor dye on the tip. When the modified NSOM probe is near the sample, however, energy transfer from the excited donor in the monolayer to the rhodamine acceptor (*dark green arrow*) on the tip leads to a new emission, shifted to the red (*red arrow*) of the donor emission (*light green arrows*). By monitoring rhodamine fluorescence, it is possible to optically probe only those structures within nanometers of the NSOM tip. Reproduced with permission from (Vickery et al. 1999) . Copyright 1999 Biophysical Society.

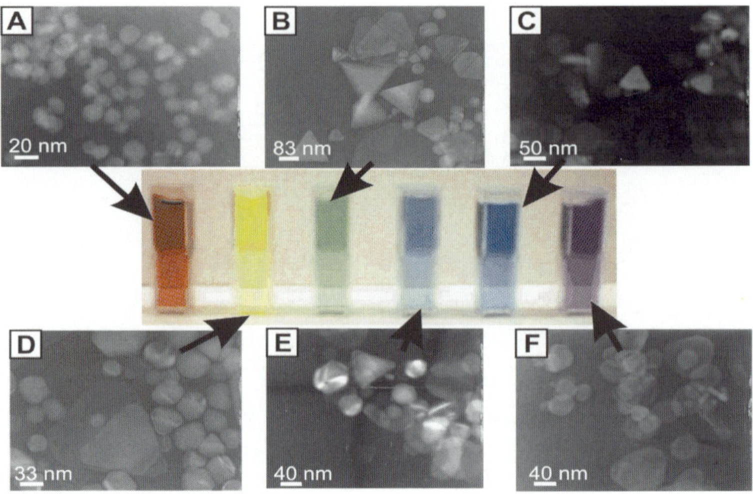

**Figure 3.3.** Solutions of nanoparticle dyes with corresponding transmission electron micrographs. (A) The red solution is made up of homogeneous Au nanospheres with D=13 nm. (B) The yellow solution is made up of inhomogeneous Ag nanoparticles: 15% trigonal prisms and 85% polygon platelets. (C) The green solution is made up of 50% trigonal prisms and 50% polygon platelets. (D) The light blue solution is made up of Ag nanoparticles including 30% trigonal prisms and 70% polygon platelets. (E) The dark blue solution is made up of 25% trigonal prisms with rounded tips and 75% polygon platelets. (F) The purple solution is made up of inhomogeneous oblong Ag nanoparticles.

**Figure 3.4.** A dark-field optical image of a field of Ag nanoparticles. The field of view is approximately 130 µm x 170 µm. The nanoparticles were fabricated by citrate reduction of silver ions in aqueous solution and drop-coated onto a glass coverslip.

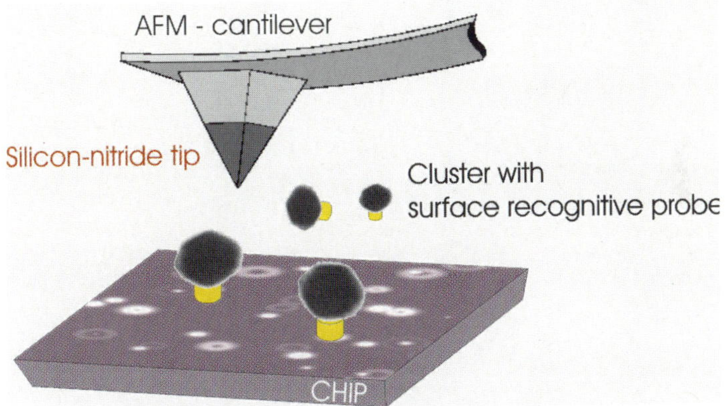


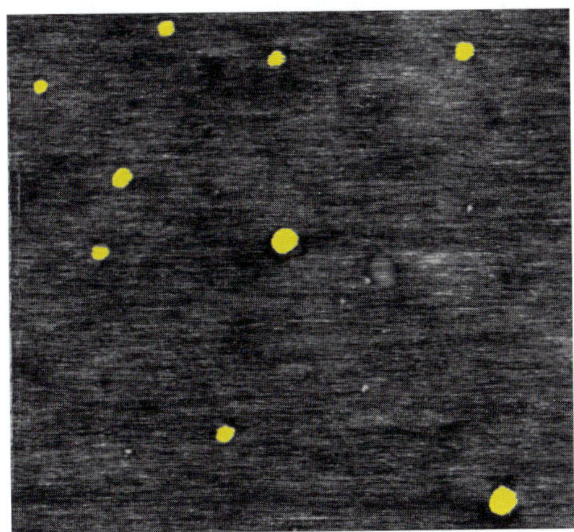

**Figure 5.14.** Bioassay via cluster counting with AFM.

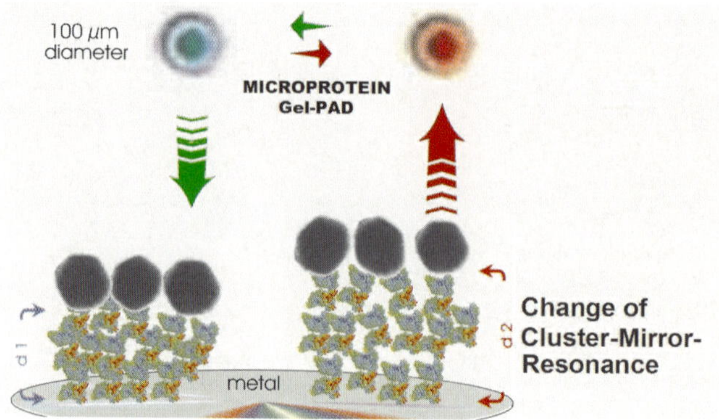



**Figure 5.23**. SEA response of a thin protein nano-pad (with a cluster top layer).

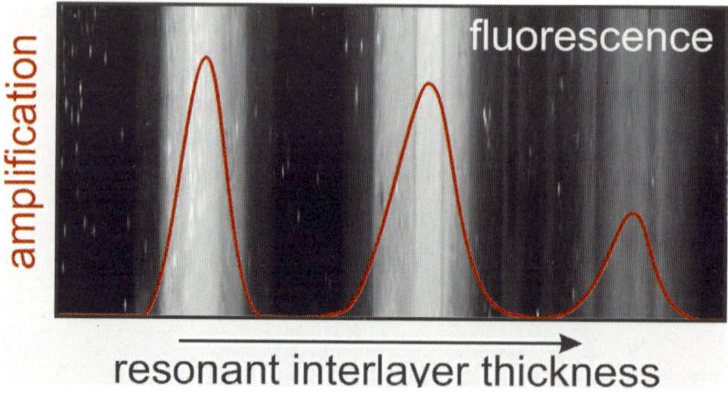

**Figure 5.26**. Fluorescence amplification in a Ag-cluster-layer – SnNx-interlayer system (steps from 0 – 700 nm) , photo (fewer steps than scan above).

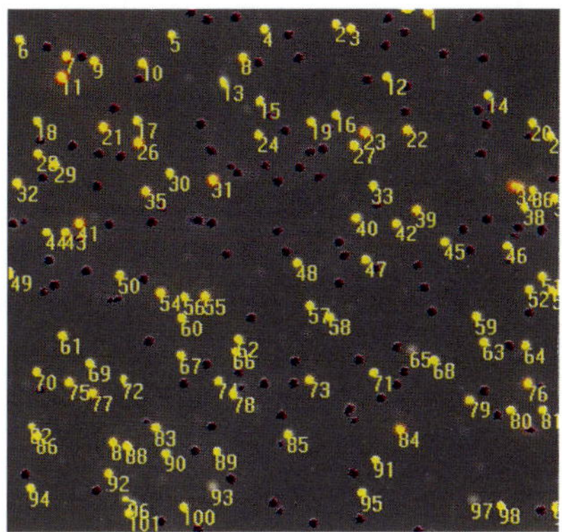

**Figure 11.12.** (Left) A typical image of a 50:50 mix of silver (blue) and gold (yellow-red) PRPs. (Right) Same image after applying automated PRP identification and counting algorithms. The identified silver particles are colored blue and the identified gold particles are colored yellow and numbered.

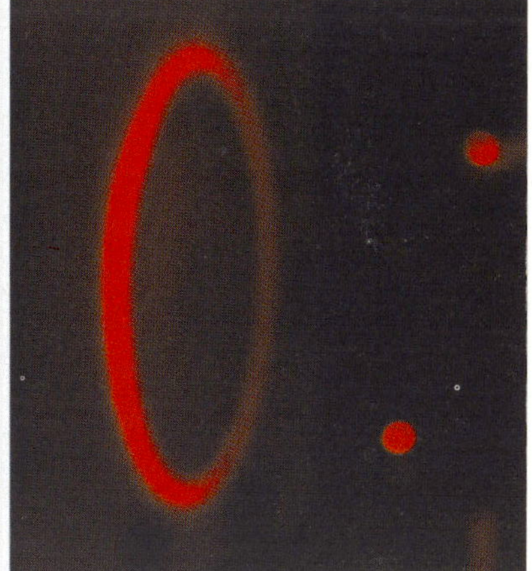

**Figure 13.17.** Cone of emission for S101 in PVA observed with a hemispherical prism and RK excitation. The emission was incident on tracing paper and photographed through a LWP 550 filter, without a notch filter (adopted from [30]).

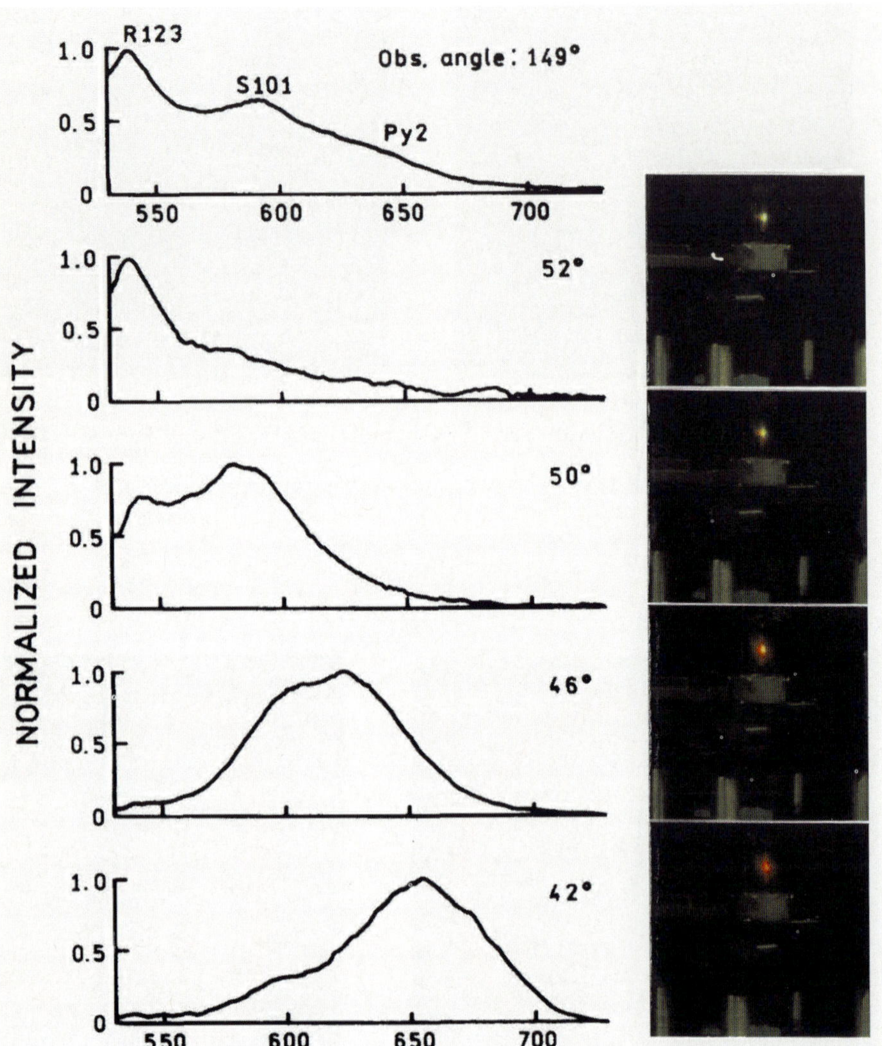



**Figure 13.19.** Emission spectra and photographs of SPCE for a three fluorophore mixture observed at different angles from the normal axis using the hemi-cylindrical prism. All concentrations of dyes in the evaporated films were about 10mM (adopted from [30]).

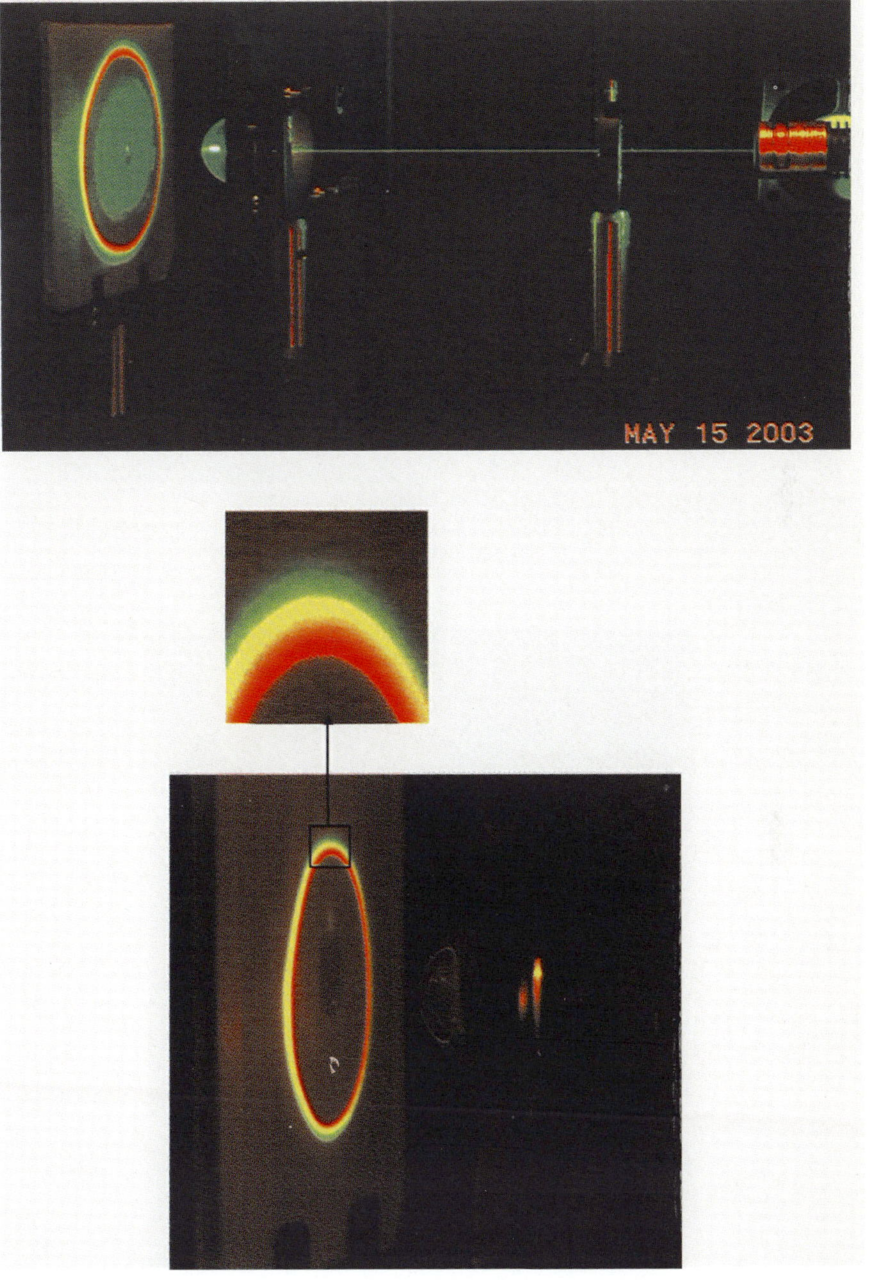

**Figure 13.20**. Photograph of SPCE from the mixture of fluorophores using RK excitation and a hemispherical prism, 532nm excitation. Top, no emission filter. Bottom, through a long pass filter but no notch filter (adopted from [30]).

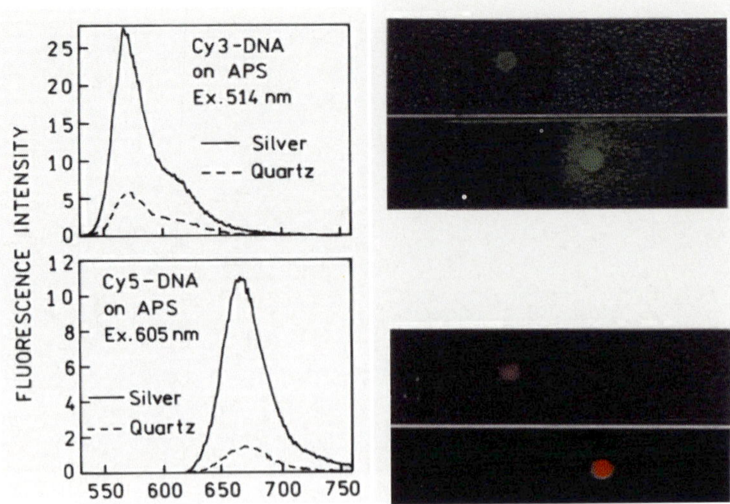



**Figure 14.9**. Emission spectra of Cy3-DNA (Top-Left) and Cy5-DNA (Bottom-Left) between quartz plates with and without SiFs. Photographs of corresponding fluorophores. Adapted from ref. 19.

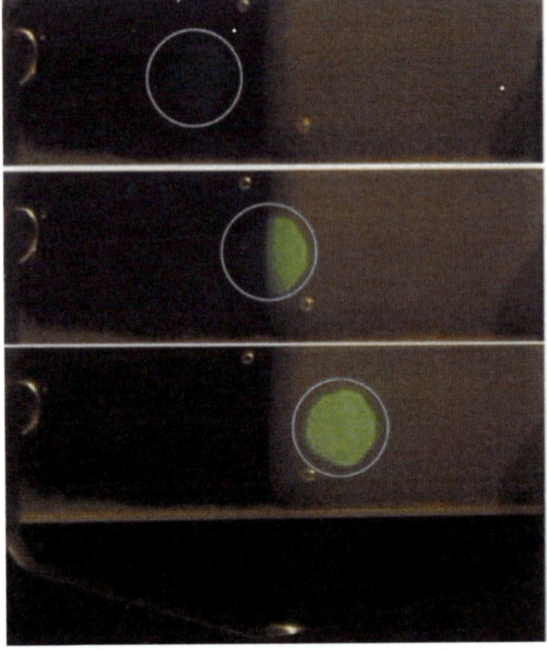

**Figure 14.20**. Photograph of fluorescein-labeled HSA (molar ratio of fluorescein/HSA =7) on quartz (left) and on SiFs (right) as observed with 430-nm excitation and a 480-nm long-pass filter. The excitation was progressively moved from the quartz side to the silver, Top to Bottom, respectively [41].

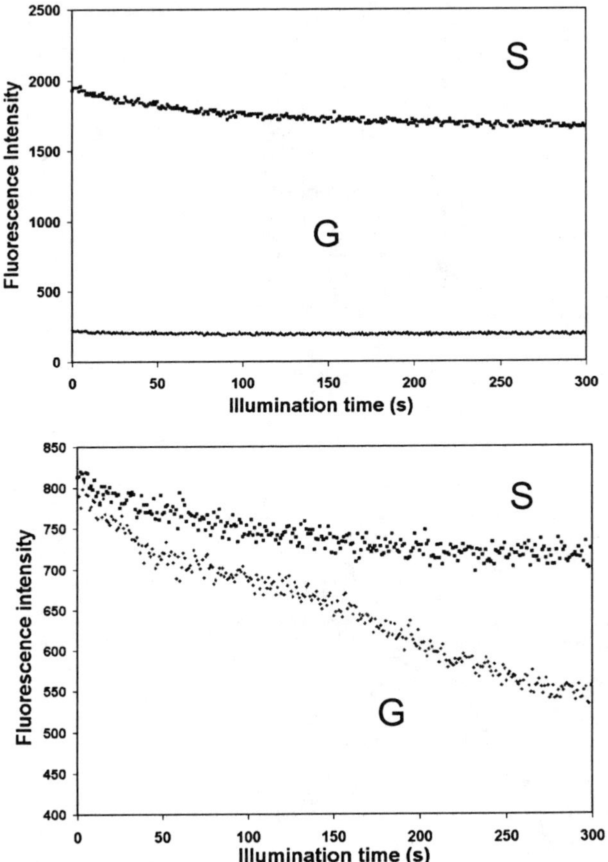


**Figure 32.** Photostability of ICG–HSA on (G) glass, and (S) laser deposited silver, measured with the same excitation power at 760 nm (Top), and with the laser power at 760 nm adjusted for the same initial fluorescence intensity (Bottom). Laser-deposited samples were made by focusing 442 nm laser light onto APS coated glass slides immersed in a AgNO₃ citrate solution for 15 min. The vzl OD of the sample was ~0.3.

                                                     

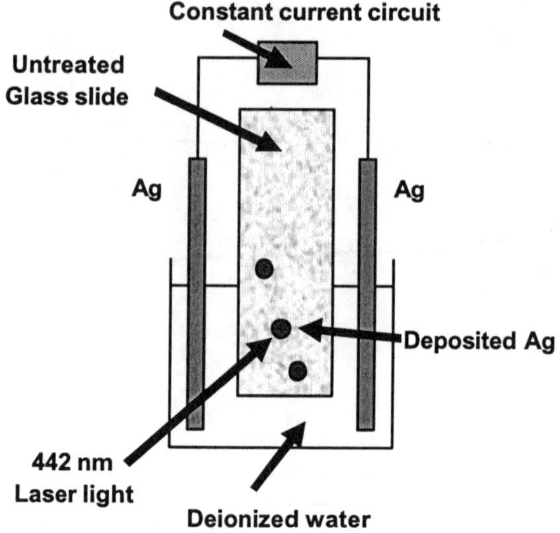


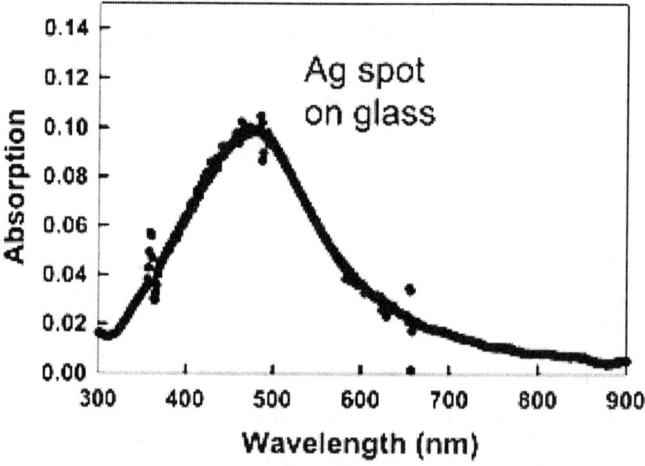


**Figure 33.** Light-deposited silver produced electrochemically. Constant current circuit (Top), absorbance spectrum of silver spot on glass (Bottom).

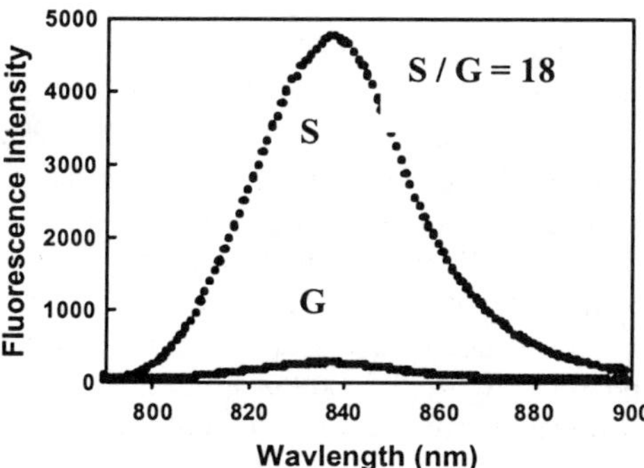



**Figure 34.** Fluorescence spectra of ICG-HSA on glass (G) and on light-deposited silver (S).

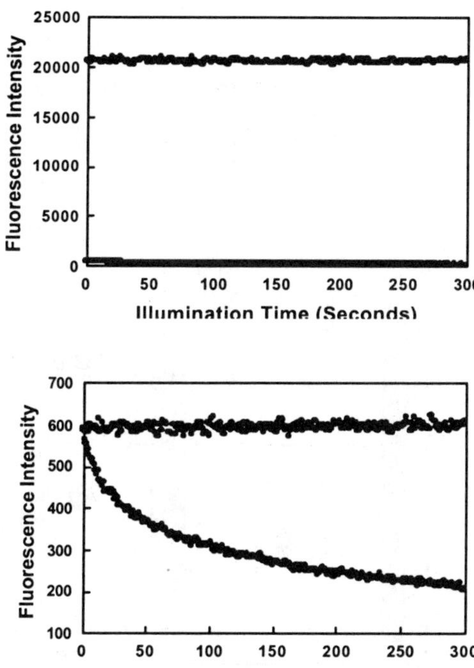



**Figure 35.** Photostability of ICG-HSA on glass and laser-deposited silver produced via electrolysis measured using the same excitation power at 760 nm (Top) and with power *adjusted* to give the same initial fluorescence intensities (Bottom). In all the measurements, vertically polarized excitation was used, while the fluorescence emission was observed at the magic angle, that is, 54.7°.

### 6.4. Electroplating of Silver on Substrates

This method is similar to the one employed in the previous section, except that the silver cathode electrode was replaced with an ITO-coated glass electrode (Figure 36). The current was again 60 μA. After a short period of time, silver readily deposited on the ITO surface (no laser illumination), the extent of which was again dependent on the exposure time. Absorption spectra of the deposited silver are shown in Figure 37 (Top). Two maxima were found on ITO, which eventually formed one large broad band. This suggests that the particles are elongated and display both transverse and longitudinal resonances (Figure 37, Top). Enhanced fluorescence emission from ICG-HSA was also found for silver particles on ITO (Figure 37, Bottom), but the enhancement was typically less and there appeared to be a small blue shift on silvered ITO.

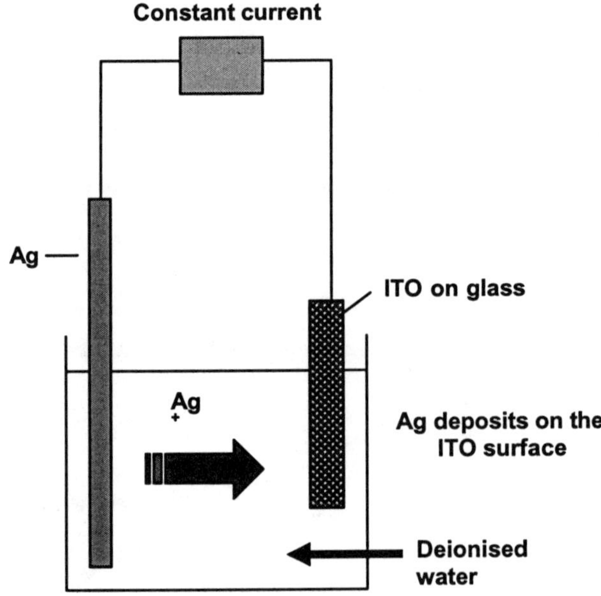


**Figure 36.** Electroplating of silver on substrates. Adapted from reference 49.

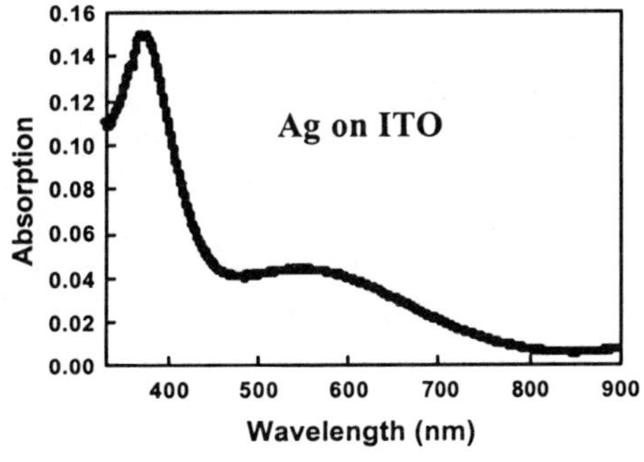


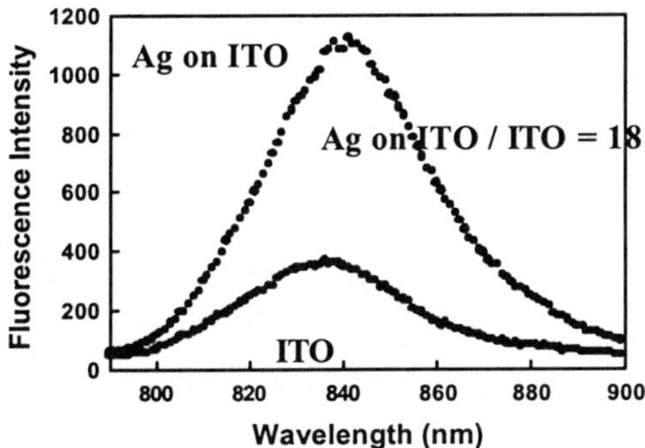


**Figure 37.** Absorption spectrum of electroplated silver on substrates (Top), fluorescence emission spectra of ICG-HSA on ITO and silver deposited on ITO (Bottom). Adapted from 48.

## 6.5. Roughened Silver Electrodes

In all of our experiments to date we have considered both the method of silver deposition and the resultant effects on metal-enhanced fluorescence (MEF). In these previous works it became apparent that for MEF to occur there is a need to localize the fluorophore in regions in close proximity to the silver surface. However, in many applications of fluorescence, it might be advantageous to have localized silver deposition for spatially selective analysis, such as lab-on-a-chip technologies. Localized silver colloid formation has been accomplished with reagents in laminar flow,[60] by nanolithography,[61-62] and by electroplating insulators.[64] Given the extensive use of roughened silver electrodes for surface-enhanced Raman scattering (SERS),[64,65] we investigated their use for spatially selective MEF due to the high surface areas of fractal-like structures.

Figure 38 shows the schematics of the method for preparing roughened electrodes. In a typical preparation, commercially available silver electrodes are placed in deionized water 10 mm apart. A constant current of 60 μA was supplied across the two electrodes for 10 minutes by a constant current generator. Figure 39 shows the time dependent growth of the silver nanostructures on the silver cathode. In comparison, the anode was relatively unperturbed.

Binding the ICG-HSA to both the silver anode and cathode after electrolysis was accomplished by soaking the electrodes in a solution of ICG-HSA overnight, followed by rinsing with water to remove the unbound material. As a control sample, an unused silver electrode was also coated with ICG-HSA. A roughened silver cathode was also dipped in $10^{-4}$ M NaCl for 1 hour before washing and then coated with ICG-HSA, so as to place our findings in context with the huge enhancements in Raman signals, typically obtained after chloride dipping the electrodes.[66-67]

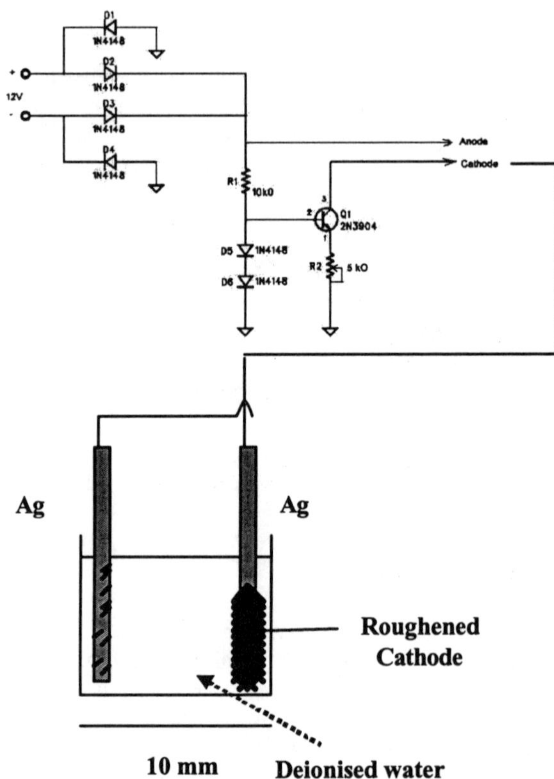


**Figure 3 8.** Experimental setup for the production of roughened silver electrodes. Adapted from reference 50.

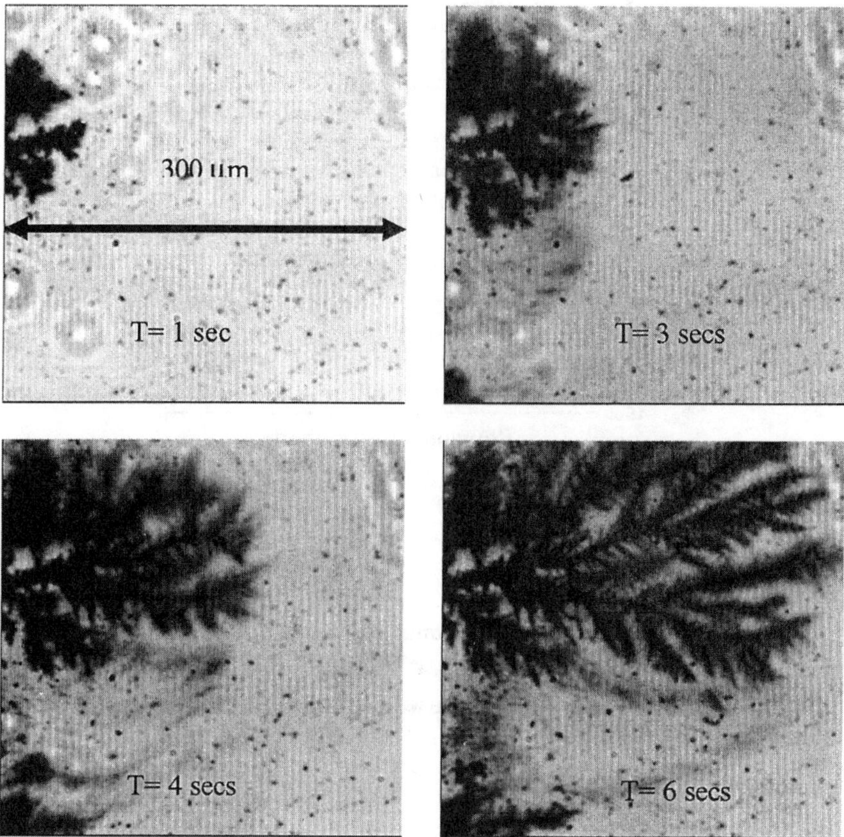



**Figure 39.** Fractal-like silver growth on the silver cathode as a function of time, visualized using transmitted light. This structure was characteristic of the whole electrode.

Three electrodes were coated with ICG-HSA and studied; the roughened cathode, the anode, and an unroughened electrode. Essentially no emission was seen from ICG-HSA on an unroughened, bright silver surface (Ag in Figure 40 (Top), the control). However, a dramatically larger signal was observed on the roughened cathode and a somewhat smaller signal was observed on the anode. In all our experiments we typically found that the roughened silver cathode was ~ 20–100 fold more fluorescent than the unroughened control Ag electrode. In comparison the Ag anode was 5–50 times more fluorescent than the Ag control. When we increased the time-for-roughening to over 1 h, the intensities of both electrodes after coating with ICG-HSA were essentially the same, but still 50-fold more fluorescent than the unroughened Ag control. The emission spectra on the two electrodes probably had the same emission maximum, where the slight shift seen in Figure 40 (Bottom) is thought to be due to the filters used to reject the scattered light. It should be noted that the amount of material coated on both surfaces was approximately the same, and the effect was not due to an increased surface area and therefore increased protein coverage on the roughened surface. The dramatic and favorable increase in fluorescence intensities shown in Figure 40 could have several explanations. Two possibilities include an increased rate of excitation, due to the enhanced incident field around the metal,[59] or increased amounts of protein bound to the fractal structure. For both eventualities the fluorescence lifetimes are expected to remain the same. Examining the intensity decay of ICG-HSA for the cathode we found that the lifetime was dramatically shortened to < 10 ps, (data not shown), in fact, so short that it was difficult to determine the absolute values with a system time-resolution of ~ 50 ps fwhm. However, a decreased lifetime with increased fluorescence intensity strongly supports an increase in the radiative decay rate, $\Gamma_m$ .

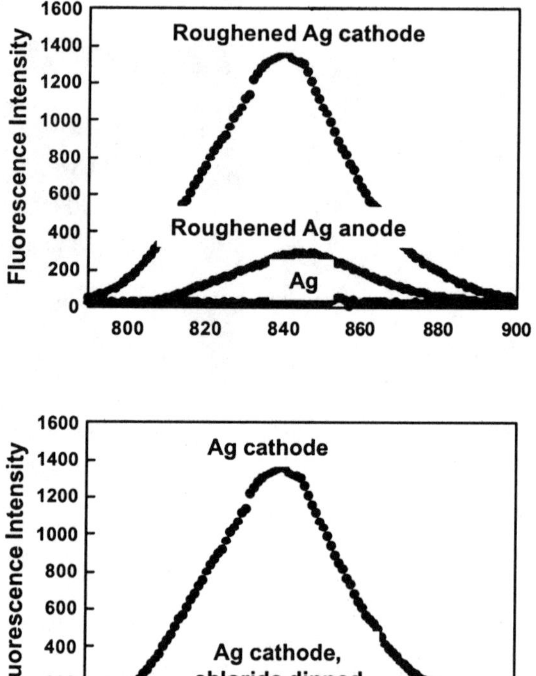


**Figure 40.** Fluorescence emission spectra of ICG-HSA on roughened silver electrodes (Top) and silver cathode (Bottom). Adapted from reference 50.

**Figure 41.** Silver fractal-like structures grown on silver electrodes.

Interestingly, Figure 41 shows and image of ICG-HSA coated silver electrodes revealing spatially-localized fluorescence *hot-spots*, which we believe are due to the areas demonstrating superior fluorescence enhancements. Indeed, this is likely to occur with most silver systems studied, and thus our measurements to date are typically spatially averaged over large excitation spot sizes.

## 6.6. Silver Fractal-like Structures on Glass

Fractal-like silver structures were also generated on glass using two silver electrodes held between two glass microscope slides, Figure 42 (Top). The electrodes were 10 x 35 x 01 mm, with about 20 mm between the two electrodes. Deionized water was placed between the slides. A direct current of 10 µA was passed between the electrodes for about 10 min, during which the voltage started near 5 V and decreased to 2 V. During the current flow, fractal silver structures grew on the cathode and then on the glass near the cathode (Figure 42), thus producing silver nanostructures on glass, compared to those on the silver electrodes as described in the previous section. Similarly to the silver electrodes, the structures grew rapidly but appeared to twist as they grew. These structures are similar to those reported recently during electroplating of insulators.[63] In addition we found that dipping the slides in 0.001 mg/dl $SnCl_2$ for 30 min, *before electrolysis*, resulted in structures that were firmly bound to the glass during working. Without the $SnCl_2$, similar structures were formed but were partially removed during washing. Following passage of the current, the silver structures on glass were soaked in 10 µM FITC-HSA overnight at 4°C, which is thought to result in a monolayer of surface bound HSA.

For the FITC-HSA-coated fractal-silver surfaces on glass, we were able to measure a fluorescence image very similar to that of the fractal silver surface alone (bright field image) using the same apparatus (Figures 43 and 44). Interestingly, regions of high and low fluorescence intensity were observed (Figure 44). This result is roughly consistent with recent SERS data, which showed the presence of intense signals that appeared to be located between clusters of particles.[66-67]

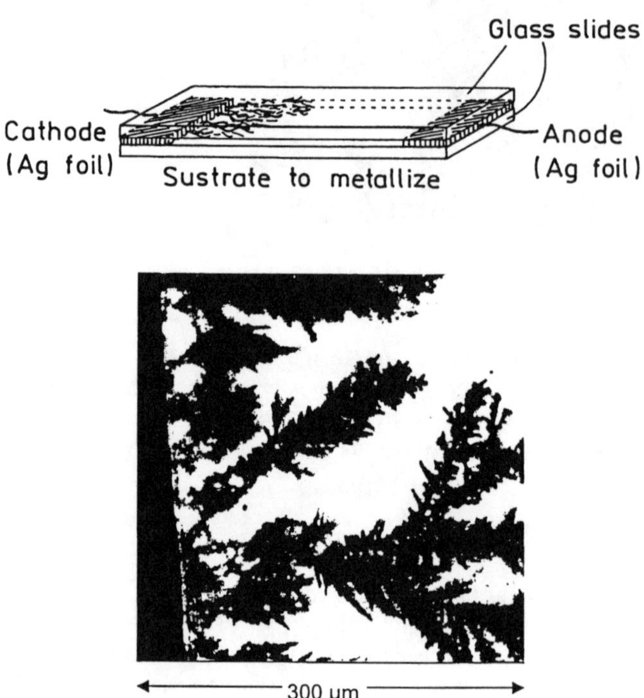


**Figure 42.** Configuration for creation of fractal-like silver surfaces on glass (Top) and bright-field image of fractal-like silver surfaces on glass.

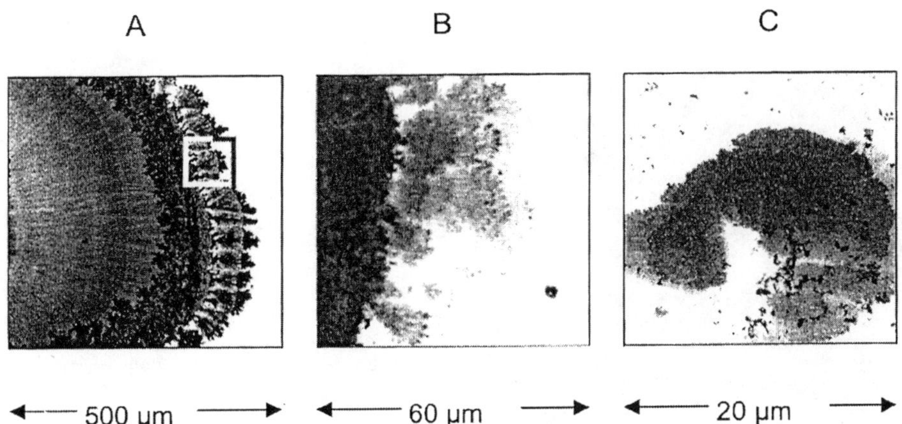



**Figure 43.** Silver nanostructures deposited on glass during electroplating (A). Panels B and C are consecutive magnification of the marked area on panel A. Bright-field image.

Emission spectra were collected from eight selected regions of varying brightness. In all cases, the emission spectra appeared to be that of fluorescein (Fig 44, Right), where the blue edge of the fluorescein emission is cut off by the emission filter. As a control the silver structures were coated with unlabeled HSA. The resulting signal was substantially lower than any of the silvered areas and lower than regions of the unsilvered glass treated with FITC-HSA. Similarly to the roughened silvered electrodes, we investigated the nature of the enhanced fluorescence intensities observed in Figure 44 (Right). If the radiative decay rate is increased, then the lifetime should decrease.[59] We measured the frequency-domain intensity decays of FITC-HSA bound to unsilvered glass and fractal silver on glass (data not shown). The amplitude-weighted lifetime of FITC-HSA bound to glass is ~80 ps, which is in agreement with previous measurements of self-quenched fluorescein on HSA.[41] On fractal silver, the amplitude weighted lifetime is dramatically reduced to about ~ 3 ps. We carefully considered whether this decrease was due to the detection of scattered light. The background signal from unlabeled HSA on fractal silver was less than 1%. The emission filter combination of a 540-nm interference filter and a solution of $CrO_4^{2-}$ / $Cr2O_7^{2-}$ were selected for low emission from the filter when exposed to scattered light from the sample. However, we do believe a small part of the increased intensity is due to the release of fluorescein self-quenching because of the silver surfaces, but this is estimated to be very small in comparison to the > 500-fold increases shown in Figure 44 (Right).

We also studied the photostability of FITC on the fractal silver surface, silver island films, and uncoated quartz. Although the relative photobleaching is higher on fractal silver, the increased rate of photobleaching is less than the increase in intensity (Figure 45). From the areas under these curves we estimate ˜ 16-fold and ˜ 160-fold more photons can be detected from the FITC-HSA on SiFs or fractal silver, respectively, relative to quartz, before photobleaching.

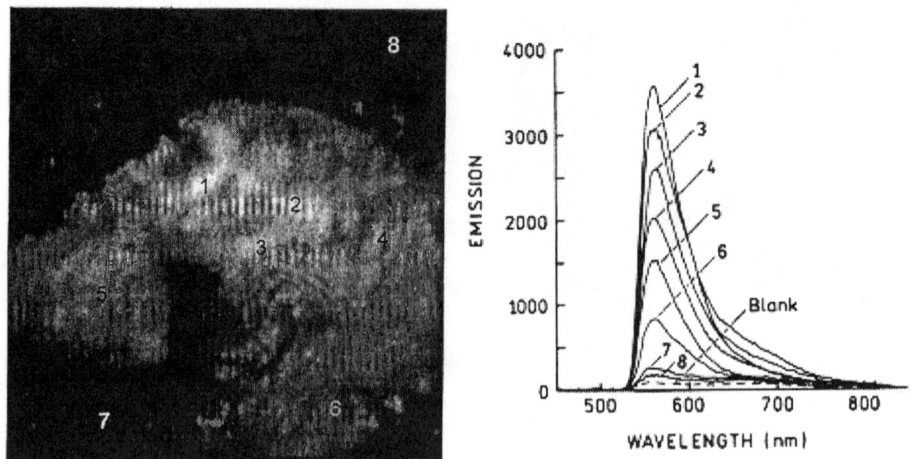

**Figure 44.** Fluorescence image of FITC-HSA deposited on the silver structure shown in Figure 43C, and the emission spectra of the numbered areas shown on the right. Adapted from reference 50.

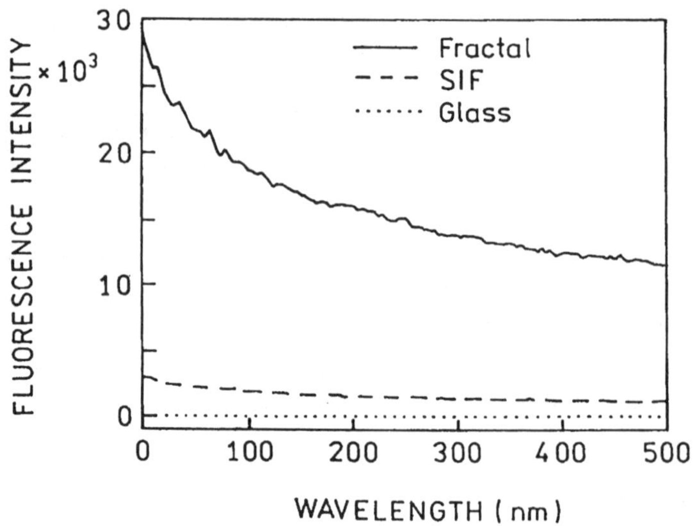

**Figure 45.** Photostability of FITC-HSA deposited surfaces. The samples were excited at 514 nm.

## 6.7. Silver Nanorods

As discussed in Section 6, elongated silver particles are predicted to enhance the emission of fluorescence due to the increased local excitation fields around the edges of the particles. Recently, we have developed a methodology for depositing silver nanorods with controlled sizes and loadings onto glass substrates. In this method, firstly, APS-coated substrates are coated with silver colloids with sizes less than 4 nm. Then, the silver seed-coated glass slides were immersed in a cationic surfactant (CTAB) solution for 5 minutes. One ml of 10 mM $AgNO_3$ and 2 ml of ascorbic acid were added. 0.4 ml of 1 M NaOH was immediately added and the solution was mixed gently to accelerate the growth process. The silver nanorods were formed on the glass slides within 10 minutes. In order to increase the loading of silver nanorods on the surface, the silver nanorods-coated glass substrates were immersed in CTAB again, and same amounts of $AgNO_3$, ascorbic acid and NaOH were added as in the first step. This process is repeated until the desired loading of silver nanorods on the glass slides was obtained. Binding of the ICG-HSA to the surfaces were accomplished by soaking the slides in a 30 µM ICG, 60 µM HSA solution overnight, followed by rinsing with deionized water to remove the unbound material.[72]

The absorption spectra of silver nanorods deposited on glass substrates are shown in Figure 47. Silver nanorods display two distinct surface plasmon peaks; transverse and longitudinal, which appear at ˜ 420 and ˜ 650 nm, respectively. In our experiments, the longitudinal surface plasmon peak shifted and increased in absorbance as more nanorods are deposited on the surface of the substrates. In parallel to these measurements, we have observed an increase in the size of the nanorods (by Atomic Force Microscopy, data not shown). In order to compare the extent of enhancement of fluorescence with respect to the extent of loading of silver nanorods deposited on the surface, we have arbitrarily chosen the value of absorbance at 650 nm as a means of loading of the nanorods on the surface. This is because the 650 nm is solely attributed to the longitudinal absorbance of the nanorods.

Figure 48 shows the fluorescence emission intensity of ICG-HSA measured from both glass and on silver nanorods, and the enhancement factor versus the loading density of silver nanorods. We have obtained up to 50-fold enhancement in emission of ICG on silver nanorods when compared to the emissions on glass. We also measured the lifetime of ICG on both glass and silver nanorods, and observed a significant reduction in the lifetime of ICG on silver nanorods, providing the evidence that an increased emissions is due to radiative rate modifications, $\Gamma_m$.

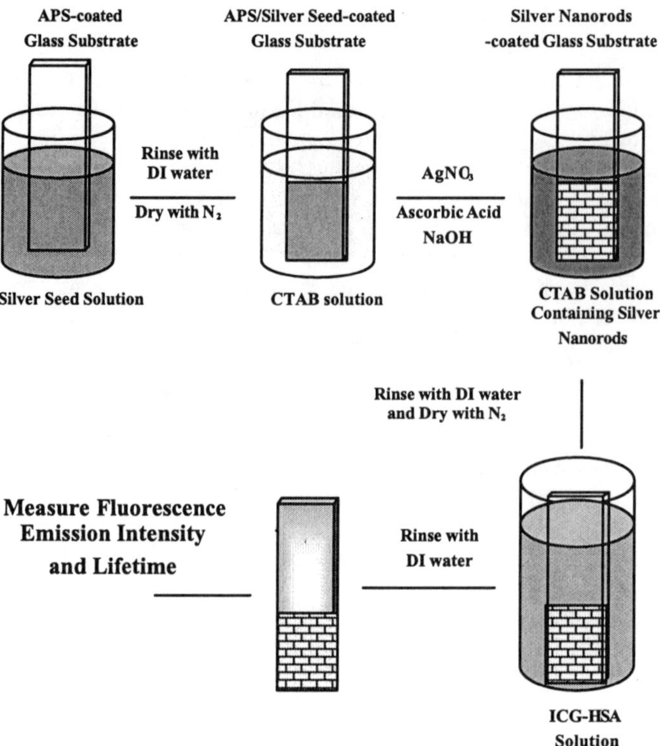



**Figure 46.** Rapid deposition of silver nanorods on glass substrates.

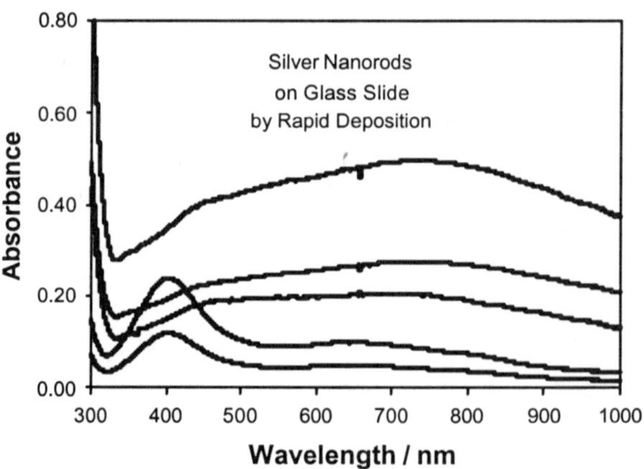



**Figure 47.** Absorption spectra of silver nanorods deposited on glass substrates.

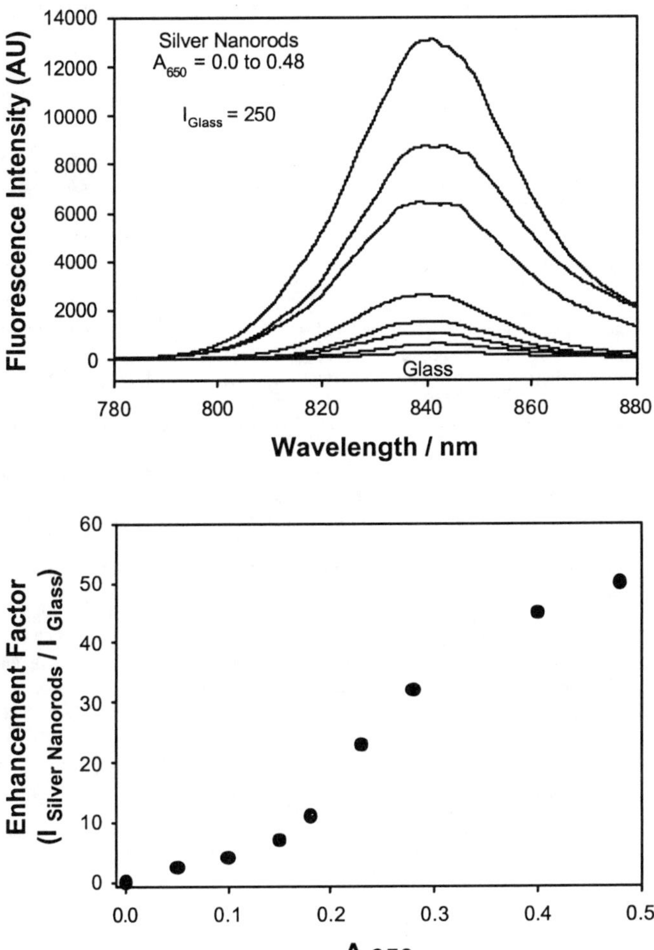


**Figure 48.** Emission spectra of ICG-HSA on silver nanorods (Top), enhancement factor versus the absorbance of silver nanorods at 650 nm (Bottom).

## 7. CLOSING REMARKS

In this review Chapter, we have shown the favorable effects of enhanced fluorescence emission intensity, (quantum yield), reduced lifetime (increased photostability) for fluorophores in close proximity to appropriately sized metallic silver nanostructures. Our findings have indeed revealed that $\gg$ 500-fold enhancement in emission intensity can be realized, which is likely to offer multifarious applications to medical diagnostics and the clinical sciences. Metal-fluorophore interactions are relatively unexplored phenomenon, but we believe will receive much academic and industrial interest in the future.


## 8. ACKNOWLEDGMENTS

Work was supported by the NIH, National Center for Research Resources, RR-01889.


## 9. REFERENCES

1. K. H. Drexhage, Interaction of light with monomolecular dye lasers. *In Progress in Optics,* edited by Wolfe, E. (North-Holland, Amsterdam, 1974), pp. 161–232.
2. R. M. Amos, and W. L. Barnes, Modification of the spontaneous emission rate of $Eu^{3+}$ ions close to a thin metal mirror, *Phys. Rev. B* **55**(11), 7249–7254 (1997).
3. W. L. Barnes, Fluorescence near interfaces: The role of photonic mode density, *J. Modern Optics* **45**(4), 661–699 (1998).
4. R. M. Amos, and W. L. Barnes, Modification of spontaneous emission lifetimes in the presence of corrugated metallic surfaces, *Phys. Rev. B* **59**(11), 7708–7714 (1999).
5. J. R. Lakowicz, Radiative decay engineering: Biophysical and biomedical applications, *Appl. Biochem.* **298**, 1–24 (2001).
6. D. A. Weitz, S. Garoff, C. D. Hanson, and T. J. Gramila, Fluorescent lifetimes of molecules on silver-island films, *Optics Letts.* **7**(2), 89–91 (1982)
7. S. Link, and M. A. El-Sayed, Spectral properties and relaxation dynamics of surface plasmon electronic oscillations in gold and silver nanodots and nanorods, *J. Phys. Chem. B* **103**, 8410–8426 (1999).
8. S. Link, and M. A. El-Sayed, Shape and size dependence of radiative, nonradiative and photothermal properties of gold nanocrystals, *Int. Rev. Phys. Chem.* **19**, 409–453 (2000).
9. U. Kreibig, M. Vollmer, and J. P. Toennies, *Optical Properties of Metal Clusters* (Springer-Verlag, Berlin, 1995).
10. J. R. Lakowicz *Principles of Fluorescence Spectroscopy* (New York: Kluwer/Plenum, 1999), p. 698
11. S. Georghiou, Nordlund, M. Thomas and A. M. Saim, Picosecond fluorescence decay time measurements of nucleic acids at room temperature in aqueous solution *Photochem. Photobiol.* **41**, 209–12 (1985)
12. S. Georghiou, T. D. Braddick, A. Philippetis and J. M. Beechem Large-amplitude picosecond anisotropy decay of the intrinsic fluorescence of double-stranded DNA *Biophys. J.* **70**, 1909–22 (1996)
13. R. F. Steiner and Y. Kubota, Fluorescent dye-nucleic acid complexes *Excited States of Biopolymers* ed R F Steiner (New York: Plenum, 1983), pp. 203–54.
14. S. Georghiou, Interaction of acridine drugs with DNA and nucleotides *Photochem. Photobiol.* **26**, 59–68 (1977).
15. C. J. Murphy, Photophysical probes of DNA sequence directed structure and dynamics *Advances in Photochemistry* (New York: Wiley, 2001), pp. 145–217.
16. I. Timtcheva, V. Maximova, T. Deligeorgiev, N. Gadjev, K. H. Drexhage and I. Petkova, Homodimeric monomethine cyanine dyes as fluorescent probes of biopolymers *J. Photochem. Photobiol. B: Biol.* **58**, 130–5 (2000).
17. J. R. Lakowicz, Y. Shen, S. D'Auria, J. Malicka, J. Fang, Z. Grcyzynski and I. Gryczynski, Radiative decay engineering 2. Effects of silver island films on fluorescence intensity, lifetimes, and resonance energy transfer, *Anal. Biochem.* **301**. 261–77 (2002).
18. J. R. Lakowicz, B. Shen, Z. Gryczynski, S. D'Auria and I., Gryczynski, Intrinsic fluorescence from DNA can be enhanced by metallic particles *Biochem Biophys. Res. Commun.* **286**, 875–879 (2001).
19. J. R. Lakowicz, J. Malicka and I. Gryczynski, I., Silver particles enhance emission of fluorescent DNA oligomers, *BioTechniques.* **34**, 62–68 (2001).
20. The human genome. *Nature* February 15, 2001, pp. 813–958.
21. The human genome. *Science* February 16, 2001, pp. 1177–1351.
22. J. Enderlein, D. L. Robbins, W. P. Ambrose, and R. A. Keller, Molecular shot noise, burst size distribution, and singlemolecule detection in fluid flow: Effects of multiple occupancy *J. Phys. Chem. A* **102**, 6089–6094 (1998)

23. A. Van Orden, N. P. Machara, P. M. Goodwin, and R. A. Keller, Single-molecule identification in flowing sample streams by fluorescence burst size and      intraburst fluorescence decay rate, *Anal. Chem.* **70**(7), 1444–1451 (1998)

24. J. V. Garcia-Ramos, and S. Sanches-Cortes, Metal colloids employed in the SERS of biomolecules: Activation when exciting in the visible and near-infrared regions. *J. Mol. Struct.* **405**, 13–28 (1997).

25. T. R. Jensen, M. D. Malinsky, C. L. Haynes, and R. P. Van Duyne, Nanosphere lithography: Tunable localized surface plasmon resonance spectra of nanoparticles. *J. Phys. Chem. B* **104**, 10549–10556 (2000).

26. M. D. Malinsky, K. L. Kelly, G. C. Schatz, and R. P. Van Duyne, Chain length dependence and sensing capabilities of the localized surface plasmon resonance of silver nanoparticles chemically modified with alkanethiol self-assembled monolayers. *J. Am. Chem. Soc.* **123**, 1471–1482 (2001).

27. D. Eck, C. A. Helm, N. J. Wagner, and K. A. Vaynberg, Plasmon resonance measurements of the adsorption and adsorption kinetics of a biopolymer onto gold nanocolloids. *Langmuir* **17**(4), 957–960 (2001).

28. M. Valina-Saba, G. Bauer, N. Stich, F. Pittner, and T. Schalkhammer, A self-assembled shell of 11-mercaptoundecanoic aminophenylboronic acids on gold nanoclusters. *Mat. Sci. Eng. C* **8–9**, 205–209 (1999).

29. J. Malicka, I. Gryczynski, Z. Gryczynski, J.R. Lakowicz, Effects of fluorophore-to-silver distance on the emission of cyanine-dye-labeled oligonucleotides, *Anal.Biochem.* **315**, 57-66 (2003).

30. J. R. Lakowicz, I. Gryczynski, Y. Shen, J. Malicka, Z. Gryczynski, Intensified fluorescence, *Photonics. Spectra*, 96–104 (2001).

31. J. Gersten, A. Nitzan, Spectroscopic properties of molecules interacting with small dielectric particles, *J. Chem. Phys.* **75**, 1139–1152 (1981).

32. H. Chew, Transition rates of atoms near spherical surfaces, *J. Chem. Phys.* **87**, 1355–1360 (1987)

33. P. C. Das, A. Puri, Energy flow and fluorescence near a small metal particle, *Phys.Rev. B* **65**, 155416–155418 (2002)

34. M. Moskovits, Surface-enhanced spectroscopy, Rev. Mod. Phys. 57 (1985) 783–826.

35. A. Wokaun, H.-P. Lutz, A.P. King, U.P. Wild, R.R. Ernst, Energy transfer in surface enhanced fluorescence, *J. Chem. Phys.* **79**, 509–514 (1983)

36. P. J. Tarcha, J. DeSaja-Gonzalez, S. Rodriquez-Llorente, R. Aroca, Surface-enhanced fluorescence on SiO2 coated silver island films, Appl. Spectrosc. 53 (1999) 43–48.

37. A. E. German, G.A. Gachko, Dependence of the amplification of giant Raman scattering and fluorescence on the distance between an adsorbed molecule and a metal surface, J. Appl. Spectrosc. 68 (2001) 987–992.

38. K. Sokolov, G. Chumanov, T.M. Cotton, Enhancement of molecular fluorescence near the surface of colloidal metal films, Anal. Chem. 70 (1998) 3898–3905.

39. G. Laczko, I. Gryczynski, Z. Gryczynski, W. Wiczk, H. Malak, J.R. Lakowicz, A 10-GHz frequency-domain fluorometer, *Rev. Sci. Instrum.* **61**, 2331–2337 (1990)

40. Th. Forster, Intermolecular energy migration and fluorescence *Ann Phys* **2** 55–75 (1948) (Transl. Knox R S, Department of Physics and Astronomy, University of Rochester, Rochester, NY 14627)

41. J. R. Lakowicz, J. Malicka, S. D'Auria and I. Gryczynski, Release of the self-quenching of fluorescence near silver metallic surfaces, *Anal. Biochem.* **320**, 13–20 (2003).

42. M. Kerker, The optics of colloidal silver: Something old and something new. *J.      Colloid      Interface Sci.* **105**, 297–314 (1985).

43. M. Faraday, The Bakerian lecture, Experimental relations of gold (and other metals) to light. *Philos. Trans.* **147**, 145–181 (1857).

44. M. R. Philpott, Effect of surface plasmons on transitions in molecules. *J. Chem. Phys.* **62**(5), 1812–1817 (1975).

45. R. R. Chance, A. Prock, and R. Silbey, Molecular fluorescence and energy transfer near interfaces. *Adv. Chem. Phys.* **37**, 1–65 (1978).

46. D. A. Weitz, S. Garoff, J. I. Gersten, and A. Nitzan, The enhancement of Raman scattering, resonance Raman scattering, and fluorescence from molecules absorbed on a rough silver surface. *J. Chem. Phys.* **78**(9), 5324–5338 (1983)

47. F. Schutt, J. Fischer, J. Kopitz, and F. G. Holz, Clin. Exp. Invest. 30(2), 110 (2002).

48. C. D. Geddes, A. Parfenov, and J. R. Lakowicz, Photodeposition of silver can result in metal-enhanced fluorescence, *Applied Spectroscopy* **57**(5), 526-531 (2003).

49. C. D. Geddes, A. Parfenov, D. Roll, J. Fang, and J. R. Lakowicz, Electrochemical and laser deposition of silver for use in metal-enhanced fluorescence, *Langmuir* **19**(15), 6236-6241 (2003).

50. C. D. Geddes, A. Parfenov, D. Roll, I. Gryczynski, J. Malicka and J. R. Lakowicz, Silver fractal-like structures for metal-enhanced fluorescence: Enhanced fluorescence intensities and increased probe photostabilities, *J. Fluoresc.* **13**(3), 267–276 (2003).

51. J. Marengo, R. A. Ucha, M. Martinez-Cartier, and J. R. Sampaolesi, *Int. Ophthalmology* **23**, 413 (2001).

52. H. Ishihara, H. Okawa, T. Iwakawa, N. Umegaki, T. Tsubo, and A. Matsuki, *Anesthesia Analgesia* **94**, 781 (2002).
53. S. G. Sakka, K. Reinhart, K. Wegscheider, and A. Meier-Hellmann, *Chest* **121**, 559 (2002).
54. P. Lanzetta, Retina. *J. Ret. VIT. Dis.* **21**, 563 (2001).
55. E. M. Sevick-Muraca, G. Lopez, J. S. Reynolds, T. L. Troy, and C. L. Hutchinson, *Photochem. Photobiol.* **66**, 55 (1997).
56. J. M. Devoisselle, S. Soulie, H. Maillols, T. Desmettre, and S. Mordon, *Proc. SPIE-Int. Soc. Opt. Eng.* **2980**, 293 (1997).
57. J. M. Devoisselle, S. Soulie, S. Mordon, T. Desmettre, and H. Maillols, *Proc. SPIE-Int. Soc. Opt. Eng.* **2980**, 453 (1997).
58. A. Becker, B. Riefke, B. Ebert, U. Sukowski, H. Rinneberg, W. Semmier, and K. Licha, *Photochem. Photobiol.* **72**, 234 (2000).
59. C. D. Geddes and J. R. Lakowicz, Metal-enhanced fluorescence, *J. Fluoresc.* **12**(2),121–129 (2002)
60. R. Keir, E. Igata, M. Arundell, W. E. Smith, D. Graham, C. McHugh, and J. M. Cooper, SERRS: In situ substrate formation and improved detection using microfluidics. *Anal. Chem.* **74**(7), 1503–1508 (2002).
61. C. L. Haynes, A. D. McFarland, M. T. Smith, J. C. Hulteen, and R. P. Van Duyne, Angle-resolved nanosphere lithography:Manipulation of nanoparticle size, shape, and interparticle spacing. *J. Phys. Chem. B* **106**, 1898–1902 (2002).
62. F. Hua, T. Cui, and Y. Lvov, Lithographic approach to pattern self-assembled nanoparticle multilayers. *Langmuir* **18**, 6712–6715 (2002).
63. V. Fleury, W. A. Watters, L. Allam, and T. Devers, Rapid electroplating of insulators, *Nature* **416**, 716–719 (2002).
64. M. Fleischmann, P. J. Hendra, and A. J. McQuillan, Raman spectra of pyridine adsorbed at a silver electrode. *Chem. Phys. Letts.* **26**(2), 163–166 (1974)
65. E. Roth, G. A. Hope, D. P. Schweinsberg, W. Kiefer, and P. M. Fredericks, Simple technique for measuring surface-enhanced fourier transform Raman spectra of organic compounds. *Appl. Spec.* **47**(11), 1794–1800 (1993)
66. A. M. Michaels, J. Jiang, and L. Brus, Ag nanocrystal junctions as the site for surface-enhanced Raman scattering of single rhodamine 6G molecules. *J. Phys.Chem. B.* **104**, 11965–11971 (2000).
67. A. M. Michaels, M. Nirmal, and L. E. Brus Surface enhanced Raman spectroscopy of individual rhodamine 6G molecules on large Ag nanocrystals. *J. Am. Chem. Soc.* **121**, 9932–9939 (1999).
68. W. C. Bell and M. L. Myrick, J. Colloid Interface Sci. **242**, 300 (2001).
69. G. Rodriguez-Gattorno, D. Diaz, L. Rendon, and G. O. Hernandez- Segura, J. Phys. Chem. B **106**, 2482 (2002).
70. J. P. Abid, A. W. Wark, P. F. Breve, and H. H. Girault, Chem. Commun. **7**, 792 (2002).
71. I. Pastoriza-Santos, C. Serra-Rodriguez, and L. M. Liz-Marzan, J. Collloid Interface Sci. **221**, 236 (2002).
72. K. Aslan, J. R. Lakowicz, and C. D. Geddes, Deposition of silver nanorods on to glass substrates and applications in metal-enhanced fluorescence, J. Phys. Chem. B (submitted).
73. J. Malicka, I. Gryczynski, C. D. Geddes, and J.R. Lakowicz, Metal-enhanced emission from indocyanince green: an new approach to in vivo imaging, *J. Biomedical Optics.* **8**(3), 472-478 (2003).

# INDEX